Frontiers in Biomaterials

(Volume 2)

Unfolding the Biopolymer Landscape

Editor

Viness Pillay

Wits Advanced Drug Delivery Platform Research Unit
Department of Pharmacy and Pharmacology
Therapeutic Sciences, Faculty of Health Sciences
University of the Witwatersrand, Johannesburg
7 York Road, Parktown, 2193
South Africa

Co-Editor

Yahya E. Choonara

Wits Advanced Drug Delivery Platform Research Unit
Department of Pharmacy and Pharmacology
Therapeutic Sciences, Faculty of Health Sciences
University of the Witwatersrand, Johannesburg
7 York Road, Parktown, 2193
South Africa

Assistant Editor

Pradeep Kumar

Wits Advanced Drug Delivery Platform Research Unit
Department of Pharmacy and Pharmacology
Therapeutic Sciences, Faculty of Health Sciences
University of the Witwatersrand, Johannesburg
7 York Road, Parktown 2193
South Africa

Frontiers in Biomaterials

Volume 2

ISSN (Online): 2352-3921

ISSN: Print: 2468-0168

Unfolding the Biopolymer Landscape

Editor: Viness Pillay

Co-Editor: Yahya E Choonara

Assistant Editor: Pradeep Kumar

ISBN (eBook): 978-1-68108-195-3

ISBN (Print): 978-1-68108-196-0

BENTHAM SCIENCE Bentham eBooks

Bentham Science Publishers Ltd.
Executive Suite Y - 2
PO Box 7917, Saif Zone
Sharjah, U.A.E.
subscriptions@benthamscience.org

to use the Work. The entire liability of Bentham Science Publishers shall be limited to the amount actually paid by you for the Work.

General:

1. Any dispute or claim arising out of or in connection with this License Agreement or the Work (including non-contractual disputes or claims) will be governed by and construed in accordance with the laws of the U.A.E. as applied in the Emirate of Dubai. Each party agrees that the courts of the Emirate of Dubai shall have exclusive jurisdiction to settle any dispute or claim arising out of or in connection with this License Agreement or the Work (including non-contractual disputes or claims).

2. Your rights under this License Agreement will automatically terminate without notice and without the need for a court order if at any point you breach any terms of this License Agreement. In no event will any delay or failure by Bentham Science Publishers in enforcing your compliance with this License Agreement constitute a waiver of any of its rights.

3. You acknowledge that you have read this License Agreement, and agree to be bound by its terms and conditions. To the extent that any other terms and conditions presented on any website of Bentham Science Publishers conflict with, or are inconsistent with, the terms and conditions set out in this License Agreement, you acknowledge that the terms and conditions set out in this License Agreement shall prevail.

Bentham Science Publishers Ltd.
Executive Suite Y - 2
PO Box 7917, Saif Zone
Sharjah, U.A.E.
subscriptions@benthamscience.org

BENTHAM SCIENCE PUBLISHERS LTD.

CONTENTS

FOREWORD

This eBook provides an expert and up-to-date overview of the use of biopolymers in the design of smart therapeutic delivery systems, tissue-engineered scaffolds, implants and devices. It also covers future trends in the design and application of biopolymers and their overall potential benefits. The versatility of biopolymer technology and modification is channelled towards the development of advanced drug delivery systems for the treatment of various diseases as well as the elaboration of nanoscale structures for scaffolding. The various chapters of this eBook with contributions from polymer chemists, pharmaceutical scientists, chemists, biomedical and chemical engineers and other allied disciplines show that multi-disciplinarity is key for spanning the frontiers in biopolymer synthesis, modification and applications. The presentation of state-of-the-art research in the biopolymer area should be most helpful not only to researchers in the field but also to graduate students. Finally, the consolidation of ideas within this eBook, evident through the sterling credentials of the Editor Prof Viness Pillay, generates a unique viewpoint for the conceptualisation of novel biopolymeric systems in future.

Dhanjay Jhurry
National Research Chair
Centre for Biomedical and Biomaterials Research (CBBR)
MSIRI Building
Réduit, Mauritius

PREFACE

The biopolymer landscape is undergoing a revolution and illuminates the propensity of certain multifunctional biomaterials to change the way biopolymers are synthesized and applied for the design of scaffolds in tissue engineering, controlling the spatial and temporal release of bioactives into the human body as well as to modulate the physicochemical and physicomechanical properties of therapeutic systems for enhanced form and function. For example at the Wits Advanced Drug Delivery Platform (WADDP) (http://www.wits.ac.za/waddp) we focus on specialized biopolymers that allow us to design therapeutic systems that can perform functions outside their range of conventional capability and into the blue skies realm, including 3D printability. The mechanized uniqueness of biopolymers is embodied not only in their macroscopic structural changes, but also in their ability to respond to varying stimuli, reversibly change their shape or even behave as *in vivo* tissue or organ surrogates. Such biopolymers have been applied in a vast array of biomedical applications that include stimuli-sensitive bioactive delivery systems, intelligent medical and surgical devices, tissue engineering, or implants for minimally invasive surgery. In this eBook, we provide a topographical view of the tortuous biopolymer landscape and elaborate on the fundamental molecular mechanisms and physical dynamics of polymer behavior that bring about their advanced function. Furthermore, we highlight the diverse stimuli that instigate biopolymer function and discuss those that have undergone modification, functionalization and customization for specialized biomedical applications. We would like to thank the experts who have made immense contributions to the various Chapters within this eBook and have emphasized the fact that certainly the biopolymer landscape is far from flat. The future is exciting.

Viness Pillay
Wits Advanced Drug Delivery Platform Research Unit
Department of Pharmacy and Pharmacology
Therapeutic Sciences, Faculty of Health Sciences
University of the Witwatersrand. Johannesburg
7 York Road, Parktown, 2193
South Africa

Yahya E. Choonara
Wits Advanced Drug Delivery Platform Research Unit
Department of Pharmacy and Pharmacology
Therapeutic Sciences, Faculty of Health Sciences
University of the Witwatersrand. Johannesburg
7 York Road, Parktown, 2193
South Africa

_&_

Pradeep Kumar
Wits Advanced Drug Delivery Platform Research Unit
Department of Pharmacy and Pharmacology
Therapeutic Sciences, Faculty of Health Sciences
University of the Witwatersrand. Johannesburg
7 York Road, Parktown 2193
South Africa

CONTRIBUTORS

Alejandro Sosnik	Group of Pharmaceutical Nanomaterials Science, Department of Materials Science and Engineering, Technion-Israel Institute of Technology, Haifa, Israel
Anjali Hirani	Department of Pharmaceutical Sciences, USF College of Pharmacy, University of South Florida, Tampa, FL 33612, USA and School of Biomedical Engineering and Sciences, Virginia Tech-Wake Forest University, Blacksburg, VA 24061
Charu Tyagi	University of the Witwatersrand, Faculty of Health Sciences, Department of Pharmacy and Pharmacology, 7 York Road, Parktown 2193, Johannesburg, South Africa and V.S.P.G.(P.G) College, C.C.S University, Meerut 250004, U.P., India
Dante M. Beltramo	Centro de Excelencia en Productos y Procesos de Córdoba (CEPROCOR); Consejo Nacional de Investigaciones Científicas y Técnicas (CONICET) and Facultad de Ciencias Químicas – Universidad Católica de Córdoba – Argentina
Divya Bijukumar	University of the Witwatersrand, Faculty of Health Sciences, Department of Pharmacy and Pharmacology, 7 York Road, Parktown 2193, Johannesburg, South Africa
Edward Suesca	Tissue Engineering Group, Department of Pharmacy, Universidad Nacional de Colombia, Bogotá, Colombia
Harpal Singh	Indian Institute of Technology Delhi (IIT D), Hauz Khas, New Delhi, India
Inderbir Singh	Department of Pharmaceutics, Chitkara College of Pharmacy, Chitkara University, Chandigarh-Patiala Highway, Rajpura-140401, Patiala, Punjab, India
Ismael D. Bianco	Centro de Excelencia en Productos y Procesos de Córdoba (CEPROCOR); Consejo Nacional de Investigaciones Científicas y Técnicas (CONICET) and Departamento de Ciencias Exactas, Físicas y Naturales - Universidad Nacional de La Rioja – Argentina
Jacques Desbrieres	Pau and Pays de l'Adour University (UPPA), Institut des Sciences Analytiques et de Physico-Chimie pour l'Environnement et les Materiaux (IPREM/EPCP), UMR CNRS 5254, Helioparc Pau Pyrenees, 2, av. President Angot, 64053 Pau Cedex 09, France
Karina L. Bierbrauer	Consejo Nacional de Investigaciones Científicas y Técnicas (CONICET), Argentina
Kuldeep Singh	Pharmaceutics Division, Department of Pharmaceutical Sciences and Drug Research, Punjabi University, Patiala, Punjab, India-147002
Lisa C. du Toit	Wits Advanced Drug Delivery Platform Research Unit, Department of Pharmacy and Pharmacology, School of Therapeutic Sciences, Faculty of Health Sciences, University of the Witwatersrand, Johannesburg, 7 York Road, Parktown, 2193, South Africa

Lomas K. Tomar	University of the Witwatersrand, Faculty of Health Sciences, Department of Pharmacy and Pharmacology, 7 York Road, Parktown 2193, Johannesburg, Gauteng, South Africa
Marcel Popa	"Gheorghe Asachi" Technical University of Iasi, Faculty of Chemical Engineering and Environmental Protection Department of Natural and Synthetic Polymers, Prof. Dr. Docent Dimitrie Mangeron Street, No. 73, 700050, Iasi, Romania
Marcelo R. Ceballos	Centro de Excelencia en Productos y Procesos de Córdoba (CEPROCOR), Argentina
Maria-Andreea Lungan	"Gheorghe Asachi" Technical University of Iasi, Faculty of Chemical Engineering and Environmental Protection Department of Natural and Synthetic Polymers, Prof. Dr. Docent Dimitrie Mangeron Street, No. 73, 700050, Iasi, Romania
Marta R. Fontanilla	Tissue Engineering Group, Department of Pharmacy, Universidad Nacional de Colombia, Bogotá, Colombia.
Navdeep Grover	Department of Chemical and Biological Engineering, University of Colorado Boulder, 3415 Colorado Avenue, Boulder, Colorado, USA
Oluwatoyin Ayotomilola Adeleke	Council for Scientific and Industrial Research, Meraka Institute, Centre for High Performance Computing, 627 Meiring Naudé Road, Pretoria 0001, South Africa
Pradeep Kumar	Wits Advanced Drug Delivery Platform Research Unit, Department of Pharmacy and Pharmacology, School of Therapeutic Sciences, Faculty of Health Sciences, University of the Witwatersrand, Johannesburg, 7 York Road, Parktown, 2193, South Africa
Riaz A. Khan	Department of Medicinal Chemistry, College of Pharmacy, Qassim University, Qassim 51452, KSA and Department of Chemistry, MRIU, Aravali Hills, National Capital Region, Faridabad, HR 121001, India
Roxana V. Alasino	Centro de Excelencia en Productos y Procesos de Córdoba (CEPROCOR) and Consejo Nacional de Investigaciones Científicas y Técnicas (CONICET), Argentina
Silvia G. Correa	Consejo Nacional de Investigaciones Científicas y Técnicas (CONICET) and Facultad de Ciencias Químicas – Universidad Nacional de Córdoba – Argentina
Silvia Vasiliu	"Petru Poni" Institute of Macromolecular Chemistry, Gr. Ghica Voda Alley No. 41A, 700487 Iasi, Romania
Sruti Chattopadhyay	Indian Institute of Technology Delhi (IIT D), Hauz Khas, New Delhi, India
Stefania Racovita	"Petru Poni" Institute of Macromolecular Chemistry, Gr. Ghica Voda Alley No. 41A, 700487 Iasi, Romania
Sunil Kamboj	Pharmaceutics Division, Department of Pharmaceutical Sciences and Drug Research, Punjabi University, Patiala, Punjab, India-147002

Swati Jain	Amity Institute of Nanotechnology, Amity University, Sec. 125, Noida, Uttar Pradesh, India
Thashree Marimuthu	Wits Advanced Drug Delivery Platform Research Unit, Department of Pharmacy and Pharmacology, School of Therapeutic Sciences, Faculty of Health Sciences, University of the Witwatersrand, Johannesburg, 7 York Road, Parktown, 2193, South Africa
Valence M.K. Ndesendo	Department of Pharmaceutics & Formulation Sciences, School of Pharmacy & Pharmaceutical Sciences, St. John's University of Tanzania, Dodoma, Tanzania
Vijaykumar B. Sutariya	Department of Pharmaceutical Sciences, USF College of Pharmacy, University of South Florida, Tampa, FL 33612, USA
Vikas Rana	Pharmaceutics Division, Department of Pharmaceutical Sciences and Drug Research, Punjabi University, Patiala, Punjab, India-147002
Vincss Pillay	Wits Advanced Drug Delivery Platform Research Unit, Department of Pharmacy and Pharmacology, School of Therapeutic Sciences, Faculty of Health Sciences, University of the Witwatersrand, Johannesburg, 7 York Road, Parktown, 2193, South Africa
Yahya E. Choonara	Wits Advanced Drug Delivery Platform Research Unit, Department of Pharmacy and Pharmacology, School of Therapeutic Sciences, Faculty of Health Sciences, University of the Witwatersrand, Johannesburg, 7 York Road, Parktown, 2193, South Africa
Yashwant Pathak	Department of Pharmaceutical Sciences, USF College of Pharmacy, University of South Florida, Tampa, FL 33612, USA
Yong W. Lee	School of Biomedical Engineering and Sciences, Virginia Tech-Wake Forest University, Blacksburg, VA 24061, USA

CHAPTER 1

Cationic Polymers for Biotechnological Applications

Roxana V. Alasino[1,2], Karina L. Bierbrauer[2], Dante M. Beltramo[1,2,3], Silvia G. Correa[2,4] and Ismael D. Bianco[1,2,5,*]

[1]Centro de Excelencia en Productos y Procesos de Córdoba (CEPROCOR), Argentina; [2]Consejo Nacional de Investigaciones Científicas y Técnicas (CONICET), Argentina; [3]Facultad de Ciencias Químicas - Universidad Católica de Córdoba - Argentina; [4]Facultad de Ciencias Químicas - Universidad Nacional de Córdoba - Argentina and [5]Departamento de Ciencias Exactas, Físicas y Naturales - Universidad Nacional de La Rioja - Argentina.

Abstract: Cationic polymers have been the subject of growing research interest in recent decades with new applications constantly appearing. They exhibit attractive physicochemical properties derived from the flexibility of the polymer chain, the formation of H-bonds, amphiphilic and electrostatic interactions as well as a great potential for being modified to obtain polymeric species for new applications. Furthermore, the different polymer architectures that are available include linear, branched and dendrimeric. According to their origin, cationic polymers can be classified in two general groups: natural or synthetic. Natural cationic polymers are interesting for therapeutic applications because they have generally a very good biocompatibility. Improvement in synthetic methods allows the preparation of cationic polymers with precise control of their properties and changes, such as the molecular weight distribution, polarity and the degradability of the chains if necessary. The reports describing antimicrobial, antioxidant, anti-tumoral and anti-inflammatory properties of cationic polymers have grown exponentially during the last years. Most of them also show interesting changes in their behaviors produced in response to stimuli such as pH and ionic strength. All these features make them even more promising for new biotechnological purposes. The wide range of uses of cationic polymers extends from industrial processes to a great variety of therapeutic, biomedical and pharmaceutical claims. In this chapter, we summarize the main properties and biotechnological applications of cationic polymers.

Keywords: Antimicrobial activity, cationic cellulose, cheese-making, chitosan, drug-delivery, Eudragit, hydrogels, immunomodulation, virus inactivation, wound healing.

INTRODUCTION

Most of the naturally existing polymers are either neutral or negatively charged. There is only one cationic polysaccharide (chitosan), the rest being proteins

[*]**Corresponding author Ismael D. Bianco:** Centro de Excelencia en Productos y Procesos de Córdoba (CEPROCOR), Ministerio de Industria, Comercio, Minería y Desarrollo Científico Tecnológico, Pabellón CEPROCOR, CP 5164, Santa María de Punilla, Córdoba - Argentina; Tel: 54-3541-489651/53 ext.143; Fax: 54-3541-488181; E-mails: ibianco@ceprocor.uncor.edu; ibianco@hotmail.com

Viness Pillay, Yahya E. Choonara and Pradeep Kumar (Eds.)

which, although having the possibility to express a net positive charge at neutral pH, are zwitterions. This is most probably the reason why cationic polymers were the last to arrive to scientific literature. Most of the research related to basic and applied properties of cationic polymers dates from the last 20 years.

In the case of proteins, their biotechnological applications are mostly based on their specific interactions that give rise to their biological functions. In this sense, they have been mostly used as active ingredients. There are many excellent reviews on protein based biotechnological applications. Therefore, we will focus this chapter mostly on cationic polysaccharides (natural and synthetic) and will include wholly synthetic cationic polymers that have been found to have interesting bioactive properties.

The presence in their structure of functional groups that allow further modification renders cationic polymers attractive for many biological applications. The initial extensive studies were mostly focused on their interaction with anionic biomolecules and proteins through electrostatic interactions. More recently, several research groups have disclosed that they may have antioxidant, anti-tumor, antimicrobial, anti-inflammatory and immunomodulatory activities as well as many different non biological properties such as stimuli responsiveness (pH- or temperature-dependent structural changes), that give them highly promising potential utility.

While some of the most widely studied cationic polymers possess intrinsic cationic charges, others have also been developed by introducing cationic groups to otherwise neutral polymers such as cationic cyclodextrins and dextrans. In this chapter, we will focus particularly on the properties and applications of three representative cationic polymers: chitosan (polysaccharide bearing primary and secondary amines), Eudragit E100 (with tertiary amine groups in its structure) and cationic cellulose SC240C (with quaternary amine groups). In the first section, we will give a brief description of the chemical nature and general properties of these cationic polymers and in the subsequent parts we will highlight recent advancements in applications, providing an overview of the interrelationships between structural and chemical properties and interactions of cationic polymers with other molecules or structures.

We expect to give the reader a broad picture of the current state of the art of cationic polymer biotechnological applications and the different avenues that research in this field has explored with its surprising findings.

Chemical Structure and General Properties of Cationic Polymers

Chitosan

Chitin is the second most abundant natural biopolymer as it is the main component of the exoskeletons of insects and crustaceans. Its chemical structure is similar to that of cellulose with monomers of N-acetylglucosamine attached *via* β(1, 4) linkages [1]. Chitosan is the low acetylated form of chitin (Fig. **1a**). It can be found in nature in the cell walls of yeasts and fungi [2]. However, due to its low abundance in nature, most of the chitosan that has been used for technological and industrial applications is prepared by partial deacetylation of chitin. The presence of primary and secondary amine groups confers chitosan water solubility at acidic pH. In this regard, chitosan deacetylated in large proportion (at least 85%) is easily soluble up to pH 6.5, but as the deacetylation degree decreases, the solubilization also declines. With an increase in the deacetylation degree more positively charged amine groups are present which gives the polymer an intrinsic possibility to interact electrostatically with negatively charged molecules. It has also been shown to interact with several metals and metal oxides, a property that has been used in several technological applications (as detailed later in the chapter).

Because of the presence of reactive amine and hydroxyl groups in its structure chitosan has been frequently modified using cationic molecules or grafting small molecules of other polymers to attain a desired function. Quaternization of chitosan has been used to provide the polymer a pH independent cationic character which results in an improvement of the stability of ionic complexes with negatively charged molecules [3, 4].

Cationic Cellulose

Cellulose is the most naturally abundant polysaccharide throughout the world. Cationic cellulose derivatives have been synthesized by several groups with the aim of designing novel biodegradable materials for cosmetic and therapeutic applications [5-7]. In this context, by reacting trimethyl ammonium substituted epoxide with a hydroxyethyl cellulose backbone, positively charged cellulose containing quaternary ammonium groups has been synthesized and is commercially available under different brand names: PQ-10, SC-230, SC-240C, Sensomer 10M, *etc* (Fig. **1b**). This cationic cellulose has been designed for use in cosmetic and hair care products. Due to its pH independent cationic character it is highly water soluble, giving clear viscous solutions.

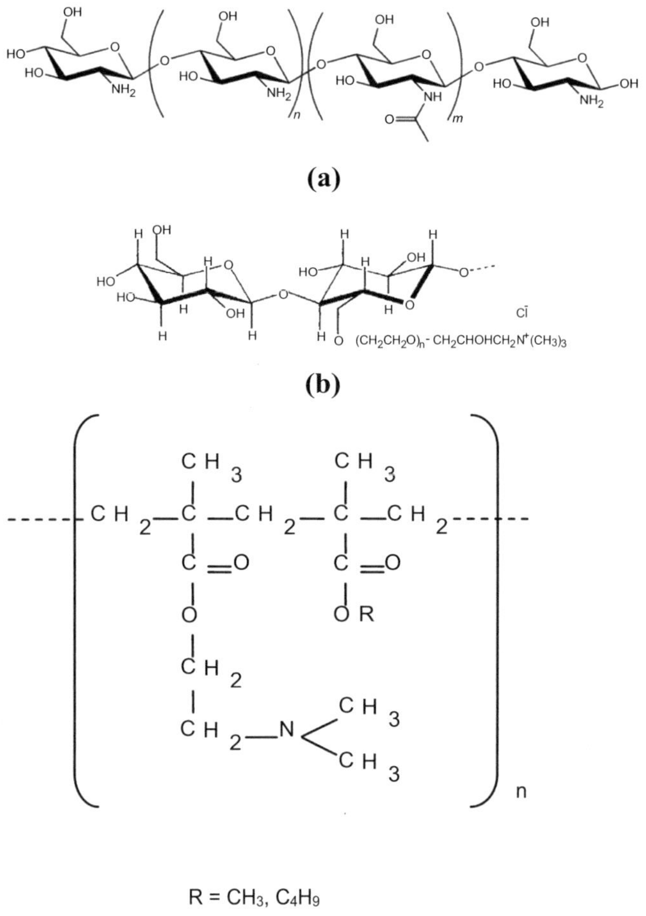

(c)

Figure 1: Chemical structure of: (**a**) chitosan; (**b**) cationic cellulose PQ-10 and (**c**) Eudragit E100.

Cationic Methacrylates

Although not strictly biopolymers, because of their wide and growing uses for biotechnological applications, we will include a brief summary of a synthetic cationic polymer derived from polyacrylates. Copolymers of methacrylates based on dimethylaminoethyl, butyl and methyl methacrylate have been prepared to be used as pharmaceutical excipients. Among these polymers, a series of derivatives known as Eudragit® have been introduced in the pharmaceutical market more than fifty years ago by Rohm GMB & Co. A series of polymers with different molecular weights are available for different applications.

As with most cationic polymers, modifications of the structure have been investigated. Several chemical modifications have been characterized in order to search for special properties, *i.e.*: inhibition of aggregate formation by grafting of PEG, improvement in pH and temperature responsiveness by grafting chitosan, improvement of biodegradability by the introduction of esther linkages and so on [8-10]. We will describe below recent applications of one of these polymers, Eudragit E100, a cationic aminoacryl methacrylate copolymer containing a tertiary amine group that was developed to be used in the pharmaceutical industry to mask flavours and to avoid drug crystallization (Fig. **1c**) [11, 12].

Biotechnological Applications

Food Applications

Chitosan

Chitosan has been shown to have interesting nutritional and physiological activities. It has been described that chitosan acts as a dietary fiber with important hypocholesterolemic effect, reducing lipid absorption and enhancing cholesterol elimination [13-15]. In the food industry, this biopolymer offers a wide range of applications including food products [16-18], food preservation [19-23], biodegradable films [24-29], recovery of waste material [30-37], water purification [38-43] juice clarification and deacidification [44-48] and its use in texture controlling, emulsification and stabilization [42]. In the dairy industry, chitosan has been used to remove milk fat, proteins and peptides from cheese whey [32, 35, 49].

Many of its properties have been related to some degree to the fact that it is a positively charged polymer. In this sense, the high content of electronegative domains in casein micelles offered an attractive model to explore not only the interactions but also on the specific properties of the complex [50-56].

As can reasonably be expected, chitosans of different mean molecular weight (MMW) promote a selective coagulation of casein micelles from whole and defatted milk. Not as obvious though, it was found that the process involves not only electrostatic but also hydrophobic interactions (Fig. **2**). The coagulation process is similar to that observed using rennet or acid coagulation methods and, although it was demonstrated that chitosan interacts and modifies the stretching of the phosphate bond in caseins, the coagulation does not depend on the presence of phosphates or milk fat [51]. On the other hand, although lipolytic enzymes only hydrolyze about 50 % of the triglycerides present in the aggregates, the complexes between chitosan and caseins are very well hydrolyzed by gastrointestinal proteases, releasing soluble peptides at a rate similar to that produced by acid or rennet, rendering the additional benefit of low fat absorption after ingestion [52].

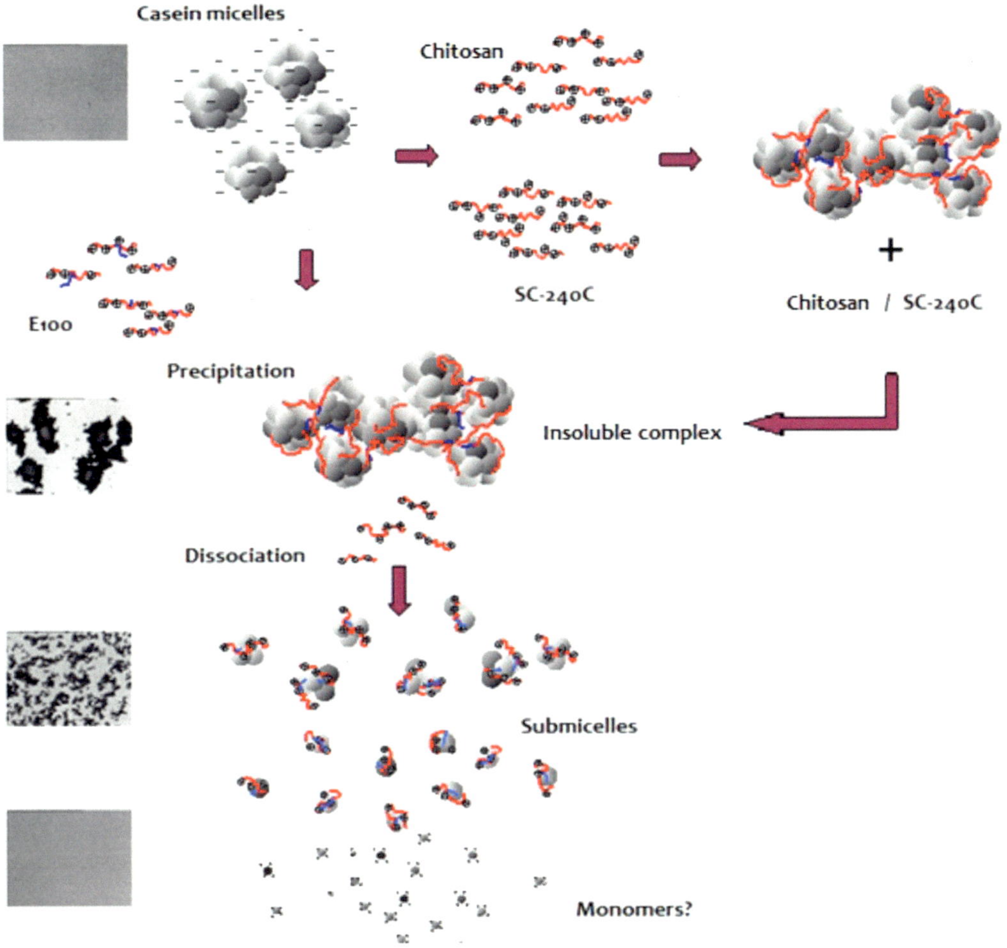

Figure 2: Schematic representation of the effect of the different cationic polymers on lactic casein. The biphasic effect of E100, with a first precipitation stage followed by a second of redissolution of the aggregates at higher concentrations, is observed. However, the addition of Chitosan or SC-240C to the micelle suspension results in formation of aggregates that do not dissociate with increasing polymer concentration.

Networks of casein micelles result from the aggregation of stable micelles caused by a reduction of repulsive forces [54, 55]. On this basis, it was demonstrated that the aggregation of caseins induced by chitosan involves electrostatic and hydrophobic interactions that contribute actively to the association [52-56].

Another applications of chitosan are based in its antimicrobial activity at pH <6 against bacteria and fungi [42, 53, 57]. Although the mode of antimicrobial action of chitosan is not completely understood, it is well established that the molecular structure of chitosan is a prerequisite for its antimicrobial activity [58, 59]. The polycationic characteristic of chitosan in acidic medium is the main factor contributing to the antimicrobial activity.

Due to the positive surface charges at acidic condition, chitosan interacts with anionic components on bacteria surface, such as negatively charged lipopolysaccharide in the outer membrane of Gram negative bacteria and peptide glycan and teichoic acid in the cell wall of Gram-positive bacteria [60, 61]. More recently, it has been shown that chitosan microparticles display antimicrobial activity even at neutral pH through the interaction *via* hydrogen bonding with the outer membrane protein OmpA [45]. The interactions between chitosan and OmpA disrupt bacterial membrane thus causing their death. Of more importance, this activity seems to be exerted against a broad-spectrum of pathogens like E.coli O157:H7, V. cholerae, S. enterica, K. pneumoniae and S. uberis [62].

In this connection, and in relation to the results described above on the interaction between chitosan with casein micelles, it is worth mentioning that, although the polymer shows a dose dependent antibacterial activity, its inhibiting effects were greatly reversed when it was incubated with milk before its exposure to bacteria. Therefore, it was possible to use the milk curds obtained using chitosan in cheese-making [53].

Eudragit E100

As with chitosan, upon addition of E100 to milk a coagulation process occurs almost immediately after addition of the polymer [63]. However, upon increasing the concentration of E100 a complete solubilization of the precipitate was observed. Upon incorporating an excess of E100 a reduction in turbidity and light scattering of milk or casein solutions confirmed that at high concentration E100 dissociates casein micelles into submicelles or casein oligomers [63]. Fig. (**2**) summarizes the events that take place upon mixing E100 with milk.

Celquat SC-240C

The cationic hydroxyethylcellulose, Polyquaternium-10 (PQ10) also known as Celquat® SC-240C (Fig. **1c**), has been also shown to destabilize and precipitate casein micelles through electrostatic interactions between its quaternary ammonium groups and the negatively charges of casein micelles [64]. However, in this case, the casein aggregates can be simply resolubilized through the addition of low ionic strength (Fig. **2**) [64]. Interestingly, the resuspended micelles retain most of the original calcium present and only around 10 % of the lactose [64]. It can also be seen as a simple method for casein concentration which retains most of the calcium content of the original milk. Furthermore, the presence of this polymer does not affect the activity of gastrointestinal lipases and proteases and growth of milk fermenting bacteria, opening the possibility of using Celquat® for preparing fermented dairy products fortified in fiber [64].

Regarding the mechanism of action, it is clear that there is a general trend from completely electrostatic associations (Celquat®) to a combination of electrostatic and hydrophobic interactions (chitosan and E100) [50]. In the case of E100, the intrinsic amphipaticity of the polymer favors a disruption of the micelles at high polymer concentrations [63]. On the other side, for the quaternary ammonium containing polysaccharide Celquat®, its association with casein micelles is basically electrostatic in nature (Fig. **2**).

Biomedical Application of Cationic Biopolymers

Hydrogels for Space Filling and Drug Delivery Applications

There is an increasing interest in biocompatible hydrogels for a variety of therapeutic uses. The most obvious is in the design of drug-delivery systems, but the possibility to attain different stress-related responses make them attractive for the development of space-filling applications. Furthermore, the possibility to exploit their bioactive properties makes them interesting as scaffolds for the design of tissue replacement products.

Natural or synthetic polymers have been used in the preparation of cationic polymer based hydrogels [65-67]. Although chitosan-based hydrogels are the most widely studied, cationic cellulose derived ones have recently been synthesized as part of a strategy to develop new biomaterials [68, 69]. In the case of hydrogels made using charged polymers there are two main preparation alternatives: a) covalent crosslinking or b) electrostatic complexation. In the latter case, reversible hydrogels are formed, which allows the preparation of an environmental multi-responsive material that changes its properties depending on the ionic strength, pH or other molecules.

Irradiation of aqueous solutions of a quaternary ammonium -substituted hydroxyethyl cellulose (Celquat SC240 from National Starch & Chem. Co.)

resulted in the formation of intermolecular crosslinks and also in the sterilization of the material [69].

Within the known hydrogels are those that arise from the combination of a cationic biopolymer with oppositely charged polymers. This combination is referred to as polyelectrolyte complexes, whom have been subject of thorough research during the last decades. These complexes have huge potential for many practical applications because they are naturally available and have a good biocompatibility. For instance, the electrostatic complexes between chitosan and negatively charged polymers have been extensively studied [70-76]. The hydrogels thus formed exhibit pH responsiveness, what makes them "rheologically smart'' for drug delivery applications. Polyelectrolyte complexes betweeen polycationic celluloses and carbopol [77] or glycosaminoglycans have been used in the development of personal care applications [78, 79].

Recently, we found that PQ-10 interacting with chondroitin 4-sulfate (C-4S) displays rheological properties that make the complex a potential candidate as a replacement o synovial fluid for osteoarticular therapy [80]. Furthermore, glycosaminoglycans (GAGs) are natural components of cartilage and have been shown to exert a positive effect on chondrocytes that promote cartilage repair [81-83].

However, although positive outcomes have been observed using intra-articular injections of hyaluronic acid or chondroitin sulfate, the unresolved problem is still that they are removed in a few months, giving rise to the need of frequent treatments. In our work, the combination of C-4S with PQ-10 would give the support for a longer residence time at the site of application (trials under course).

Polycations are also employed in ophthalmic drug delivery. Chitosan, among others have been frequently used to prepare highly viscous ophthalmic solutions, that also take advantage of its strong adhesion to epithelial tissues and its known activity as a penetration enhancer [84-87].

Various synthetic cationic polymers have been also studied as positively charged nanoparticles that electrostatically interact with the ocular surface [84]. Several groups demonstrated that positively charged acrylate copolymers are well tolerated and do not induce inflammation and/or discomfort in the rabbit eye [85, 87, 88]. Another example of polymeric nanoparticle suspensions is Flurbiprofen (FLU) loaded in Eudragit RS 100 and RL 100 which have been shown to be effective in preventing myosis induced during extracapsular cataract surgery [86].

Soluble inserts are also made with chitosan and Eudragit RS for veterinary applications [89].

Due to the possibility to interact electrostatically with nucleic acids, cationic polymers have been extensively studied as potential gene delivery systems. In this section, we will review some recent reports in their use for different therapeutic applications.

Metal Complexion Activity of Chitosan- its Application in a Mouthwash

As stated above, the presence of primary amine groups in chitosan confers upon it the possibility to form stable complexes with several metal ions. This property has been used to stabilize a redox couple formed by ascorbic acid and Cu^{2+}. Numerous studies support the antimicrobial activity of ascorbic acid [90, 91]. It is also known that this antimicrobial activity is increased by the presence of Cu^{2+} which not only catalyzes the oxidation of ascorbic acid but also increases its antimicrobial activity with a synergic effect. The antimicrobial activity of chitosan on *Streptococcus mutans* added to its ability to form complexes with Cu^{2+} were the basis for the development of a mouthwash [92]. Curiously, a formulation containing Cu^{2+}, ascorbic acid and chitosan also displayed antimicotic activity against *Candida albicans,* the most frequent pathogen that induces oral micosis [92]. It is well known that aqueous solutions of $CuSO_4$ have a long stability. On the other side, ascorbic acid is stable as a solid and is rapidly hydrolyzed when dissolved in water. Our results demonstrate that in the presence of 0.1 % (w/w) chitosan aqueous solutions of ascorbic acid retain their antimicrobial activity for at least 2 months. Furthermore, one of the main concerns of using a redox couple between ascorbic acid and Cu^{2+} was the extremely low stability caused by the catalytic activity of Cu^{2+} on ascorbic acid oxidation. As expected, complexation of Cu^{2+} with chitosan increased the stability of ascorbic acid from less than 1 min to more than 1 h [92].

Virus Inactivation

The thermodynamic characterization of the interaction between E100 and casein micelles previously described, led to the suggestion that the polymer could have amphipathic properties [93]. This behavior was observed only with Eudragit E100, a cationic polymer with tertiary ammonium groups but not with other polymers with primary and secondary amine groups (chitosan), and quaternary ammonium compounds like cationic cellulose (see above).

Studies using model membrane systems revealed that the hydrophilic / lipophilic balance of the molecule E100 is an essential factor in the amphipathic behavior of the polymer and that the interaction of E100 with different biological and synthetic structures involves electrostatic and hydrophobic forces. The balance of hydrophilic and hydrophobic residues in E100 allows this polymer to interact with different supramolecular structures [94]. Furthermore, the strong interaction of E100 with biological membranes produces a dramatic disruption of membrane permeability. These results were the basis for studies aimed to evaluate if E100 could have an effect on the infectivity of enveloped virus. The basic strategy (outlined in Fig. **3**) was based on the idea that if the polymer-virus interaction is strong enough but there are still positively charged groups it would be possible to design a simple protocol to remove the polymer from the solution either by its interaction with a cation exchange chromatography column or by its precipitation from the solution after its positively charged groups are neutralized by a subtle change in pH (Fig. **3**). The results revealed that after treatment with E100, infectious titers were reduced to values below the detection limit of herpes virus (VHS), vesicular stomatitis virus (VSV), bovine viral diarrhea virus (BVDV) and measles virus (MV) using both remotion strategies (Table **1**). In addition, molecular biology studies with the immunodeficiency virus (VIH) and hepatitis C virus (VHC) confirmed that the interaction of E100 with virus allows removing the virus particles from the media (Table **1**). Interestingly, the treatment of plasma components or biopharmaceuticals with E100 did not affect significantly the biological activity of most proteins studied (Table **2**) [93, 94].

Modulation of Inflammatory and Immune Response

Wound Healing Activity

From its discovery, chitosan has been widely used in the treatment of wounds because of its hemostatic properties and also because it stimulates healing, has antimicrobial activity and is harmless and biodegradable. Many *in vitro* studies addressed applications of chitosan in wounds and burns. Formulations studied include chitosan and its derivatives with other molecules. For these applications, chitosan was processed as fibers, membranes, gels, beads, and sponges [95-105].

Chitin and chitosan promote wound healing and a good-looking skin surface through the activation of neutrophils and macrophages present at wounded sites [106-108].

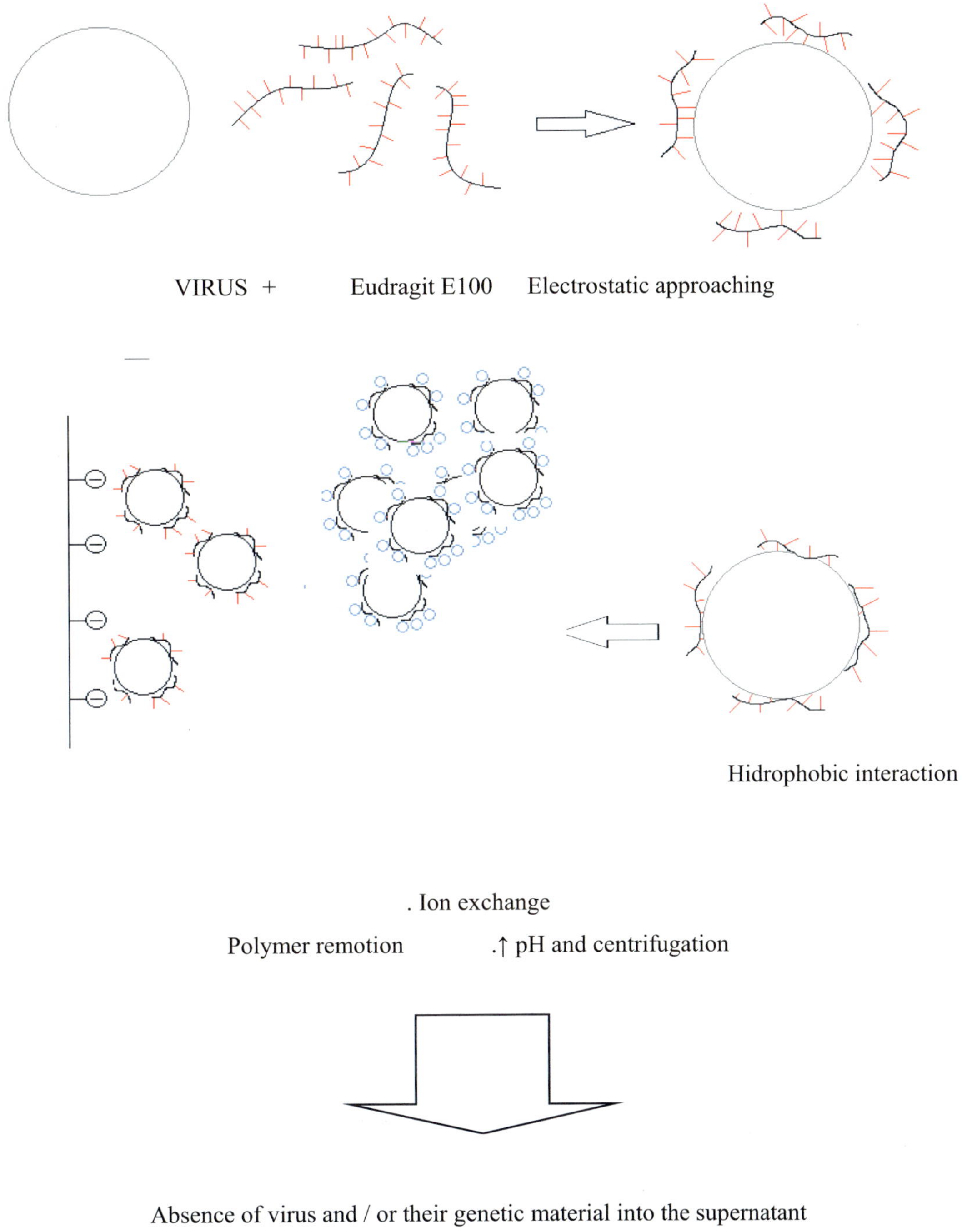

Figure 3: Summary of E100 interactions with virus and their inactivation and remotion by either ion exchange chromatography or precipitation by pH change (for details see Alasino *et al.* [93,94]).

Table 1. Summary of viral reduction by E100 and cation exchange chromatography.

Virus	Virus Characteristics			Virus Inactivation (log_{10})		
	Family	Genome	Size(nm)	E100	CEC	TR
HSV	Herpes	dsDNA	120-200	≥ 7.2[(a)]	≥ 5.0[(a)]	≥ 12.2[(a)]
BVDV	Flavi	+ssRNA	50-70	≥ 6.5[(a)]	≥ 5.0[(a)]	≥ 11.5[(a)]
VSV	Rhabdo	-ssRNA	70-170	≥ 7.7[(a)]	≥ 5.0[(a)]	≥ 12.7[(a)]
MV	Paramyxo	-ssRNA	150-300	≥ 5.0[(a)]	nd	____
HCV	Flavi	+ssRNA	40-60	bdl[(b)]	bdl[(b)]	____
HIV	Retro	+ssRNA	80-100	bdl[(c)]	nd	____

E100: Incubation with eudragit E100 (10 mg/ml) for 60 min at 37 °C and removal of the polymer by rising pH to 8.5 and centrifugation a 5000 x g during 10 min.
CEC: Cation Exchange Chromatography.
TR: Total Reduction of combined treatments.
nd: Not determined bdl: Below Detection Limit
(a) $TCID_{50}$/ml (b) 600 IU/ml (c) 50 copies/ml

Table 2. Concentration of lipids and proteins in human plasma treated with E100.

	Level in plasma (Control without E100)	Level in plasma treated with E100 - 10 mg/mL)	Percentage remaining
Cholesterol (mg/dl)	149 +/- 9	51 +/- 4	34 %
Triglyceride (mg/dl)	174 +/- 10	63 +/- 5	36 %
Protein (mg/dl)	6.1 +/- 0.5	6.0 +/- 0.4	98 %
Glutamic oxalacetic transaminase GOT (IU/dl)	13 +/- 1	13 +/- 1	100 %
Glutamic pyruvic transaminase GPT (IU/dl)	11 +/- 1	10 +/- 1	91 %
Lactate dehydrogenase LDH (IU/dl)	292 +/- 26	282 +/- 27	97 %
Alkaline phosphatase (IU/ml)	154 +/- 14	137 +/- 13	89 %
Anti-D titre (IU/dl)	1:32	1:32	100%
Antithrombin III (recovery post treatment)	90	88	98%

In this connection, the activation of mannose receptors of neutrophils and macrophages is responsible for the primary inflammatory reaction [107, 109-111]. Afterwards, they increase the production of the anti-inflammatory cytokine transforming growth factor β (TGF [112, 113].

At the molecular level, it is also known that L-Arginine metabolic products are important modulators of the wound healing process. Two distinct enzymes participate in this process. On the one side, arginase converts arginine into urea and ornithine, the precursor for polyamines which promote proliferation in various cell types acting as a positive stimulus for wound healing. On the other side, inducible nitric oxide synthase (iNOS) that catalyzes the formation of nitric oxide (NO) which is an important signal for collagen accumulation and wound strength [114]. In this context, we showed that LMW chitosan enhances arginase and iNOS activity in resident macrophages, with a more marked effect on the activation of arginase in inflammatory macrophages [108]. It has been demonstrated that the hydrolysis of L-arginine by arginase is the main pathway in maturating wounds [115]. Therefore, an enhanced arginase activity could be an important effector of the healing activity of positively charged polymers like chitosan [108].

These results clearly demonstrate that chitosan, and other cationic polymers, can be used for the treatment of wound and burn infections not only because of their ability to deliver extrinsic antimicrobial agents and growth factors to wounds and burns, but also by virtue of their intrinsic properties to modulate the main physiological pathways involved in healing. Altogether, these results indicate that cationic polysaccharides will certainly be used in the development of new technological products for the management of wounds and burns.

Another polycation of synthetic origin that has begun to be used in patches is E100, in combination with cohesion promoters, and secondary polymers such as Eudragit RL or RS which are also polycations. These facilitate the release of the drug incorporated for controlling anaerobic infections in chronic wounds [116].

Modulation of Mucosal Immunity

As stated above, one of the most studied applications of biopolymers is on drug delivery. Cationic polymers have been used as hydrogels, drug conjugates (*via* covalent hydrolysable bonds) or polyelectrolyte complexes [65-67]. It has been proposed that complexes with protein/peptide APIs (active pharmaceutical ingredients) chitosan prevents their degradation by intestinal proteases [117].

Despite all the claimed benefits of as drug delivery vehicles, it has been recently established that biopolymers, can exert modulating activities on gut associated immune system [117-120]. In this regard, microspheres of chitosan have shown adjuvant properties against diphtheria and *Bordetella bronchiseptica* increasing antibody levels after intranasal or oral administration [121, 122].

Although it is generally accepted that the oral route is tolerogenic, oral administration of an antigen can result in local and systemic priming or tolerance, and the basis of this dichotomy is poorly understood [123, 124].

Recently, we found that after a single oral dose of chitosan, either alone or combined with proteins, there is an increase in the production of anti-inflammatory cytokines at the level of mucosal sites [117, 118]. We demonstrated that upon feeding, the polysaccharide is taken by cells present at Peyer's patches (PP) and mesenteric lymph nodes (MLNs) [119]. We also demonstrated that the co-administration with type II collagen enhanced the oral and systemic tolerance toward this antigen, reducing the symptoms in experimental arthritis [119, 124]

Rats fed type II collagen (CII) in association with chitosan had reduced levels of immunoglobulin G anti-CII, a limited proliferation in draining lymph nodes and a lower release of interferon- γ (IFN-γ) after re-stimulation with CII.

In summary, oral co-administration of chitosan with proteins can promote their uptake and distribution modifying the profile of cytokines and chemokines not only PP and MLN but also in the spleen of treated animals [117].

Curiously, chitosan itself elicited signals at the epithelial lining that stabilized the homeostatic non-inflammatory microenvironment in gut mucosa [125]. Upon chitosan feeding the epithelium maintained the constitutive expression of injury markers suggesting that, at least orally, chitosan is not an inflammatory stimulus [125]. Moreover, following chitosan administration, the levels of the anti-inflammatory cytokines IL-10, IL-6 and TGF-β increased. Altogether, after chitosan feeding, a mild activation of IECs occurs which is followed by an increase in the production of regulatory factors that are responsible for its immunomodulatory effects (Fig. **4**).

From all what is known, it is clear that gut epithelium is an active component in the modulation of intestinal microbiota and homeostasis through the production of mucus and peptidic factors and also through a complex signaling mechanism that involves the mucosal immune system [126, 127]. In this context, detection and response to the presence of microbes are very important in maintaining gut homeostasis. Recognition of microbial pathogen-associated patterns is performed by Toll-like receptors (TLRs). These receptors bind to molecules associated to threats, most of which are negatively charged molecules like lipopolysaccharides (LPS) and nucleic acids. As expected, cationic polymers can interact electrostatically with these molecules independently of their sequence, therefore acting as a kind of "molecular scavenger" with anti-inflammatory activity [128, 129].

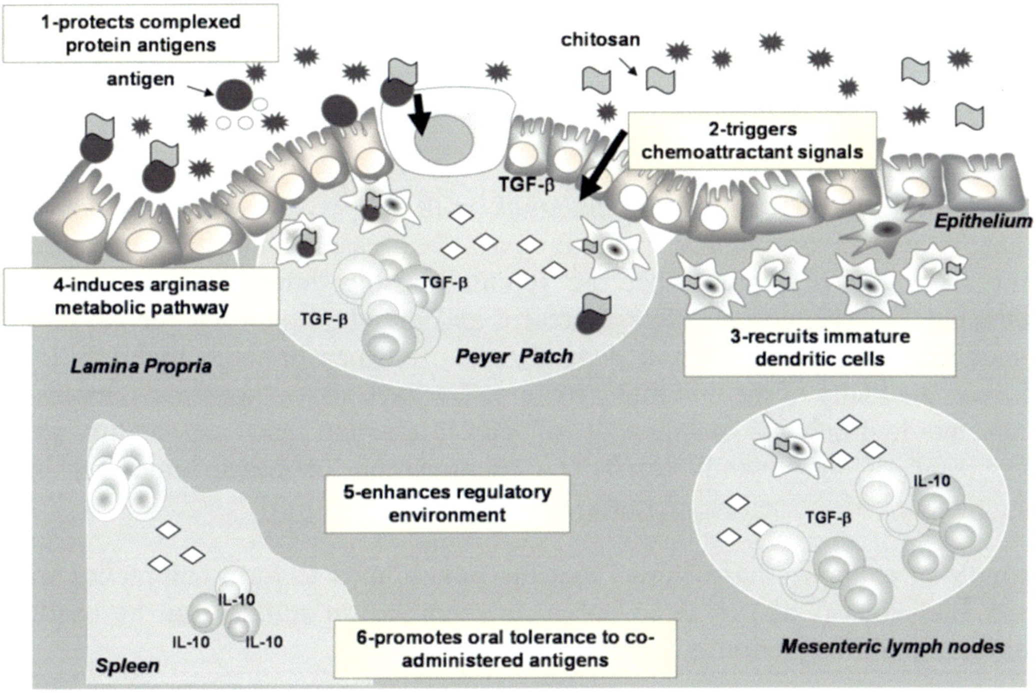

Figure 4: Chitosan main effects on the epithelian lining and immune cells located beneath enhance the regulatory environment. The polysaccharide protects complexed protein antigens from protease degradation (#1), promoting the induction of oral tolerance (#6). Chitosan is able to stimulate the epithelial lining inducing arginase activity (#4) and triggering chemoattractant signals (#2) that recruits immature dendritic cells (#3). The biological activity of this cationic polymer elicits a stream of regulatory signals (#5) that contribute to the intestinal homeostasis.

CONCLUDING REMARKS AND FUTURE PERSPECTIVES

From the countless biotechnological applications in which cationic biopolymers have been studied we selected just a few to stress several of the major points that should be kept in mind when attempting to use any of them, either alone or combined with other molecules, for the design of novel products or processes.

When dissolved cationic polymers generally increase the viscosity of the solution, a property shared with neutral and negatively charged polymers. Although for some applications this is a desired effect, viscosity should be carefully controlled when preparing a solution of the polymer as an intermediate step in a process that requires further interactions of the cationic polymer with other molecules. The presence of several positive charges within a single molecule is the dominant force for establishing electrostatic interactions, which take place extremely fast. Therefore, it is frequently seen that complete mixing of components is difficult and should be carefully controlled. Otherwise, the structures formed will be unstable and not easily reproducible in their properties (water retention, degradability, *etc*).

When studied in applications in which the most obvious effects would be those derived from the electrostatic interactions with negatively charged structures (either individual molecules or complexes like casein micelles or cellular membranes) the first point that the reported results stress is that although electrostatic interactions are always present and may be important for the initial approaching the final result will depend on the balance between hydrophilicity and hydrophobicity of the polymer structure. We have shown several examples in which the hydrophobic component of the interaction ends up acting as a determinant of the overall result (*i.e.* dissociation of casein micelles into monomers or lysis of cell membranes).

A couple of examples have shown that the possibilities to form complexes with either small ions (Cu^{2+}) or large polymers (chondroitin sulfate) can be used to stabilize and prolong a desired effect.

The possibility that cationic polymers could have an intrinsic bioactivity has generally been mistreated, ignored or simply denied. We have shown herein the numerous effects that can be induced just by a single oral dose of chitosan, either alone or associated with a protein. It is thus extremely important to stress the point that cationic polymers should never be considered as an inert, biocompatible component of biotechnological applications aimed to interact with living systems (drug delivery systems or hydrogels for space filling applications). In this context, it should also be emphasized that biocompatibility, biodegradability and polymer bioactivities not only depend on their chemical nature but also on the physical structure in which they are presented to biological systems.

In conclusion, the versatility of properties and the diversity of functions attainable with cationic polymers make it reasonably to assume that we will see more and more biotechnological applications of these marvelous molecules in the near future.

ACKNOWLEDGEMENTS

R.V. Alasino, K.L. Bierbrauer, D.M. Bettramo, S.G. Correa and I.D. Bianco are staff members of CONICET. We wish to acknowledge Andrew Gee for his technical assistance with English grammar and spelling.

CONFLICT OF INTEREST

The authors declare no conflict of interest. All the work has been supported with grants from CEPROCOR and CONICET.

ABBREVIATIONS

APIs	=	active pharmaceutical ingredients
BVDV	=	bovine viral diarrhea virus
CII	=	type II collagen
CD4+	=	lymphocytes
C-4S	=	chondroitin 4-sulfate
DCs	=	mucosal dendritic cells
E100	=	Eudragit, cationic aminoacryl methacrylate copolymer containing a tertiary amine group
GAGs	=	sulfated glycosaminoglycans
ICAM-1	=	Intercellular Adhesion Molecule 1
IECs	=	isolated intestinal epithelial cells
IFN-γ	=	interferon- γ
(Ig)G2a anti-CII	=	immunoglobulin
IL-	=	interleukin

iNOS	=	inducible nitric oxide synthase
LMW	=	low molecular weight
LPS	=	lipopolysaccharides
MHC	=	histocompatibility complex
MLNs	=	mesenteric lymph nodes
MMW	=	mean molecular weight
mRNA	=	messenger RNA
MV	=	measles virus
NO	=	nitric oxide
OmpA	=	outer membrane protein
PEG	=	Polyethylene Glycol
PMN	=	polymorphonuclear leukocytes cells
PP	=	Peyer's patches
PQ-10	=	Polyquaternium-10 cationic hydroxyethylcellulose, also known as Celquat® SC-240C, containing quaternary ammonium group
SC-230	=	another brand name for PQ-10
SC240C	=	*see* PQ-10
TFFs	=	trefoil factors
TGF-β	=	transforming growth factor β
TLRs	=	Toll-like receptors
VHC	=	hepatitis C virus

VHS = herpes virus

VIH = immunodeficiency virus

VSV = vesicular stomatitis virus

REFERENCES

[1] Aranaz I, Mengíbar M, Harris R, *et al.* Functional Characterization of Chitin and Chitosan. Current Chemical Biology 2009, 3: 203-30.

[2] Peniche C, Argüelles-Monal W, Goicoolea FM. Chitin and Chitosan: Major Sources, Properties and Applications. Belgacem MN, Gandini A, Eds.; Monomers, Polymers and Composites from Renewable Resources; Elsevier: Amsterdam, 2008, chapter 25, 517-42.

[3] Stepnova E A, Tikhonov VE, Babushkina TA, *et al.* New approach to the quaternization of chitosan and its amphiphilic derivatives. Eur Polym J 2007, 43: 2414-21.

[4] Belalia R, Grelier S, Benaissa M, *et al.* New bioactive biomaterials based on quaternized chitosan. J Agric. Food Chem 2008; 56: 1582-8.

[5] Song Y, Zhang L, Gan W, *et al.*Self-assembled micelles based on hydrophobically modified quaternized cellulose for drug delivery. Colloids Surf B 2011; 83: 313-20.

[6] Xu FJ, Ping Y, Ma J, *et al.* Comb-Shaped Copolymers Composed of Hydroxypropyl Cellulose Backbones and Cationic Poly((2-dimethyl amino)ethyl methacrylate) Side Chains for Gene Delivery. Bioconj Chem 2009; 20: 1449-58.

[7] Fayazpour F, Lucas B, Alvarez-Lorenzo C, *et al.* Physicochemical and transfection properties of cationic Hydroxyethylcellulose/DNA nanoparticles. Biomacromolecules 2006; 7: 2856-62.

[8] Verbaan FJ, Oussoren C, Snel CJ, *et al.* A comparative study of different cationic transfection agents for *in vivo* gene delivery after intravenous administration. J Gene Med 2004; 6: 64-75.

[9] Guo S, Huang Y, Wei T, *et al.* Amphiphilic and biodegradable methoxy polyethylene glycol-block-(polycaprolactone-graft-poly(2-(dimethylamino)ethyl methacrylate)) as an effective gene carrier. Biomaterials 2011; 32: 879-89.

[10] Zhang Y, Zheng M, Kissel T, *et al.* Design and biophysical characterization of bioresponsive degradable poly(dimethylaminoethyl methacrylate) based polymers for *in vitro* DNA transfection. Biomacromolecules 2012; 13: 313-22.

[11] Kotiyan PN, Vavia PR. Eudragits: role as crystallization inhibitors in drug-in-adhesive transdermal systems of estradiol. Eur J Pharmaceutics Biopharm 2001; 52: 173-80.

[12] Horisawa E, Danjo K, Haruna M. Physical properties of solid dispersion of a nonsteroidal anti-inflammatory drug (M-5011) with Eudragit E. Drug Dev Ind Pharm 2000; 26: 1271-78.

[13] Ebihara K, Schneeman BO. Interaction of bile acids, phospholipids, cholesterol and triglyceryde with dietary fibers in the small intestine of rats. J Nutr 1989; 119: 1100-6.

[14] Ormrod DJ, Holmes CC, Miller TE. Dietary chitosan inhibits hypercholesterolaemia and atherogenesis in the apolipoprotein E-deficient mouse model of atherosclerosis. Atherosclerosis 1998; 138: 329-34.

[15] Ventura P. Lipid lowering activity of chitosan, a new dietary integrator. Muzarelli R.A.A., Ed. Chitin Enzimology. Atec Edizioni, Italy. 1996, vol. 2: 55-62

[16] Shahidi F, Synowiecki J. Isolation and Characterization of Nutrients and Value-Added Products from Snow Crab (Chinoecetes Opilio) and Shrimp (Pandalus Borealis) Processing Discards. J Agric Food Chem1991; 39: 1527-32.

[17] Carroad PA, Tom RA. Bioconversion of Shellfish Chitin Wastes: Process Conception and Selection of Microorganisms. J Food Sci 1978; 43: 1158-61.

[18] Revah-Moiseev S, Carroad A. Conversion of the enzymatic hydrolysate of shellfish waste chitin to single-cell protein. Biotechnol Bioeng 1981; 23, 1067-78.

[19] Papineau AM, Hoover DG, Knorr D *et al.* Antimicrobial Effect of Water-Soluble Chitosans with High Hydrostatic Pressure. Food Biotechnol1991; 5: 45-57.

[20] El Ghaouth A, Arul J, Asselin A *et al.* Antifungal Activity of Chitosan on Post-Harvest Pathogens: Induction of Morphological and Cytological Alterations in Rhizopus Stolonifer. Mycol Res 1992; 96: 769-79.

[21] Sudharshan NR, Hoover DG, Knorr D. Anti-bacterial Action of Chitosan. Food Biotechno. 1992; 6: 257-72.

[22] Fang SW, Li CF, Shih DYC. Antifungal Activity of Chitosan and its Preservative Effect on Low-Sugar Candied Kumquat. J Food Protection 1994; 56: 136-40.

[23] Chen C, Liau W, Tsai G. Antibacterial Effects of N-Sulfonated and N-Sulfobenzoyl Chitosan and Application to Oyster Preservation. J Food Protection 1998; 61: 1124-8.

[24] Wong DWS, Gastineau FA, Gregorski KA, *et al.* Chitosan-Lipid Films: Microstructure and Surface Energy. J Agric Food Chem 1992; 40: 540-4.

[25] Uragami T, Matsuda T, Okuno H *et al.* Structure of hemically Modified Chitosan Membranes and their Characteristics of Permeation and Separation of Aqueous Ethanol Solutions. J Membrane Sci 1994; 88, 243-51.

[26] Butler BL, Vergano PJ, Testin RF, *et al.* Mechanical and Barrier Properties of Edible Chitosan Films as Afected by Composition and Storage, J Food Sci 1996; 61: 953-6.

[27] Chen RH, Hwa H. Effect of Molecular Weight of Chitosan with the Same Degree of Deacetylation on the Thermal, Mechanical, and Permeability Properties of the Prepared Membrane. Carbohydr Poly 1996; 29: 353-8.

[28] Hoagland PD, Parris N. Chitosan/Pectin Laminated Films. J Agric Food Chem 1996; 44: 1915-9.

[29] Kittur FS, Kumar KR, Tharanathan RN. Functional Packaging Properties of Chitosan Films. Z Lesbensm Unters Forsch A 1998; 206: 44-7.

[30] Bough WA. Reduction of Suspended Solids in Vegetable Canning Waste Efluents by Coagulation with Chitosan. J Food Sci 1975; 40: 297-301.

[31] Bough WA. Chitosan a Polymer from Seafood Waste, for Use in Treatment of Food Processing Wastes and Activated Sludge. Proc Biochem 1976; 11: 13-6.

[32] Bough WA, Landes DR. Recovery and Nutritional Evaluation of Proteinaceous Solids Separated from Whey by Coagulation with Chitosan. J Dairy Sci 1976; 59: 1874-80.

[33] Latlief SJ, Knorr D. Effect of Chitin as Coagulating Aid on Protein Yield, Composition and Functionality of Tomato Seed Protein Concentrates. J Food Sci 1983; 48: 1587-90.

[34] Senstad C, Mattiasson B. Purification of Wheat Germ Agglutinin Using Afinity Flocculation with Chitosan and a Subsequent Centrifugation or Flotation Step. Biotechnol Bioeng 1989; 34: 387-93

[35] Hwang D, Damodaran S. Selective Precipitation and Removal of Lipids from Cheese Whey Using Chitosan. J Agric Food Chem 1995; 43, 33-7.

[36] Sun W, Payne GF. Tyrosinase-Containing Chitosan Gels: A Combined Catalyst and Sorbent for Selective Phenol Remova.l Biotechnol Bioeng 1996; 51: 79-86.

[37] Pinotti A, Bevilacqua A, Zaritzky N. Optimization of the Flocculation Stage in a Model System of a Food Emulsion Waste Using Chitosan as Polyelectrolyte. J Food Eng 1997; 32: 69-81.

[38] Jeuniaux C. Chitosan as a Tool for the Purification of Waters in Chitin in Nature and Technology. Muzzarelli, RAA, Jeuniaux C, Gooday GW, Eds, Plenum Press, New York, USA 1986, 551-70.

[39] Micera G, Deiana S, Dessi A, *et al.* Copper and Vanadium Complexes of Chitosan in Chitin in Nature and Technology, Muzzarelli RAA, Jeuniaux C, Gooday GW, Eds, Plenum Press, New York, USA 1986, 565-7.

[40] Muzzarelli RAA, Weckx M, Fillipini O. Removal of Trace Metal Ions From Industrial Waters, Unclear Efluents and Drinking Water, with the aid of Cross-Linked N-Carboxymethyl Chitosan. Carbohydr Poly 1989; 11: 293-6.

[41] Deans JR, Dixon BG. Bioabsorbents for Waste-water Treatment in Advances in Chitin and Chitosan. Brine CJ, Sandford PA, Zikakis JP, Eds., Elsevier Applied Science, Oxford, UK 1992 648-56.

[42] Allan CR, Hadwiger LA. The fungicidal chitosan on Fungi of varying cell wall composition. Exp Mycol 1979; 3: 285-7.

[43] Bianco ID, Balsinde J, Beltramo DM, *et al.* Chitosan-induced phospholipase A_2 activation and arachidonic acid mobilization in $P388D_1$ macrophages. FEBS Lett 2000; 466: 292-4.

[44] Imeri AG, Knorr D. Effect of Chitosan on Yield and Compositional Data of Carrot and Apple Juice. J Food Sci 1988; 53: 1707-9.

[45] Soto-Perlata NV, Muller H, Knorr D. Effect of Chitosan Treatments on the Clarity and Color of Apple Juice. J Food Sci1989; 54: 495-6.

[46] Chen Y, Li C. Studies on the Application of Chitosan to Clarification of Grapefruit Juice. Food Sci 1996; 23: 617-28.

[47] Rwan J, Wu J. Deacidification of Grapefruit Juice with Chitosan. Food Sci 1996; 23: 509-19.

[48] Spagna G, Pieri PG, Rangoni C, *et al.* The Stabilization of White Wines by Adsorption of Phenolic Compounds on Chitin and Chitosan. Food Res Intern 1996; 29: 241-8.

[49] Fernandez M, Fox PF. Fractionation of cheese nitrogen using chitosan. Food Chem 1997; 58: 319-22.

[50] Ausar SF, Bianco ID, Badini RG, *et al.* Characterization of casein micelle precipitation by chitosans. J Dairy Sci 2001; 84: 361-9.

[51] Fernández C, Ausar SF, Badini RG, *et al.* An FTIR spectroscopy study of the interaction between α-casein-bound phosphoryl groups and chitosan. Int Dairy J 2003; 13: 897-901.

[52] Ausar SF, Landa CA, Bianco ID, *et al.* Hydrolysis of chitosan-induced milk aggregates by pepsin, trypsin and pancreatic lipase. Biosci Biotechnol Biochem 2001; 65: 2412-8.

[53] Ausar SF, Passalacqua N, Castagna LF, *et al.* Growth of milk fermentative bacteria in the presence of chitosan. Int Dairy J 2002; 12: 899-906.

[54] Dalgleish DA. Casein micelle as colloids: Surface structures and stabilities. J Dairy Sci 1998; 81: 3013-8.

[55] de Kruif CG. Casein micelle interactions. Int Dairy J 1999; 9: 183-8.

[56] Narambuena CF, Ausar SF, Bianco ID, *et al.* Aggregation of casein micelles by interactions with chitosans: A study by Monte Carlo simulations. J Agric Food Chem 2005; 53: 459-63.

[57] Vinsova J, Vavrikova E. Recent advances in drugs and prodrugs design of chitosan. Curr Pharm Des 2008; 14, 1311-26.

[58] Kong M, Chen XG, Xing K, *et al.* Antimicrobial properties of chitosan and mode of action: A state of the art review. Int J Food Microbiol 2010; 144: 51-63.

[59] Rabea EI, Badawy ME, Stevens CV, *et al.* Chitosan as antimicrobial agent: Applications. Biomacromolecules 2003; 6: 1457-65.

[60] Davydova VN, Bratskaya SY, Gorbach VI, *et al.* Comparative study of electrokinetic potentials and binding affinity of lipopolysaccharides-chitosan complexes. Biophys Chem 2008; 136: 1-6.

[61] Xing K, Chen XG, Liu CS, *et al.* Oleoyl-chitosan nanoparticles inhibits Escherichia coli and Staphylococcus aureus by damaging the cell membrane and putative binding to extracellular or intracellular targets. Int J Food Microbiol 2009; 132: 127-33.

[62] Jeon SJ, Oh M, Yeo WS, *et al.* Underlying Mechanism of Antimicrobial Activity of Chitosan Microparticles and Implications for the Treatment of Infectious Diseases. PLoS ONE 2014; 9(3): e92723.

[63] Ausar SF, Bianco ID, Castagna LF, *et al.* Interaction of a cationic acrylate polymer with caseins: biphasic effect of Eudragit E100 on the stability of casein micelles. J Agric Food Chem 2003; 51: 4417-23.

[64] Ausar SF, Bianco ID, Castagna LF, *et al.* Reversible precipitation of casein micelles with a cationic hydroxyethylcellulose. J Agric Food Chem 2005; 53: 9031-8.

[65] Prego C, Torres D, Alonso MJ. The potential of chitosan for the oral administration of peptides. Expert Opin Drug Deliv 2005; 2: 843-54.

[66] Şenel S. Potential applications of chitosan in oral mucosal delivery. J Drug Deliv Sci Technol 2010; 20: 23-32.

[67] Rodrigues S, Dionisio M, Remuñán López C, *et al.* Biocompatibility of chitosan carriers with application in drug delivery. J Funct Biomater 2012; 3, 615-41.

[68] Favre E, Girard S. Release kinetics of low molecular weight solutes from mixed cellulose hydrogels: a critical experimental study. Eur Polym J 2001; 37: 1527-32.

[69] Perez RD, Rubio M, Perez CA, *et al.* Applications of SR-XRF to characterize biogels obtained by irradiation of dilute aqueous biopolymers solutions. Radiation Phys Chem 2004; 71: 999-1000.

[70] Sakiyama T, Takata H, Kikuchi M, *et al.* Polyelectrolyte complex gel with high pH-sensitivity prepared from dextran sulfate and chitosan. J Appl Polym Sci 1999; 73 (11): 2227-33.

[71] Argüelles Monal W, Cabrera G, Peniche C, *et al.* Conductimetric study of the interpolyelectrolyte reaction between chitosan and polygalacturonic acid. Polymer 2000; 41(7): 2373-8.

[72] Lee JW, Kim SY, Kim SS, *et al.* Synthesis and characteristics of interpenetrating polymer network hydrogel composed of chitosan and poly(acrylic acid). J Appl Polym Sci 1999; 73 (1): 113-20.

[73] Denuziere A, Ferrier D, Domard A. Chitosan-chondroitin sulfate and chitosan-hyaluronate polyelectrolyte complexes. Physico-chemical aspects. Carbohydr Polym 1996; 29 (4): 317-23.

[74] Sui W, Huang L, Wuang J, *et al.* Preparation and properties of chitosan chondroitin sulfate complex microcapsules. Colloids Surf B 2008; 65 (1): 69-73.

[75] Fajardo AR, Piai JF, Rubira AF, *et al.* Time- and pH-dependent self-rearrangement of a swollen polymer network based on polyelectrolytes complexes of chitosan/chondroitin sulfate. Carbohydr Polym 2010; 80 (3): 934-43.

[76] Lawrie G, Keen I, Drew B, *et al.* Interactions between alginate and chitosan biopolymers characterized using FTIR and XPS. Biomacromolecules 2007; 8 (8): 2533-41.

[77] Rodríguez R, Alvarez-Lorenzo C, Concheiro A. Rheological evaluation of the interactions between cationic celluloses and carbopol 974P in water. Biomacromolecules 2001; 2 (3): 886-93.

[78] Brode GL, Band PA, Goddard ED, *et al.* Glycosaminoglycan and cationic polymer combinations. US Patent 1988; No. 4767463.

[79] Brode GL, Band PA, Goddard ED, *et al.* Processes for managing keratinous material using glycosaminoglycan and cationic polymer combinations. US Patent 1990; No. 4913743.

[80] Bierbrauer KL, Alasino RV, Strumia MC, *et al.* Cationic cellulose and its interaction with chondroitin sulfate Rheological properties of the polyelectrolyte complex. European Polymer J 2014; 50: 142-9.

[81] Sawitzke AD. Clinical efficacy and safety of glucosamine, chondroitin sulphate, their combination, celecoxib or placebo taken to treat osteoarthritis of the knee: 2-year results from GAIT. Arthritis Rheum 2010; 69 (8): 1459-64.

[82] Erickson JM, Messer TM. Glucosamine and chondroitin sulfate treatment of hand osteoarthritis. J Hand Surg Am 2013; 38 (8): 1638-40.

[83] Frisbie DD, McIlwraith CW, Kawcak CE, *et al.* Evaluation of intra-articularhyaluronan, sodium chondroitin sulfate and N-acetyld-glucosamine combination *versus* saline (0.9% NaCl) for osteoarthritis using an equine model. Vet J 2013; 197 (3): 824-9.

[84] Sasaki H, Yamamura K, Nishida K, *et al.* Delivery of drugs to the eye by topical application. Prog Retinal Eye Res 1996; 15 (2): 583-620.

[85] Ludwig A. The use of mucoadhesive polymers in ocular drug delivery. Adv Drug Delivery Rev 2005; 57: 1595- 1639.

[86] Sahoo SK, Dilnawaz F and Krishnakumar S. Nanotechnology in ocular drug delivery. Drug Discov Today 2008; 13 (3-4): 144-51.

[87] Chang JN. Recent Advances in Ophthalmic Drug Delivery. Handbook of Non-Invasive Drug Delivery Systems 2010; 7: 165-92.

[88] Pignatello R, Bucolo C, Ferrara P, *et al.* Eudragit RS100 nanosuspensions for the ophthalmic controlled delivery of ibuprofen. Eur J Pharm Sci 2002; 16: 53-61.

[89] Baeyens V, Percicot C, Zignani M, *et al.* Ocular drug delivery in veterinary medicine. Adv Drug Delivery Rev 1997; 28: 335-61.

[90] McCormick WJ. Vitamin C in the prophylaxis and therapy of infectious diseases. Arch Pediatr 1951; 68:1-9.

[91] Vilchèze C, Hartman T, Weinrick T, *et al.* *Mycobacterium tuberculosis* is extraordinarily sensitive to killing by a vitamin C-induced Fenton reaction. Nat Commun 2013; 4: 1881-1904.

[92] Beltramo DM, Ausar SF, Virga C, *et al.* Argentinean Patent 010101270 Composición para la prevención y el tratamiento de las enfermedades bucales, producto farmacéutico, kit que lo contiene, uso del mismo y método de preparación.

[93] Alasino RV, Ausar SF, Bianco ID, *et al.* Amphipathic and membrane-destabilizing properties of the cationic acrylate polymer Eudragit E100. Macromol Biosci 2005; 5: 207-13.

[94] Alasino RV, Bianco ID, Vitali MS, *et al.* Characterization of the inhibition of enveloped virus infectivity by the cationic acrylate polymer Eudragit E100. Macromol Biosci 2007; 7: 1132-8.

[95] Azad AK, Sermsintham N, Chandrkrachang S, *et al.* Chitosan membrane as a wound-healing dressing: Characterization and clinical application. J Biomed Mater Res 2004; 69B: 216-22.

[96] Dai T, Tanaka M, Huang Y-Y, *et al.* Chitosan preparations for wounds and burns: antimicrobial and wound-healing effects. Expert Rev Anti Infect Ther 2011; 9 (7): 857-79.

[97] Archana D, Dutta J, Dutta P.K. Evaluation of chitosan nano dressing for wound healing: Characterization, *in vitro* and *in vivo* studies. Int J Biol Macromol 2013; 57: 193-203.

[98] Sanandam M, Salunkhe A, Shejale K, *et al.* Chitosan bandage for faster blood clotting and wound healing. Int J Adv Biotechnol Res 2013; 4 (1): 47-50.

[99] Jayakumar R, Prabaharan M, Kumar PTS, *et al.* Chitin and Chitosan Materials in Wound Dressing, Biomedical Engineering. Trends in Materials Science, Mr Anthony Laskovski (Ed.), ISBN: 978-953-307-513-6, InTech, 2011. Available from: http://www.intechopen.com/books/biomedical-engineering-trends-in-materials-science/novel-chitin-andchitosan-materials-in-wound-dressing

[100] Halim AS and Lim CK Biomedical-Grade Chitosan in Wound Management and Its Biocompatibility *In vitro*, Biopolymers, MagdyElnashar (Ed.), ISBN: 978-953-307-109-1, InTech, 2010. Available from: http://www.intechopen.com/books/biopolymers/biomedical-grade-chitosan-in-wound-management-andits-biocompatibility-*in vitro*

[101] Giri TK, Thakur A, Alexander A, *et al.* Modified chitosan hydrogels as drug delivery and tissue engineering systems: present status and applications. Acta Pharmaceutica Sinica B 2012; 2 (5): 439-49.

[102] Aziz A, Cabral JD, Brooks HJL, *et al.* Antimicrobial Properties of a Chitosan Dextran-Based Hydrogel for Surgical Use. Antimicrob Agents Chemother 2012; 56 (1): 280-7.

[103] Akncbay H, Senel S, Ay ZY. Application of chitosan gel in the treatment of chronic periodontitis. J Biomed Mater Res B Appl Biomater 2007; 80 (2): 290-6.

[104] Elsabeea MZ, Naguiba HF, Morsib RE. Chitosan based nanofibers. Review Mat Sci Eng C 2010; 32 (7): 1711-26.

[105] Jayakumar R, Prabaharan M, Nair SV, *et al.* Novel chitin and chitosan nanofibers in biomedical applications. Biotechnol Adv 2010; 28(1): 142-50.

[106] Li Q, Dunn ET, Grandmaison EW, *et al.* Application and properties of chitosan. M.F.A. Goosen, Ed. Applications of Chitin and Chitosan, Technomic Publishing Co. Lancaster PA, USA 1997, 3-30.

[107] Peluso G, Petillo O, Ranieri M, *et al.* Chitosan-mediated stimulation of macrophage function. Biomaterials 1994; 15: 1215-20.

[108] Porporatto C., Bianco ID, Riera CM, *et al.* Chitosan induces different L-arginine metabolic pathways in resting and inflammatory macrophages. Biochem Biophys Res Commun 2003; 304: 266-72.

[109] Usami Y, Okamoto Y, Takayama T, *et al.* Chitin and chitosan stimulate canine polymorphonuclear cells to release leukotriene B4 and prostaglandin E2 J. Biomed Mater Res 1998; 42: 517-22.

[110] Mori T, Okumura M, Matsuura M, *et al.* Effects of chitin and its derivatives on the proliferation and cytokine production of fibroblasts *in vitro*. Biomaterials 1997; 18: 947-51.

[111] Shibata Y, Metzger WJ, Myrvik QN. Chitin particle-induced cell-mediated immunity is inhibited by soluble mannan: mannose receptor-mediated phagocytosis initiates IL-12 production. J Immunol 1997, 159: 2462-7.

[112] Ueno H, Nakamura F, Murakami M, *et al.* Evaluation effects of chitosan for the extra-cellular matrix production by fibroblasts and the growth factors production by macrophages. Biomaterials 2001; 22: 2125-30.

[113] Nishimura K, Ishihara C, Ukei S, *et al.* Stimulation of cytokine production in mice using deacetylated chitin. Vaccine 1986; 4: 151-6.

[114] Schaffer MR, Tantry U, Gross SS, *et al.* Nitric oxide regulates wound healing. J Surg Res 1996; 63: 237-40.

[115] Albina JE, Mills CD, Henry WL Jr, *et al.* Temporal expression of different pathways of L-arginine metabolism in healing wounds. J Immunol 1990; 144: 3877-30.

[116] Elgindy N, Samy W. Evaluation of the mechanical properties and drug release of cross-linked Eudragit films containing metronidazole. Int J Pharm 2009; 376: 1-6.

[117] Porporatto C, Bianco ID, Cabanillas AM, *et al.* Early events associated to the oral coadministration of type II collagen and chitosan: induction of anti-inflammatory cytokines. Int Immunol 2004; 16: 433-41.

[118] Porporatto C, Bianco ID, Correa SG. Local and systemic activity of the polysaccharide chitosan at lymphoid tissues after oral administration. J Leukoc Biol 2005; 78 (1): 62-9.

[119] Porporatto C, Canali MM, Bianco ID, *et al.* Ability of the polysaccharide chitosan to inhibit proliferation of CD4+ lymphocytes from mucosal inductive sites, *in vitro* and *in vivo*. Cell Prolif 2009; 42 (6): 780-7.

[120] Porporatto C, Canali MM, Bianco ID, *et al.* The biocompatible polysaccharide chitosan enhances the oral tolerance to type II collagen. Clin Exp Immunol 2009; 155 (1): 79-87.
[121] Van der Lubben I, Kersten G, Fretz MM, *et al.* Chitosan microparticles for mucosal vaccination against diphtheria: oral and nasal efficacy studies in mice. Vaccine 2003; 21 (13-14): 1400-8.
[122] Kang ML, Kang SG, Jiang HL, *et al. In vivo* induction of mucosal immune responses by intranasal administration of chitosan microspheres containing Bordetella bronchiseptica DNT. Eur J Pharm Biopharm 2006; 63 (2): 215-20.
[123] Schipper N GM, Olsson S, Hoogstraate JA, *et al.* Chitosans as absorption enhancers for poorly absorbable drugs 2: mechanism of absorption enhancement. Pharm Res 1997; 14: 923-9.
[124] Weiner HL. Oral tolerance: immune mechanisms and treatment of autoimmune diseases. Immunol Today 1997; 18: 335-43.
[125] Canali MM, Porporatto C, Aoki MP, *et al.* Signals elicited at the intestinal epithelium upon chitosan feeding contribute to immunomodulatory activity and biocompatibility of the polysaccharide. Vaccine 2010; 28: 5718-24.
[126] Muller CA, Autenrieth IB, Peschel A. Innate defenses of the intestinal epithelial barrier. Cell Mol Life Sci 2005; 62 (12): 1297-307.
[127] Neutra MR, Mantis NJ, Kraehenbuhl JP. Collaboration of epithelial cells with organized mucosal lymphoid tissues. Nat Immunol 2001; 1 (11): 1004-9.
[128] Chen H, Li P, Yin Y, *et al.* The promotion of type 1 T helper cell responses to cationic polymers *in vivo via* toll-like receptor-4 mediated IL-12 secretion. Biomaterials 2010; 31: 8172-80.
[129] Lee J, Sohn JW, Zhang Y, *et al.* Nucleic acid-binding polymers as anti-inflammatory agents. Proc Natl Acad Sci USA 2011, 108: 14055-60.

CHAPTER 2

Collagen Substrates for Soft Tissue Engineering

Edward Suesca[1], Alejandro Sosnik[2] and Marta R. Fontanilla[1,*]

[1]Tissue Engineering Group, Department of Pharmacy, Universidad Nacional de Colombia, Bogotá, Colombia and [2]Group of Pharmaceutical Nanomaterials Science, Department of Materials Science and Engineering, Technion-Israel Institute of Technology, Haifa, Israel

Abstract: A main challenge for soft tissue regeneration is to develop products and therapies that minimize the fibrotic scarring characteristic of tissue repair. Scaffolds are three-dimensional structures in which cells can attach, grow and differentiate to form *ex vivo* or *in vivo* artificial tissue. They provide signals that trigger cell migration from the wound bed, as well as cell differentiation and cell secretion of extracellular matrix constituents. Scaffolds are made of natural or synthetic materials, with proteins from the collagen family being among the most used natural polymers. Collagen type I is a major component of the complex extracellular network of proteins that form the matrix of mammal tissues. Besides having cell-binding sequences, this protein is biodegradable, biocompatible, and exhibits a haemostatic effect when placed in open wounds. The aforementioned properties have made this compound a widely used natural material to produce scaffolds for tissue engineering skin and oral mucosa substitutes. This chapter reviews some of the parameters that influence the bioactivity of scaffolds emphasizing on collagen I scaffolds and their major applications in soft tissue engineering.

Keywords: Collagen I, artificial tissue, biomaterial, biocompatibility, extracellular matrix, scaffold microstructure, soft tissue regeneration, regenerative medicine, tissue engineering, wound healing.

1. INTRODUCTION

Early in human development, the capacity of various tissues and organs to regenerate in response to tissue damage or tissue turnover is lost [1, 2]. Blood, liver parenchyma, epithelium, hair, nails, and bone marrow are among the tissues that maintain the ability to regenerate [3]. Conversely, when the wound is larger than a critical size that depends on the tissue (*e.g.*, cartilage, retina, muscle, spinal cord, alveoli, pancreas, myocardium, mucous connective tissue and dermis) is replaced by scar tissue, which is structurally and functionally different to the

***Corresponding author Marta R. Fontanilla:** Head of the Tissue Engineering Group, Laboratory 318, Department of Pharmacy, Universidad Nacional de Colombia. Av. Carrera 30 # 45-10, Bogotá, Colombia; Tel: +57-1-3165000 ext. 14664, 14668 and 16960; Fax: +57-1-3165000 ext. 16959; E-mail: mrfontanillad@unal.edu.co original one [4]. Some vertebrates grouped as Urodela within the order Caudata, have kept the extraordinary ability to regenerate tissues, appendages and body parts when these structures are lost or injured [5, 6]. Cellular dedifferentiation underlies this capability because differentiated cells present in the injured site dedifferentiate to progenitor cells that will then re-differentiate and synthesize the new tissue required to replace the lost one [7]. Regeneration involves three stages: acute response to the injury, formation of progenitor cells, and morphogenesis of tissues and organs that will replace the missing structure [8]. During regeneration, removal of damaged tissue is accompanied by synthesis and remodeling of new tissue that replaces the damaged or lost one [9]. For this reason, only the formation of new tissues and organs should be regarded as regeneration. Despite this, it is common to consider the compensatory hyperplasia that occurs in mammal glands such as thyroids, adrenals, and liver as regeneration [10].

Wound healing in adult mammals does not exhibit cell dedifferentiation and morphogenesis [11]. Except epithelia and bone, most mammal tissues do not regenerate like the salamander´s tissues do, instead upon injury they heal with scarring [4, 11]. This process leads to the formation of non-functional contracted tissue that lacks the structure of the injured one [4, 12]. Due to the limited human capacity to regenerate, the most commonly used treatments to address the loss of tissues and organs are transplants, autografts, homografts, and prosthesis. Usual drawbacks of these approaches are the scarceness of graft tissue, creation of wounds in donor areas, and immune rejection. Tissue engineering approaches have been developed to overcome them [13]. Despite their limitations, artificial tissues and organs developed by tissue engineering are becoming an option to promote regeneration rather than fibrotic tissue repair [14].

The scaffold that serves as an extracellular matrix (ECM) analog is an important component of soft engineered tissues [15]. It has been demonstrated that scaffolds effectively delay wound contraction and induce regeneration when they are bioactive [16]. In recent years, several types of scaffolds have been designed and produced using different biomaterials and methods that influence their bioactivity. In fact, data gathered during *in vivo* evaluation of engineered tissues indicate that the outcome of wound healing is associated with scaffold composition, biodegradability, mechanical properties, and microstructure.

Biomaterials used to manufacture scaffolds must be biocompatible, biodegradable and, in most of the cases, pro-angiogenic. They could be natural or synthetic polymers, and can be used alone or in combination [17]. Natural biomaterials are mainly animal proteins constituents of the ECM of mammal tissues that display characteristics that promote cell adhesion, proliferation and migration. Besides being present in the connective tissue, collagen type I underlie the vascular endothelium and its exposure upon endothelium injury is a primary initiator of the coagulation cascade, being this the reason for its use as a haemostatic agent [18]. Due to its biocompatibility, biodegradability, and ease of tailoring in a 3D assembly, it has been extensively investigated as scaffolding material for accelerated tissue regeneration [16, 19-21]. The above mentioned advantages have triggered development of substitutes for skin [22], oral mucosa [23], cartilage, tendon, nerve, and vocal fold regeneration based on collagen type I scaffolds. This chapter initially reviews some of the parameters that influence scaffold bioactivity emphasizing on collagen type I scaffolds. Then the major clinical applications of collagen I substrates in soft tissue engineering are discussed.

2. SCAFFOLDS FOR SOFT TISSUE ENGINEERING

Tissue major components are cells and the ECM secreted by them during tissue morphogenesis. The ECM consists of a complex network of proteins, proteoglycans, and polysaccharides that offer attachment sites for cells and play an important role in cell differentiation, cell migration, cell signaling, regeneration, and repair [24-26]. It contributes to the mechanical integrity of the tissue by increasing its resistance to tensile stresses [27], and facilitates transport of nutrients and wastes [28, 29]. Most cells are constantly changing the ECM to sculpt their microenvironment in a dynamic reciprocal communication that conveys intracellular signaling [24, 25], influences the phenotype of cell, induces tissue formation [29], and controls homeostasis [30]. Consequently, cell-ECM interactions are essential in morphogenesis and wound healing.

In tissue engineering, scaffolds are used as analogs of the ECM during the healing of tissues and organs that have been damaged or lost. Increasing evidence has demonstrated that scaffolds seeded or unseeded with cells, when implanted, can promote tissue regeneration. In fact, we have shown in a lagomorphic model of oral mucosa wound that partial thickness mucosal wounds heal faster and better when grafted either with autologous artificial tissue made from collagen type I scaffolds or with acellular collagen type I scaffolds than those closed by secondary intention [31].

Ideally, tissue engineered products should promote tissue regeneration over tissue repair; to fulfill this purpose, scaffolds must be designed considering the characteristics of the site to be implanted. In other words, a scaffold intended for soft tissue application will demand a different design than a scaffold that will be employed in hard tissue [32, 33]. Chemical composition, degradation rate, mechanical properties, and microstructure are the four parameters that must account in designing pro-regenerative scaffolds. These characteristics are interconnected and determine scaffold bioactivity, *i.e.*, whether or not regeneration rather than repair will occur when a scaffold is implanted (Fig. **1**).

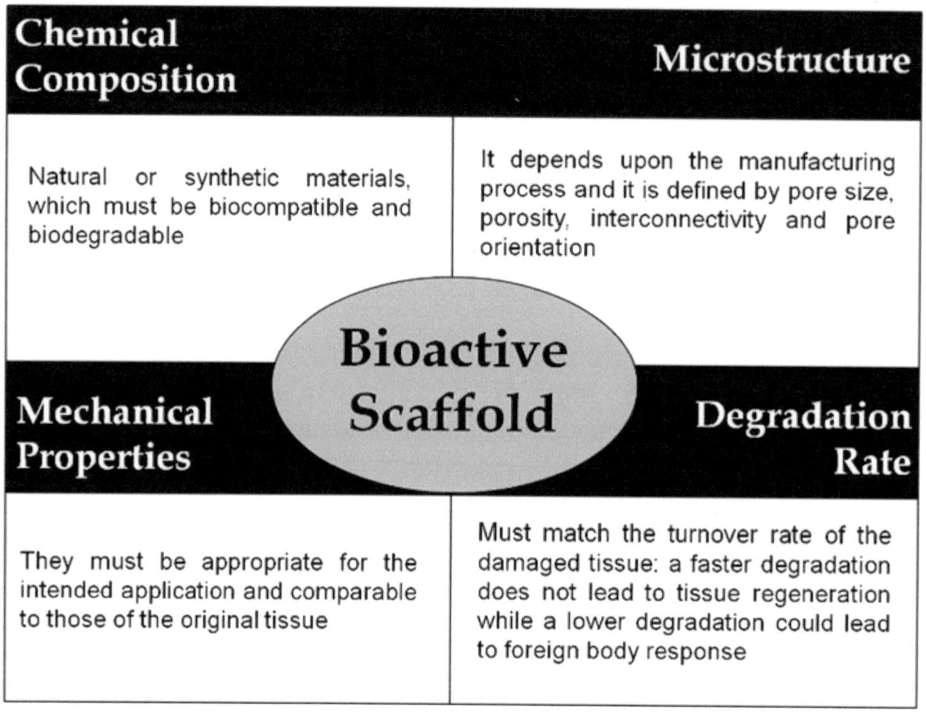

Figure 1: Scaffold parameters that influence tissue regeneration. The chemical composition, microstructure, mechanical properties, and degradation rate of the scaffold must be designed to inhibit wound contraction in the specific implantation site.

The biomaterials used to manufacture scaffolds are natural and synthetic polymers and their modification or combination (Table **1**). In general, they are easy to tailor in a 3D assembly exhibiting the strength, degradation time, and microstructure demanded by the zone to be implanted.

One critical feature of the biomaterials used in tissue engineering is biocompatibility; they do not induce adverse responses in animal beings and can interact properly with the living environment to induce tissue regeneration [34]. Scaffolds and artificial tissues synthesized with them are not intended to be permanent in the grafted site [35]. The own cells of the host degrade and replace these products with new tissue similar to the one that has been lost [36]. Consequently, the scaffold material must be biodegradable and the resulting degradation products should be easy to eliminate from the body. They also must be pro-angiogenic because seeded cell survival crucially depends on the rapid establishment of blood supply (except in avascular tissues such as cartilage and the transparent structures of the eye) [37].

Table 1. Polymers used to manufacture scaffold for soft tissue engineering.

	Biomaterial	Reference
Natural	*Decellularized tissues* Heart valves, human skin, human nerves, porcine intestinal submucosa and bladder	[38-42]
	Proteins Collagen, fibrin, elastin and silk	[31, 43, 44]
	Polysaccharides Glycosaminoglycan, chitin, chitosan (chitin derivative), starch, cellulose, agar and alginate	[45, 46]
	Natural polyesters Polyhydroxybutyrate, poly(hydroxybutyrate-co-valerate)	[47, 48]
Synthetic	*Polyesters* Poly (epsilon-caprolactone), poly(lactic acid), poly(glycolic acid) and their copolymers.	[49, 50]
	Polyanhydrides Polymers and copolymers of 1,6-bis (p-carboxyphenoxy) hexane; 1,8-bis (p-carboxyphenoxy) -3,6-dioxaoctane; 1,6-bis (p-carboxyphenoxy) propane; (o-carboxyphenoxy) p-xylene	[51-53]
	Polyalkylcarbonates Structures based on 1,3-trimethylene carbonate	[54, 55]
	Polyethers Polyethylene glycol	[56, 57]
	Synthetic polypeptides Self-assembling peptides	[58]

Collagen type I fulfills all the above mentioned requirements for a biomaterial with tissue engineering application: one of its major roles is to provide tensile strength, it has cell binding sequences (*e.g.*, RGD motifs recognized by cell membrane integrins), it is degraded by cell secreted collagenases, and it is a highly conserved protein within mammals characterized for being a poor immunogenic biomaterial [59]. In fact, studies in mice and rabbits have shown immune responses to collagen only when immunizations are carried out with adjuvants, and after several expositions during an extended period of time [60].

The main natural sources of this protein are bovine, porcine, murine and human tissues. Currently, one of the most important sources of human collagen type I is the human placenta [61]. Collagen type I clinical application in soft tissue engineered is versatile and widespread. In this chapter we will not discuss decellularized mammal tissues as collagen substrates for soft tissue engineering because they have unknown quantities of other ECM components (e.g., glycosaminoglycans, proteoglycans, and growth factors) [62].

3. MAIN CHARACTERISTICS OF COLLAGEN SCAFFOLDS

3.1. Degradation Rate and Mechanical Properties

Tissue engineering scaffolds are temporary devices that facilitate tissue healing and regeneration. After implantation they become degraded and undergo changes leading to a decrease in their mechanical properties [63, 64]. Therefore, scaffold degradation rate might be tuned with tissue turnover to ensure that injured tissue is replaced by new one. Some soft tissue engineering applications such skin and mucosa may require relatively rapid degradation times [16], while other applications like vascular grafts and heart valves may require low degradation rates to maintain their mechanical integrity for months or even years [65]. The lack of tuning between degradation rate and tissue turnover may elicit a foreign body response in which a fibrous layer is formed at the interface between material and tissue.

Wound healing comprises a sequentially organized series of events that involves inflammation, formation of granulation tissue, and remodeling of the ECM [66]. Ideally, scaffolds must be degraded *in vivo* when granulation tissue is being formed and ECM is being remodeled. In skin wound models, dermis regeneration is observed when scaffold degradation and new tissue synthesis are approximately equal [66]. Fast degradable collagen-glycosaminoglycan scaffolds do not delay wound contraction therefore they do not improve scarring healing [67]. Degradation rate can be controlled through scaffold physicochemical properties such as hydrophilicity, glass transition temperature, solid content, and porosity. Ideally, scaffolds must undergo surface erosion rather than mass hydrolysis (*e.g.*, polyesters) because the latest process results in the rapid loss of mechanical strength and structure. In this context, scaffold degradation and remodeling by cell produced enzymes is more likely to result in the generation of new tissue.

Collagen type I degradation rate is controlled through cross-linking, a process in which free lysine amino groups are involved. Several physical and chemical agents are used in the cross-linking treatments. Dehydrothermal treatment and ultraviolet and gamma irradiation are the most used physical procedures while glutaraldehyde, carbodiimides, genipin are the reagents commonly used for chemical procedures. Collagen type I cross-linking not only influences the scaffold degradation rate, it also modifies the scaffold mechanical properties and efficacy in the healing process. For example, the effectiveness of collagen scaffolds to support chondrogenesis of human mesenchymal stem cells has been related to their degradation rate, more than to their chemical composition. Scaffolds made with cross-linked collagen that degraded slowly supported cell differentiation and matrix deposition while the uncross-linked that collapsed rapidly did not [68].

Scaffolds must provide temporary mechanical support until the regenerated tissue can bear mechanical loads by itself. This means that scaffolds must match the mechanical properties of the biological environment, for instance, a scaffold should not be stiffer than the surrounding tissue because it can cause adverse physiologic responses, such as tissue resorption [69]. Conversely, poor mechanical behavior might lead to rapid scaffold failure. The mechanical properties of a scaffold decrease as a result of degradation. Therefore, an *in vivo* evaluation can be useful in establishing whether the mechanical properties of a scaffold are appropriate or not for a particular application in tissue engineering. In other words, *in vitro* experiments cannot replace *in vivo* ones although, it is worth stressing that *in vitro* experiments are very relevant from an ethical perspective because they enable the selection of the best candidates and the minimization of the use of animals.

Mechanical properties of the scaffolds can be modified by changing parameters like biomaterial concentration, porosity, and cross-linking approach and extent [70]. For collagen hydrogels, increased biomaterial concentration results in greater mechanical properties that better withstand mechanical stress [64]. Conversely, the mechanical properties of thermoplastic polymers might be adjusted by means of molecular weight, thermal properties, and crystallinity. Tissue engineered scaffolds exhibit high porosity and low material content, hence their mechanical properties are often dictated primarily by their microstructure [70, 71]. Another approach to achieve this goal is the control of the cross-linking density [72]. Most of the studies on cross-linking influence on the mechanical properties of collagen type I scaffolds for soft tissue applications have been focused on comparing the effect of various cross-linking procedures on the physicochemical properties of the scaffold [73-76]. However, studies on the influence of the mechanical properties of collagen type I scaffolds on their *in vivo* performance are rather sparse.

The studies reported on this matter have shown that tuning the mechanical properties with a cross-linking procedure affects the outcome of applying a scaffold. To improve stem cell engraftment in cardiac tissue, scaffolds made of bovine collagen type I with different degrees of cross-linking were evaluated *in vitro* and *in vivo* in their mechanical properties and biocompatibility. Cross-linked and non-cross-linked scaffolds showed statistical differences in their mechanical characteristics which were also influenced by the degree of cross-linking. Conversely, *in vitro* evaluation of adipose derived stem cell adhesion and proliferation showed that these biological properties were neither affected by cross-linking nor by the degree of cross-linking. The implantation of the collagen scaffolds in a rat model of myocardial infarction indicated the opposite: the non-cross-linked ones were the scaffolds that adhered to the heart [77]. These results pointed out again the relevance of conducting *in vivo* experiments and the limitations of the *in vitro* models. A different work assessing ECM gene expression, secretion profile and growth of endothelial cells seeded into denatured collagen type I matrices cross-linked with different concentrations of carbodiimides, found that the highest growth occurred into scaffolds with intermediate cross-linking. In general, scaffold stiffness affects inhibitory control factors but not the proliferative ones *i.e.* endothelial cells grown on stiffer scaffolds inhibited T cell but not smooth muscle cell proliferation and ECM production except for elastin [78]. The above mentioned results are contradicted by data from a work that evaluated growth of endothelial cells into collagen type I scaffolds exhibiting different cross-linking degrees attained with a non-enzymatic-glycation procedure which does not affect scaffold microstructure. This study found that endothelial cell spreading and angiogenic sprouts increased in number and length proportionally to the stiffness of the scaffolds [79].

3.2. Scaffold Microstructure

A major challenge in tissue engineering is mimicking the complex cellular organization and function of the tissues found in the human body [80]. Tissue structure and function are highly interrelated. For example, in tendon, fibrils are highly aligned following a parallel pattern to provide tensile strength [81]. In the skin, the wickerwork pattern of collagen fibers confers mechanical strength [82], while in the lung elastic fibers form a cuplike matrix that shapes the alveoli and sustains the process of gas exchange [80], and collagen fibers confer the function and polygonal arrangement of hepatocytes in the liver [83].

Microstructure of engineered scaffolds is highly dependent on the fabrication process selected and on the material used. Recent advances have made it possible to create scaffolds with controlled architecture, that is, with defined porosity, pore size, interconnectivity, and orientation. Collagen type I scaffolds that resemble the lung, tendon and skin architecture have been manufactured using different freezing conditions during the freeze-drying procedure [80]. Below we will discuss some of the considerations that several researchers have been made regarding these properties in artificial analogs of extracellular matrix of the connective tissue, stressing in those made with collagen type I.

3.2.1. Porosity

Porosity is the measure of the void volume fraction within the scaffold and it defines the amount of solid material. High porosity, typically greater than 90%, is needed for cell seeding and in-growth [84]. However, this property is inversely related to the surface area of the scaffold that defines the total surface of the scaffold interacting with the cells. Consequently, there must be a balance between porosity and total surface to ensure that cell binding sites and mechanical properties of the scaffold are well adjusted and maintained [85]. Porosity also affects the diffusion of nutrients and waste products that allows cell survival inside the scaffold [86]. Overall, data is pointing out that scaffolds with high-porosity are likely to support greater cellular metabolism, cell migration and differentiation, and ECM formation compared to low-porosity scaffolds [87].

Fibroblast cells preference to grow over porous structures was demonstrated by Daskalova *et al.* [88]. These researchers used laser beams to create porous rows in a region of a collagen-elastin film while left the rest of the film untreated to have a non-porous surface. After seeding fibroblasts in both porous and non-porous regions, it was found that cells selectively adhered, migrated and proliferated into the porous zone rather than into the non-porous one. Spreading and migration of human foreskin fibroblast have been studied in scaffolds manufactured with 1 to 4 mg/mL collagen suspensions exhibiting porosities between 3.7 and 0.9 μm^2. Cell migration increased in function of collagen concentration (lower porosity) and cells on the surface of the scaffolds change from dentritic to flattened and polarized morphology. Results were similar when the scaffolds were cross-linked with gluraldehyde, suggesting that porosity rather than stiffness influences fibroblast migration and morphology [89].

3.2.2. Pore Size

Studies of collagen type I based scaffolds have shown that pore size must be within a critical range depending upon the type of cell used and the tissue being engineered [16]. The size of the pores must allow cells to migrate into the scaffold, and to adhere to the fibers surface. Therefore, the lower pore size of a scaffold depends upon the size of the cells seeded into it. The upper limit is defined by the surface area accessible to cells for attachment. Major studies carried out with scaffolds based on collagen established out a range of pore size favoring tissue formation: collagen-glycosaminoglycan scaffolds used as skin substitutes had shown not to favor regeneration when the pore size was out of the 20-120 μm range [22]. Other authors have demonstrated that scaffold pore size affects cell attachment, growth rate and viability. A study evaluating cell attachment and viability in relationship with the scaffold structure has shown that collagen-glycosaminoglycan scaffolds exhibiting different pore sizes present different viable cell attachment. The same study also found that cell attachment decreased when the pore size of the scaffolds increased [85]. Moreover, scaffold pore size has been shown to be a critical determinant of the rate of vessel in-growth, which is significantly highest in the scaffold with the largest pores [90]. Pore microstructure affects cell binding and cell migration *in vitro* and influences cell in-growth into the scaffold *in vitro* and *in vivo*. The size of the pore requirements also depend on the biomaterial use for scaffold manufacturing. The critical size of cylindrical pores in scaffolds used for fibrovascularization was found to be above 500 μm in poly(L-lactic acid) (PLLA). Scaffolds for dermis regeneration made of cross-linked collagen and glycosaminoglycans need a pore range between 20 and 120 μm [22], and the range of pore diameters required for collagen scaffolds used in peripheral nerve regeneration is 5-10 μm [91].

3.2.3. Interconnectivity

Non-interconnected porosity exhibits membrane-like faces between adjacent pores, sealing the environment of one pore from its neighbors. Therefore, the porosity of scaffolds for tissue engineering applications must be interconnected to allow in-growth of cells, vascularization, cell interaction, and diffusion of nutrients and waste material within the construct. The influence of scaffold interconnectivity on the rate of tissue in-growth was studied comparing scaffolds based on poly(ethylene glycol terephthalate)-poly(butylene terephthalate) (PEGT-PBT) copolymer and hybrid scaffolds composed of PEGT-PBT and lyophilized collagen [92]. Due to better seeding efficiency PEGT-PBT scaffolds with similar porosity (75-80%) and pore size range (90–350 μm) and three different interconnectivities were chosen to be studied. Formation of homogenous connective tissue throughout the complete void space was observed in the scaffolds with intermediate interconnectivity (160 ± 56 μm). The scaffolds with the lowest interconnectivity showed formation of tissue but several pores did not present tissue formation, while the ones with the highest interconnectivity (191 ± 69 μm) showed localized formation of dense tissue and many empty areas. In general, it is accepted that scaffold interconnectivity must be not to low and not too high to allow tissue to grow homogeneously. In this revision, similar studies on interconnectivity for collagen-based scaffolds were not found.

3.2.4. Scaffold Fiber Orientation

Some tissues like tendon, cornea, and nerve are aligned to display their mechanical performance or physiology. Therefore, fabrication of unidirectional scaffolds has been investigated with regard to regeneration of peripheral nerves [93-95], spinal cord [93-95], tendon [96, 97], cornea [98], and bone [99, 100]. It has been demonstrated *in vivo* that nerve guidance matrices with longitudinally oriented micro-channels favor regeneration over repair in the same way that an autograft does [95, 101]. Another study showed that scaffold microstructure had an effect on corneal fibroblast cell phenotype and that oriented collagen fibers down-regulated alpha-smooth muscle actin expression in rabbit corneal fibroblasts [98]. These results shown that the phenotype of corneal cells can be controlled in the tissue-engineered environment.

Commonly, when cells are seeded into scaffolds tissue is formed on the outer edge as a result of limitations on cell migration and poor exchange of nutrients and waste products with the culture medium. Collagen type I scaffolds that favor cell migration and nutrient diffusion to their inner part have been developed incorporating aligned channels into them to overcome this issue. Also, it was found that when compared with random pores, unidirectional pores promoted in-growth of surrounding tissues into the pores [102]. Another interesting example of scaffold designing is the development of modular constructs coated with endothelial cells to provide a more blood-compatible and non-thrombogenic surface [103].

We have developed oral artificial connective tissue by seeding oral fibroblasts into collagen I scaffolds with multi and unidirectional oriented fibers (Fig. **2**). The assessment of protein secretion profile of the artificial tissue showed that secretion of factors that modulate wound healing was higher when fibroblasts were seeded into the unidirectional scaffolds than when they were seeded into the multidirectional ones. Additionally, the histological analysis carried out show that cells exhibit different alignment and morphology in both types of scaffolds [104].

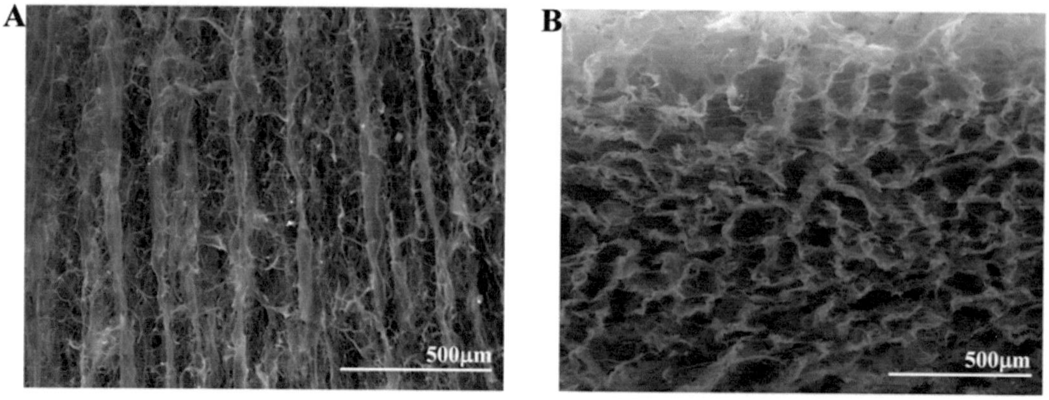

Figure 2: Collagen type I scaffolds exhibiting **(A)** unidirectional and **(B)** multidirectional oriented fibers. Collagen type I suspensions (5 mg/mL) were frozen following protocols already described to obtain scaffolds (10 cm x 10 cm) with different fiber orientation. (Scale bar: 500 µm).

4. CLINICAL APPLICATION OF COLLAGEN TYPE I BASED PRODUCTS

Preclinical grafting of artificial tissue made using collagen substrates have been widely reported. Those studies had shed light into the *in vivo* performance of these products and their impact in the quality of wound healing [31, 105-107]. However, the challenge of tissue-engineered products is to demonstrate their efficacy in clinical studies and to get the approval of the regulatory authorities. For that reason, this chapter is focused on describing the most relevant products based on collagen type I, alone or combined with other materials that are clinically available for application in soft tissues.

4.1. Skin Substitutes Based on Collagen Scaffolds

Although autologous skin transplantation remains the most common form of treatment in patients with significant skin loss, over the last three decades, development of skin substitutes based on tissue engineering has offered huge potential in treating skin-related disorders like burns and chronic wounds [108]. Different approaches have been pursued to generate skin substitutes that fulfill at least some of the functions of this organ. They involve autologous or heterologous artificial substitutes made using fibroblasts, keratinocytes, and their combination, to acellular substitutes made using biomaterials molded into scaffolds. However, to date, no skin substitute able to replace all the vital functions of native tissue has been developed. The reason for this is that they lack cells (melanocytes,

Langerhans cells, macrophages, lymphocytes) and structures (blood vessels, hair follicles and sweat glands) that are present in the skin.

The first attempt to produce a skin substitute was performed in 1975 by Rheinwald *et al.* who cultured human epidermal keratinocyte sheets [109]. Later on, the clinical use of cultured keratinocytes in the treatment of burn wounds was published [110]. However, an epidermal sheet alone often fails to restore the skin structure in full thickness wounds because of problems like fragility and poor attachment of the graft, wound contraction, and scar formation [111]. Preclinical and clinical evidence has demonstrated that effective wound healing always requires the presence of the dermis layer in skin substitutes [107]. Hence, one crucial factor for dermis restoration is the construction of a three-dimensional scaffold that serves as the template that guides the restructuration of cells and subsequent host infiltration of thc skin graft.

Collagen type I scaffolds, alone or combined with other biomaterials, have been used as dermal substitutes for the treatment of full thickness wounds [112]. In fact, the first skin-engineered product approved by the Food and Drug Administration of the United States (US-FDA) was Integra® (Integra LifeSciences Corp.) a template of cross-linked bovine tendon collagen mixed with chondroitin-6-sulphate, a glycosaminoglycan, coated with a semi-permeable silicone layer that functions as a semipermeable barrier [113, 114]. This product is considered as a device type III and it is indicated for the treatment of chronic and traumatic wounds that need dermis formation. After Integra® has supported the migration, proliferation and differentiation of dermal cell required for remodeling damaged tissue, the silicone layer is surgically removed and the newly formed dermis is covered by epidermal cells that migrate from the wound bed. In the case of large wounds the silicon layer is replaced by thin skin autografts.

Within the group of dermal substitutes clinically used is also Nevelia®, a bilaminar medical device that received its CE approval in 2013. Formerly known as Renoskin® is a porous scaffold of cross-linked bovine collagen type I covered with a silicone layer that has been designed and manufactured to promote dermis formation by recruiting cells from the wound bed [115]. As Integra®, it requires an additional surgical procedure to remove the silicone layer and to cover the wound with split-thickness skin autografts. The application of both products, Integra® and Nevelia®, requires two surgical procedures therefore they are considered two-step procedures

There have been also developed dermal substitutes that can be applied in one surgical procedure such as Matriderm®. It is a biodegradable device made of a mix of collagens (types I, III and V) and elastin, that serves as scaffold for dermal tissue repair [116]. It has been successfully used under split-thickness skin autografts on the hands [116, 117], face [118], and other soft tissue burn wounds [115] with good aesthetic results. This acellular three-dimensional matrix increases skin elasticity parameters after its application [119].

This chapter aims to describe products that have been approved for human use, and not to conclude about their clinical efficacy. However, it is important to discuss data from clinical studies that compare dermal substitutes to the conventional split-thickness autograft in order to understand the real impact of dermal substitute application in patients. A clinical trial carried out intra-individual comparisons of 42 paired burn wounds and 44 paired scar reconstructions treated with a dermal substitute made from bovine collagen and elastin-hydrolyzate. The study evaluated elasticity, scar contraction, and scar tissue characteristics according the Vancouver Scar Scale in the wounds treated with the engineered product and in the ones treated with the autograft. Results showed that no significant differences were observed between both treatments one year after surgery [120].

A comparison of the performance of Integra®, Renoskin® (actually known as Nevelia®), Matriderm® in a published preclinical study, showed no significant differences between treatments and between treatments and control (secondary healing) in long-term wound retraction. Because the lack of beneficial effects in the animal model, the authors of the study suggested that results of human testing are required to conclude about the benefits of dermal substitutes [115]. We also believe that clinical data is required to conclude on the efficacy and safety of each of dermal-engineered products. The mentioned study also included ProDerm® described as a 2 mm thick equine matrix made of collagen type I, chitosan and chondroitin sulfate. However, we could not find recent information about this product even in the web site of the manufacturer [121].

Another group of skin substitutes is those made by culturing fibroblasts into a collagen type I based scaffold. An example of them is TransCyte® (Advanced Tissue Sciences) that was the first cellular skin substitute used in excised full-thickness and partial-thickness burn management approved by the US-FDA. It is made of human neonatal skin fibroblasts cultured into a nylon mesh coated with porcine collagen type I which is covered by a silicone layer. The fibroblasts within the nylon mesh proliferate and differentiate secreting ECM components and growth factors that remain after freezing the product in order to eliminate any cellular metabolic activity. TransCyte® needs to be stored between -70 and -20 °C and must be unfrozen directly before use. After 7 to 14 days of grafting, the TransCyte® will begin to lift off by itself [122, 123].

The approach of replacing both dermis and epithelium of the skin has led to develop dermo-epidermal substitutes that upon application should function as a barrier to control water loss, infection, and pain. Apligraf® (Organogenesis) was the first product of this nature approved by US-FDA; it is manufactured from human skin fibroblasts and keratinocytes using a collagen I scaffold and it is considered a Class III medical device. This product dermal layer is the result of growing the skin fibroblasts within a collagen type I matrix that then is remodeled and replaced by the extracellular matrix that the seeded cells secrete. The epidermal layer is formed after the artificial dermis has been obtained by culturing keratinocytes onto the dermis and by promoting *in vitro* epithelium differentiation. Apligraf® is deprived of other cells of the human skin, blood vessels, hair follicles and sweat glands [124]. It is a prescription product not available in pharmacies or online and that must be applied by medical professionals in a Wound Care Center with experience in its management [125]. Apligraf® is indicated for treatment of diabetic foot ulcers and venous leg ulcers that have no responded to conventional therapy [126].

OrCel® (Ortec International, Inc.; currently Forticell Bioscience), a product similar to Apligraf®, contains living allogeneic human keratinocytes and fibroblasts, but instead of using a collagen gel as a scaffold this product uses a cross-linked collagen asymmetric scaffold. One side of the scaffold is coated with a thin film of collagen to close surface pores where keratinocytes are cultured, while fibroblasts are grown within the porous side of the scaffold. Orcel® received FDA approval in 2001 for treatment of autograft donor sites in burned patients and for patients with epidermolysis bullosa [127].

4.2. Oral Mucosa Substitutes Based on Collagen Scaffolds

Commonly used treatments of oral mucosa discontinuities are autografts and pedicle flaps. Tissue grafting is often limited by shortage of donor areas and generation of wounds that heal by secondary intention resulting in scare formation and reduction of donor tissue for further interventions. These drawbacks have launched the development of oral mucosa engineered tissue to treat large wounds.

Manufacturing of oral mucosa substitutes based on collagen type I platforms, alone or in combination with other materials, have been described as well as preclinical studies carried out in order to test their safety and efficacy [128, 129]. Our research group has extensively investigated the development of accessible artificial connective tissue that promotes the regeneration processes of oral mucosa defects through modulating contraction, inflammatory response, cell migration, and cell proliferation. Using rabbits as animal models, an autologous artificial oral connective tissue (AACT) was developed by seeding second-third passage autologous oral fibroblasts into porous cross-linked bovine collagen type I scaffolds. The healing process of oral mucosa wounds covered with AACT, acellular collagen type I scaffolds (CS), or left to heal by secondary intention was determined. It was found that AACT secretes pro-angiogenic, tissue-remodeling, and immune-modulating factors that promote tissue regeneration and wound healing to a greater extent than CS. Also, that the use of CS improved wound closure when compared with secondary intention [31].

Clinical assays of oral mucosa connective tissue substitutes based on collagen type I also are well-known. Rastogi *et al.* evaluated the use of cross-linked porous collagen scaffolds in 60 patients as a dressing material for surgical mucosa defects [23]. In most cases, collagen scaffolds promoted hemostasis, relieved pain, induced granulation, assisted in rapid epithelialization at the wound site and prevented infection, contracture and scarring. All collagen scaffolds, on the seventh postoperative day, slowly underwent collagenolysis and were eventually sloughed off.

As in the skin, there are collagen type I-based substitutes that replace both, connective tissue and epithelium [130, 131]. In fact, skin engineered substitutes have been evaluated clinically in humans to treat oral lesions. A case series of four patients examined the clinical and histological performance of a tissue-engineered bilayered skin substitute in oral defects requiring mucogingival surgery. All treated sites healed with no adverse effects and the tissue synthesized matched the texture and color of the healthy surrounding tissue. The clinical goal of increasing the width of keratinized gingival tissue was achieved in all patients avoiding the use of palatal grafting [132].

Gintiut® is a product made of allogenic fibroblasts and keratinocytes cultured in bovine collagen type I that followed the same approach of Apligraf® and that has been developed by the same company. This device is applied topically in a surgically created vascular wound bed for the treatment of adult mucogingival defects [133]. Its use is not indicated to provide root coverage and the product must not be used in infected wounds or when the patient present allergy to bovine collagen.

Other challenges in oral mucosal defects are the reduced width of keratinized gingiva and gingival recession that could lead to exposure of teeth roots. The presence of a suitable width of keratinized gingiva is important for effective homecare, resist frenum pull, and potentially reduce further gingival recession [134]. The absence of adequate keratinized mucosa has been associated with high plaque accumulation and gingival inflammation that can lead to gingival recession. In order to augment the width of keratinized gingiva, a free gingival graft is commonly used. Because of the similarities of healing between cutaneous and oral mucosa wounds, it was thought that a skin substitute may deliver better results than the standard free gingival graft in terms of the color and texture of the gingiva while providing an adequate amount of new keratinized gingiva. A series of studies evaluating Apligraf® as a tissue substitute to augment the width of keratinized gingiva were carried out and they showed that the graft was safe and able to produce keratinized gingiva although not as much as the free gingival graft [135-137]. In 2012, the US-FDA authorized Organogenesis, the company manufacturing Apligraf®, to market this product as Gintiut® to treat mucogingival conditions [133].

4.3. Nerve Guides Based on Collagen Scaffolds

Nerve repair is still a major challenge in regenerative medicine. The axons of detached peripheral nerves can regenerate but will not be functional unless the nerve stumps are surgically reconnected. Despite the advances seen in surgical techniques over the last years, functional recovery after a severe lesion of a major nerve is not successful yet [138]. Functional recovery of an injured peripheral nerve depends on the precise regeneration of axons to their original target tissues. The use of nerve guides in surgical procedures to repair acute peripheral nerve injuries is increasing as a consequence of the proliferation of nerve devices, and also because their clinical acceptance. Despite regeneration distance is commonly limited to 20-25 mm [139], several studies have shown improvements in nerve regeneration distance when nerve guides include coatings, internal scaffolds, topographical cues, or the delivery of support cells [140-142].

In gaps up to 20 mm collagen type I nerve guides have shown to induce similar regeneration than autografts. NeuraGen® (Integra LifeSciences Corp.) is a degradable collagen tube used as an interface between the nerve and the surrounding tissue and to serve as a channel for axonal growth across a nerve gap [143]. Neuromatrix® (Collagen Matrix, Inc.) is a degradable collagen-based tubular scaffold that protects the peripheral nerve after injury and creates a conduit for axonal growth across a nerve gap. This product is made from collagen type I dense fibers that confer mechanical strength. At the same time, it slowly resorbs *in vivo* [144].

4.4. Tendon Substitutes Based on Collagen Scaffolds

Severe tendon injury is treated through surgery to repair or replace the damaged tendon with autografts, allografts, xenografts, or prosthetic devices. However, the clinical outcomes remain unsatisfactory due to limitations including donor site morbidity, high failure rates, risk of injury recurrence, and limited long-term function recovery [145]. Therefore, tendon substitutes require scaffolds that provide mechanical support to the healing tissue until endogenous cells have produced native ECM, thereby, preventing re-rupture. In addition, collagen type I scaffolds can improve healing by inducing cell proliferation, promoting ECM synthesis, and organizing the matrix into functional tendon tissues. Because tendons are composed mostly of parallel collagen fibers, orientation is an important microstructure characteristic to mimic in tendon tissue engineering. The topographical cues provided by scaffolds must be carefully considered when engineering tendon constructs.

Some collagen scaffolds used for tendon treatment include TenoGlide® (Integra LifeSciences Corp.), a porous collagen-glycosaminoglycan scaffold [146], and BioBlanket® (Kensey Nash Corp.; currently DSM), a porous cross-linked collagen scaffold that was approved for reinforcement and repair of a variety of soft tissues, including rotator cuff tear and designed to degrade slowly and it is expected to last up to one year *in vivo* [147]. However, further information on the latter product is unavailable.

CONCLUSION

Soft tissue lesions still represent one of the commonest challenges in regenerative medicine. In a great percentage of the cases, they are life threatening and represent a significant burden to health systems and economy worldwide. In this framework, a cornucopia of natural and synthetic biomaterials and devices have been explored over the last four decades to provide haemostasis, reduce local pain and microbial colonization and promote regeneration over reparatory cicatrization. Among the biomaterials, collagen type I, a major component of the

ECM, has emerged as one of the most extensively used and has reached the market in the form of a variety of acellular and cellularized wound dressings. Regardless of the beneficial features displayed by these products, they are costly and their affordability in developing countries and constrained-setting environments is extremely limited. In this scenario, the local development of products or new more affordable strategies is a fundamental step to ensure the access of less favored patients to these therapies. The recent contributions of our group [31, 102, 148, 149] are intended to offer locally soft-tissue engineered products based on collagen type I isolated with standardized and validated procedures from by-products of cattle slaughter.

Despite extensive research on collagen I uses the effect of many of the fundamental structural aspects of collagen type I scaffolds (*e.g.*, fiber directionality, supplementation with natural extracts that induce tissue growth) on cell adhesion, proliferation, migration and cell secretory activity that are crucial for the attainment of the expected *in vivo* performance remain unrevealed. This understanding is necessary to fine-tune the properties of collagen type I scaffolds and to enable the extension of their use to other applications.

Collagen type I abundance in the connective tissue of mammals is the best indicative of its role in the regeneration of organs and tissues. Even though the extensive search for new biomaterials that is being conducted and the development of new technologies for boosting organ and tissue regeneration, this biomaterial will remain to be used in the manufacturing of tissue-engineered products. Finally, a mandatory stage to maximize the chances of bench-to-bedside translation is the development of reproducible, traceable, scalable and cost-effective production processes. Our investigations are currently oriented in this direction and we envision the conduction of extensive clinical trials in the near future.

ACKNOWLEDGMENTS

The authors thank Colciencias-SENA (grants 110145221387 and 110152128661), the ‘Red Iberoamericana de Nuevos Materiales para el Diseño de Sistemas Avanzados de Liberación de Fármacos en Enfermedades de Alto Impacto Socioeconómico’ (RIMADEL) of the Iberoamerican Program CYTED (2011-4) and the Universidad Nacional de Colombia.

CONFLICT OF INTEREST

The authors confirm that this chapter contents have no conflict of interest.

REFERENCES

[1] Hantash BM, Zhao L, Knowles JA, Lorenz HP. Adult and fetal wound healing. Frontiers in bioscience: a journal and virtual library. 2008;13:51-61.

[2] Yannas IV, Kwan MD, Longaker MT. Early fetal healing as a model for adult organ regeneration. Tissue engineering. 2007;13(8):1789-98.

[3] Stocum DL. Regenerative biology and engineering: strategies for tissue restoration. Wound repair and regeneration: official publication of the Wound Healing Society [and] the European Tissue Repair Society. 1998;6(4):276-90.

[4] Carlson BM. Some principles of regeneration in mammalian systems. Anatomical record Part B, New anatomist. 2005;287(1):4-13.

[5] Slack JM. Regeneration research today. Developmental dynamics: an official publication of the American Association of Anatomists. 2003;226(2):162-6.

[6] Odelberg SJ. Cellular plasticity in vertebrate regeneration. Anatomical record Part B, New anatomist. 2005;287(1):25-35.

[7] Han M, Yang X, Taylor G, Burdsal CA, Anderson RA, Muneoka K. Limb regeneration in higher vertebrates: developing a roadmap. Anatomical record Part B, New anatomist. 2005;287(1):14-24.

[8] Kumar A, Godwin JW, Gates PB, Garza-Garcia AA, Brockes JP. Molecular basis for the nerve dependence of limb regeneration in an adult vertebrate. Science. 2007;318(5851):772-7.

[9] Vinarsky V, Atkinson DL, Stevenson TJ, Keating MT, Odelberg SJ. Normal newt limb regeneration requires matrix metalloproteinase function. Developmental biology. 2005;279(1):86-98.

[10] Martinez-Hernandez A, Amenta PS. The extracellular matrix in hepatic regeneration. FASEB journal: official publication of the Federation of American Societies for Experimental Biology. 1995;9(14):1401-10.

[11] Harty M, Neff AW, King MW, Mescher AL. Regeneration or scarring: an immunologic perspective. Developmental dynamics: an official publication of the American Association of Anatomists. 2003;226(2):268-79.

[12] Snowden JM, Cliff WJ. Wound contraction. Correlations between the tension generated by granulation tissue, cellular content and rate of contraction. Q J Exp Physiol. 1985;70(4):539-48.

[13] Fuchs JR, Nasseri BA, Vacanti JP. Tissue engineering: a 21st century solution to surgical reconstruction. The Annals of thoracic surgery. 2001;72(2):577-91.

[14] Samadikuchaksaraei A. Scientific and industrial status of tissue engineering. African Journal of Biotechnology. 2007;6(25).

[15] O'Brien FJ. Biomaterials & scaffolds for tissue engineering. Materials Today. 2011;14(3):88-95.

[16] Yannas I, Tzeranis D, Harley B, So P. Biologically active collagen-based scaffolds: advances in processing and characterization. Philosophical Transactions of the Royal Society A: Mathematical, Physical and Engineering Sciences. 2010;368(1917):2123-39.

[17] Ng KW, Khor HL, Hutmacher D. *In vitro* characterization of natural and synthetic dermal matrices cultured with human dermal fibroblasts. Biomaterials. 2004;25(14):2807-18.

[18] Bos JJ, Hunink MG, Mali WPTM. Use of a collagen hemostatic closure device to achieve hemostasis after arterial puncture: a cost-effectiveness analysis. Journal of Vascular and Interventional Radiology. 1996;7(4):479-86.

[19] Abou Neel EA, Bozec L, Knowles JC, Syed O, Mudera V, Day R, *et al.* Collagen—emerging collagen based therapies hit the patient. Advanced drug delivery reviews. 2013;65(4):429-56.

[20] Lee CH, Singla A, Lee Y. Biomedical applications of collagen. International journal of pharmaceutics. 2001;221(1):1-22.

[21] Glowacki J, Mizuno S. Collagen scaffolds for tissue engineering. Biopolymers. 2008;89(5):338-44.

[22] Yannas IV. Studies on the biological activity of the dermal regeneration template. Wound Repair and Regeneration. 1998;6(6):518-23.

[23] Rastogi S, Modi M, Sathian B. The efficacy of collagen membrane as a biodegradable wound dressing material for surgical defects of oral mucosa: a prospective study. Journal of Oral and Maxillofacial Surgery. 2009;67(8):1600-6.

[24] Geiger B, Bershadsky A, Pankov R, Yamada KM. Transmembrane crosstalk between the extracellular matrix and the cytoskeleton. Nature Reviews Molecular Cell Biology. 2001;2(11):793-805.

[25] Juliano R, Haskill S. Signal transduction from the extracellular matrix. Journal of Cell Biology. 1993;120:577-.

[26] Meredith J, Fazeli B, Schwartz M. The extracellular matrix as a cell survival factor. Molecular biology of the cell. 1993;4(9):953-61.
[27] Chiquet M, Koch M, Matthisson M, Tannheimer M, Chiquet-Ehrismann R. Regulation of extracellular matrix synthesis by mechanical stress. Biochemistry and cell biology. 1996;74(6):737-44.
[28] Midwood KS, Williams LV, Schwarzbauer JE. Tissue repair and the dynamics of the extracellular matrix. The international journal of biochemistry & cell biology. 2004;36(6):1031-7.
[29] Kleinman HK, Philp D, Hoffman MP. Role of the extracellular matrix in morphogenesis. Current opinion in biotechnology. 2003;14(5):526-32.
[30] Nelson CM, Bissell MJ. Of extracellular matrix, scaffolds, and signaling: tissue architecture regulates development, homeostasis, and cancer. Annual review of cell and developmental biology. 2006;22:287.
[31] Espinosa L, Sosnik A, Fontanilla MR. Development and preclinical evaluation of acellular collagen scaffolding and autologous artificial connective tissue in the regeneration of oral mucosa wounds. Tissue Engineering Part A. 2010;16(5):1667-79.
[32] Karande TS, Ong JL, Agrawal CM. Diffusion in musculoskeletal tissue engineering scaffolds: design issues related to porosity, permeability, architecture, and nutrient mixing. Annals of biomedical engineering. 2004;32(12):1728-43.
[33] Yuksel E, Choo J, Wettergreen M, Liebschner M, editors. Challenges in Soft Tissue Engineering. Seminars in Plastic Surgery; 2005: Thieme Medical Publishers.
[34] Williams DF. On the mechanisms of biocompatibility. Biomaterials. 2008;29(20):2941-53.
[35] Valentin JE, Stewart-Akers AM, Gilbert TW, Badylak SF. Macrophage participation in the degradation and remodeling of extracellular matrix scaffolds. Tissue Engineering Part A. 2009;15(7):1687-94.
[36] Boruch AV, Nieponice A, Qureshi IR, Gilbert TW, Badylak SF. Constructive remodeling of biologic scaffolds is dependent on early exposure to physiologic bladder filling in a canine partial cystectomy model. Journal of Surgical Research. 2010;161(2):217-25.
[37] Folkman J, Hochberg M. Self-regulation of growth in three dimensions. The Journal of experimental medicine. 1973;138(4):745-53.
[38] Bloch O, Golde P, Dohmen PM, Posner S, Konertz W, Erdbrugger W. Immune response in patients receiving a bioprosthetic heart valve: lack of response with decellularized valves. Tissue engineering Part A. 2011;17(19-20):2399-405.
[39] Juhasz I, Kiss B, Lukacs L, Erdei I, Peter Z, Remenyik E. Long-term followup of dermal substitution with acellular dermal implant in burns and postburn scar corrections. Dermatology research and practice. 2010;2010:210150.
[40] Karabekmez FE, Duymaz A, Moran SL. Early clinical outcomes with the use of decellularized nerve allograft for repair of sensory defects within the hand. Hand. 2009;4(3):245-9.
[41] Atala A, Bauer SB, Soker S, Yoo JJ, Retik AB. Tissue-engineered autologous bladders for patients needing cystoplasty. Lancet. 2006;367(9518):1241-6.
[42] Bejjani GK, Zabramski J, Durasis Study G. Safety and efficacy of the porcine small intestinal submucosa dural substitute: results of a prospective multicenter study and literature review. Journal of neurosurgery. 2007;106(6):1028-33.
[43] Liao J, Guo X, Grande-Allen KJ, Kasper FK, Mikos AG. Bioactive polymer/extracellular matrix scaffolds fabricated with a flow perfusion bioreactor for cartilage tissue engineering. Biomaterials. 2010;31(34):8911-20.
[44] Decaris ML, Binder BY, Soicher MA, Bhat A, Leach JK. Cell-derived matrix coatings for polymeric scaffolds. Tissue engineering Part A. 2012;18(19-20):2148-57.
[45] Salgado AJ, Coutinho OP, Reis RL. Novel starch-based scaffolds for bone tissue engineering: cytotoxicity, cell culture, and protein expression. Tissue engineering. 2004;10(3-4):465-74.
[46] Mi FL, Tan YC, Liang HC, Huang RN, Sung HW. *In vitro* evaluation of a chitosan membrane cross-linked with genipin. Journal of biomaterials science Polymer edition. 2001;12(8):835-50.
[47] Nair LS, Laurencin CT. Polymers as biomaterials for tissue engineering and controlled drug delivery. Advances in biochemical engineering/biotechnology. 2006;102:47-90.
[48] Chen GQ, Wu Q. The application of polyhydroxyalkanoates as tissue engineering materials. Biomaterials. 2005;26(33):6565-78.

[49] Agrawal CM, Niederauer GG, Athanasiou KA. Fabrication and Characterization of PLA-PGA Orthopedic Implants. Tissue engineering. 1995;1(3):241-52.

[50] Kumarasuriyar A, Jackson RA, Grondahl L, Trau M, Nurcombe V, Cool SM. Poly(beta-hydroxybutyrate-co-beta-hydroxyvalerate) supports *in vitro* osteogenesis. Tissue engineering. 2005;11(7-8):1281-95.

[51] Petersen LK, Oh J, Sakaguchi DS, Mallapragada SK, Narasimhan B. Amphiphilic polyanhydride films promote neural stem cell adhesion and differentiation. Tissue engineering Part A. 2011;17(19-20):2533-41.

[52] Griffin J, Carbone A, Delgado-Rivera R, Meiners S, Uhrich KE. Design and evaluation of novel polyanhydride blends as nerve guidance conduits. Acta biomaterialia. 2010;6(6):1917-24.

[53] Burkoth AK, Burdick J, Anseth KS. Surface and bulk modifications to photocrosslinked polyanhydrides to control degradation behavior. Journal of biomedical materials research. 2000;51(3):352-9.

[54] Song Y, Wennink JW, Poot AA, Vermes I, Feijen J, Grijpma DW. Evaluation of tubular poly(trimethylene carbonate) tissue engineering scaffolds in a circulating pulsatile flow system. The International journal of artificial organs. 2011;34(2):161-71.

[55] Pego AP, Siebum B, Van Luyn MJ, Gallego y Van Seijen XJ, Poot AA, Grijpma DW, *et al.* Preparation of degradable porous structures based on 1,3-trimethylene carbonate and D,L-lactide (co)polymers for heart tissue engineering. Tissue engineering. 2003;9(5):981-94.

[56] Buxton AN, Zhu J, Marchant R, West JL, Yoo JU, Johnstone B. Design and characterization of poly(ethylene glycol) photopolymerizable semi-interpenetrating networks for chondrogenesis of human mesenchymal stem cells. Tissue engineering. 2007;13(10):2549-60.

[57] Patel PN, Gobin AS, West JL, Patrick CW, Jr. Poly(ethylene glycol) hydrogel system supports preadipocyte viability, adhesion, and proliferation. Tissue engineering. 2005;11(9-10):1498-505.

[58] Kumada Y, Zhang S. Significant type I and type III collagen production from human periodontal ligament fibroblasts in 3D peptide scaffolds without extra growth factors. PloS one. 2010;5(4):e10305.

[59] Lynn A, Yannas I, Bonfield W. Antigenicity and immunogenicity of collagen. Journal of Biomedical Materials Research Part B: Applied Biomaterials. 2004;71(2):343-54.

[60] Peng YY, Glattauer V, Ramshaw JA, Werkmeister JA. Evaluation of the immunogenicity and cell compatibility of avian collagen for biomedical applications. Journal of Biomedical Materials Research Part A. 2010;93(4):1235-44.

[61] Liu B, Xu Z, Yu R, Wang J, Wang Z, Harrell CR, editors. The use of type I and type III injectable human collagen for dermal fill: 10 years of clinical experience in China. Seminars in Plastic Surgery; 2005: Thieme Medical Publishers.

[62] Gilbert TW, Sellaro TL, Badylak SF. Decellularization of tissues and organs. Biomaterials. 2006;27(19):3675-83.

[63] Wu L, Ding J. *In vitro* degradation of three-dimensional porous poly (d, l-lactide-< i> co</i>-glycolide) scaffolds for tissue engineering. Biomaterials. 2004;25(27):5821-30.

[64] Pek Y, Spector M, Yannas I, Gibson L. Degradation of a collagen–chondroitin-6-sulfate matrix by collagenase and by chondroitinase. Biomaterials. 2004;25(3):473-82.

[65] Schmidt CE, Baier JM. Acellular vascular tissues: natural biomaterials for tissue repair and tissue engineering. Biomaterials. 2000;21(22):2215-31.

[66] Diegelmann RF, Evans MC. Wound healing: an overview of acute, fibrotic and delayed healing. Front Biosci. 2004;9(1):283-9.

[67] Yannas I, Lee E, Orgill D, Skrabut E, Murphy G. Synthesis and characterization of a model extracellular matrix that induces partial regeneration of adult mammalian skin. Proceedings of the National Academy of Sciences. 1989;86(3):933-7.

[68] Hofmann S, Knecht S, Langer R, Kaplan DL, Vunjak-Novakovic G, Merkle HP, *et al.* Cartilage-like tissue engineering using silk scaffolds and mesenchymal stem cells. Tissue engineering. 2006;12(10):2729-38.

[69] Sturm S, Zhou S, Mai Y-W, Li Q. On stiffness of scaffolds for bone tissue engineering—a numerical study. Journal of biomechanics. 2010;43(9):1738-44.

[70] Gibson LJ. Biomechanics of cellular solids. Journal of biomechanics. 2005;38(3):377-99.

[71] Harley BA, Leung JH, Silva EC, Gibson LJ. Mechanical characterization of collagen–glycosaminoglycan scaffolds. Acta biomaterialia. 2007;3(4):463-74.
[72] Haugh MG, Murphy CM, McKiernan RC, Altenbuchner C, O'Brien FJ. Crosslinking and mechanical properties significantly influence cell attachment, proliferation, and migration within collagen glycosaminoglycan scaffolds. Tissue Engineering Part A. 2011;17(9-10):1201-8.
[73] Charulatha V, Rajaram A. Influence of different crosslinking treatments on the physical properties of collagen membranes. Biomaterials. 2003;24(5):759-67.
[74] Duan X, Sheardown H. Crosslinking of collagen with dendrimers. Journal of Biomedical Materials Research Part A. 2005;75(3):510-8.
[75] Chan B, Hui T, Chan O, So K-F, Lu W, Cheung K, *et al.* Photochemical cross-linking for collagen-based scaffolds: a study on optical properties, mechanical properties, stability, and hematocompatibility. Tissue engineering. 2007;13(1):73-85.
[76] Veríssimo D, Leitão R, Ribeiro R, Figueiró S, Sombra A, Góes J, *et al.* Polyanionic collagen membranes for guided tissue regeneration: Effect of progressive glutaraldehyde cross-linking on biocompatibility and degradation. Acta biomaterialia. 2010;6(10):4011-8.
[77] Araña M, Peña E, Abizanda G, Cilla M, Ochoa I, Gavira JJ, *et al.* Preparation and characterization of collagen-based ADSC-carrier sheets for cardiovascular application. Acta biomaterialia. 2013;9(4):6075-83.
[78] Murikipudi S, Methe H, Edelman ER. The effect of substrate modulus on the growth and function of matrix-embedded endothelial cells. Biomaterials. 2013;34(3):677-84.
[79] Mason BN, Starchenko A, Williams RM, Bonassar LJ, Reinhart-King CA. Tuning three-dimensional collagen matrix stiffness independently of collagen concentration modulates endothelial cell behavior. Acta biomaterialia. 2013;9(1):4635-44.
[80] Faraj KA, Van Kuppevelt TH, Daamen WF. Construction of collagen scaffolds that mimic the three-dimensional architecture of specific tissues. Tissue engineering. 2007;13(10):2387-94.
[81] Caliari SR, Harley BA. The effect of anisotropic collagen-GAG scaffolds and growth factor supplementation on tendon cell recruitment, alignment, and metabolic activity. Biomaterials. 2011;32(23):5330-40.
[82] Annaidh AN, Ottenio M, Bruyère K, Destrade M, Gilchrist M, editors. Mechanical properties of excised human skin. 6th World Congress of Biomechanics (WCB 2010) August 1-6, 2010 Singapore; 2010: Springer.
[83] Kim Y, Rajagopalan P. 3D hepatic cultures simultaneously maintain primary hepatocyte and liver sinusoidal endothelial cell phenotypes. PloS one. 2010;5(11):e15456.
[84] Freyman T, Yannas I, Gibson L. Cellular materials as porous scaffolds for tissue engineering. Progress in Materials Science. 2001;46(3):273-82.
[85] O'Brien FJ, Harley B, Yannas IV, Gibson LJ. The effect of pore size on cell adhesion in collagen-GAG scaffolds. Biomaterials. 2005;26(4):433-41.
[86] Hu Y, Topolkaraev V, Hiltner A, Baer E. Measurement of water vapor transmission rate in highly permeable films. J Appl Polym Sci. 2001;81(7):1624-33.
[87] Zeltinger J, Sherwood JK, Graham DA, Müeller R, Griffith LG. Effect of pore size and void fraction on cellular adhesion, proliferation, and matrix deposition. Tissue engineering. 2001;7(5):557-72.
[88] Daskalova A, Nathala CS, Bliznakova I, Stoyanova E, Zhelyazkova A, Ganz T, *et al.* Controlling the porosity of collagen, gelatin and elastin biomaterials by ultrashort laser pulses. Applied Surface Science. 2014;292:367-77.
[89] Miron-Mendoza M, Seemann J, Grinnell F. The differential regulation of cell motile activity through matrix stiffness and porosity in three dimensional collagen matrices. Biomaterials. 2010;31(25):6425-35.
[90] Druecke D, Langer S, Lamme E, Pieper J, Ugarkovic M, Steinau HU, *et al.* Neovascularization of poly (ether ester) block-copolymer scaffolds *in vivo*: Long-term investigations using intravital fluorescent microscopy. Journal of Biomedical Materials Research Part A. 2004;68(1):10-8.
[91] Yannas IV. Emerging rules for inducing organ regeneration. Biomaterials. 2013;34(2):321-30.
[92] Wang H, Pieper J, Peters F, van Blitterswijk CA, Lamme EN. Synthetic scaffold morphology controls human dermal connective tissue formation. Journal of Biomedical Materials Research Part A. 2005;74(4):523-32.

[93] Stokols S, Tuszynski MH. The fabrication and characterization of linearly oriented nerve guidance scaffolds for spinal cord injury. Biomaterials. 2004;25(27):5839-46.
[94] Chamberlain LJ, Yannas IV, Hsu HP, Strichartz G, Spector M. Collagen-GAG substrate enhances the quality of nerve regeneration through collagen tubes up to level of autograft. Experimental neurology. 1998;154(2):315-29.
[95] Hu X, Huang J, Ye Z, Xia L, Li M, Lv B, *et al.* A novel scaffold with longitudinally oriented microchannels promotes peripheral nerve regeneration. Tissue engineering Part A. 2009;15(11):3297-308.
[96] Liu H, Ge Z, Wang Y, Toh SL, Sutthikhum V, Goh JC. Modification of sericin-free silk fibers for ligament tissue engineering application. Journal of biomedical materials research Part B, Applied biomaterials. 2007;82(1):129-38.
[97] Louie LK, Yannas IV, Hsu H-P, Spector M. Healing of Tendon Defects Implanted with a Porous Collagen-GAG Matrix: Histological Evaluation. Tissue engineering. 1997;3(2):187-95.
[98] Wray LS, Orwin EJ. Recreating the microenvironment of the native cornea for tissue engineering applications. Tissue engineering Part A. 2009;15(7):1463-72.
[99] Silva MM, Cyster LA, Barry JJ, Yang XB, Oreffo RO, Grant DM, *et al.* The effect of anisotropic architecture on cell and tissue infiltration into tissue engineering scaffolds. Biomaterials. 2006;27(35):5909-17.
[100] Yunoki S, Ikoma T, Tsuchiya A, Monkawa A, Ohta K, Sotome S, *et al.* Fabrication and mechanical and tissue ingrowth properties of unidirectionally porous hydroxyapatite/collagen composite. Journal of biomedical materials research Part B, Applied biomaterials. 2007;80(1):166-73.
[101] Yannas IV. Facts and theories of induced organ regeneration. Advances in biochemical engineering/biotechnology. 2005;93:1-38.
[102] Bustos RH, Suesca E, Millán D, González JM, Fontanilla MR. Real-Time Quantification of Proteins Secreted by Artificial Connective Tissue Made from Uni-Or Multidirectional Collagen I Scaffolds and Oral Mucosa Fibroblasts. Analytical chemistry. 2014;86(5):2421-8.
[103] McGuigan AP, Sefton MV. Vascularized organoid engineered by modular assembly enables blood perfusion. Proceedings of the National Academy of Sciences. 2006;103(31):11461-6.
[104] Suesca E. Optimización de la obtención de soportes de colágeno y estudio del efecto de su microestructura en el desarrollo de mucosa oral artificial [Doctoral Thesis]. Bogotá Universidad Nacional de Colombia; 2013.
[105] Izumi K, Feinberg SE, Terashi H, Marcelo CL. Evaluation of transplanted tissue-engineered oral mucosa equivalents in severe combined immunodeficient mice. Tissue engineering. 2003;9(1):163-74.
[106] Butler CE, Orgill DP, Yannas IV, Compton CC. Effect of keratinocyte seeding of collagen-glycosaminoglycan membranes on the regeneration of skin in a porcine model. Plastic and reconstructive surgery. 1998;101(6):1572-9.
[107] Chamberlain L, Yannas I, Hsu H, Strichartz G, Spector M. Collagen-GAG substrate enhances the quality of nerve regeneration through collagen tubes up to level of autograft. Experimental neurology. 1998;154(2):315-29.
[108] Kopp J, Jeschke MG, Bach AD, Kneser U, Horch RE. Applied tissue engineering in the closure of severe burns and chronic wounds using cultured human autologous keratinocytes in a natural fibrin matrix. Cell and tissue banking. 2004;5(2):89-96.
[109] Rheinwald JG, Green H. Formation of a keratinizing epithelium in culture by a cloned cell line derived from a teratoma. Cell. 1975;6(3):317-30.
[110] O'Connor N, Mulliken J, Banks-Schlegel S, Kehinde O, Green H. Grafting of burns with cultured epithelium prepared from autologous epidermal cells. The Lancet. 1981;317(8211):75-8.
[111] Cooper ML, Andree C, Hansbrough JF, Zapata-Sirvent RL, Spielvogel RL. Direct comparison of a cultured composite skin substitute containing human keratinocytes and fibroblasts to an epidermal sheet graft containing human keratinocytes on athymic mice. Journal of investigative dermatology. 1993;101(6):811-9.
[112] Ruszczak Z. Effect of collagen matrices on dermal wound healing. Advanced drug delivery reviews. 2003;55(12):1595-611.
[113] Jeng JC, Fidler PE, Sokolich JC, Jaskille AD, Khan S, White PM, *et al.* Seven years' experience with Integra as a reconstructive tool. Journal of burn care & research. 2007;28(1):120-6.

[114] Heimbach DM, Warden GD, Luterman A, Jordan MH, Ozobia N, Ryan CM, *et al.* Multicenter postapproval clinical trial of Integra (R) dermal regeneration template for burn treatment. Journal of burn care & research. 2003;24(1):42-8.
[115] Philandrianos C, Andrac-Meyer L, Mordon S, Feuerstein J-M, Sabatier F, Veran J, *et al.* Comparison of five dermal substitutes in full-thickness skin wound healing in a porcine model. Burns. 2012;38(6):820-9.
[116] Haslik W, Kamolz L-P, Nathschläger G, Andel H, Meissl G, Frey M. First experiences with the collagen-elastin matrix Matriderm< sup>®</sup> as a dermal substitute in severe burn injuries of the hand. Burns. 2007;33(3):364-8.
[117] Ryssel H, Germann G, Kloeters O, Gazyakan E, Radu C. Dermal substitution with Matriderm< sup>®</sup> in burns on the dorsum of the hand. Burns. 2010;36(8):1248-53.
[118] Cent JTOM. One—Step Reconstruction of Severe Burn Injury on the Face of A Child by Using the Collagen-Elastin Matrix.
[119] Ryssel H, Gazyakan E, Germann G, Öhlbauer M. The use of MatriDerm< sup>®</sup> in early excision and simultaneous autologous skin grafting in burns—A pilot study. Burns. 2008;34(1):93-7.
[120] Van Zuijlen P, Vloemans J, Van Trier A, Suijker MH, van Unen E, Groenevelt F, *et al.* Dermal substitution in acute burns and reconstructive surgery: a subjective and objective long-term follow-up. Plastic and reconstructive surgery. 2001;108(7):1938-46.
[121] Laboratories G. http://www.laboratoires-genevrier.com/pages/?page=469&idl=21 2014 [cited 2014 October 17].
[122] Noordenbos J, Doré C, Hansbrough JF. Safety and Efficacy of TransCyte* for the Treatment of Partial-Thickness Burns. Journal of burn care & research. 1999;20(4):275-81.
[123] Amani H, Dougherty WR, Blome-Eberwein S. Use of Transcyte< sup>®</sup> and dermabrasion to treat burns reduces length of stay in burns of all size and etiology. Burns. 2006;32(7):828-32.
[124] Eaglstein WH, Falanga V. Tissue engineering and the development of Apligraf< sup>®</sup>, a human skin equivalent. Clinical therapeutics. 1997;19(5):894-905.
[125] Zaulyanov L, Kirsner RS. A review of a bi-layered living cell treatment (Apligraf®) in the treatment of venous leg ulcers and diabetic foot ulcers. Clinical interventions in aging. 2007;2(1):93.
[126] Falanga V, Sabolinski M. A bilayered living skin construct (APLIGRAF®) accelerates complete closure of hard-to-heal venous ulcers. Wound Repair and Regeneration. 1999;7(4):201-7.
[127] Bioscience F. http://www.forticellbioscience.com/orcel.html 2014 [cited 2014 October 17].
[128] Moharamzadeh K, Brook I, Van Noort R, Scutt A, Thornhill M. Tissue-engineered oral mucosa: a review of the scientific literature. Journal of dental research. 2007;86(2):115-24.
[129] Moharamzadeh K, Colley H, Murdoch C, Hearnden V, Chai W, Brook I, *et al.* Tissue-engineered oral mucosa. Journal of dental research. 2012;91(7):642-50.
[130] Kinikoglu B, Auxenfans C, Pierrillas P, Justin V, Breton P, Burillon C, *et al.* Reconstruction of a full-thickness collagen-based human oral mucosal equivalent. Biomaterials. 2009;30(32):6418-25.
[131] Izumi K, Neiva RF, Feinberg SE. Intraoral grafting of tissue-engineered human oral mucosa. The International journal of oral & maxillofacial implants. 2012;28(5):e295-303.
[132] Nevins ML. Tissue-engineered bilayered cell therapy for the treatment of oral mucosal defects: a case series. The International journal of periodontics & restorative dentistry. 2010;30(1):31-9.
[133] Schmidt C. Gintuit cell therapy approval signals shift at US regulator. Nature biotechnology. 2012;30(6):479-.
[134] Kassab MM, Cohen RE. The etiology and prevalence of gingival recession. Journal of the American Dental Association (1939). 2003;134(2):220-5.
[135] McGuire MK, Nunn ME. Evaluation of the safety and efficacy of periodontal applications of a living tissue-engineered human fibroblast-derived dermal substitute. I. Comparison to the gingival autograft: a randomized controlled pilot study. Journal of periodontology. 2005;76(6):867-80.
[136] McGuire MK, Scheyer ET, Nevins ML, Neiva R, Cochran DL, Mellonig JT, *et al.* Living cellular construct for increasing the width of keratinized gingiva: results from a randomized, within-patient, controlled trial. Journal of periodontology. 2011;82(10):1414-23.
[137] Wilson Jr TG, McGuire MK, Nunn ME. Evaluation of the safety and efficacy of periodontal applications of a living tissue-engineered human fibroblast-derived dermal substitute. II. Comparison to the subepithelial connective tissue graft: a randomized controlled feasibility study. Journal of periodontology. 2005;76(6):881-9.

[138] Gordon T, Tyreman N, Raji MA. The basis for diminished functional recovery after delayed peripheral nerve repair. The Journal of Neuroscience. 2011;31(14):5325-34.
[139] Bell JH, Haycock JW. Next generation nerve guides: materials, fabrication, growth factors, and cell delivery. Tissue Engineering Part B: Reviews. 2011;18(2):116-28.
[140] Bozkurt A, Lassner F, O'Dey D, Deumens R, Böcker A, Schwendt T, *et al.* The role of microstructured and interconnected pore channels in a collagen-based nerve guide on axonal regeneration in peripheral nerves. Biomaterials. 2012;33(5):1363-75.
[141] Beigi MH, Ghasemi-Mobarakeh L, Prabhakaran MP, Karbalaie K, Azadeh H, Ramakrishna S, *et al.* *In vivo* integration of poly (ε-caprolactone)/gelatin nanofibrous nerve guide seeded with teeth derived stem cells for peripheral nerve regeneration. Journal of Biomedical Materials Research Part A. 2014.
[142] Liu T, Houle JD, Xu J, Chan BP, Chew SY. Nanofibrous collagen nerve conduits for spinal cord repair. Tissue Engineering Part A. 2012;18(9-10):1057-66.
[143] Lifesciences I. http://www.ilstraining.com/NTC Solutions/NeuraGen/NeuraGen_00.html 2014 [cited 2014 October 17].
[144] Collagen-Matrix.Inc. http://www.collagenmatrix.com/products-neurological.htm 2014 [cited 2014 October 17].
[145] Sharma P, Maffulli N. Biology of tendon injury: healing, modeling and remodeling. Journal of Musculoskeletal and Neuronal Interactions. 2006;6(2):181.
[146] Lifescience I. http://www.ilstraining.com/NTC Solutions/TenoGlide/TenoGlide_00.html 2014 [cited 2014 October 17].
[147] Coons DA, Barber FA. Tendon graft substitutes—rotator cuff patches. Sports medicine and arthroscopy review. 2006;14(3):185-90.
[148] Fontanilla MR, Espinosa LG. *In vitro* and *in vivo* assessment of oral autologous artificial connective tissue characteristics that influence its performance as a graft. Tissue Engineering Part A. 2012;18(17-18):1857-66.
[149] Bello SA, Pereira Ebratt R, Fontanilla Duque MR. Elaboración de tejido conectivo artificial autólogo de mucosa oral y evaluación de su desempeño como cobertura biológica en lesiones mucosas inducidas en conejos. Rev Fed Odontol Colomb. 2004(20):12-28.

CHAPTER 3

Gums and Mucilages Based Mucoadhesive Biopolymers

Inderbir Singh[1,*] and Pradeep Kumar[2]

[1]Department of Pharmaceutics, Chitkara College of Pharmacy, Chitkara University, Chandigarh-Patiala Highway, Rajpura-140401, Patiala, Punjab, India and [2]Wits Advanced Drug Delivery Platform Research Unit, Department of Pharmacy and Pharmacology, School of Therapeutic Sciences, Faculty of Health Sciences, University of the Witwatersrand, Johannesburg, 7 York Road, Parktown, 2193, South Africa

Abstract: In recent years, there have been extensive developments associated with the application of natural, derivatized and semi-synthetic biopolymers for the localized and prolonged delivery of active pharmaceutical ingredients at the mucosal interface. Naturally-derived gums and mucilages are biomacromolecular assemblies employed for traditional (films and tablets) as well as advanced (nanomedicine and conjugated systems) bioactive delivery paradigms. These natural biomaterials are stable, easily available, economical, non toxic and associated with less regulatory issues as compared to their synthetic counterpart. Additionally, these biopolymers could be easily tailored *via* graft-polymerization, functionalization, conjugation, and polyelectrolyte formation in order to render specific properties such as mucoadhesivity. In this chapter, gums and mucilages along with their modified derivatives have been discussed accompanied by gums- and mucilages-derived mucoadhesive drug delivery systems. Additionally, the mechanistic phenomena dictating their mucoadhesive performance will be described with special reference to the constituent functional groups and their role in muco-tethering and -penetration. A representative list of gums and mucilages include, but not limited to, alginic acid, agar, carrageenans, and laminarin (marine origin); gum arabic, gum karaya, locust bean gum, gum ghatti, khaya gum, tragacanth, albizia gum, guar gum, starch, cellulose, larch gum and pectin (plant origin); and curdian, pullulan, xanthan, dextran, zanflo, emulsan, Baker's yeast glycan, lentinan, schizophyllan, scleroglucan and krestin (microbial origin).

Keywords: Gums, mucilages, biopolymers, mucoadhesion, drug delivery systems, Kondagogu Gellan gum, Tamarind seed polysaccharide, Psyllium mucilage, Linum (linseed) mucilage

INTRODUCTION

Pharmaceutical excipients play an important role in terms of quality and performance of a pharmaceutical formulation. Recent trends towards the use of

***Corresponding author Inderbir Singh:** Department of Pharmaceutics, Chitkara College of Pharmacy, Chitkara University, Chandigarh-Patiala Highway, Rajpura-140401, Patiala, Punjab, India; Tel: +91-98550-24140; E-mail: inderbir.singh@chitkara.edu.in excipients of natural origin have posed tremendous competition to their synthetic counterparts because of low toxicity, low price biocompatibility, biodegradability, easy availability and environment friendliness.

Gums are abnormal products usually formed in some trees because of wounding, infection or adverse weather conditions. The formation of gums is a pathological condition resulting from injury or fungal or bacterial infection of the injured tree. Mucilages are the normal product produced without any injury to the plant and are formed within the cell. The applicability and usefulness of plant based polymers has been confirmed in various drug delivery archetypes such as buccal films, monolithic matrix systems, microspheres, nanoparticles, solutions, suspensions, implants. These biopolymers have been employed for wider applications in the form of viscosity enhancers, binders, stabilizers, emulsifiers, suspending agents, solubilizers, gelling agents, disintegrants and bioadhesives [1, 2].

Extraction is a vital procedure for recovering of target compounds from plants. Plant material may be dried in sunlight or oven followed by treatment with chloroform/petroleum ether for the removal of chlorophyll and pigments. Final isolated mucilage should be air- or oven-dried carefully below 50°C or should preferably be vacuum dried so that its properties remain unchanged. In order to minimize moisture uptake and subsequent hydrolytic degradation, the above dried natural material should be stored in desiccators. Some recent studies reported microwave energy as an innovative technique as a simple, efficient, fast and eco-friendly method for the extraction of phytoconstituents. Although the microwave assisted method follows principle similar to maceration or percolation, this method is more efficient in terms of time and solvent and allows higher extraction rate with better quality at lower cost.

In simple terms, "gums and mucilages are polysaccharide-containing sugars". The presence of different sugars can be confirmed by chromatographic techniques (TLC/HPLC) whereas structural elucidation may be conducted by FTIR, mass, and NMR spectroscopic techniques. Specific analytical tests for steroids, alkaloids, glycosides, carbohydrates, terpenes, flavonoids, saponins, amino acids, oils, fatty constituents, tannins and phenols may be performed for ensuring the of purity of gums and mucilages. Estimation of physicochemical properties such as colour, odour, texture, taste, shape, solubility, pH, loss on drying, swelling index, hygroscopicity and powder properties such as angle of repose, bulk and true densities, porosity, and surface tension can also be conducted. The presence of specific pathogens and microbial load are the indicator of the quality of natural materials. As gums and mucilages are viscous in nature, therefore rheological properties are considered as an important criterion for deciding their commercial use and applicability.

CLASSIFICATION OF NATURAL POLYSACCHARIDES

Natural polysaccharides including natural gums and mucilages could be classified on the basis of surface charge, source, shape and monomeric units in chemical structure. These classifications are being presented in Table **1** along with suitable examples.

Table 1. Classification of gums and mucilages on the basis of charge, source, shape and monomeric units present in the chemical structure.

Basis	Classification	Examples
Charge	Anionic Polysaccharides	Natural: Alginic acid, Xanthan gum, Gum Arabic, Gum Karaya, Gum Tragacanth Semi-Natural: Carboxymethyl, Chitin, Cellulose gum
	Cationic Polysaccharides	Natural: Chitosan Semi-Natural: Cationic Guar gum
	Nonionic Polysaccharides	Natural: Starch, Dextrins, Guar gum Semi-Natural: Cellulose Ethers (*e.g.* hydroxyethyl cellulose, Methylcellulose, Nitrocellulose).
	Amphoteric Polysaccharides	Semi-Natural: Carboxymethylchitosan, N-hydroxyl-Dicarboxyethylchitosan, Modified Potato starch
Source	Marine origin	Agar, Carrageenans, Alginic acid, Laminarin
	Shrubs/tree exudates	Gum Arabica, Ggum Ghatti, Gum Karaya, Gum Tragacanth, Khaya and Albizia gums
	Seed gums	Guar Gum, Locust bean Gum, Starch, Amylose, Cellulose
	Extracts	Pectin, Larch gum
	Tuber and roots	Potato starch
	Animal origin	Chitin and chitosan, Chondroitin sulfate, Hyaluronic acid
	Microbial origin (bacterial and fungal)	Xanthan, Dextrin, Curdian, Pullulan, Zanflo, emulsan, Baker's yeast glycan, schizophyllan, lentinan, krestin, scleroglucan
Shape	Linear	Algins, Amylose, Cellulose, pectins
	Short branches	Xanthan, Xylan, Galactomannans
	Branch-on-branch	Amylopectin, Gum Arabic, Tragacanth
Monomeric units	Homoglycans	Amylose, Arabinanas, Cellulose
	Diheteroglycans	Algins, Carragennans, Galactomannans
	Tri-heteroglycans	Arabinoxylans, Gellan, Xanthan
	Tetra-heteroglycans	Gum Arabic, Psyllium seed gum
	Penta-heteroglycans	Ghatti gum, Tragacanth

ADVANTAGES AND DISADVANTAGES OF NATURAL GUMS AND MUCILAGES

Plant based natural materials have number of advantages over synthesis and semi synthetic materials. Being non-toxic, biodegradable and biocompatible, they have no adverse impact on human health and are environment friendly. Low cost and easy availability reduces the production costs to greater extent. High quality natural materials could be collected in excess quantities owing to simple production procedures.

The equilibrium moisture content present in plant based materials is normally 10% or more, which is a major cause of microbial contamination. However, this may be prevented by use of appropriate preservatives and proper handling of the material. Another matter of concern is the batch to batch variation and uncontrolled hydration of the natural materials which is probably due to collection of natural materials at different times, region, species, and climate. Moreover, changes in rheological properties of the product containing natural materials has to be taken care off during the stability testing of the pharmaceutical formulations. Hence, there is an urgent need to develop suitable monographs of available gums and mucilages indicating their quality attributes at large.

MUCOADHESION

Bioadhesion may be defined as "the binding of a polymer (bio-based or artificial) to a biological substrate". The term used is mucoadhesion, when the substrate is a mucous layer. The mucosal membrane is present in various regions of the human body including the urogential tract, gastrointestinal tract, airways, eye, ear and nose. Hence, mucoadhesive drug delivery systems may be designed for potentially targeting oral, buccal, rectal, vaginal, ocular and nasal routes of administration. Transmucosal products including sprays, gels, tablets, lozenges, suppositories, pessaries, vaginal rods and gels have already been used worldwide. Mucoadhesive drug delivery systems offer a number of advantages including enhanced residence time of drug at the site of application, increased permeability of drug into the systemic circulation and enhanced bioavailability of the drug [3]. Mucoadhesive polymers besides being non-toxic, non-absorbable, biocompatible and economical must possess some molecular attributes required to develop a good mucoadhesive drug delivery system. These attributes includes: presence of hydrogen bond forming groups, sufficiently low surface free energy to achieve wetting of the mucosal surface, sufficient flexibility to go through and intermingle/penetrate within the mucin network, strong anionic charge to support the electrostatic interaction with the mucin, and high molecular weight for maximizing the adhesion through mechanical entanglements and bonding through van der Waals forces. Moreover, polymeric chains must have high segmental mobility assists rapid and deep penetration into the mucosal tissue substrate [4]. In general, the mucoadhesion phenomenon could be described in three steps: first step involves wetting of the tissue by the polymer followed by swelling of the polymer for an intimate contact with the biological tissue; second, interpenetration and/or entanglement of the polymeric chains with the mucin; and finally, the chemical bond formation leads to adhesion of the polymer to the mucosal surface. Certain theories of mucoadhesion have suggested that it might occur *via* physical entanglement and/or interpenetration of polymer chains and mucus chains (diffusion theory), ability of the mucoadhesive polymer to spread and build up intimate contact with the mucous layer (Wetting theory), surface forces: chemical bonds, strong primary forces: covalent bonds, weak secondary forces: hydrogen bonds and van der Waals forces (Adsorption theory), attractive electrostatic forces between the mucoadhesive polymer network and glycoprotein mucin chains (Electronic theory) [5]. The mucoadhesive polymers are vital components of drug delivery systems as it is the type and properties of the mucoadhesive polymers which is responsible for the development of mucoadhesive bonds (mucoadhesion) and controlling the release of the drug from the delivery system.

A number of *in vitro*, *in vivo* and *ex vivo* methods are reported in literature for the assessment of mucoadhesive property of the drug delivery systems (tablets, microspheres, nanoparticles, films, *etc.*) Commonly employed *in vitro*/*ex vivo* methods for the evaluation of mucoadhesiveness of the drug delivery system include shear strength measurement, tensile strength measurement and chip based systems. For studying the mechanism of mucoadhesion, texture analysis, rheological measurements, atomic force microscopy, wetting measurements, and zeta potential measurements have been successfully used [6]. Fig. (**1**) represents some commonly employed techniques for the assessment of mucoadhesion phenomenon of drug delivery systems.

Mucoadhesive polymers have sufficient affinity to develop a bond or intermingle with the mucin of mucosal surface. These polymers could be used alone or in combination for sufficient mucoadhesion of the drug delivery system. Non-mucoadhesive pharmaceutical formulation could be made mucoadhesive by incorporating mucoadhesive polymer in the drug delivery device. Chemical modification of polymers is another approach to enhance the mucoadhesive capacity of the delivery system containing the polymer. Kondagogu gum, gellan gum, okra gum, Tamarind seed polysaccharide, psyllium mucilage, linum (linseed) mucilage, fenugreek seed mucilage and various other gums and mucilages have been discussed explicating the mucoadhesive potential and pharmaceutical applications of these biopolymers.

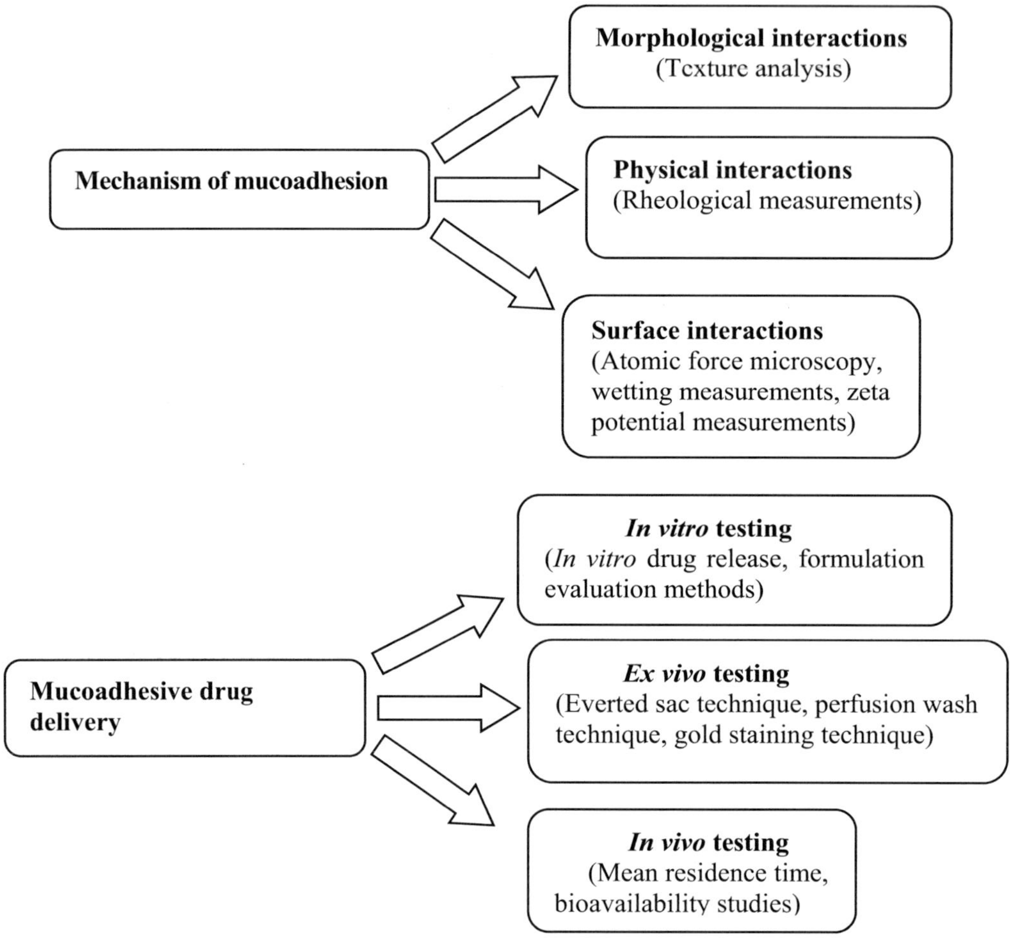

Figure 1: Schematic representation of various mucoadhesion assessment techniques.

KONDAGOGU GUM

Gum kondagogu is a natural gum exudate obtained from branches and stems of *Cochlospermum gossypium* of family Bixaceae. Gum kondagogu is negative charged and is composed of high-molecular weight complex polysaccharides. It contains glucuronic acid, galacturonic acid, rhamnose, D galactopyranose, D-glucose, galactose, mannose, arabinose, and fructose. Various pharmaceutical formulations like mini-matrices, hydrophilic matrices, transdermal patches and microcapsules have been developed using gum kondagogu as biopolymer [7]. The applicability of gum kondagogu is based on its viscosity. Putta *et al.* 2010, formulated mucoadhesive microcapsules of glipizide with gum Kondagogu by emulsification ionotropic gelation process. Evaluation of microcapsules was conducted in terms of hausner ratio, Carr's index, encapsulation efficiency, *in vitro* drug release, surface characteristics; mucoadhesive properties and *in vivo* hypoglycemic activity. Good mucoadhesive property was exhibited in the *in vitro* wash-off test [8]. Krishna and Murthy, 2010, also prepared mucoadhesive microspheres of glipizide using gum kondagogu as the mucoadhesive polymer. The microspheres showed good mucoadhesive property. The *in vitro* drug release studies indicated that gum kondagogu had more potentiality to retard the drug release compared to other gums and concentrations [9]. To enhance the performance and functionality of gum kondagogu, Kumar and Ahuja (2012) carried out carboxymethylation of gum kondagogu and evaluated it as a mucoadhesive polymer. For carboxymethylation gum kondagogu was reacted with monochloroacetic acid under controlled alkaline conditions. The said reaction resulted in increase in surface roughness and degree of crystallinity, reduction in viscosity and improvement in mucoadhesive properties of the natural polymer. Further, carboxymethyl gum kondagogu was used to prepare beads by ionic cross linking method. *Ex vivo* bioadhesion study conducted on isolated chick-ileum revealed more than 80% bioadhesion over a period of 24h [10]. In another study, the researchers carried out graft co-polymerization of N-vinyl-2-pyrrolidone on gum kondagogu with an aim to enhance its mucoadhesive potential. Characterization of the graft co polymer was done by FTIR, DSC and SEM studies. Buccal discs of metronidazole were prepared using grafted polymer and were evaluated for mucoadhesive properties. Buccal discs formulated using grafted gum exhibited better *ex vivo* bioadhesion as compared to the discs formulated using pristine gum. It was concluded that grafting of gum kondagogu with N-vinyl-2-pyrrolidone enhances its mucoadhesive property without having significant effect on the release behaviour [11].

GELLAN GUM

Gellan gum is an extracellular polysaccharide which is produced by *Pseudomonas elodea* bacterium. The natural form of gellan gum is a linear anionic hetero polysaccharide based on a tetrasaccharide unit of rhamnose, glucose and glucuronic acid in the molar ratio of 1:2:1. Natural form of gellan gum is partially acetylated with L glyceryl and acetyl groups located on the glucose residues. The degree to which the glucuronic acid is neutralized with various salts determines the molecular formula of gellan gum. It has been widely used in the formulation and development of *in situ* gelling ophthalmic as well as oral controlled release preparations because of characteristic property of cation induced gelation of the gum. Gellan gum is a promising mucoadhesive polymer for the development of nasal formulations. Residence time of 4-5 hrs in the nasal cavity has been reported after employing gellan gum in the formulations [12].

Narayana *et al.*, formulated *in situ* forming vaginal gels of secnidazole using gellan gum at concentrations ranging from 0.1 to 0.75% w/v. Formulations were evaluated for mucoadhesive force and *in vitro* drug release studies. The gel strength and mucoadhesion (as determined using texture analyzer) were found to be 6.5 g and 4 g, respectively. The gels showed potential to improve the drug administration at the site of infection thereby decreasing the dosing frequency [13]. The localised mucosal adherence of gellan gum was further explored by Narkar *et al.,* 2010, wherein the gastric mucus was used as the site of action. The controlled release of amoxicillin trihydrate from mucoadhesive gellan beads was employed for stomach specific delivery for *Helicobacter pylori* eradication. *In vitro* and *in vivo* studies showed good mucoadhesivity of beads and it was established that after 7 h more than 85% beads remained adhered to stomach mucosa of albino rats. Complete eradication of *H. Pylori* was depicted in growth inhibition study. It was concluded that controlled release mucoadhesive beads of amoxicillin were effective for the treatment of *H. pylori* infections [14]. In addition to vaginal and gastric mucosal surfaces, nasal route provides an interesting opportunity for site specific delivery of therapeutic molecules. Shah *et al.,* 2010, prepared mucoadhesive microspheres of gellan gum for intranasal delivery of sildenafil citrate. The microspheres were formulated by spray drying method and were evaluated for various properties like particle size, incorporation efficiency, swelling, zeta potential, *in vitro* and *ex vivo* mucoadhesion, and *in vitro* drug release. The formulation exhibited good mucoadhesive strength as determined by falling film technique. Good swelling index of the microspheres further confirmed enhanced mucoadhesive property of the formulation [15].

In line with the modification of gum kondagugu, Ahuja *et al.,* developed carboxymethyl gellan gum (by reaction with monochloroacetic acid) for its application as a bioadhesive polymer. The modified gellan gum showed an increase in mucoadhesive strength as compared to pure gellan gum (2.71 folds). Biocompatible and non-irritant property of the modified gum was established through *ex vivo* ocular tolerance and cytotoxic screening [16]. The researchers further conjugated gellan gum with thioglycollic acid for improving the mucoadhesive properties of gellan gum. The SH stretch in the fourier transform infrared spectra (FTIR) at 2571 cm^{-1} confirmed the formation of conjugates. 13.92 mM of thiol groups were reported to be present per gram of the conjugate. Thiolation of gellan gum resulted in increase in the degree of crystallinity and sensitivity to Ca^{2+}induced gelation of the gum. Metronidazole gels prepared using thiolated gellan gum as exhibited 1.82 folds higher mucoadhesive strength as compared to pure gellan gum formulation. Moreover, the drug released followed first-order and Higuchi's square-root release kinetics from both the formulations. Gellan thioglycollic acid conjugate or thiolated gellan gum was proved to be a promising bioadhesive excipient for various pharmaceutical formulations [17].

OKRA GUM

Okra gum is obtained from the pods of *Abelmoschus esculentus* (Malvaceae). Okra gum contains random coiled polysaccharides consisting of rhamnose, galactose and galacturonic acid. The gum contains repeating units of (1-2) rhamnose and (1-4) galacturonic acid residues with disaccharide side chains with a degree of acetylation. When extracted in water, these polysaccharides produce highly viscous solution. The highly viscous property of okra gum may be useful as a retarding polymer in the formulation of sustained release tablets. It has been investigated as a binding agent to produce tablets with good hardness, friability and drug release profiles. Its application as mucoadhesive agent has also been reported by many researchers. In one of the first studies, Attama *et al.,* 2003, used okra gum as bioadhesive polymers in the formulation of mucoadhesive tablets of indomethacin. Different batches of the tablet compacts were formulated based on different combination ratios of the polymers. The bioadhesive properties of the tablets were studied using a tensiometer [18]. The oral drug delivery potential of a similar gum, taro gum, was studied by Arora *et al.,* 2011, through the formulation of oral controlled release mucoadhesive matrix tablets of domperidone. Taro is a common name for the corms and tubers of *Colocasia esculenta* belonging to family Araceae. Tablets prepared by direct compression were evaluated for bioadhesive strength and *in vitro* dissolution performance. A 2 factors and 3 levels central composite design was employed for studying the effect of significant formulation variables like amount of PVP K 30 and taro gum on tensile strength, mucoadhesive strength, release exponent (n) and t50 (time for 50% drug release). Taro gum was found to be an effective natural material for the preparation of mucoadhesive tablets [19].

In a recent study, the hydrogelling potential of okra gum was evaluated in detail by Sharma *et al.* wherein a nasal gel of Rizatriptan benzoate was developed employing okra gum. The formulation was characterized for mucoadhesion, pH, viscosity and *in vitro* permeation. Formulation showed good properties in terms of mucoadhesion, pH and viscosity. It was concluded that okra mucilage is a safe gel forming natural material that can be developed to serve as an alternative to currently used synthetic gel formers for various drug delivery applications [20].

TAMARIND SEED POLYSACCHARIDE

Biological source of tamarind seedpolysaccharide is *Tamarindus indica* (Fabaceae). It is composed of (1-4) β-D-glucan backbone substituted at the O-6 position of glucopyranosyl residues with α-D-xylopyranose side chains. The xylose residues are substituted at O-2 position with β-D-galactosylate. Xyloglucan is the major structural polysaccharide present in the cell wall and is composed of glucosyl: xylosyl:galactosyl in the ratio of 3:2:1. Tamarind seed polysaccharide has been employed as an excipient in the formulation of hydrogels and various mucoadhesive drug delivery systems. Pharmaceutically, it is used as emulsifier, binding agent, release retardant, suspending agent and mucoadhesive agent. Its applications as a stabilizing agent and thickening in food industry have also been reported [21-23]. Mucoadhesive potential of tamarind seed polysaccharide could be enhanced by thiol fictionalization and co blending. Applications of tamarind seed polysaccharide as mucoadhesive biopolymer are depicted in Fig. (**2**).

In an interesting and unique study, Ghelardi *et al.* (2004), demonstrated that tamarind seed polysaccharide can significantly enhance the intra-aqueous penetration of rufloxacin in both infected and uninfected eyes leading to significant reduction in levels of *P. aeruginosa* and *S. aureus* in the cornea in comparison to rufloxacin alone. It was suggested that the tamarind seed polysaccharide prolongs the precorneal residence time of the formulation thereby enhancing the availability of the antibiotics for preconeal absorption. Tamarind seed polysaccharide appears to be a promising natural mucoadhesive polymer for the development of formulations for the treatment of bacterial keratitis [24]. Since the start of this decade, several researchers have tested the mucoadhesive efficacy of tamarind seed polysaccharide. Jana *et al.* (2010) prepared mucoadhesive buccal patches of meteronidazole using tamarind seed polysaccharide. The patches were cross linked with epichlorohydrin. The drug release depends on the cross-linking efficiency; lower level of crosslinker showed high drug permeation. The mucoadhesive strength and buccal residence time were better studied by the *ex vivo* permeation study. Tamarind seed mucilage was found to be a promising excipient for fabricating mucoadhesive buccal patches [25].

The physicomechanical properties of tamarind seed mucilage can be modulated *via* simplistic techniques such as crosslinking and thiolation. Patil *et al.* (2011) studied that tamarind seed mucilage, when cross-linked with epichlorohydrin, can be used for the formulation of novel wound dressing films, after loading with povidone iodine solution by soaking method. Tamarind seed mucilage showed ideal elasticity and tensile strength depending on the thickness and extent of cross-linking. The *in vivo* efficiency of antibacterial activity and wound healing activity studies by excision wound model of albino rats established significant results. Tamarind seed mucilage films treated groups showed greater rates of wound contraction and faster epithelialisation with significantly increased tensile strengths and collagen content of the regenerated tissue [26]. Kaur *et al.* (2012) studied thiol-functionalization of tamarind seed polysaccharide. Thioglycolic acid induced esterification of the polysaccharide was confirmed by FTIR, DSC and XRD studies. Thiolated tamarind seed polysaccharide showed 6.85 times higher mucoadhesive strength as compared to pristine tamarind seed polysaccharide [27]. Bhalekar *et al.,* 2013, also demonstrated the improved mucoadhesive potential of xyloglucan polymer *via* the covalent attachment of cysteine. Xyloglucan is a natural mucoadhesive polymer derived from the seeds of *Tamarindus indica* and used as an additive in the paper industry. Refined xyloglucan is used as a stabilizing, thickening and gelling agent in the food industry. Due to the introduction of thiol group within xyloglucan, its mucoadhesive potential was greatly improved. It can be used as a new generation mucoadhesive polymer for prolonging the residence time of drug delivery systems on various mucosal tissues [28].

Tamarind seed polysaccharide additionally offers favourable blending opportunities with other polymers, especially anionic polysaccharides. Uccello-Barretta *et al.,* 2013, developed aggregates of tamarind seed polysaccharide and hyaluronic acid. The aggregates exhibited enhanced mucoadhesive properties and were employed as artificial tears in dry eye syndrome for ophthalmic applications. A minimum concentration of tamarind seed mucilage (1.5mg/mL) and hyaluronic acid was required for NMR detectable mucoadhesive properties [29]. In another study involving polymer blending, Saikia *et al.,* 2013, developed mucoadhesive nanoparticles of tamarind seed mucilage for sustained delivery of anticancer drug irinotecan. Nanoparticles were prepared by ion gelation method and the polysaccharide was cross-linked with sodium alginate in different ratios. Stable nanoparticles with average particle size of 405±25.2 nm and polydispersity index of 0.497±0.02 were obtained. It was concluded that tamarind seed polysaccharides may be suitable for formulation of mucoadhesive nanoparticles for better efficacy and sustained delivery of anticancer drugs with reduced toxicity [30]. Nayak *et al.* in two different studies described the development and evaluation of mucoadhesive beads for oral delivery of metformin using tamarind seed polysaccharide-gellan gum and tamarind seed polysaccharide-calcium pectinate blend. The mucoadhesive formulation showed zero order drug release patter along with pH-dependent swelling and good mucoadhesivity with biological mucosal membrane. A significant hypoglycemic effect over prolonged period was observed in alloxan-induced diabetic rats after oral administration of mucoadhesive beads [31, 32].

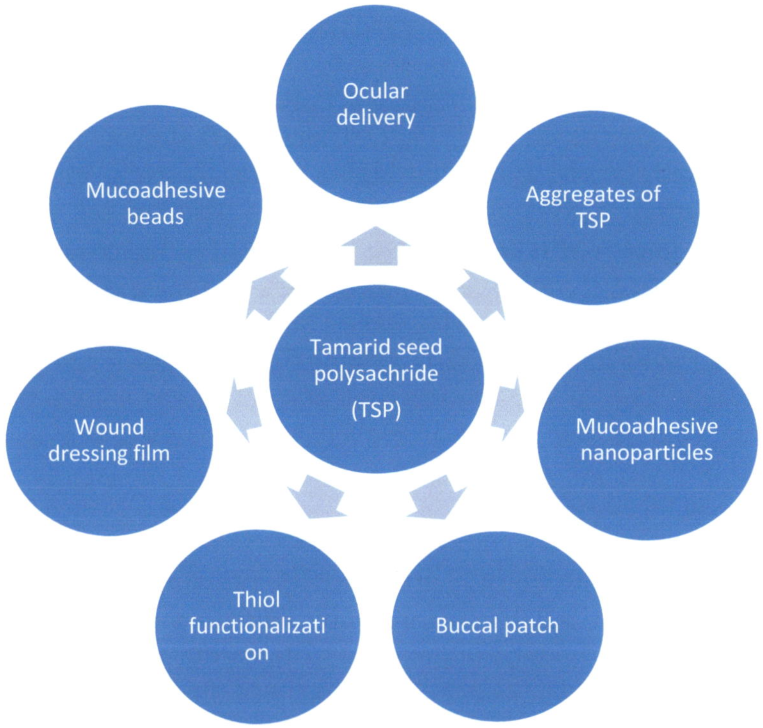

Figure 2: Applications of tamarind seed polysaccharide (TSP) as mucoadhesive biopolymer.

PSYLLIUM MUCILAGE

Psyllium also known as ispaghula is comprised of seed husks of *Plantago ovata* (Plantaginaceae).The laxative and long acting action of *Plantago* husk has been attributed to the gel forming fraction of the husk. Psyllium mucilage obtained from psyllium husk is white fibrous hydrophilic material that forms clear colorless mucilaginous gel by absorbing water. The polysaccharides extracted from the husk of *Plantagoovata* have been chemically characterized to contain a high proportion of hemicellulose which is the alkali soluble fraction of the husk. It consists of highly branched acidic arabinoxylan comprising of xylan backbone chain with xylose and arabinose forming the side chains [33]. Psyllium and its chemically modified derivatives have been explored for modification of water absorbing, gelling, and release properties. It is very well known from the times of ayurveda and is widely used as laxative to relive constipation. It also used for the treatment of diabetes, high cholesterol diarrhea, colon cancer, crohns disease (inflammatory bowel ulcerative colitis disease), obesity in children and adolescents. Psyllium seed husk has been successfully evaluated as binder, disintegrant and release retardant for microspheres and hydrogels formulations.

As described earlier in the chapter, thiolation and polymer blending provide enhanced mucoadhesion functionalities to the gums and mucilages. Bhatia and Ahuja, 2013, carried out thiol modification of psyllium with an aim to enhance its mucoadhesive potential. Thiolated psyllium was found to contain 3.282 m moles of thiol groups/g of the polymer. Mucoadhesive applications of thiolated psylium were explored by formulating gels using metronidazole as the model drug. On comparative evaluation thiolated psyllium gels showed 3-fold higher mucoadhesive strength than the psyllium gels [34]. Similar to tamarind seed polysaccharide, anionic polysaccharide (such as pectin and gellan gum) blending of psyllium has been reported by several researchers. Nayak *et al.,* 2014, prepared mucoadhesive beads of metformin HCl using polymer blends of ispaghula husk mucilage and pectin. The optimized beads exhibited pH-dependent swelling and good mucoadhesivity when tested with goat intestinal mucosa. Moreover, after oral administration in alloxan induced diabetic rats, a significant hypoglycemic effect over prolonged period of time was observed [35]. The researchers further employed response surface methodology based on 3(2) factorial design for the development of ispaghula mucilage-gellan gum mucoadhesive beads containing metformin HCl. Cumulative drug release after 10 h and drug encapsulation efficiency were selected as dependent variables. The *in vitro* drug release exhibited zero-order controlled release pattern (super case-II transport) over 10 h. The optimized beads showed good mucoadhesive potential and pH-dependent swelling [36].

LINUM (LINSEED) MUCILAGE

Linum usitatissimum (Linaceae) seed mucilage is a water soluble heterogeneous polysaccharide composed of glucose, galactose, arabinose, xylose, galacturonic acid, fucose and rhamnose. It is commonly known as flax or linseed. Flax is grown for its use as a nutritional supplement, edible oil and as an ingredient in many wood finishing products. Three natural phenolic glucosides ferulic acid, secoisolariciresinol diglucoside and p-coumaric acid are found in flax seeds. Flax seeds contain high levels of dietary fibers, lignans and omega-3 fatty acids.

In one of the first representative studies, Hayter and O'mullane (1993) filed a patent on compositions comprising of linseed mucilage for topical applications to the skin for potential having therapeutic and cosmetic utility. Linseed mucilage may be used as artificial mucus and has good mucoadherent properties. In another aspect, the invention provides the use of a composition of matter comprising linseed mucilage as a cosmetic preparation for topical application to the skin and/or mucous membranes of the human body. Amongst the many and varied uses of linseed mucilage as a mucosal or mucous-adherent are included its use as an artificial mucus and/or lubricant for application to the skin surface, the occular, nasal, oral, vaginal and anal cavities; its use as a mucoadherent in the gastrointestinal tract; and its use as a cytoprotective agent [37].

In an interesting study, Basu *et al.,* 2009, isolated mucilage from the seeds of *Linum usitatissimum* L and used it in the preparation of mucoadhesive nasal gels containing midazolam hydrochloride. Characterization of the isolated mucilage was done in terms of pH, swelling property, viscosity and mucoadhesive strength. Nasal gels of midazolam were formulated with and without penetration enhancers and the permeation studies were performed using excised goat nasal mucosa. In a comparative study, the mucilage isolated from *Linum usitatissimum* L. seeds exhibited improved mucoadhesive property as compared to the synthetic polymers like HPMC and carbopol 934 [38]. Mahant *et al.,* 2011, blended, formulated and evaluated mucoadhesive tablets of ranitidine hydrochloride using linseed mucilage and dried flax seed mucilage. It was concluded from the study that incorporation of linseed mucilage provides mucoadhesive as well as release retardant property to the formulation [39].

Nerkar and Gattani recently carried out a series of formulation studies involving linseed mucilage and targeted nasal and buccal mucosal surfaces employing microspheres and mucoadhesive gels. Firstly, buccal mucoadhesive microspheres of venlafaxine were developed using linseed mucilage as mucoadhesive biopolymer. Linseed mucilage was found to be a promising mucoadhesive biopolymer for the production of mucoadhesive microspheres [40]. Later on, prepared buccal gels of Venlafaxine using *Lepidium sativum* seed mucilage were prepared and were evaluated for rheological and mechanical (gel strength, adhesiveness) properties. A combination of the *Lepidium sativum* seed mucilage and carbopol 934 P exhibited prolonged delivery of venlafaxine when administered *via* buccal route [41]. Recently, Nerkar and Gattani (2013) extracted mucilage from linseed for its use as mucoadhesive gelling agent for the administration of venlafaxine by buccal route in the gel form. Linseed mucilage based buccal mucoadhesive gel preparations in combination with chitosan, carbopol 934P, carboxy methylcellulose and polyvinyl pyrrolidone were formulated and evaluated for gel strength, viscosity, *in vitro* diffusion and percentage mucoadhesion. Formulation, which contained linseed mucilage (2 %) and chitosan (0.5 %), showed the highest percentage of mucoadhesion, gel strength and sustained drug diffusion. It was concluded that the combination of linseed mucilage and chitosan can be used to formulate buccal mucoadhesive gel and increase the bioavailability of venlafaxine [42].

FENUGREEK MUCILAGE

Trigonella foenum-graecum (Fabaceae) seeds are used to flavor many foods, spice blends and pickles. It is commonly known as fenugreek and methi. Fresh fenugreek leaves are used as flavouring agent in Indian curries and its sprouted seeds are used in salads. Pharmaceutical applications of fenugreek seed mucilage includes gelling agent, granulating agent, binding agent, disintegrating agent, suspending agent, release retardant and mucoadhesive polymer. Datta *et al.* (2005) developed nasal gel for the delivery of diazepam using fenugreek mucilage as mucoadhesive agent. The gel formulation was evaluated for mucoadhesive strength, viscosity, gelling property and *in vitro* drug release study. Fenugreek mucilage was found to exhibit better mucoadheive strength when compared with standard mucoadhesive polymers [43]. Recently, Nayak *et al.* (2013) investigated the optimization and development of fenugreek seed mucilage-alginate based mucoadhesive beads containing metformin HCl prepared by ionotropic gelation technique. The average size of the beads was found to be ranging between 0.92 ± 0.05 to 1.30 ± 0.14 mm. It was reported that pH of the test medium had a significant influence on degradation and swelling of mucoadhesive beads. The beads exhibited good mucoadhesivity in wash-off test [44]. The fenugreek mucilage when evaluated for its bucco-adhesive properties by Gondaliya and c-workers wherein buccal gels of verapamil HCl were formulated. The mucosal permeation of drug from the formulation was evaluated using Franz diffusion cell, goat buccal mucosa as semi-permeable membrane. The drug release was sustained up to 6 hrs by optimizing concentration of polymers [45].

MISCELLANEOUS GUMS

Various other gums have also been reported to possess mucoadhesive potential for the delivery of active pharmaceutical ingredient through different pharmaceutical formulations like tablets, gels *etc*. Table **2** enlists various natural gums explored for the development of different drug delivery systems. Some of the representative studies involving various "miscellaneous gums" are detailed as follows:

1. Alur *et al.* (1999) prepared buccal tablets of chlorpheniramine maleate using gum obtained from *Hakea gibbosa*. The natural gum exhibited excellent mucoadhesive properties. Relaxation of the polymeric chains of the hydrated hakea gum was the proposed mechanism of drug release from the formulation. The mucoadhesive strength could be modulated by altering the amount of gum in the tablet formulation [46].

2. Arora *et al.* (2011) prepared controlled release matrix mucoadhesive tablets of domperidone employing *Salvia plebeian* gum as natural bio polymer. Different concentrations (5, 10, 15 and 20% w/w) of the natural polymer were used for preparing matrix tablets. Both diffusion and polymer erosion in combination were reported to be responsible for the release of drug from the formulation. The findings indicate the potential of *Salvia plebeian* gum as binder, mucoadhesive and release retardant natural material in tablet formulations [47].

3. Nep and Conway (2011) compared the mucoadhesive performance of *Grewia mollis* gum with guar gum, hydroxypropyl methylcellulose, carboxymethylcellulose and carbopol 971P. Compacts and gels were the prepared formulations. Better mucoadhesive properties were observed with Grewia polysaccharide gum when compared with guar gum, carboxymethylcellulose, carbopol 971P and hydroxypropyl methylcellulose. Grewia polysaccharide gum is a suitable biopolymer for the development of mucoadhesive drug delivery devices [48].

4. Gowthamarajan *et al.* (2012) used *Anacardium occidentale* gum from cashewnut trees for preparing bilayered, buccoadhesive tablets of curcumin. The drug release kinetics from the tablets was non-Fickian or anomalous diffusion. Cashewnut tree gum was proposed to be a potential mucoadhesive polymer for the development of

buccoadhesive tablets for improving the bioavailability of curcumin [49].

5. Mahakalkar and Upadhye (2013) developed *in situ* mucoadhesive nasal gel formulation of zolmitriptan using a natural mucoadhesive polymer obtained from bark of *Sterculia foetida*. The formulation was found to be beneficial for the treatment of migraine for long period of time. Improved bioadhesion by using *sterculia foetida* gum enhanced nasal residence time of the drug delivery system owing to increased viscosity of gel in the nose which maybe advantageous to protect the drug from draining out. Moreover the natural polymer also exhibits a permeation enhancing effect thereby increasing the bioavailability of drug. Bioadhesive force, gel strength and sustained release effects were found to increase with the increase in concentration of *sterculia foetida* gum in the formulation [50].

6. Mylangam *et al.* (2014) formulated buccoadhesive tablets of metoprolol succinate using natural gum obtained from *Terminalia catappa* as a buccoadhesive polymer. Mucoadhesive properties of the gum were found to increase with the increase in concentration of the biopolymer in the formulation. This was evident form the detachment force measurement, *ex vivo* residence time, and swelling studies. There was increase in the bioavailability of metoprolol succinate from *Terminalia catappa* formulation which was evident from high AUC and MRT values compared with the commercial formulation [51].

Table 2. Miscellaneous natural gums explored for developing various drug delivery systems.

Natural gum	Formulation	Drug	Ref.
Hakea gibbosa	Buccal tablets	Chlorpheniramine maleate	[46]
Salvia plebeian	Matrix mucoadhesive tablets	Domperidone	[47]
Grewia mollis	Compacts and gels	-	[48]
Anacardium occidentale	Buccoadhesive tablets	Curcumin	[49]
Sterculia foetida	Nasal gel	Zolmitriptan	[50]
Terminalia catappa	Buccoadhesive tablets	Metoprolol succinate	[51]

MISCELLANEOUS MUCILAGES

Table **3** enlists some mucilages from different biological sources employed in the formulation and development of mucoadhesive drug delivery systems for the delivery of wide variety of bioactives and are described below:

1. Metia and Bandyopadhyay (2008) employed mucilage isolated from edible *Diospyros peregrina* and *Dillenia indica* fruits to formulate mucoadhesive buccal tablets of oxytocin. Freshly excised bovine buccal mucosas was used for evaluating shear, tensile and peel strengths of the formulation. A 27% drug bioavailability without damaging the buccal mucosa was exhibited by the mucoadhesive buccal tablets suggesting its suitability as an alternative to noninvasive administration of oxytocin [52, 53].

2. Sudhakar and Bandyopadhyay (2008) evaluated mucilage from whole seeds of *Sinapis alba* as mucoadhesive polymer for the development of buccal tablets with aim to improve the bioavailability of diltiazem. Three stage process using specially fabricated punches was employed to prepare novel buccal adhesive tablets in cup and core fashion to achieve unidirectional release towards mucosa. The adhesive cups were evaluated for shear, tensile, and peel strengths using specially designed apparatus employing freshly excised ruminant and porcine buccal mucosa as substrates. *Ex vivo* permeation studies in a Franz diffusion cell were conducted through porcine buccal mucosa. It was concluded that the novel formulation prevented salivary scavenging of drug thereby exhibited improved bioavailability [54].

3. Basu and Bandyopadhyay (2010) prepared *in situ* mucoadhesive nasal gels of midazolam hydrochloride with mucilage isolated from *Ficus carica* fruits. The formulated gels were evaluated for texture profile analysis, viscosity and mucoadhesive strength. *In vitro* drug permeation studies were conducted using excised goat nasal mucosa on gels prepared with and without sodium taurocholate (permeation enhancer). Histological study of nasal mucosa before and after the permeation study was also performed. *In vivo* experiments in rabbits confirmed better bioavailability of midazolam from mucilage formulation as compared to gels prepared from synthetic mucoadhesive polymers. Moreover, it was observed that no damage of the nasal mucosa was observed during the permeation study [55].

4. Moghimipour *et al.* (2012) used mucilage of *Cordia myxa* as a natural mucoadhesive polymer to formulate buccal tablets of chlorhexidine. Effect of mucilage concentration on various tablet parameters like dissolution and muco-adhesiveness strength was studied [56].

5. Sabale *et al.* (2012) isolated the mucilage from the fresh fruits of *Artocarpus heterophyllus* Lam. (Moraceae) commonly known as jackfruit. Buccal mucoadhesive tablets for the controlled release of chlorpheniramine maleate were developed employing jackfruit mucilage. The drug release kinetics showed that tablets containing natural mucilage were the best formulations because they showed a prolonged drug release with linear kinetics when compared to Carbopol 974P [57].

6. Sabale *et al.* (2014) isolated *Calendula officinalis* mucilage as a mucoadhesive agent and to formulate controlled release buccoadhesive tablets. The tablets were characterized for various physicochemical parameters and mucoadhesive property. The results of *in vitro* release studies indicated controlled drug release from the formulation. *Calendula* mucilage was reported to be a potential mucoadhesive and controlled release biopolymer [58].

Table 3. Miscellaneous natural mucilages explored for developing various drug delivery systems.

Natural Mucilage	Formulation	Drug	Ref.
Diospyros peregrina	Buccal tablets	Oxytocin	[52]
Dillenia indica	Buccal tablets	Oxytocin	[53]
Sinapis alba	Buccal tablets	Diltiazem	[54]
Ficus carica	Nasal gel	Midazolam	[55]
Cordia myxa	Tablets	Chlorhexidine	[56]
Artocarpus heterophyllus	Buccal tablets	Chlorpheniramine maleate	[57]
Calendula officinalis	Buccal tablets	Chlorpheniramine maleate	[58]

CONCLUSION

Natural biopolymers such as gums and mucilages present exciting opportunities for the pharmaceutical research and industry in terms of cost-effectiveness, availability, and biocompatibility. In addition, the derivatisation potential of the gums and mucilages discussed in this chapter provide an interesting paradigm for the design and development of novel semi-synthetic biomaterials overcoming the non-reproducibility disadvantage of gums and mucilages to some extent. Furthermore, the inherently bioadhesive nature of these excipients is an added advantage with no additional chemical efforts as compared to extensive chemical processes applied to synthetic acrylate-based mucoadhesive polymers. The biomucoadhesivity of these excipients has been frequently employed to formulate specialized delivery systems such as gastroretentive, buccal, nasal, vaginal, and transdermal delivery systems for local as well as systemic drug transport. However, there are certain areas such as – extraction methods and associated costs, microbial contamination and susceptibility, and environmental stability and storage concerns that need immediate attention to make these biopolymers more acceptable at commercial level.

ACKNOWLEDGMENTS

Declared none.

CONFLICT OF INTEREST

The authors confirm that this chapter content have no conflict of interest.

REFERENCES

[1] Wallis T.E. Text Book of Pharmacognosy. 5th Ed, CBS Publishers and Distributors, Delhi, India, 2005; pp. 465,472, 474-7.

[2] Evans W.C. Treese and Evans Pharmacognocy.15th Ed., Elsevier, 2007; pp. 205,206,209.

[3] Woodley J. Bioadhesion: New possibilities for drug administration. Clin Pharmacokinet 2001; 40: 77-84.

[4] Roy S, Pal K, Anis A, Pramanik K, Prabhakar B. Polymers in mucoadhesive drug delivery system: A brief note. Design Monomer Polym 2009; 12: 483-495.

[5] Madsen F, Eberth K, Smart J. A rheological assessment of the nature of interactions between mucoadhesive polymers and a homogenised mucus gel. Biomaterials 1998; 19: 1083-92.

[6] Singh I, Rana V. Techniques for the assessment of mucoadhesion in drug delivery systems: An overview. J Adhesion Sci Technol 2012; 26: 2251-67.

[7] Janaki B, Sashidhar RB. Physico-chemical analysis of gum kondagogu (*Cochlospermum gossypium*): a potential food additive. Food Chem 1998; 61(1/2): 231-6.

[8] Putta SK, Kumar A, Kumar A. Formulation and *in vitro* evaluation of mucoadhesive microcapsules of glipizide with gum kondagogu. J Chem Pharm Res 2010; 2(5): 356-64.

[9] Krishna RR, Murthy TEGK. Preparation and evaluation of mucoadhesive microcapsules of glipizide formulated with gum kondagogu: *In vitro* and *in vivo.* Acta Pharmaceutica Sciencia 2010; 52(3): 335-44.

[10] Kumar A, Ahuja M. Carboxymethyl gum kondagogu: Synthesis, characterization and evaluation as mucoadhesive polymer. Carbohydr Polym 2012; 90(1): 637-43.

[11] Malik S, Kumar A, Ahuja M. Synthesis of gum kondagogu-g-poly(N-vinyl-2-pyrrolidone) and its evaluation as a mucoadhesive polymer. Int J Biol Macromol 2012; 51(5): 756-62.

[12] Jansson PE, Lindberg B, Sanford PA. Structural studies of gellan gum an extra cellular polysaccharide elaborated by Pseudomonas eloda. Carbohydr Res 1983; 124: 135-9.

[13] Narayana RC, Harish NM, Gulzar A M, Prabhakara P, Singh AK, Subrahmanyam EV. Formulation and *in vitro*evaluation of in situgelscontainingsecnidazole for vaginitis. Yakugaku Zasshi 2009; 129(5): 569-74.

[14] Narkar M, Sher P, Pawar A. Stomach-specific controlled release gellan beads of acid-soluble drug prepared by ionotropic gelation method. AAPS PharmSciTech 2010; 11(1): 267-77.

[15] Shah V, Sharma M, Parmar V, Upadhyay U. Formulation of sildenafil citrate loaded nasal microspheres: An *in vitro, ex vivo* characterization. Int J Drug Del 2010; 2: 213-20.

[16] Yadav S, Ahuja M, Kumar A, Kaur H. Gellan-thioglycolic acid conjugate: synthesis, characterization and evaluation as mucoadhesive polymer. Carbohydr Polym 2014; 99: 601-7.

[17] Ahuja M, Singh S, Kumar A. Evaluation of carboxymethyl gellan gum as a mucoadhesive polymer. Int J Biol Macromol 2013; 53: 114-21.

[18] Attama AA, Adikwu MU, Amorha CJ.Release of indomethacin from bioadhesive tablets containing carbopol 941 modified with Abelmuschus esculentus (okra) gum. Boll Chim Farm 2003; 42(7): 298-302.

[19] Arora G, Malik K, Singh I. Formulation and evaluation of mucoadhesive matrix tablets of taro gum: optimization using response surface methodology. Polim Med 2011; 41(2): 23-34.

[20] Sharma N, Kulkarni GT and Sharma A. Development of Abelmoschus esculentus (Okra) based mucoadhesive gel for nasal delivery of rizatriptan benzoate. Trop J Pharm Res 2013; 12(2): 149-53.

[21] Datta R, Bandyopadhyay AK. A new nasal drug delivery system for diazepam using natural mucoadhesive polysaccharide obtain from tamarind seed. Saudi Pharm J 2006; 14: 115-9.

[22] Khanna M, Nandi RC, Sarin JP. Standardisation of tamarind seed polyosacchride for pharmaceutical use. Indian Drugs 1987; 24: 268-9.

[23] Kulkarni GT, Gowthamarajan K, Dhobe RR, Yohanan F, Suresh B. Development of controlled release spheroids using natural polysaccharide as release modifier. Drug Deliv 2005; 12(4): 201-6.

[24] Ghelardi E, Tavanti A, Davini P, Celandroni F, Salvetti S, Parisio E, Boldrini E, Senesi S, Campa M. A mucoadhesive polymer extracted from tamarind seed improves the intraocular penetration and efficacy of rufloxacin in topical treatment of experimental bacterial keratitis. Antimicrob Agents Chemother 2004; 48(9): 3396-401.

[25] Jana S, Lakshman D, Sen KK, Basu SK. Development and evaluation of epichlorohydrin cross-linked mucoadhesive patches of tamarind seed polysaccharide for buccal application. Int J Pharm Sci Drug Res. 2010; 2:193-8.

[26] Patil BS, Mastiholimath V, Kulkarni AR. Novel wound dressing material derived from crosslinked tamarind seeds polysaccharide: *in vitro* characterization and wound healing activity. Trends Biomater Artif Organs 2011; 25:149-53.

[27] Kaur H, Yadav S, Ahuja M, Dilbaghi N. Synthesis, characterization and evaluation of thiolated tamarind seed polysaccharide as a mucoadhesive polymer. Carbohydr Polym 2012; 90(4): 1543-9.

[28] Bhalekar M, Sonawane S, Shimpi S. Synthesis and characterization of a cysteine xyloglucan conjugate as mucoadhesive polymer. Brazilian J Pharm Sci 2013; 49(2): 285-92.

[29] Uccello-Barretta G, Balzano F, Vanni L, Sansò M. Mucoadhesive properties of tamarind-seed polysaccharide/hyaluronic acid mixtures: A nuclear magnetic resonance spectroscopy investigation. Carbohydr Polym 2013; 91(2): 568-72.

[30] Saikia P, Sahu BP, Dash SK. Mucoadhesive nanoparticles from tamarind seed polysaccharides for sustained delivery of anticancer drug irinotecan. Asian J Pharm 2013;7: 163-9.

[31] Nayak AK, Pal D, Santra K. Tamarind seed polysaccharide-gellan mucoadhesive beads for controlled release of metformin HCl. Carbohydr Polym 2014; 99: 601-7.

[32] Nayak AK, Pal D, Santra K. Development of calcium pectinate-tamarind seed polysaccharide mucoadhesive beads containing metformin HCl. Carbohydr Polym 2014; 101: 220-30.

[33] Kennedy JF, Sandhu JS, Southgate DAT Structural data for the carbohydrate of ispaghula husk ex Plantago ovate forsk. Carbohydr Res 1979; 75, 265-74.

[34] Bhatia M, Ahuja M. Thiol modification of psyllium husk mucilage and evaluation of its mucoadhesive applications. Scientific World J 2013; Article ID 284182.

[35] Nayak AK, Pal D, Santra K. Development of pectinate-ispagula mucilage mucoadhesive beads of metformin HCl by central composite design. Int J Biol Macromol 2014; 66: 203-11.

[36] Nayak AK, Pal D, Santra K. Ispaghula mucilage-gellan mucoadhesive beads of metformin HCl: Development by response surface methodology. Carbohydr Polym 2014; 107: 41-50.

[37] Hayter IP and O'mullane JE. Linseed mucilage. Application No.PCT/GB1993/000343, Publication No.WO1993016707 A1. (Publication date 2 Sep' 1993).

[38] Basu S, Chakrabatorty S, Bandyopadhyay AK.Development and evaluation of a mkucoadhesive nasal gel of midazolam prepared with *Linum usitatissimum* L. seed mucilage. Sci Pharm 2009; 77: 899-910.

[39] Mahant S, Khurana N, Dua S, Takhar P and Bakshi IS Formulation and evaluation of mucoadhesive tablets using dried flax seed mucilage. J Pharm Biomed Sci 2011; 6(8): 1-7.

[40] Nerkar PP, Gattani S. *In vivo, in vitro* evaluation of linseed mucilage based buccal mucoadhesive microspheres of venlafaxine. Drug Deliv 2011; 18(2): 111-21.

[41] Nerkar PP, Gattani SG. Cress seed mucilage based buccal mucoadhesive gel of venlafaxine: *in vivo, in vitro* evaluation. J Mater Sci Mater Med 2012; 23(3): 771-9.

[42] Nerkar PP, Gattani SG. Oromucosal delivery of venlafaxine by linseed mucilage based gel: *in vitro* and *in vivo* evaluation in rabbits. Arch Pharm Res 2013; 36 (7):846-53.

[43] Datta R and Bandyopadhyay AK.Development of a new nasal drug delivery system of diazepam with natural mucoadhesive agent from *Trigonella foenum-graecum* L. J Sci Ind Res 2005; 64, 973-7.

[44] Nayak AK, Pal D, Pradhan J, Hasnain MS. Fenugreek seed mucilage-alginate mucoadhesive beads of metformin HCl: Design, optimization and evaluation. Int J Biol Macromol 2013; 54: 144-54.

[45] Gondaliya RP, Shah PD, Nerkar PP, Mahajan HS, Ige PP. Buccal gel of verapamil HCl based on fenugreek mucilage and xanthan gum: *in vitro* evaluation. Am J PharmTech Res 2015 (Article in Press).

[46] Alur HH, Pather SI, Mitra AK, Johnston TP. Evaluation of the gum from Hakea gibbosa as a sustained-release and mucoadhesive component in buccal tablets. Pharm Dev Technol 1999; 4(3): 347-58.

[47] Arora G, Malik K, Singh I, Arora S, Rana V. Formulation and evaluation of controlled release matrix mucoadhesive tablets of domperidone using *Salvia plebeian* gum. J Adv Pharm Technol Res 2011; 2(3):163-9.

[48] Nep EI, Conway BR. Grewia gum 2: mucoadhesive properties of compacts and gels. Tropical J Pharm Res 2011; 10(4): 393-401.

[49] Gowthamaraian K, Jawahar N, Wake P, Jain K, Sood S. Development of buccal tablets for curcumin using *Anacardium occidentale* gum.Carbohydr Polym 2012; 88(4): 1177–83.

[50] Mahakalkar NG, Upadhye KP. Zolmitriptan nasal i*n-situ* gel using *Sterculia Foetida* Linn gum as natural mucoadhesive polymer. Int J Pharm Sci Rev Res 2013; 22(2): 206-13.

[51] Mylangam CK, Beeravelli S, Medikonda J, Pidaparthi JS, Kolapalli VR. Badam gum: a natural polymer in mucoadhesive drug delivery. Design, optimization, and biopharmaceutical evaluation of badam gum-based metoprolol succinate buccoadhesive tablets. Drug Deliv 2014 (Article In Press).

[52] Metia PK, Bandyopadhyay AK. *In vitro* and *in vivo* evaluation of a novel mucoadhesive buccal oxytocin tablet prepared with *Dillenia indica* fruit mucilage. Pharmazie 2008; 63(4): 270-4.

[53] Metia PK, Bandyopadhyay AK. *In vitro* evaluation of novel mucoadhesive buccal tablet of oxytocin prepared with *Diospyros peregrina* fruit mucilages. Yakugaku Zasshi 2008; 128(4): 603-9.

[54] Sudhakar Y, Bandyopadhyay AK. Novel buccal adhesive tablets using Aloe vera L and Sinapis alba--a promising option for improved bioavailability of diltiazem hydrochloride. PDA J Pharm Sci Technol 2008; 62(2): 97-110.

[55] Basu S, Bandyopadhyay AK. Development and characterization of mucoadhesive *in situ* nasal gel of midazolam prepared with *Ficus carica* mucilage. AAPS PharmSciTech 2010; 11(3): 1223-31.

[56] Moghimipour E, Aghel N, Adelpour A. formulation and characterization of oral mucoadhesive chlorhexidine tablets using cordia myxa mucilage. Jundishapur J Nat Pharm Prod 2012; 7(4): 129-33.

[57] Sabale V, Patel V, Paranjape A. Isolation and characterization of jackfruit mucilage and its comparative evaluation as a mucoadhesive and controlled release component in buccal tablets. Int J Pharm Invest 2012; 2(2): 61-9.

[58] Sabale V, Patel V, Paranjape A. Evaluation of *Calendula* mucilage as a mucoadhesive and controlled release component in buccal tablets. Res Pharm Sci 2014; 9(1): 39-48.

CHAPTER 4

Natural Polymers for Ophthalmic Drug Delivery

Anjali Hirani[1,2], Yong W. Lee[2], Vijaykumar B. Sutariya[1] and Yashwant Pathak[1,*]

[1]Department of Pharmaceutical Sciences, USF College of Pharmacy, University of South Florida, Tampa, FL 33612, USA and [2]School of Biomedical Engineering and Sciences, Virginia Tech-Wake Forest University, Blacksburg, VA 24061, USA

Abstract: Recent developments in polymer technology have created new opportunities for ocular drug delivery, which has unique challenges due to specific attributes of the eye. The unique requirements of ocular drug delivery pose several difficulties for a formulator to select an appropriate polymer and use it in ODD. Nanotechnology has opened up new vistas as the nano particle becomes a choice of drug delivery in ODD. Layers of tissue, blood barriers, choroidal flow, lymphatic drainage, and lacrimation are some of the factors that limit therapeutic concentrations of drugs from reaching diseased parts of the eye. Currently, research is being directed towards the identification and discovery of novel drug delivery systems using newly introduced polymers, nanoparticles, liposomes, and adhesive gels to circumvent barriers and obtain sustained levels of therapeutics at target ocular sites. The ocular route of drug delivery is determined based on disease application and each method presents its own challenge: topical (uveitis, conjunctivitis), systemic (cytomegalovirus retinitis), intravitreal (macular degeneration, diabetic reinopathy), subconjunctival (glaucoma), subtenon (retinal vein occlusion). Formulation factors such as drug loading, drug stability, drug excipient interaction, carrier biocompatibility and drug carrier biodegradability need to be considered when designing ocular drug delivery systems. Ocular biocompatibility and ocular biodegradability are one of the few important characteristics which dictate the selection of such polymeric materials to be used as ODD carriers. This chapter covers various aspects of the ocular delivery and discusses the characteristic of the nano biomaterials to be used as ocular drug delivery carriers.

Keywords: Nanoparticles, Ocular Drug delivery, Liposomes, Mucoadhesive gels, Age-related macular degeneration, Diabetic retinopathy, Glaucoma, Conjunctivitis, Polysaccharide, Chitosan.

1. INTRODUCTION

Approximately 285 million people worldwide are estimated to be visually impaired. 37 million people have ocular disorders that have resulted in complete

***Corresponding author Yashwant Pathak:** Department of Pharmaceutical Sciences, USF College of Pharmacy, University of South Florida, Tampa, FL 33612, USA; Tel: 813-974-1029; E-mail: ypathak1@health.usf.edu

Viness Pillay, Yahya E. Choonara and Pradeep Kumar (Eds.)

blindness [1]. The prevalence of ocular disorders will increase with the worldwide aging population. The global causes as a percentage of total blindness are depicted in Fig. (**1**).

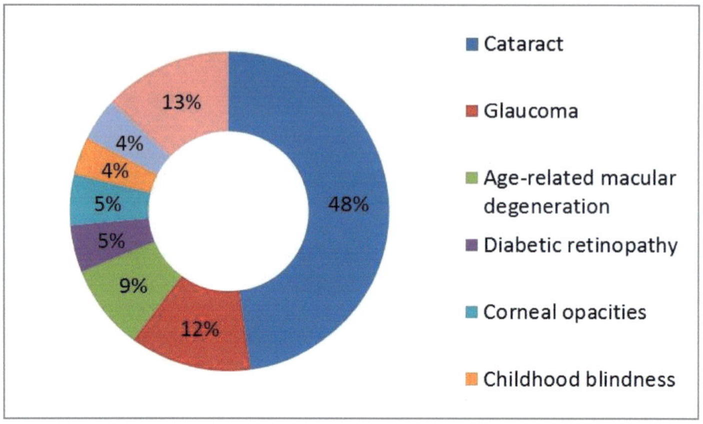

Figure 1: Global causes of blindness as reported by the World Health Organization [2].

Cataract is responsible for almost half of the reported cases of vision impairment in the world; however, in developed countries, age-related macular degeneration is the reigning source of blindness [2]. Although current treatments exist, there is a need for more efficient therapies that can halt the progression and even reverse damage of ocular diseases. Recent developments in polymer technology have created new opportunities to discover new drug targets, enhance ocular drug discovery, and reposition current drugs for more efficacious therapy.

2. CHALLENGES IN OCULAR DRUG DELIVERY

2.1. Static and Dynamic Barriers of the Eye

The eye is protected from foreign substances by its unique natural anatomic barriers. The two main static barriers that shield the front and the back of the eye are the cornea and the blood-retina barrier, respectively. Addition static barriers include the sclera, retina and blood aqueous barrier [3]. These barriers restrict administration of potential therapeutics. After circumventing the immediate physiological barriers, therapeutics are prevented from localizing at target sites due to the dynamic nature of the eye; this includes blood flow in the conjunctiva and choroid, lymphatic clearance, nasolacrimal drainage and efflux pumps. Design of drug delivery systems to overcome these barriers as well as allowing for simple patient administration while not irritating ocular tissue are some of the basic requirements needed for more effective ocular therapeutics [3].

2.2. Routes of Drug Administration

Therapeutic agents can be delivered to the eye by a variety of routes including topical, periocular injection, intravitreal injection, and systemic administration. The route of drug administration is an important consideration that can promote or hinder drug efficacy. Ideally, the drug should be delivered to the site of action at a rate and concentration that minimizes side effects and maximizes therapeutic effect; however, the delivery and targeting barriers are so great, most drugs are rendered ineffective. Fig. (**2**) depicts the common modalities of drug delivery.

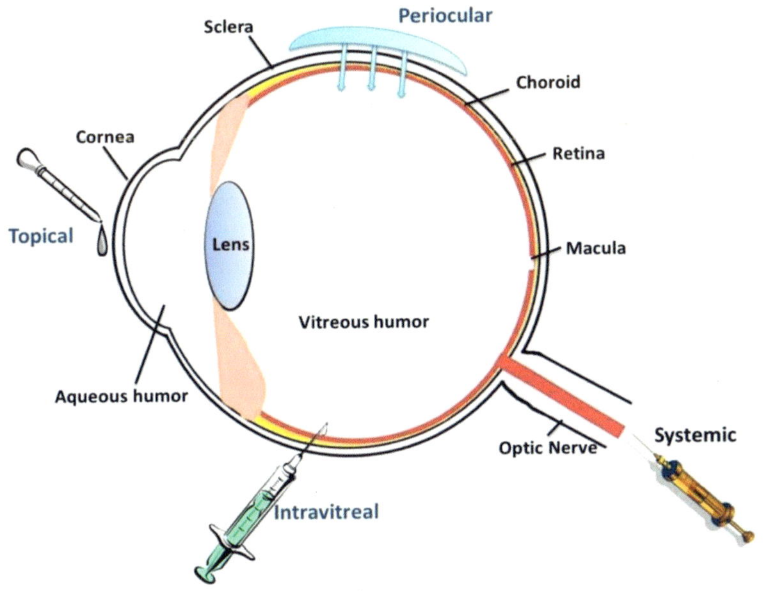

Figure 2: Structure of the eye showing drug delivery modalities for treating ocular disease.

The ocular route of drug delivery is determined based on disease application and each method presents its own challenge: topical (uveitis, conjunctivitis), systemic (cytomegalovirus retinitis), intravitreal (macular degeneration, diabetic retinopathy), subconjunctival (glaucoma), subtenon (retinal vein occlusion). Table **1** describes these common ocular disorders, current treatments utilized clinically, and the route of drug administration utilized in detail.

2.2.1. Topical

The topical route is a convenient method of drug delivery to the anterior segment of the eye because it is noninvasive and can be self-administered which leads to high patient compliance. However, topical medications are stopped from reaching the target site due to the eye's natural protective surface with the cornea acting as a barrier and the presence of efflux pumps [4]. Additionally, there is poor bioavailability due to nasolacrimal drainage, tear turnover, and nonproductive conjunctival uptake [5]. A transient diffusion model described in the literature has shown that less than 5% of a lipophilic drug and 0.5% of a hydrophilic drug actually reach the anterior chamber [6]. The amount of available drug transported is further diminished across every subsequent layer: the sclera, choroid, and retinal pigment epithelium [7]. Permeability of cationic and lipophilic solutes is limited across the sclera, and the tight intercellular junctions of the retinal pigment epithelium prevent transport of hydrophilic molecules [7]. Finally, the lymphatic system, blood vessels and active transporters all work to clear drugs administered through transscleral routes.

2.2.2. Systemic

To achieve sufficient concentrations of drug to the eye, medications can be administered systemically. Patient compliance is higher as it is a generally non-invasive route of administration; however, systemic administration of drugs requires high doses that are potentially toxic to obtain a therapeutic concentration across the blood aqueous and blood retinal barriers [8, 9]. These dosage requirements can result in toxicity [4] and the drug can still be quickly removed by eye's circulatory process requiring frequent injections.

2.2.3. Intravitreal

Intravitreal injections circumvent physiological barriers and maintain therapeutic doses without damage to bystander tissues. Dose efficacy can be affected by clearance mechanisms *via* diffusion across both the anterior and posterior segment. Additionally, the retinal pigment epithelium acts as a barrier for drugs to affect the retina. The most prevalent complications that occur with frequent injections include retinal detachment, cataract, increase in ocular pressure, and hemorrhage [10]. These side-effects and invasiveness of the administrations leads to low patient compliance [4].

2.2.4. Periocular: Subconjunctival and Subtenon

Periocular injections allow for specific delivery to the anterior and posterior segment with high vitreal drug levels. It is relatively noninvasive, and leads to fewer complications unlike intravitreal delivery. However, it is still a victim to clearance by conjunctival and choroidal circulation, and can lead to subconjunctival hemorrhage [4]

Given the presence of these physiological barriers, the development of therapies utilizing biopolymers and nanotechnology that efficiently deliver drugs and extend drug release to the eye would be beneficial to the progression of ocular disease treatment.

Table 1. List of major ocular diseases, descriptions, treatments, and route of drug delivery administration.

Disease	Symptoms	Treatment	Route of Drug Administration	Refs.
Uveitis	Inflammation in the uvea of the eye (ciliary body, iris, choroid)	Corticosteroids and immunosuppressive agents	Topical	[11]
Conjunctivitis	Inflammation of the conjunctiva, redness, irritation	Anti-histamines, antibiotics, corticosteroids	Topical	[11]
Cytomeglovirus	Inflammation of the retina, retina detachment, complete blindness	Cidofovir, ganciclovir, foscarne, corticosteroids	Systemic	[11]
Age-related Macular Degeneration (AMD)	Disease that destroys sharp, central vision. It affects the central region of the retina known as the macula, which is responsible for fine vision. It is characterized by the presence of drusen and area of hyperpigmentation on the retinal pigment epithelium (RPE). Advanced AMD is characterized by choroidal neovascularization (CNV), the growth of abnormal blood vessels beneath the RPE or between the RPE and the retina. It is accompanied by fluid and blood rupturing Bruch's membrane into the subretinal space, leading to irregularities of the retina.	Three anti-angiogenic drugs are currently being used in the treatment of AMD. Ranibizumab (Lucentis) is a human recombinant antibody fragment that displays high binding affinity towards all VEGF isoforms. Bevacizumab (Avastin) has been used more recently as an 'off label' therapy. It is a full-length human recombinant monoclonal antibody, which binds all VEGF isoforms. Pegaptanib (Macugen) is a pegylated aptamer that acts as an anti-VEGF agent. It binds the VEGF165 isoform and inhibits angiogenesis.	Intravitreal	[12-14]
Diabetic Retinopathy (DR)	DR is a consequence of diabetes, usually developing within 5-10 years. The disease is associated with progressive retinal ischemia. It is characterized by loss of pericytes in retinal capillaries and subsequent thickening of capillary basement membranes and local ischemia. This induces the expression of VEGF, which can cause vessel leakage and formation of abnormal blood vessels, similar to AMD. The majority of vision loss from	Current therapies include photocoagulation therapy and pharmacological agents that block VEGF signaling, such as ruboxistaurin mesylate which is a protein kinase C-β inhibitor. Additionally, corticosteroids are also being evaluated for their angiostatic, antipermeable and antifibrotic properties. Triamcinolone acetonide and dexamethasone are used in combination with other treatments for posterior ocular disorders. They act by binding steroid receptors in cells to induce or repress targeted genes, thereby inhibiting inflammatory symptoms like edema	Intravitreal	[15-21]

Table 1: contd…

	DR is due to macular edema. Other factors such as proliferation of fibrovascular membranes can lead to retinal detachment.	and vascular permeability. Corticosteroids act on VEGF by inhibiting VEGF secretion and inhibiting cytokine production. Corticosteroids can also inhibit basic fibroblast growth factor and ICAM-1 expression and decrease VEGF levels that are evident in neovascularization.		
Glaucoma	Glaucoma is caused by progressive damage to the trabecular meshwork, optic nerve head, and retinal ganglion cells; however, the exact mechanism behind the nerve damage is yet unknown. Glaucoma can usually be diagnosed as open- or closed-angle and primary or secondary based on the degree of ocular hypertension. Elevated intraocular pressure (IOP) is characteristic of primary open-angle glaucoma (POAG) and is the most common target of treatment. A number of risk factors have been identified for glaucoma, including hypertension, family history, and age, among others.	Prostaglandin analogs, β-adrenergic receptor agonists, α2-adrenergic agonists, parasympathomimetics, carbonic anhydrase inhibitors.	Periocular – subconjunctival	[11, 22, 23]
Retinal Vein Occlusion	Blockage of small veins that carry blood away from retina. It often occurs in locations where arteries have been hardened by atherosclerosis and place pressure on retinal vein.	Corticosteroids, anti-VEGF agents.	Periocular - subtenon	[24, 25]

3. CONSIDERATIONS IN FORMULATING OCULAR DRUG DELIVERY SYSTEMS

Drug delivery systems are utilized to overcome limitations of conventional treatments to enhance drug bioavailability, the rate at which drug is available for absorption at site of action, and reduce side effects. In theory, ocular drug delivery systems should prevent nonspecific uptake, premature metabolism and excretion of the drug. The difficulty in gaining FDA approval for new therapies also lends itself to the demand of newly repositioned drugs for ocular disease. This would expand current disease therapy and repurpose currently used drugs, thereby extending their patent life.

In the preparation of drug/biopolymer ophthalmic formulations, drug release can be controlled over an extended duration [26]. For any drug delivery system therapy to be effective, it should fulfill three major goals: (1) the therapeutic agent must be targeted to the site of action; (2) the drug release should be controlled at an optimal rate to reduce toxicity; and (3) the drug should be maintained at therapeutic dosage levels to eliminate need for repeated administration [27].

The unique physiology of the eye, and dynamic and static barriers are few of the issues that need to be considered when designing ocular drug delivery systems. From a production standpoint, preparation of the treatment also needs to include specifications on critical dose of drug, sterility of the product, and protecting the eye from sensitivity to irritants and toxicants [28]. Therefore, formulation factors such as drug loading, drug stability, drug excipient interaction, carrier biocompatibility and drug carrier biodegradability need to be considered when designing ocular drug delivery systems. The bioavailability of the drug is dependent on the disease state; therefore, the ability of a compound to permeate barriers is dependent on the delivery system properties, such as size, charge, and solubility. The interaction of the drug delivery system with the dynamic barriers of eye is also pertinent. This section covers various aspects of ocular delivery and discusses the characteristics of the biopolymers to be used as ocular drug delivery carriers.

3.1. Drug Carrier Biocompatibility and Biodegradability

Biocompatibility and biodegradability in the ocular environment are few of the important characteristics that will dictate the selection of polymeric materials to be used as ophthalmic drug delivery system carriers. Biocompatibility refers to the delivery system's ability to perform the desired function without eliciting toxic or immunogenic responses [3]. Biodegradable polymers allow for release of the drug *via* polymer erosion; the polymeric system degrades chemically or enzymatically by the physiological environment [3]. Examples of biodegradable polymers include polysaccharides and protein-based polymers are from natural sources [29].

However, if the materials used contain any toxic chemicals or byproducts, due to poor degradation, there may be a buildup in tissues, causing blockages and leading to increased intraocular pressure. Retina and optic neural tissue is highly sensitive to toxins. The most common forms of toxicity in ocular tissue are expressed as oxidative stress and inflammation, and counteractions with cell membranes [28, 30].

3.2. Drug Loading

One goal for formulators is to prepare delivery systems with optimized drug loading and release properties [31]. Drug loading can be incorporated during or adsorbed after production [32]. The drug loading capacity of polymeric delivery systems is affected by the hydrophobic effect between the system and the active pharmacological ingredient. Hydrophobic interactions are heavily relied upon in the case of loading poorly soluble drugs. Drug loading can be affected by other interactions like hydrogen bonding, dipole-dipole interaction and electrostatic interactions. The efficiency also depends on the miscibility between the polymer and the drug [33].

3.3 Drug Excipient Interaction

Excipients are generally considered to be inert; however, they can interact with the drugs thereby lowering the efficacy and performance of a medication. Interactions between the excipient and active pharmacological ingredient are classified as either chemical or physical. Chemical interactions can cause the drug to degrade, resulting in reduced concentration. Additionally, by products of the chemical reaction can lead to ocular irritation and sensitivity for the patient. Physical interactions between the drug and excipient can alter the drug release rate and even the ability to administer the drug. Preparing the drug delivery system with the consideration of potential impurities and modes of drug decomposition is of great importance in ocular drug delivery systems [34, 35].

3.4. Stability and Sustained Release

Another factor considered in the preparation of drug delivery systems is stability and sustained release of the drug. Stability is significant, because even if an adequate amount of drug is loaded, it is useless if the drug is lost in systemic circulation [33]. Increased drug residence time in the eye provides drug absorption for a longer duration and reduces the need for consistent dosing [28].

Many variables can affect the release rate of the drug. Biodegradable polymers are broken down by degradation of the polymeric matrix; water penetrates into matrix at a higher rate than polymer degrades. Bulk and surface diffusion and erosion can also affect drug release [36]. These values can be varied based on type of polymer, ratio, and molecular weight.

Additional strategies to control the release of ophthalmic drug delivery systems include prolonging drug action by reducing lacrimal drainage using viscosity-enhancing agents, suspensions, emulsions, and erodible matrices. The use of water-soluble matrices increases precorneal retention time and duration of action. Hydrophobic polymers can also reduce the penetration of water into the polymeric matrix, thereby slowing the drug release rate [3].

4. IDENTIFICATION AND DISCOVERY OF NOVEL DRUG DELIVERY SYSTEMS

Currently, research is being directed towards the identification and discovery of novel drug delivery systems using newly introduced polymers, nanoparticles, liposomes, and adhesive gels to circumvent barriers and obtain sustained levels of therapeutics at target ocular sites.

Utilization of nanoparticles and biomaterials can increase treatment efficacy overall by enhancing corneal penetration, prolonging contact time with ocular tissue, and reducing patient irritation. Drug delivery systems can also serve to solve solubility issues, overcome poor bioavailability of drugs, decrease dosing frequency, diminish toxicity, and resistance to clearance and degradation. The following section lists current materials that are being tested or used to improve drug delivery to the eye.

4.1. Polymers

Polymers are widely used to achieve controlled release, stability and improved bioavailability. As mentioned previously, the rate of release depends on the initial drug concentration and the swelling or erosion of the polymer matrix. The rate of drug release can be modified by membrane thickness, pore structure, fillers, and the use of plasticizers [29].

Ocular delivery of drugs can be improved by use of naturally occurring biopolymers such polysaccharides, proteins and nucleic acids [37]. Naturally derived biomaterials are known for their biocompatibility and biodegradability, as well as their low immunogenic nature [38]. Biopolymers are advantageous because they can be produced with well defined degradation kinetics and mechanical properties. Proteins offer the benefit was being metabolized by enzymes into peptides [38]. Some common natural polymers include cellulose, chitosan, and alginate.

Chitosan is a biopolymer produced by the deacetylation of chitin. It boasts many suitable characteristics for ocular drug delivery such as low toxicity, good ocular tolerability, and ability to enhance permeation. A chitosan solution was studied for its ability to enhance corneal permeability after topical application of ofloxacin. The pharmacokinetic data demonstrated a significant increase in corneal permeability [39].

N-trimethylchitosan (TMC), a derivative of chitosan, possesses the benefit of solubility in tear fluid at physiology pH. It has been tested on a rat model to enhance intraocular delivery of dexamethasone [39].

In situ gels can be administered as a solution and transition to gels with specific stimuli, such as pH or temperature [40, 41]. The gel can be loaded with bioactive macromolecules and pharmacological agents irrespective of their solubility properties. Since the gel forms quickly *in vivo*, it can be used to achieve localized and sustained release. They are attractive due to their ease of administration and improved bioavailability. Chitosan and TMC were also evaluated as promoters for drug absorption in combination with polyethylene glycol *in situ* gels. The presence of these biopolymers significantly increased the presence of dexamethasone, which the authors suggest is indicative of the biopolymer affecting the transcellular penetration pathway [39].

A polysaccharide derived from tamarind seeds, xyloglucan, has been evaluated as an *in situ* gelling system. It does not exhibit thermoreversible properties; however, upon partial degradation by β-galactosidase, it exhibits a solution to gel transition at 40°C. Researchers have studied xyloglucan for the delivery of pilocarpine to treat glaucoma [39].

4.2. Nanoparticles

Nanoparticles have been researched as carriers in ocular pharmaceuticals [4, 22, 42, 43]. They can be prepared from biodegradable, polymeric materials in which the drug can be dissolved, entrapped, or adsorbed [44]. Nanoparticles range from around 100 nm and allow targeted and controlled delivery of hydrophilic or hydrophobic drugs. Surface modification with targeting moieties can increase specificity and uptake to desired site of action [32].

Benefits for using nanoparticles in ophthalmic drug delivery include: (1) smaller particles are well tolerated in the eye; (2) nanometer sized particles demonstrate higher solubility, surface area, and drug dissolution; (3) nanoparticles can be manipulated by polymer weight and hydrophilicity to sustain drug release; (4) small size facilitates administration *via* injection. The application of biodegradable nanoparticles can be used to enhance current methods of therapy by significantly enhancing the quality of life for patients suffering from degenerative eye disorders. A recent study reported the localization of nanoparticles approximately 200 nm in size in the retinal pigment epithelium within 6 hours after intravitreal injection. Cytoplasmic concentrations of the nanoparticles remained elevated for up to 4 months [45].

Chitosan is a polysaccharide commonly used for nanoparticle formulation. It is hydrophilic and is degraded by lipases and proteases. Recent studies have demonstrated the use of chitosan nanoparticles for the delivery of cyclosporine A to the ocular mucosa. In 48 hours, the drug was concentrated in the cornea and conjunctiva and has shown promise for treatment in conjunctivitis and dry eye disease [38].

Another study reports the incorporation of chitosan in a dexamethasone-loaded poloxamer nanomicelle. Incorporation of the chitosan raised the zeta potential and increased membrane permeability and ocular bioavailability [46].

Dorzolamide-loaded chitosan nanoparticles were studied for their efficacy in the treatment of increased intraocular pressure in glaucoma. As shown in Fig. (**3**), an *ex vivo* corneal permeation study demonstrated a 35.80% permeation of nanoparticles in 2 hours, whereas the drug alone showed a 75.30% permeation due to the sustained release of drug from the nanoparticle formulation. Additionally, gamma scintigraphic displayed higher corneal retention [47].

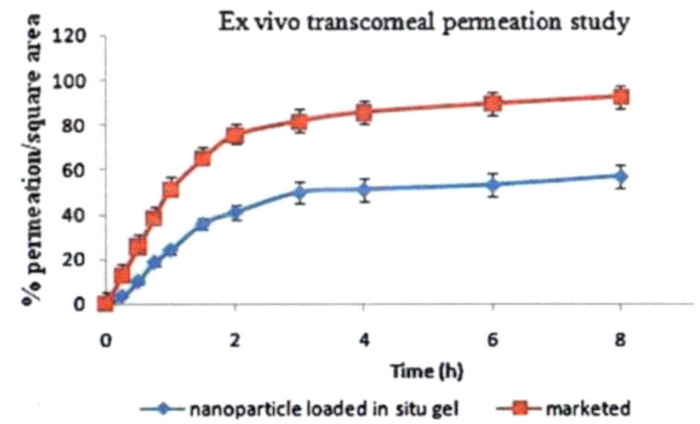

Figure 3: *Ex vivo* transcorneal penetration study. Nanoparticle loaded drug showed lower permeation due to sustained release properties. (Reproduced with permission from [47]).

Studies using nanoemulsions and polymeric micelles containing anti-VEGF bevacizumab result in sustained delivery. Furthermore, pDNA encapsulated by micelles can ameliorate choroidal neovascularization seen in age-related macular degeneration [48]. They offer more effective drug delivery by their ability to maintain therapeutic concentrations of the active drug molecule over a longer duration than previous methods.

4.3. Liposomes

Liposomes are vesicles consisting of phospholipid bilayers that surround an aqueous core. Encapsulation in lipid vesicles improves solubility and transport across the corneal barrier and improves controlled drug delivery.

A drug delivery system consisting of liposomes complexed with chitosan nanoparticles was designed to study its potential use as an alternative to topical eye drops. Preliminary studies showed strong cellular uptake in an *in vivo* rabbit model [49]. Fluorescence microscopy of rabbit corneal and conjunctival tissue of sham, control, and liposome-chitosan nanoparticles (LCS-NP) are shown in Fig. (**4**). Sham control rabbit eyes show no fluorescence; the experimental control showed weak signals and a strong fluorescent signal was viewed in LCS-NP treated eyes.

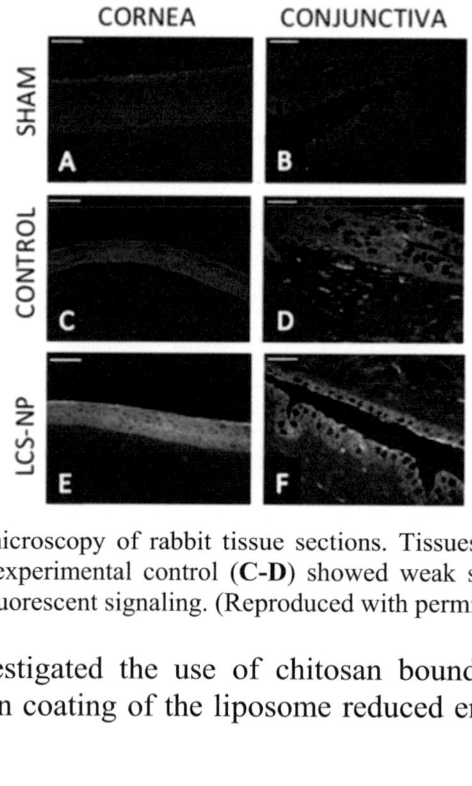

Figure 4: Fluorescence microscopy of rabbit tissue sections. Tissues from sham control (**A-B**) showed no fluorescence, experimental control (**C-D**) showed weak signals, and LCS-NP (E-F) treatment showed strong fluorescent signaling. (Reproduced with permission from [49]).

A similar model investigated the use of chitosan bound liposomes to deliver ciprofloxacin. Chitosan coating of the liposome reduced encapsulation efficiency of ciprofloxacin; however, the chitosomes exhibited enhanced anti-bacterial activity *in vitro*. *Ex vivo* studies showed enhanced corneal penetration across a rabbit cornea due to the absorption enhancing nature of the chitosan [50].

4.4. Adhesive Gels

Bioadhesive polymers can be utilized to increase the concentration of drugs at the site of action, thereby increasing drug absorption. These polymers bind to biological substrates such as mucosal membranes and are characterized by their hydrophilic nature to form strong adhesive bonds with membranes, as well as their ability to maintain surface tension to spread onto the epithelial cell surface. Preparations that contain polymers that can attach to the mucin are effective at maintaining sustained drug release into the eye [51].

Arabinogalactan is a natural polysaccharide that displays mucoadhesive properties. Preliminary studies have shown that its prolonged retention time on the corneal surface will have potential applications for the treatment of dry eye and corneal injuries [39].

A derivative of cellulose, hydroxypropylmethyl cellulose (HPMC), has also been studied for its gelation properties. An ophthalmic delivery system for ketorolac tromethamine was prepared with metylcellulose and HPMC. While gelation occurred at 60ºC, viscosity increased at 37ºC and sustained drug release was maintained for up to 4 hours, thereby lending evidence that it could be used as a better alternative to traditional eye drops for the treatment of post-operative inflammation [52].

HPMC *in situ* gels have also been used to optimize delivery of sinomenine hydrochloride for the treatment of uveitis. In a study reported by Song *et al.*, a gel composed of 0.1% (w/v) Carbopol and 0.4% (w/v) HPMC provided sustained release of sinomenine hydrochloride over 8 hours. The gel was studied on an experimental autoimmune anterior uveitis model in female Lewis rats [53].

As depicted in Fig. (**5**), the group pretreated with the gel developed uveitis later and with lower inflammation symptoms that the blank or control group. By the fifteenth day, all signs of inflammation were gone.

CONCLUSION

Better therapeutics for ocular drug delivery is a continuous challenge for formulators. Innovative drug delivery systems comprised of biopolymers and nanomaterials work to enhance bioavailability and sustain therapeutic doses. A variety of biopolymers in different formulations have been presented for their ability to maintain the stability and sustain drug delivery of certain ocular therapeutics. The optimal biopolymer for each application is largely dependent on the disease state and route of drug delivery. Parameters can be adjusted to enhance and facilitate the mode of delivery and lower the risk of ocular irritation or toxicity. Development of novel therapeutics will work to reduce vision impairment and blindness that afflicts the global population.

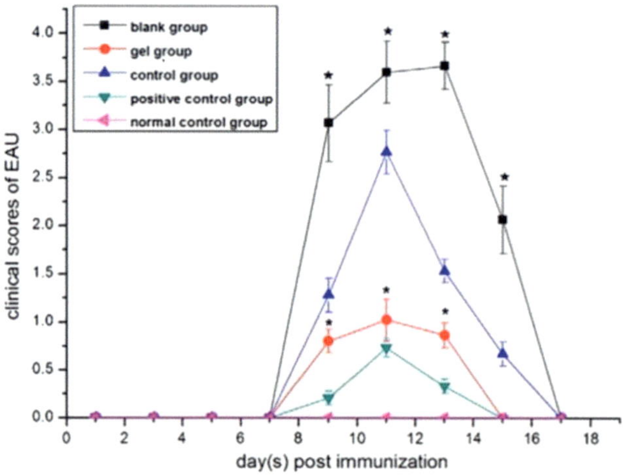

Figure 5: Disease severity scoring of the experimental autoimmune anterior uveitis model. Severity was assessed on a scale of 0 (no disease, normal) to 4 (maximum disease symptoms). Groups were treated accordingly: Normal control – no drops; blank- PBS; control – 0.5% drug; gel – 0.5% drug in gel; positive control - prednisolone acetate. Reproduced with permission from [53].

ACKNOWLEDGEMENTS

Declared none.

CONFLICT OF INTEREST

The authors confirm that this chapter content have no conflict of interest.

REFERENCES

[1] Organization, W.H. *Prevention of Blindness and Visual Impairment.* 2014 [cited 2014 09/01/14].

[2] Resnikoff, S., *et al., Global data on visual impairment in the year 2002.* Bull World Health Organ, 2004. **82**(11): p. 844-51.

[3] *Drug Delivery and Targeting*, in *Martin's Physical Pharmacy and Pharmaceutical Sciences: Physical Chemical and Biopharmaceutical Principles in the Pharmaceutical Sciences*, P.J. Sinko and Y. Singh, Editors. 2011, Lippincott Williams & Wilkins: Philadelphia, PA. p. 594-645.

[4] Gaudana, R., *et al.*, *Ocular drug delivery.* AAPS J, 2010. **12**(3): p. 348-60.

[5] Gambhire, S., K. Bhalerao, and S. Singh, *In situ hydrogel: Different approaches to ocular drug delivery.* Int. J. Pharm. Pharm Sci., 2013. **5**(2): p. 27-36.

[6] Zhang, W., M.R. Prausnitz, and A. Edwards, *Model of transient drug diffusion across cornea.* J Control Release, 2004. **99**(2): p. 241-58.

[7] Urtti, A., *Challenges and obstacles of ocular pharmacokinetics and drug delivery.* Adv Drug Deliv Rev, 2006. **58**(11): p. 1131-5.

[8] Geroski, D.H. and H.F. Edelhauser, *Drug delivery for posterior segment eye disease.* Invest Ophthalmol Vis Sci, 2000. **41**(5): p. 961-4.

[9] Sigurdsson, H.H., *et al.*, *Topical and systemic absorption in delivery of dexamethasone to the anterior and posterior segments of the eye.* Acta Ophthalmol Scand, 2007. **85**(6): p. 598-602.

[10] Peyman, G.A., E.M. Lad, and D.M. Moshfeghi, *Intravitreal injection of therapeutic agents.* Retina, 2009. **29**(7): p. 875-912.

[11] Gaudana, R., *et al.*, *Recent perspectives in ocular drug delivery.* Pharm Res, 2009. **26**(5): p. 1197-216.

[12] Vadlapudi, A.D., *et al.*, *Recent Patents on Emerging Therapeutics for the Treatment of Glaucoma, Age Related Macular Degeneration and Uveitis.* Recent Pat Biomed Eng, 2012. **5**(1): p. 83-101.

[13] Rosenfeld, P.J., *et al.*, *Ranibizumab for neovascular age-related macular degeneration.* N Engl J Med, 2006. **355**(14): p. 1419-31.

[14] Bylsma, G.W. and R.H. Guymer, *Treatment of age-related macular degeneration.* Clinical and Experimental Optometry, 2005. **88**(5): p. 322-334.

[15] Clark, A.F. and T. Yorio, *Ophthalmic drug discovery.* Nat Rev Drug Discov, 2003. **2**(6): p. 448-59.

[16] Samudre, S.S., *et al.*, *Comparison of topical steroids for acute anterior uveitis.* J Ocul Pharmacol Ther, 2004. **20**(6): p. 533-47.

[17] Wang, Y., V.M. Wang, and C.C. Chan, *The role of anti-inflammatory agents in age-related macular degeneration (AMD) treatment.* Eye (Lond), 2011. **25**(2): p. 127-39.

[18] Sherif, Z. and U. Pleyer, *Corticosteroids in ophthalmology: past-present-future.* Ophthalmologica, 2002. **216**(5): p. 305-15.

[19] Wu, W.S., *et al.*, *Dexamethasone induction of keloid regression through effective suppression of VEGF expression and keloid fibroblast proliferation.* J Invest Dermatol, 2006. **126**(6): p. 1264-71.

[20] Penfold, P.L., *et al.*, *Triamcinolone acetonide modulates permeability and intercellular adhesion molecule-1 (ICAM-1) expression of the ECV304 cell line: implications for macular degeneration.* Clin Exp Immunol, 2000. **121**(3): p. 458-65.

[21] Wang, Y.S., *et al.*, *Inhibitory effects of triamcinolone acetonide on bFGF-induced migration and tube formation in choroidal microvascular endothelial cells.* Graefes Arch Clin Exp Ophthalmol, 2002. **240**(1): p. 42-8.

[22] Zhou, H.Y., *et al.*, *Nanoparticles in the ocular drug delivery.* International journal of ophthalmology, 2013. **6**(3): p. 390.

[23] Clark, A.F. and T. Yorio, *Ophthalmic drug discovery.* Nature Reviews Drug Discovery, 2003. **2**(6): p. 448-459.

[24] Gurram, M.M., *Comparison of Posterior Sub-Tenon Triamcinolone with Intravitreal Bevacizumab for Macular Edema in Non-Ischemic Retinal Vein Occlusions.* International Journal of Current Research and Review, 2013. **5**(12): p. 61-71.

[25] Karia, N., *Retinal vein occlusion: pathophysiology and treatment options.* Clinical Ophthalmology, 2010. **4**: p. 809-816.

[26] Pahuja, P., S. Arora, and P. Pawar, *Ocular drug delivery system: a reference to natural polymers.* Expert Opin Drug Deliv, 2012. **9**(7): p. 837-61.

[27] Yasukawa, T., *et al.*, *Drug delivery systems for vitreoretinal diseases.* Prog Retin Eye Res, 2004. **23**(3): p. 253-81.

[28] Missel, P.J., *et al.*, *Design and Evaluation of Ophthalmic Pharmaceutical Products*, in *Modern Pharmaceutics*, A.T. Florence and J. Siepmann, Editors. 2009, Informa Healthcare USA, Inc.: New York, NY. p. 101-189.

[29] *Pharmaceutical Polymers*, in *Martin's Physical Pharmacy and Pharmaceutical Sciences: Physical Chemical and Biopharmaceutical Principles in the Pharmaceutical Sciences*, P.J. Sinko and Y. Singh, Editors. 2011, Lippincott Williams & Wilkins: Philadelphia, PA. p. 492-515.

[30] Grover, A., *et al.*, *Ocular Toxicity of Nanoparticles*, in *Biointeractions of Nanomaterials*, V. Sutariya and Y. Pathak, Editors. 2015, CRC Press: Boca Raton, FL. p. 347-352.
[31] Gaurav Tiwari, *et al.*, *Drug delivery systems: An updated review.* International Journal of Pharmaceutical Investigations, 2012. **2**(1): p. 2-11.
[32] Danhier, F., *et al.*, *PLGA-based nanoparticles: An overview of biomedical applications.* Journal of Controlled Release, 2012. **161**: p. 502-522.
[33] Sungwon Kim, *et al.*, *Overcoming the barriers in micellar drug delivery: loading efficiency, in vivo stability, and micelle–cell interaction.* Expert Opin Drug Deliv, 2010. **7**(1): p. 49-62.
[34] Vranic, E., *Basic principles of drug-excipients interactiosn.* Bosnia Journal of Basical Medical Science, 2004. **4**(2): p. 56-58.
[35] Crowley, P. and L. Martini, *Drug-Excipient Interactions.* Pharmaceutical Technology Europe, 2001. **13**(3): p. 26-34.
[36] Makadia, H.K. and S.J. Siegel, *Poly Lactic-co-Glycolic Acid (PLGA) as Biodegradable Controlled Drug Delivery Carrier.* Polymers, 2011. **3**(3): p. 1377-1397.
[37] Nitta, S.K. and K. Numata, *Biopolymer-Based Nanoparticles for Drug/Gene Delivery and Tissue Engineering.* International Journal of Molecular Science, 2013. **14**: p. 1629-1654.
[38] Singh, A.V., *Biopolymers in Drug Delivery: A Review* Pharmacologyonline, 2011. **1**: p. 666-674.
[39] Zambito, Y. and G.D. Colo, *Polysaccharides as Excipients for Ocular Topical Formulations*, in *Biomaterials Applications for Nanomedicine*, R. Pignatello, Editor. 2011, InTech.
[40] Nirmal, H.B., S.R. Bakliwal, and S.P. Pawar, *In-Situ gel: New trends in Controlled and Sustained Drug Delivery System.* Int J PharmTech Res, 2010. **2**(2): p. 1398-1408.
[41] Choi, B.G., *et al.*, *In situ thermal gelling polypeptide for chondrocytes 3D culture.* Biomaterials, 2010. **31**(35): p. 9266-72.
[42] Patel, A., *et al.*, *Ocular drug delivery systems: An overview.* World J Pharmacol., 2013. **2**(2): p. 47-64.
[43] Sabzevari, A., *et al.*, *Polymeric triamcinolone acetonide nanoparticles as a new alternative in the treatment of uveitis: in vitro and in vivo studies.* Eur J Pharm Biopharm, 2013. **84**(1): p. 63-71.
[44] Bu, H.Z., *et al.*, *Ocular disposition, pharmacokinetics, efficacy and safety of nanoparticle-formulated ophthalmic drugs.* Curr Drug Metab, 2007. **8**(2): p. 91-107.
[45] Bourges, J.L., *et al.*, *Ocular drug delivery targeting the retina and retinal pigment epithelium using polylactide nanoparticles.* Invest Ophthalmol Vis Sci, 2003. **44**(8): p. 3562-9.
[46] Vaishya, R., *et al.*, *Controlled ocular drug delivery with nanomicelles.* WIREs Nanomed Nanobiotechnol, 2014. **6**: p. 422-437.
[47] Katiyar, S., *et al.*, *In situ gelling dorzolamide loaded chitosan nanoparticles for the treatment of glaucoma.* Carbohydr Polym, 2014. **102**: p. 117-24.
[48] Ong, F.S., *et al.*, *Personalized Medicine in Ophthalmology: From Pharmacogenetic Biomarkers to Therapeutic and Dosage Optimization.* J Pers Med, 2013. **3**(1): p. 40-69.
[49] Diebold, Y., *et al.*, *Ocular drug delivery by liposome–chitosan nanoparticle complexes (LCS-NP).* Biomaterials 2007. **28**: p. 1553-1564.
[50] Mishra, G.P., *et al.*, *Recent Applications of Liposomes in Ophthalmic Drug Delivery.* Journal of Drug Delivery, 2011. **2011**: p. 1-14.
[51] Shaikh, R., *et al.*, *Mucoadhesive Drug Delivery Systems.* Journal of Pharmacy and BioAllied Sciences, 2011. **3**(1): p. 89-100.
[52] Bhowmik, M., *et al.*, *Study of Thermo-Sensitive In-Situ Gels for Ocular Delivery.* Scientia Pharmaceutica, 2011. **79**(2): p. 351-358.
[53] Song, J., *et al.*, *Preparation and evaluation of sinomenine hydrochloride in situ gel for uveitis treatment.* Int Immunopharmacol, 2013. **17**(1): p. 99-107.

CHAPTER 5

Customized Shape-Memory Biopolymers

Lisa C. du Toit, Thashree Marimuthu, Pradeep Kumar, Yahya E. Choonara and Viness Pillay*

Wits Advanced Drug Delivery Platform Research Unit, Department of Pharmacy and Pharmacology, School of Therapeutic Sciences, Faculty of Health Sciences, University of the Witwatersrand, Johannesburg, 7 York Road, Parktown, 2193, South Africa

Abstract: The technological revolution which is smart biopolymers illuminates the propensity of certain multifunctional materials - shape biopolymers - to change form and move on exposure to a stimulus, and in so doing, perform functions which furnish them with a range of capabilities. Thus, the mechanized uniqueness of these polymers is embodied not only in their macroscopic structural changes, but in their reversible shape change. There is a binary classification of shape biopolymers: shape-memory and shape-changing biopolymers. The focus of this chapter is on shape-memory biopolymers (SMBPs), which possess dual-shape competence. In terms of biomedical applications, SMBPs could have diverse applications in stimuli-sensitive drug delivery systems, intelligent medical and surgical devices, tissue engineering, or implants for minimally invasive surgery. In this chapter, we firstly elaborate on the fundamental molecular mechanisms culminating in the macroscopic dynamics of shape-memory systems bringing about their pertinent action and the recovery of their original shape. Further, we highlight the diverse stimuli instigating the polymeric response such as thermal stimuli, light, magnetism, mechanical stress or moisture. In addition, we discuss the biopolymers exhibiting potential shape-memory capabilities, as well as various modifications, functionalizations and reinforcements to these to customize their biomedical applications. An expert summation of the application of these biopolymers in the design of smart delivery systems, implants and devices, as divulged through recent investigations, is presented. The chapter culminates in current and future trends in the design and application of SMBPs and their overall potential benefits.

Keywords: Biopolymers, external stimulus activation, intelligent drug delivery systems, medical devices, shape-memory polymers, stimuli-responsive systems.

INTRODUCTION

The technological revolution which is smart biopolymers illuminates the propensity of certain multifunctional materials - shape biopolymers - to change

***Corresponding author Viness Pillay:** Department of Pharmacy and Pharmacology, University of the Witwatersrand, Johannesburg, Wits Medical School, 7 York Road, Parktown 2193, South Africa; Tel: +27-11-717-2274; Fax: +27-86-553-4733; E-mail: viness.pillay@wits.ac.za

Viness Pillay, Yahya E. Choonara and Pradeep Kumar (Eds.)

form and move on exposure to a stimulus, and in doing so, perform functions which furnish them with a range of capabilities. Thus, the mechanized uniqueness of these polymers is embodied not only in their macroscopic structural changes, but in their reversible shape change. There is a binary classification of shape biopolymers: shape-memory and shape-changing biopolymers. The focus of this chapter is on shape-memory biopolymers (SMBPs), which possess dual-shape competence. In terms of biomedical applications, SMBPs could have diverse applications in stimuli-sensitive drug delivery systems, intelligent medical and surgical devices, tissue engineering, or implants for minimally invasive surgery [1, 2].

Biopolymers: Nature's Shape-Changing Polymers

To differentiate biopolymers from synthetic polymers, it must be emphasized that biopolymers have a natural origin, being produced by living organisms, as proteins, polysaccharides, or nucleic acids (*i.e.* polymeric biomolecules). Pertinent biopolymers, having application as SMBPs, include:

1. *Cellulose*

   Cellulose is a highly crystalline and rigid structural polysaccharide comprising a linear chain of β(1→4) linked D-glucose units [3]. Its natural abundance and renewability furnish advantages such as biodegradability and low cost; it also possesses low density and high stiffness [4].

2. *Alginate*

   The structure of alginate is a polymeric uronic belonging to the family of unbranched binary copolymers. Alginate has a widely varying composition and sequence of (1→4) linked β-D-mannuronic acid and α-L-guluronic acid residues [5]. It is widely employed in the form of crosslinked matrices and scaffolds for drug delivery and tissue engineering applications.

3. *Chitosan*

   Chitosan (CHT) (poly(*N*-acetyl-d-glucosamine-co-d-glucosamine)) is a partially N-deacetylated derivative of chitin (poly(N-acetyl-d-glucosamine) (isolated from crustacean shells), which is the second most abundant biopolymer in nature following cellulose, with diverse biomedical applications. Its advantages include low cost, good natural availability, antimicrobial activity and non-toxicity, together with being inherently biodegradable and biocompatible. Chitosan is diversely formulated into matrices, nanosystems and scaffolds for drug delivery and tissue engineering [6, 7].

4. *Starch*

Starch is essentially composed of two homopolymers of D-glucose, namely, amylose, a linear α-D(1,4')-glucan, and branched amylopectin, possessing the backbone structure of amylose with the inclusion of α-1, 6'-linked branch points. It is a sustainable material with complete biodegradability and low cost [8].

5. *Collagen (gelatin is denatured collagen)*

Collagen exists in all animals as an abundant structural protein. The structure of collagen is based on three parallel polypeptide strands existing as a left-handed, polyproline II-type helical conformation coil with a one-residue stagger creating a right-handed triple helix [9].

6. *Casein (milk protein)*

Casein forms part of a group of phosphoproteins and includes four individual gene product components *i.e.* αs1-, αs2-, β- and κ-casein, contrasting in primary structure and type and modification degree following translation [10].

7. *Polyesters*

Polyesters are one of the most extensively studied classes of biopolymers, containing the ester functional group in their main chain. They are produced by bacteria, and can be manufactured commercially on large scales through fermentation processes. The chemical structure predicts whether the polyester is a thermoplastic or thermoset [11].

During the past century we have seen the institution of polymers in elementary drug delivery functions for the creation of controlled release systems, but with the dawn of the new millennium, progressive drug delivery strategies are the key for the enhanced therapeutic management of the patient - we want polymeric drug delivery systems that are intelligent and capable of adapting to their environment to ensure an optimization of their behavior or that of the system of which they form an intrinsic component.

Nature is constantly providing solutions; various natural systems embody stimulus-responsive mechanisms, which we as pharmaceutical scientists ultimately embrace. A case in point: employing sea cucumber skin for design inspiration, scientists have developed a novel material for purported application in Parkinson's disease, stroke and spinal cord injury treatment (Fig. **1**), as published in the journal *Science* by Capadona and co-workers [12]. The material, which resembles the skin of the sea cucumber and can switch between rigid and flexible states (with a factor of 10 change in modulus), may purportedly be used in the future in biomedical implants "as a protective sheath around brain microelectrodes, which would be stiff when implanted but could then become flexible, reducing the impact on surrounding tissue" according to AAAS, publisher of *Science* [12]. They possess a composite structure in their inner dermis, which is reinformed with collagen fibrils that undergo reversible interactions [3].

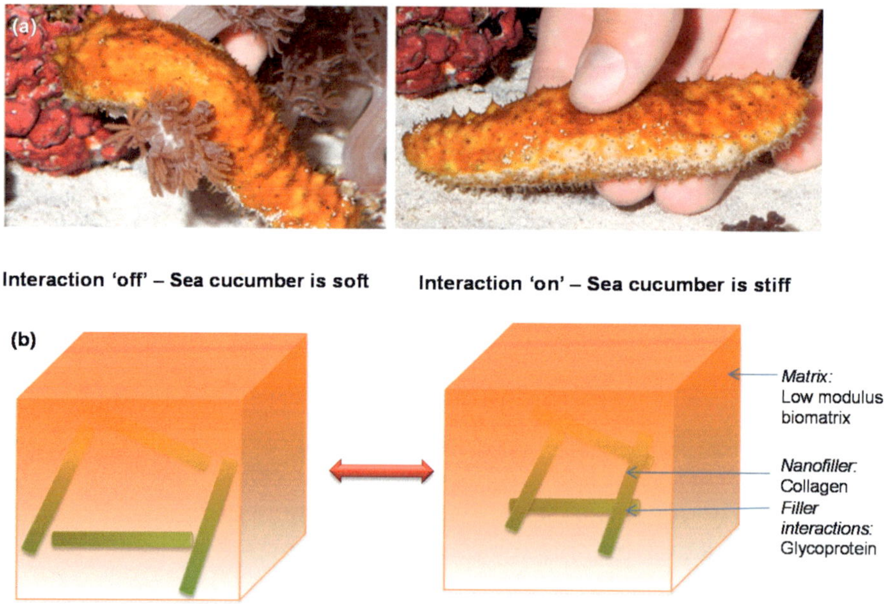

Figure 1: Sea cucumbers inspired the design of stimuli-responsive polymer nanocomposites with adaptive mechanical properties [12]. (**a**) Photographs depicting the soft and stiff state of the sea cucumber. (**b**) Biological nanocomposite model of the sea cucumber dermis (adapted from ref [13]).

THE INTELLIGENT POLYMER PRINCIPLE

To further elaborate on the stimulus responsive approach introduced in the section above, one can delve into the 'Degrees of Intelligence':

1. A basic response implicates a polymer receiving a stimulus from its environment that involves the adaptation of one of its properties according to the intensity of the stimulus.

2. The sophistication of an intelligent system supercedes this through interpretation of the property changes of the smart polymer. It may even comprise a stimulus-emitting material and an adaptive material that controls the adaptations of a certain property to counterbalance the stimulus and to maintain system stability [14].

The universal intelligent polymer does not exist. However, in this chapter we discuss several approaches based on the intelligent polymer principle, as depicted for enabling drug release (Fig. **2**), that were applied to generate innovative SMBP systems.

Figure 2: The intelligent polymer principle for responsive drug release.

An increasingly broad panoply of stimulus-adaptation couples (Table **1**) are under development to fulfill all the biotechnological and medical sectors. Well-established and employed couples include heat-shrinkage or heat-color, whereas others possess very specific applications or are still in early phases of development leading to more concrete uses [14].

Table 1. Coupled stimuli and responses (adapted from ref [14]).

Stimulus	Response/ Adaptive Property
Temperature	Color
	Shape
	Rheology
	Swelling/shrinkage in a liquid
	Adhesive strength
Chemical	Deformation
	Color
	Release of active product
Mechanical Stress	Colour
	Shape
	Chemical property
	Electrical property
	Solubility of an aqueous solution
	Rheology
Electric	Shape
	Color
	Rheology
Light, UV	Colour
	Deformation
	Dichroism
	Shape
	Permeability
Magnetic	Mechanical properties
	Rheology

In recent years, intelligent polymers have been promulgated as valuable tools in biotechnological applications including controlled drug release, immunodiagnostics, and gene therapy. Intelligent or stimuli-responsive polymers mimic biological systems in a rudimentary manner in that they exhibit notable reversible physical changes in response to an external stimulus such as temperature, pH, ionic strength, solvents, electrical or magnetic field, radiation, redox potential, or concentration of biomolecules (*e.g.* electrolytes, glucose). These polymers may be in solid hydrogel form, immobilized on solid surfaces, or existing as water-soluble chains. A stimulus, proposedly an increase in temperature over a certain critical value causes a response, which can be manifold: dissolution/ precipitation, degradation, drug release, altered hydration state, swelling/ collapsing, adjustment of hydrophilic/ hydrophobic balance (modification of the polymeric chains from water-soluble coils to water-insoluble globules in aqueous media), shape/ conformation change, and micellization; often with ensuing release of an incorporated bioactive molecule [15, 16].

MECHANISM OF OPERATION OF SHAPE-MEMORY BIOPOLYMERS

Lendlein and co-workers [17] have provided a detailed description of the molecular *modus operandi* of SMBPs. SMBPs have the propensity to be stabilized in a temporary secondary shape when as external stress is applied, followed by fixation. Stability of the temporary shape is maintained up to the point at which the required stimulus is applied to the fixed shape, with reversal of the mechanical deformation enabling the ultimate recovery of the original shape (Fig. **2**). The basis of the SMBP is a molecular biopolymeric network architecture coupled with a processing and programming technology, or shape-memory creation process. The network architecture of biopolymers exhibiting this shape-memory effect incorporate netpoints (visualized as spheres in Fig. **3**) connected by chain segments as molecular switches (represented by lines in Fig. **3**) with sensitivity to an external stimulus. The permanent shape of the SMBP network is dependent on the netpoints which may be a result of covalent bonding (chemical) or intermolecular interactions (physical). In order to be deformed, this biopolymeric network must possess adequate elastic deformation. Crosslinking chemistry attains covalent points, whereas physical interactions necessitate a segregated domain morphology, being soft (amorphous) and hard (crystalline), with the hard domains possessing the highest thermal transition temperature. The growing length and flexibility of the chain segments predict the deformability of the network for allowing the attainment of a specific orientation. Entropy drives the chain recoil when the material is in the rubbery state, for permanent shape retrieval. Fixation in the temporary deformed shape can be enhanced by the inclusion of reversible netpoints (either chemically based or by solidification of switching domains, represented as the encircled portion in the middle schematic of Fig. **3**) undergoing on-demand formation and cleavage [17].

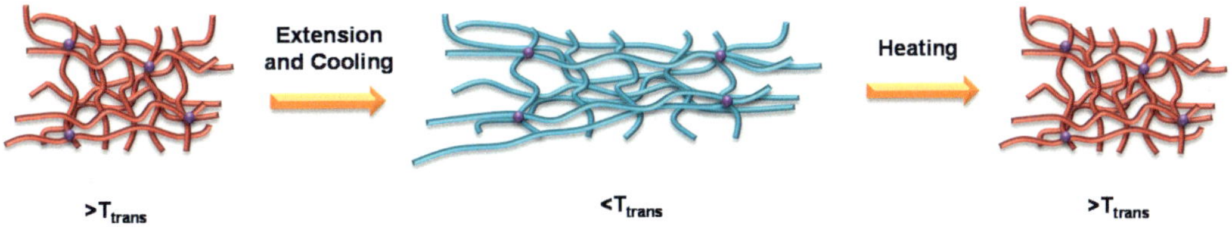

Figure 3: Molecular mechanism of the thermally-induced shape-memory effect of a covalently crosslinked polymer where the T_{trans} is the T_m.

Most commonly, SMBPs respond to a thermal stimulus; classification is in accordance with the solidification of the polymer chain segments, which act as the molecular switches arising from their correlating thermal transition temperatures (T_{trans}). For fixing of the temporary deformed shape, reversible crosslinks form between the switching domains. Solidification with crystallization on cooling emanates in physical crosslinking. Reheating above T_{trans} melts the crystallites or returns the glassy polymer chain segments to a viscous state with shape recovery. Thermoplastic SMBPs thus have the biopolymer chain segments associated with the second highest T_{trans} as the switching domains. When chemical crosslinks serve as the switches, two functional groups will form a reversible chemical bond, which is cleaved by a suitable stimulus. In the case of physical crosslinking, the T_{trans} can be differentiated as the glass transition (T_g), melting transition (T_m), or liquid crystalline transition (T_c) temperatures [17]. SMBP express recovery of close to 100% following deformation [18].

BIOMEDICAL APPLICATIONS OF SMBPs

The application of SMBPs in the biomedical field is diverse, extending to controlled drug delivery systems, surgical instruments, wound covers and extracorporeal devices; each requiring implicit biomaterial properties and functions. This is a result of their biocompatibility, as well as broad range of tunable mechanical properties (including elasticity, stiffness, large deformation and recovery), tailorable transition temperatures and rapid actuation [6]. When considering the design of an implant, it needs to fulfill a number of requirements, including certain mechanical requirements (*e.g.* wound chips and hernia meshes). Substitution of the function of tissues or organs by the biopolymeric device necessitates biofunctionality. In the instance of devices in contact with the blood (*e.g.* stents, artificial valves, and extracorporeal tubes), hemocompatibility of the surface is essential. Past innovations in biomaterial functions have enabled forward strides in medical engineering, such as the hydrolytic cleavage which is evident in biodegradable scaffolds; this enables wound healing or regeneration. In formulating a controlled release drug delivery system, the diffusional and erosional behavior of the biomaterial must be considered Wischke *et al.,* [19] highlighted sequential points for consideration in developing a SMBP system for controlled bioactive release applications:

1. Aqueous environment effect on shape-memory and thermomechanical properties

2. Ascertainment of conditions for enhanced drug loading

3. Bioactive incorporation effects on the thermomechanical properties of the dry and wet SMBP system

4. Drug release kinetics, erosional, and mechanical behavior considering the effect of a cycle of programming and recovery

Because of progress in the application of novel surgical techniques, including the evolution of minimally invasive surgery, the demands from implants become more intricate. It has become an increasingly important goal to customize an implant to the patient-specific requirements, by for example, adapting the shape to the individual geometry of the patient. This is where the shape-memory functionality of biopolymers becomes pertinent, where a predefined change in shape can be induced on exposure to a stimulus [17]. In the case of thermally-responsive SMBPs for biomedical applications, the insertion of a large/ bulky implant through a small incision as a compressed or elongated temporary shape can be achieved if said implant possesses for example, a T_{trans} existing between room temperature and body temperature. The initial shape may also be customized to the individual patient requirements. Once the implant reaches body temperature following insertion, it reverts to its bulky functional shape. Additional functionality may need to be built into the implant, such as biodegradability (excluding the need for a secondary explantation surgery, and implicit with the use of biopolymers) or drug release [17]. The extensive biomedical applications of SMBPs are discussed for specific systems in the ensuing section; some additional uses are provided in Table **2**. The described SMBP systems display favorable and diversely applicable shape-memory properties, often with simultaneous bioactive release. However, it must be noted that a number of these systems are in the early stages of development, with progression only to *in vitro* and *ex vivo* evaluation. The true clinical potential of these biopolymeric systems can only be realized following *in vivo* assessment.

Table 2. Diverse biomedical application of SMBP systems (adapted from ref [6]).

SMBP System	Biomedical Application	Reference
Biodegradable shape-memory sutures	Surgical applications	[20]
Shape-memory foams	Weight loss in obese patients	[21]
Shape-memory stents and foams	Gastric and intestinal disorders	[22, 23]
Shape-memory appliance	Orthodontic applications	[24]

SPECIFIC SHAPE-MEMORY BIOPOLYMERIC SYSTEMS AND THEIR DIVERSE BIOMEDICAL AND DRUG DELIVERY APPLICATIONS

Hydration-Induced Chitosan Shape-Memory Biopolymeric Scaffolds

Correia and Mano [7] developed chitosan-based porous scaffolds, both semi-crystalline non-crosslinked (CHT0) and genipin-crosslinked (CHT1), where the shape-memory effect was triggered by hydration. Environmental conditions (*e.g.* humidity) can dramatically affect the shape-memory capabilities of certain biopolymers through effects on the conformational mobility of macromolecular chains, which can be exploited for the design of a moisture-triggered SMBP. This is evident for SMBPs possessing glass transitions as switching transitions where water absorption affects their physicochemical and mechanical properties [25]. Hydrogen bond disruption by water creates a plasticizing effect, this reducing the T_g and enabling room temperature activation [26].

Previous investigations have demonstrated that CHT can undergo a glass transition *via* hydration [27]. Correia *et al.* [28] also investigated the shape-memory capability of CHT membranes induced by hydration and their mechanical properties when exposed to water/ethanol solutions, demonstrating that an increased water content, decreased the Young's and storage modulus. They reported that the limitation of a single permanent shape can be overcome through fixation of the membrane *via* a crosslinking agent (*e.g.* genipin), yielding varied permanent shapes.

Correia and Mano [7] set out to combine this observation with the maintenance of the structural integrity of the system (achieved through crosslinking) to design three-dimensional devices that could achieve geometrical recovery upon hydration. Novel hydromechanical compressive tests and dynamic mechanical analysis were employed for evaluation of the shape-memory effect, with immersion in differing water-ethanol mixtures. Dehydration with increased amounts of ethanol enabled the vitreous-like nature of the amorphous component of CHT to become fixated in the temporary shape. In the presence of water, there is disruption of the inter-molecular hydrogen bonds enabling enhanced segmental mobility of the CHT chains on encountering the glass transition, with recovery of the permanent shape of the pre-deformed scaffold. The shape-memory characteristics of chitosan were as follows: a glass transition at 32.3%v/v of water in water-ethanol mixtures for CHT0 and 40%v/v for CHT1; fixity ratio above 97.2% for CHT0 and above 99.2% for CHT1; recovery ratio above 70.5% for CHT0 and 98.5% for CHT1. CHT1 thus demonstrated enhanced mechanical properties and recoverability even at high deformations. The shape recovery of the scaffolds at various strains is shown in Fig. (**4**). The potential of these scaffolds as multifunctional drug delivery devices was also demonstrated *via in vitro* drug release studies, with the porous structures enabling controlled release. Their application in minimally invasive surgery and tissue regeneration was also noted. The resultant scaffolds could be applied as 3D devices in empty hydrated regions such as bone or cartilage defects, with the natural hydrated environment promoting the recovery of the desired shape, and the pressure created supporting fixation of the implant [7].

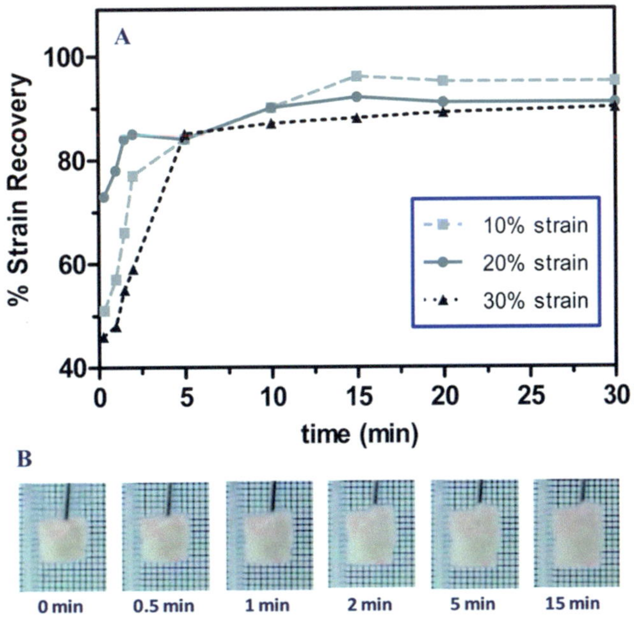

Figure 4: (**A**) CHT0 shape recovery along time after different maximum strains (ε_m). (**B**) Series of photographs demonstrating the shape recovery process for CHT0; the refereed times indicate the immersion time in water upon deformation at $\varepsilon_m = 30\%$ and dehydration. [7].

Biodegradable Chitosan/Poly(l-Lactide) Composites as Shape-Memory Biopolymeric Systems

Poly(l-lactide) (PLLA) is a biodegradable and biocompatible linear aliphatic biopolymer of the polyester class with derivation from completely renewable resources (*e.g.* corn and sugar beets). Hydrolytic cleavage under mild conditions yields lactic acid, a common biodegradable organic acid in the human body. It has been applied in numerous biomedical applications such as implants, drug delivery systems, scaffolds for tissue engineering and bone fixation, and surgical sutures [6].

Investigators have previously characterized the shape-memory effect of PLLA, reporting that the shape recovery of PLLA decreased and approached steady state as thermomechanical cyclic tensile testing number increased [29, 30]. There have been attempts to chemically and physically compound PLLA and CHT to create novel functionality, and owing to the fact that pure PLA undergoes fast biodegradation with evolution of lactic acid, and hydrophobicity. CHT-PLLA composites, however, have demonstrated hydrophilicity, slower biodegration, better maintenance of shape and mechanical properties, as well as a favourable architecture for cell growth [31].

Meng and co-workers [6] exploited the shape-memory characteristics of PLLA and CHT/PLLA composites as novel biodegradable and biocompatible materials. They found that CHT exerted a minimal effect on the T_g and T_m of PLLA. Shape-memory effects were observed for both the pure PLLA and CHT/PLLA composites due to the viscoelastic properties of PLLA, composed of semi-crystalline structures. Increasing CHT content in the composite reduced the shape recovery ratio due to the enhanced immiscibility between PLLA and CHT. The preferred shape-memory effect was observed at a CHT content of 15%$^{w}/_{w}$.

Poly(D,L-Lactide)/Hydroxyapatite Composites as Shape-Memory Biopolymeric Systems

Calcium phosphate ceramics including hydroxyapatite (HA), such as HA($Ca_{10}(PO_4)_6$ $(OH)_2$), possess commendable biocompatibility and osteoconductivity, but poor mechanical properties (*e.g.* low strength and resistance to fracture, thus limiting their hard tissue engineering application. Besides its known biocompatibility and biodegradability, poly(D,L-lactide) (PDLLA) possesses thermally-induced shape-memory behavior. In order to build on their advantages and tackle their individual limitations, the compounding of PDLLA and HA for the attainment of shape-memory capabilities was undertaken by Zheng *et al.*, [32]. Serial imposing shape-memory composites were formulated varying the shape, composite ratios and sample thicknesses. The physicochemical attributes of these systems were subsequently assessed *via* scanning electron microscopy for analysis of the surface morphology, differential scanning calorimetry for determination of the T_g and dynamic mechanical analysis for assessment of the shape-memory behaviors. The enhanced disperse morphology of HA grains, and a specific PDLLA/HA composite ratio (between 2.0 and 2.5), elaborated an improved shape-memory effect with rapid recovery time, compared to pure PDLLA. Fig. (**5**) demonstrates this rapid shape recovery and excellent shape-memory effect of the composite system, which were with other advantages, such as biodegradation, biocompatibility, easier availability of PDLLA, and osteoconductivity of HA. This highlighted the propensity of HA particles to improve the shape-memory effect of the composite for biomedical applications such minimally invasive surgery and tissue repair [32].

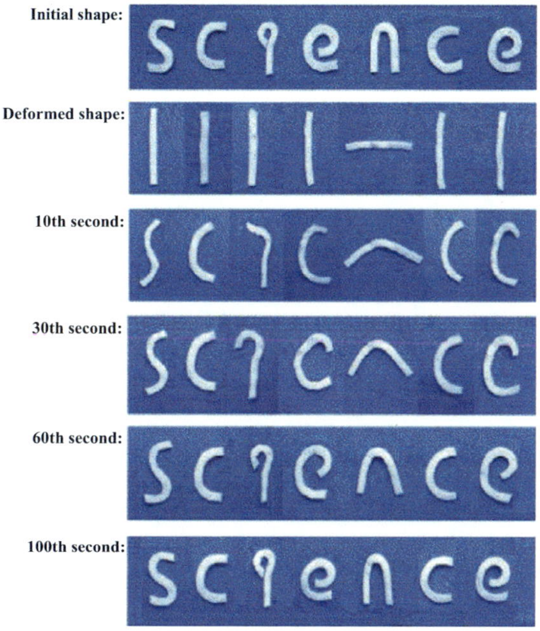

Figure 5: Digital images depicting the process of shape-memory recovery of the word 'science' fashioned from the PDLLA /HA composite.

Chitosan-Functionalized Poly(Lactic-co-Glycolic Acid) Microspheres as an Ultrasound-Modulated Shape-Memory Biopolymer System

Generally shape-memory effects are instigated *via* direct heating, which can lack practicality *in vivo*. Thus a number of indirect stimuli have been investigated, such as magnetic field, electrical resistance heating, laser, and light. This, however, necessitates additional components, such as magnetic particles, with subsequent biodegradability and biocompatibility concerns arising. Thus, in addition to looking for safer SMBPs, the need for an effective stimulus is also a pertinent consideration for triggering the required shape change. The application of high-intensity-focused ultrasound (HIFU) in clinical settings is on the increase as a non-invasive surgical tool [33]. The specific *modus operandi* of ultrasound is that the energy absorbed by tissues on its application causes molecular vibrations. The resultant heat energy causes a rise in temperature in the focal zone, with the surrounding tissues minimally affected. Focusing the acoustic waves on polymeric chains within the required area will thus elicit the shape change recovery effect of the SMBP. Additionally, ultrasound is a powerful modality for spatiotemporal control of 'on-off' pulsatile drug delivery [34]. Ultrasound application in achieving the shape-memory effect with integrated drug release, has however, received limited attention.

Bao and co-workers [35] highlighted the need for multifunctional minimally invasive implants and scaffolds, which change from a small to voluminous structure on application of a stimulus, with spatially-controlled bioactive release, employing ultrasound as a stimulus. They attempted to achieve this through the design of CHT functionalized poly(lactic-*co*-glycolic acid) (PLGA) microspheres incorporating the model bioactive, lysozyme. CHT was included for modification of the physical, chemical and biological attributes of the PLGA microspheres. A water-in-oil-in-water emulsion method was used for microsphere preparation and these were subsequently sintered in a mold to form a cylindrical rod. The microspheres thus served as building blocks with substantial "spatiotemporal designability" for construction of a macroscopic scaffold. The rod was then exposed to HIFU to elicit shape-memory effects with lysozyme release. The inclusion of CHT in the PLGA microspheres slightly increased the T_g, ultimately impacting the shape-memory ration of the 3D cylindrical structure. Furthermore, addition of CHT served to minimize the acidic degradation inherent to PLGA. It was demonstrated, *via* shape-memory analysis and drug release testing, that HIFU modulated the shape recovery process and synchronized the release kinetics of lysozyme in a switchable manner, which could be varied through adjustment of the acoustic power and duration of insonation. The designed implant possessed compressive properties falling within the range of trabecular bone, as ascertained *via* mechanical testing before and after ultrasound irradiation. The study demonstrated the favourable ultrasound-responsive shape change of the device with resultant bioactive release.

The shape-memory potential of PLGA has been previously described by Wong *et al.* [36]. Following physical adsorption of the CHT onto the surface of the PLGA, its influence on the macroscopic shape recovery of the microspheres needed to be investigated. This was done *via* the bending test, which involved folding samples to 180° shape, and the focus of the HIFU beam (at 300 W) was subsequently directed to the water-immersed scaffold to trigger the shape recovery process for ~150 s. The resultant shape recovery ratio *(R)* was calculated as follows:

$$R = (180° - \theta)/180° \times 100\% \qquad \text{(Equation 1)}$$

where θ refers to the final angle, as a quantitative measure of the ability of a material to memorize its permanent shape (shape-memory capability).

Bao *et al.* [35] found that the shape recovery ratio reached the maximum of ~97.6% when the PLGA/CHT ratio was 2.5:1. Pure PLGA had a lower *R* value of ~96%. The anchoring of CHT to the microsphere surface *via* physical crosslinking between CHT and PLGA macromolecular chains creates a stationary phase constraining random motion of molecular chain segments of PLGA, resulting in an increased shape recovery potential. In Fig. (**6**), the synchronized shape recovery with consequent lysozyme release from the PLGA/CHT rod is visualized. HIFU of 300W was applied for 30 seconds, instigating a shape recovery from a deformed 'V' configuration with coordinated lysozyme release. Shape recovery immediately desisted when the HIFU stimulus was removed, with drug release tapering off, thus resulting in pulsatile release kinetics. The initial rod shape was regained after exposure to 4 cycles of HIFU. Bao and co-workers [35] thus propose their system as favorable for biomedical applications, such as minimally invasive surgery, owing to the simultaneous shape recovery and drug release.

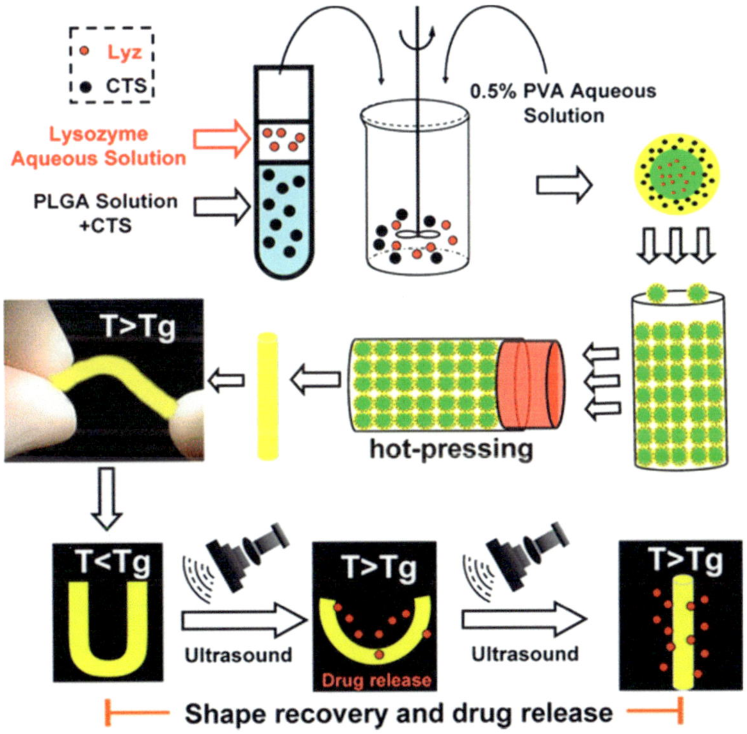

Figure 6: Schematic highlighting the formulation and subsequent shape recovery and drug release behavior from the CHT-functionalized PLGA microsphere implant [35].

Collagen-Like Transient Nodes as Shape-Memory Biopolymer Networks

Polyurethane multiblock copolymers are widely used SMBPs, with alteration in their properties and T_g achieved through modifying the ratio of the soft and hard domains. A high fraction of hard domains renders crystallization of the soft domains more difficult and decreases the melting temperature. Hydrogels are of particular interest as SMBPs, as their softness and water uptake behavior renders them more compatible with biological tissues [18].

Skrzeszewska and team [18] thus pursued the development of hydrogels employing collagen-like triple helical assemblies as 'pins' for the temporary shape of a lightly chemically crosslinked polymeric network. They formed the permanent network from chemically crosslinked monodisperse biodegradable telechelic polypeptides (TR4T), which comprised of identical end blocks (trimer-forming or T) and a random coil-like middle block (R4). The T domain was formed from nine amino acid triplets (Pro-Gly-Pro), which undergo folding on cooling as collagen-like triple helices. In contrast, the R4 domain behaves as a random coil over a wide temperature range. Addition of gluteraldehyde as a crosslinking agent served to attain the permanent shape through linking the lysine residues of the R4 domain. The triple helices were fixed into a temporary shape with cooling in the presence of applied deformation. Following exposure to the melting temperature of the triple helices of ~45°C, the system returned to its original shape (Fig. 7) [18].

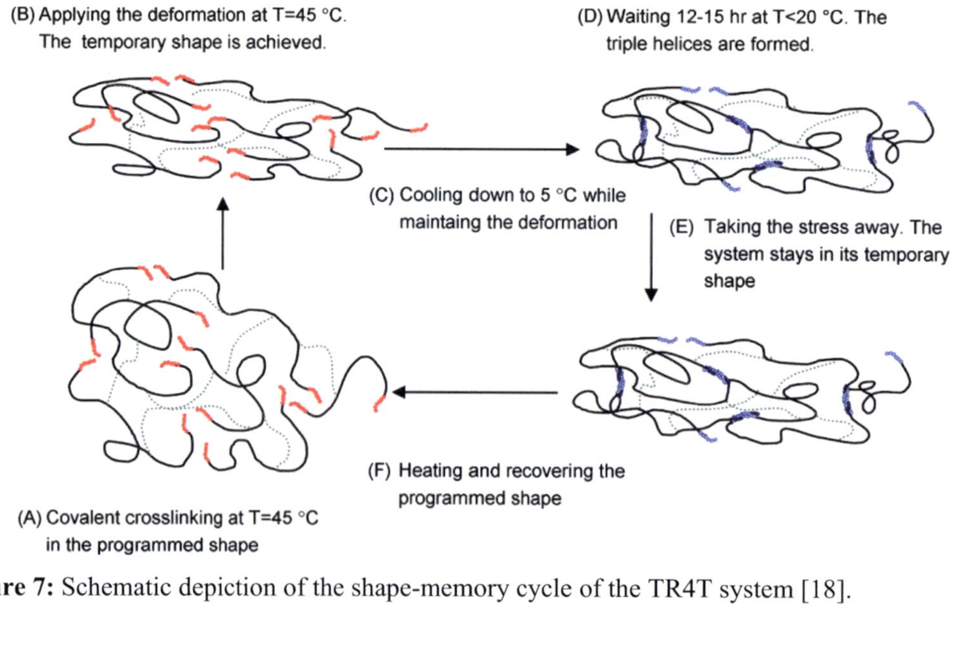

Figure 7: Schematic depiction of the shape-memory cycle of the TR4T system [18].

Cellulose/Elastomer Nanocomposites as Shape-Memory Biopolymeric Systems

When looking at alternative mechanisms for initiating shape change, other than thermally (direct or indirect) or light-induced, investigators [25] have identified the water-sensitive/driven programmable group of SMBPs. Yang *et al.* [25] found that water immersion for hours followed by drying at ~180°C were required for shape recovery, with water removal for fixing behavior. Bonded water has a pertinent influence on the T_g, rendering it very stable in these SMBPs, with removal by heating to at least 140°C, in order to raise the T_g above ambient temperature required for new shape fixing. This limits applicability as rapid shape fixing and shape recovery cannot be achieved through a mild programming cycle. However, a water-sensitive SMBP would have pertinent applications (*e.g.* in medical devices), as it would be preferred to use these for shape recovery, as these would only require exposure to body fluids. Rather than using indirect thermal triggering, Zhu *et al.* [4] utilized water to soften their novel SMBP system and provide it a temporary shape *via* rapid removal of water (within minutes). This was followed by triggering of shape recovery of the programmed SMBP at ambient temperatures through addition of water, thus avoiding a long response time with potential damage or adverse effects as a result of exposure to heat or light.

In the design of their SMBP system, Zhu *et al.* [4] identified a novel phenomenon where a water-activated rapidly switchable shape-memory effect was demonstrated through the reversible formation and disruption of a cellulose nano-whisker (CNW) percolation network contained in an elastomeric thermoplastic polyurethane matrix (TPU). CNWs were employed based on their high aspect ratio and good mechanical strength for reinforcement. Earlier in this chapter, we introduced the adaptability of the sea cucumber, and cellulose whiskers possess a similar architecture and mechanical adaptability to this marine animal, which is under stimulus-responsive selective and reversible control *via* coupling and decoupling of a three-dimensional nanofiber network. Zhu and team [4] developed a model which demonstrated that exposure to water enabled the CNW percolation network to break up enabling deformation of the flexible elastomer matrix into the required shape. Following programmed drying, the CNW percolation network reformed to create a fixed force facilitating the temporary shape. The TPU matrix then allowed fast recovery of the shape due to elastic entropy when the CNW percolation network is disrupted by subsequent wetting. The shape-memory effect of this system is thus completely thermally-independent, using a mild programming cycle to attain the mechanism *via* chemo-mechanical adaptation, as compared to other systems discussed where water or other solvents act as plasticizers that lower the T_g of the SMBP, allowing the shape recovery at room temperature or less; thus providing an innovative advance in SMBP systems. Samples achieved shape recovery following only 10 minutes of water immersion and rapid shape fixing with drying at 75 °C for 10 minutes. Systems of this type could possess important application as medical devices triggered by body fluids [4].

Fig. (**8**) highlights the shape-memory mechanism of the CNW/TPU composite, based on the chemo-mechanical adaptability of the CNW and the entropic elasticity of the elastomer**.** There is initially water molecule-induced softening of the CNW/TPUs, directed at the nano-whisker hydrogen bonds. Transformation into the temporary shape is enabled by removal of the water molecules, with fixation on drying as a hydrogen-bonded 3-D network of nano-whiskers. When the external stimulus (wetting) is applied, decoupling of the nano-whisker network occurs with recovery of the programmed shape. Zhu *et al.* [4] propose that additional hydrogen-bonding competitive agents may also be used as alternative triggers, further broadening the biomedical applications of this system.

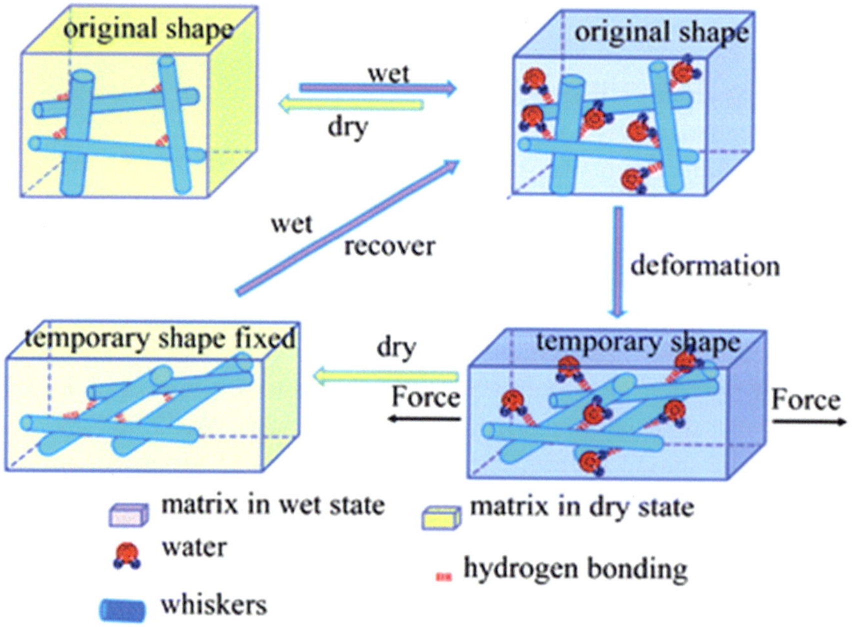

Figure 8: Proposed rapidly switchable water-sensitive shape-memory mechanism for the CNW/TPUs comprising a CNW percolation network in an elastomeric matrix [4]

Chitosan-Modified Cellulose Whisker/Elastomer Composites as Mechanically-Adaptive Shape-Memory Biopolymeric Systems

Following the investigations of Zhu *et al.* [4], Wu and team [3] also exploited the concept of cellulose whiskers, again referring to the stimuli-responsive mechanically-adaptive properties which creates stiffness in echinoderms (*e.g.* sea cucumbers), as a defense mechanism. The properties of cellulose whiskers (CWs) as crystals of cellulose which provide reinforcement have been described in this chapter, through interaction of hydroxyl groups *via* hydrogen bonding, which may be regulated by a competing hydrogen-bonding agent such as water, emanating in the hard vs. soft behavior. In order to customize the material for specific biomedical applications, hard and soft states are tuneable by variation of the polymeric matrix and the content of CWs [37]. For water-active mechanically adaptive shape-memory systems, an elastomer is generally employed as the matrix. The interest in these systems for biomedical applications (*e.g.* smart surgical sutures, minimally invasive medical devices, artificial skin, and wound closure) is increasing.

With these latest advances in water-active mechanically adaptive shape-memory systems, there is still limited research into stimuli-responsive behavior on exposure to different pH environments. It has been demonstrated that the surface of CWs have an abundant presence of sulfate ester groups ($O\text{-}SO_3H$), originated by sulfuric acid hydrolysis. Subsequent pH-sensitive surface dissociation of sulfate ester and hydroxyl groups is anticipated to be responsible for a pH-responsive wettability of CWs. Wu *et al.* [3] thus investigated the potential of modifying CWs with CHT (as the stimulus-response reinforcing phase) in order to develop bio-inspired CHT-modified CW (CHT-CW)/ TPU composites that would exhibit water-responsive characteristics in varied pH milieus. The positive charge of CHT under acidic conditions enables it to undergo surface adsorption onto CWs through electrostatic interactions and hydrogen bonding. They subsequently investigated the effect of this surface modification of the water absorption capacity, and mechanically adaptive shape-memory behavior at different pHs [3].

The shape-memory effect of a CHT-CW/TPU sample is visualized in Fig. (**9**), where the strip was prewetted in the presence of water, stretched and dried under ambient conditions. Following deformation, it was maintained in a stress-free state for 24 h, with no length change, indicating stability of the temporary fixed shape. The strip then recovered its original length following immersion in water, highlighting its water-responsive shape-memory performance [3].

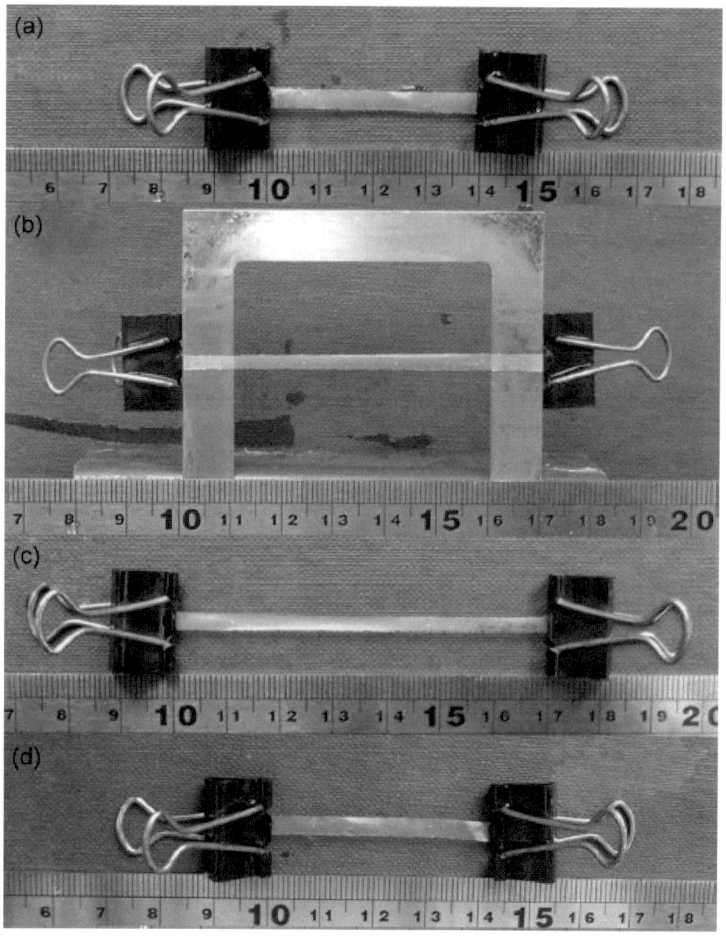

Figure 9: Water-active CHT-CW/TPU strip showing shape-memory behavior. (**a**) Pre-treatment by immersion in water (1 h). (**b**) Stretched wet sample. (**c**) Deformed sample following drying at ambient temperature (4 h). (**d**) Shape recovery following immersion in water (1 h) [3].

The water absorption was notably affected by surface modification with CHT, and varied with the solution pH, with a greater degree of absorption at acidic pH's compared to the non-CHT-modified CW/TPUs. The pH values also affected the mechanical properties of the CHT-modified strips in relation to unmodified strips, with increased mechanically-adaptive behavior under acidic conditions (greater decrease in modulus), due to CHT's acid responsive properties, and a maintained modulus in basic media. Wu *et al.* [3] thus achieved a simplified approach for customizing the pH-dependent mechanically adaptive properties of CW/TPU composites through surface modification with a pH-dependent biopolymer, expanding on the study of Zhu *et al.* [4].

Cellulose Nanocrystals for Modifying Castor-Oil based Polyurethanes and Shape-Memory Biopolymeric Systems

Segmented polyurethanes have found application as biomaterials owing to their good physical properties and fairly good biocompatibility [38]. Segmented thermoplastic polyurethanes (STPU) are technically described as block copolymers comprised of alternating amorphous (soft) and crystalline (hard) segments or domains, with manipulatable composition. In the STPU, on a molecular level, the soft segment is the switching domain undergoing a transition at a certain temperature, and is responsible for shape fixity through crosslinks, whereas the hard segment provides the network points that link these soft segments for establishing the permanent shape. Recently, there has been application of components from renewable sources for polyurethane synthesis, such as the synthesis of polyols from vegetable oil. The use of STPU in thermally-actuated shape-memory applications has received increasing attention, with potential uses in self-healing and health-monitoring purposes, and in biomedical devices. In combination with this, it has already been noted in this chapter that a number of the nanoreinforcements are biopolymers from renewable sources such as the cellulose nanocrystals/ whiskers, chitin nanocrystals, as well as bacterial cellulose. These ultimately improve the shape-memory characteristics of the nanocomposite, while increasing the carbon level from renewable sources.

In this study, Saralegi *et al.* [39] employed *in situ* polymerization to synthesise cellulose nanocrystal/ polyurethane bionanocomposites with a high renewable source content. The crystals were covalently bonded to polyurethane chains, using a castor oil- based diol and a corn-sugar-based chain extender as raw materials. The effect of the cellulose nanocrystals on the shape-memory properties were ascertained in response to thermal activation. Two specific thermal transitions were noted, being the melting temperatures of the soft (affecting shape fixity) and hard segments (affecting the shape recovery), respectively. Unmodified polyurethane possessed poor shape-memory behavior; but the addition of cellulose nanocrystals enhanced the crystallinity (a hard phase with increasing network points), improving the shape recovery to almost 100%. As intelligent biomaterials, these bionanocomposites are potential candidates for thermally-activated applications [39].

Porous Alginate Scaffolds as Shape-Memory Biopolymeric Systems

In an attempt to promote regeneration of annulus fibrosis (AF) occurring with disc herniation following degeneration or traumatic injury, Guillaume and co-workers [40] formulated a porous alginate scaffold possessing shape-memory attributes that can be delivered with minimal invasion with shape recovery on hydration. The hydrogels were covalently crosslinked employing carbodiimide chemistry with subsequent lypholization, creating a porous system. The resultant scaffolds demonstrated crosslinker concentration-dependent shape-memory recovery and mechanical characteristics. Repeated compression and expansion was supported by the scaffold, enabling direct delivery to the traumatized AF tissue. Cytocompatibility of the scaffolds with potential for cell-seeding and proliferation was demonstrated under *in vitro* conditions simulated the intervertebral disc milieu (low oxygen and glucose conditions). AF cells were enabled to secrete extracellular matrix rich in collagen type I and glycosaminoglycan, which filled the porous scaffold. The matrix network was, however, not sufficiently mature after *in vitro* culturing over 21 days to possess the mechanical integrity of native AF tissue. A pertinent observation following *ex vivo* studies in a bovine tail intervertebral disc (IVD) model with a 5mm defect in AF tissue was the facilitation of immediate AF defect filling following injection of the alginate scaffold. The shape-memory effects achieved, depicted in Fig. (**10**), were thus promising as a regenerative approach [40].

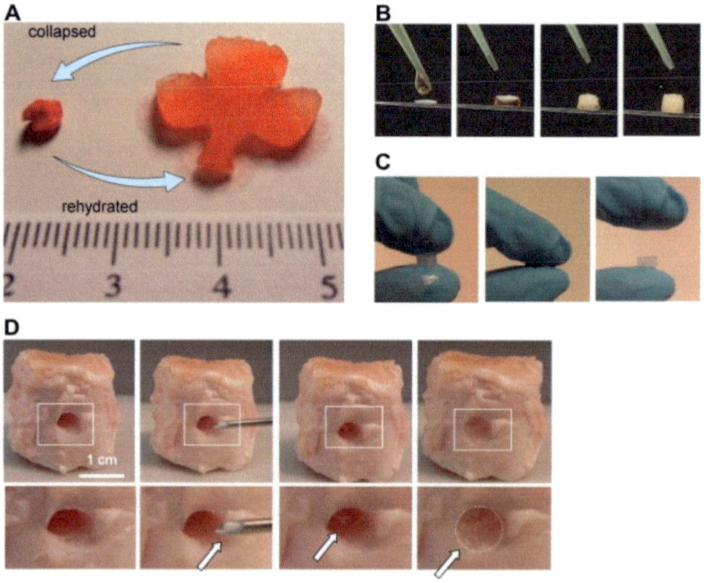

Figure 10: Digital photographic documentation of the shape-memory characteristics of covalently crosslinked porous alginate scaffolds. (**A**) Shape-memory properties of porous alginate scaffolds. (**B**) Deformation properties in dry and rehydrated state in association with a cell suspension. (**C**) Elasticity of the hydrated scaffold elasticity. (**D**) *Ex vivo* demonstration of injection of the porous alginate scaffold with ensuing rehydration in the defect created in AF tissue of a bovine caudal IVD. The scaffold is indicated by the white arrows and the white dashed line circle delineates the scaffold position after rehydration. Scale bar: = 1 cm [40].

Crosslinked Alginate Scaffolds as Shape-Memory Biopolymeric System for Cell and Growth Factor Delivery

Following cell and growth factor delivery, the limitations of notable cell loss and fast growth factor dissipation following implantation retards their benefits in tissue repair. Researchers have endeavoured to address these concerns *via* developing covalently crosslinked porous alginate systems for enhancing cell survival and growth factor release; however open-wound surgical procedures have been necessitated for their insertion, and their *in vivo* degradation was limited [41, 42]. Wang *et al.* [43] thus designed a partially crosslinked alginate scaffold for enhanced biodegradation with the required shape-memory characteristics for minimally invasive surgery. Covalent crosslinking employing carbodiimide chemistry of high and low molecular weight arginylglycylaspartic acid (RGD) peptide-modified oxidized alginate. The low molecular weight (LMW) alginate was produced by irradiation, they applied oxidation to enhance biodegration, while RGD modification increased cell adhesion. Compressibility of the scaffold was significant – 11 times smaller than the original size – it then returned to the original shape on hydration. The scaffold also demonstrated a preferred *in vitro* biodegradability of 85% after 28 days. Scaffolds exceeding 90% porosity provided *in vitro* recombinant growth factor insulin-like growth factor-1 release over several days, which would enable survival of skeletal muscle, and increase satellite cell proliferation and migration from the scaffold over 28 days. The scaffold has potential suitability as a shape change system for temporary delivery of cells and growth factors for tissue repair. Wang and co-workers [43] reported that they would subsequently undertake *in vivo* analyses in a muscle injury model *via* implantation of the scaffold with myogenic cells and growth factors. The efficacy of the scaffolds compared to bolus delivery of these agents would ultimately be elucidated [43].

Photoresponsive Polyesters as Shape-Memory Biopolymeric Systems

As highlighted, the application of SMBPs becomes limited when a thermal transition is targeted, because of the requirement of being set near body temperature, which does not enable significant tailoring of other properties. Employment of a light-induced mechanism overcomes this difficulty and enables shape change to occur at ambient temperatures. Previous investigators have disclosed their approach for achieving thermally-independent light-induced shape change.

For example, Wu and co-workers [44] formulated innovative multiblock polyesterurethanes possessing soft and hard domains with pendant cinnamamide moieties. This was *via* a two-step polyaddition reaction employing *N,N*-bis(2-hydroxyethyl) cinnamamide (BHECA), biodegradable PLLA, and poly(ε-caprolactone) (PCL) diols as raw materials and hexamethylene diisocyanate (HDI) as coupling agent. Reversible [2+2] cycloaddition crosslinking occured. Behavior of the copolymers showed typical thermoplastic elastomeric behavior as well as an ambient temperature light induced shape-memory effect as a result of the pendant cinnamamide group functioning as a photoresponsive molecular switch. There could, however, be no modification of the thermal or mechanical properties of the platforms without alteration of the cinnamamide group concentration. PLLA and PCL also demonstrate undesirably slow degradation rates.

Rochette *et al.* [45] thus proposed the design of biodegradable amorphous poly(ester urethane) (PEU) prepolymers, with each unit possessing a photoresponsive group, thus providing an even distribution. The resulting thermoset polymers could undergo a light-induced shape-memory change, with modification of their thermomechanical properties for addressing a broad range of biomedical applications, as well as fairly tuneable rapid degradation. Tuning of thermal and mechanical properties was achieved *via* interchanging the two diester photomonomers and cyclic/linear diols. Fixation was achieved at a UV wavelength of 302nm, whereas further irradiation at λ = 254nm attained the original shape when the temporary crosslinks were cleaved. Biocompatibility studies revealed minor cytotoxicity for HeLa cells. These systems could have further applications as triple shape-memory biomaterials incorporating light and thermal triggers in a single platform, which could then exhibit multiple shape changes and complex movements, with application as biomedical devices and implants [45].

Potato Starch as a Shape Change Biopolymeric System

Recently the shape-memory effects of non-modified potato starch have been reported [46] and has potential advantages in that it does not contain residual components (*e.g.* proteins or lipids). Potato starch is a polyanhydroglucose comprised of linear amylose (average content) and branched amylopectin as homologous polymers. In its native form, it is granular and semi-crystalline, with an amorphous state achieved *via* thermomechanical processing. On encountering moisture, the hydrophilic starch experiences a decrease in T_g, emanating in residual stress relaxation. Water uptake can thus activate a shape reduction of chemically grafted starch-based films [47].

The study of Véchambre and co-workers [47] focused rather on the shape-memory capabilities of non-modified potato starch prepared *via* extrusion and activated *via* humidity. Glycerol was employed as a T_g depressor. The recovery ratio of amorphous natural starch exceeds 90% to achieve 200% deformation. Extrusion cooking was executed with stabilization below the T_g in order to fix the amorphous starch in an unstressed state and create the initial permanent shape. The second shape was achieved by thermomolding above the T_g, which caused the appearance of residual stresses, but stability was maintained with cooling below the T_g. Relaxation of the residual stresses above the T_g enabled return to the initial unstressed equilibrium shape. It was found that water sorption at 20°C triggered shape recovery. Starch plasticized with 10% glycerol exhibited a recovery ratio ~50% due to crystallization limiting shape recovery. This study is a trailblazer for the use of an unmodified biopolymer for creating innovative environmentally safe systems with biomedical applications [47].

Resorbable Starch-Based Stents as a Shape-Memory Biopolymeric System

Stents are commonly employed in sialendocopy for maintaining the salivary duct open following surgery and need to be removed thereafter. Beilvert and his team [48] opted to develop a novel starch-based shape-memory material. A device based on this biopolymer could be a potential deployable resorbable stent. Formulations were prepared employing potato and high amylose content maize starch, in the presence and absence of a plasticizer. The degradability of the stents in simulated saliva conditions was ascertained together with its performance in a pig model following insertion in the submandibular duct. Plasticized starch stents possessed the necessary shape-memory behavior for application as a deployable system; however, they were rapidly hydrolyzed in simulated saliva. Stents did not adversely affect the epithelium following their insertion in pigs. The designed stents could have enhanced biomedical applications in pathologies occurring in environments with a lower α- amylase content [48].

Copolyesters as a Shape-Memory Biopolymeric System for Controlled Drug Release

The investigation of Wischke *et al.*, [19] employed an amorphous biodegradable shape-memory polymer for controlled drug release. Synthesis of a copolyester urethane network was undertaken by crosslinking tetrahydroxy telechelics of oligo[(*rac*-lactide)-*co*-glycolide] with an aliphatic diisocyanate. Wischke *et al.* [19] selected enoxacin (EN), nitrofurantoin (NF) and ethacridine lactate (EL) as three model drugs with varying physicochemical properties. These were incorporated within the SMBP *via* swelling in saturated drug solutions. The SMBP employed is as an intelligent multifunctional matrix material incorporating a crystallisable segment. In this investigation, they explored the suitability completely amorphous shape-memory networks vitrified below body temperature as intelligent matrices. The rationale for this was based on the advantages of amorphous materials, which include more maintained mechanical properties due to more homogenous biodegradation compared to semi-crystalline biopolymers, a wider range of copolymer variation for water uptake control, and potentially high drug loading. The biodegradability of linear polyesters prepared from dilactides and diglycolide (PLGA) is known. Polyurethanes involve copolyester sequences from lactic and glycolic acid and include copolyester urethanes, which have encouraging mechanical properties with an adjustable rate of hydrolysis (depending on glycolide content and overall molecular weight). The designed SMBP was evaluated for its usefulness in controlled drug delivery.

The SMBP demonstrated favorable thermal and mechanical properties and shape recovery could be *via* a one- or multi-step methodology (Fig. **11**). The shape-memory capabilities of the system were not negatively affected by drug loading. The SMBP network enabled long-term controlled drug release over 4 months, with release profiles demonstrated four phases. All compounds exhibited a burst release, but predictably, the water soluble EL showed more rapid release, especially in the first 40 days. In phase 2, a lag phase was observed for EN compared to controlled diffusional release of EL and NF. Phase 3 was characterized by accelerated release as a result if the collapse of the network structure with biodegradation. During phase 4, the remaining amount of entrapped drug (that was possibly strongly associated/ interacting with the polymer) was gradually released. The effect of the shape-memory functionality on the drug release rates was minimal. Contrary to anticipated, a lower drug release was observed from the stretched network, potentially due to the greater linearity of polymer chain orientation slowing diffusion of drug. The resultant drug loaded SMBP system ultimately functions as a novel multifunctional device, combining shape-memory effect, drug release and biodegradation; that can be inserted through a small incision into the body. It consequently unfolds to create mechanical stabilization, provides controlled release of drugs, and ultimately degrades, thus avoiding secondary surgery for implant removal [19].

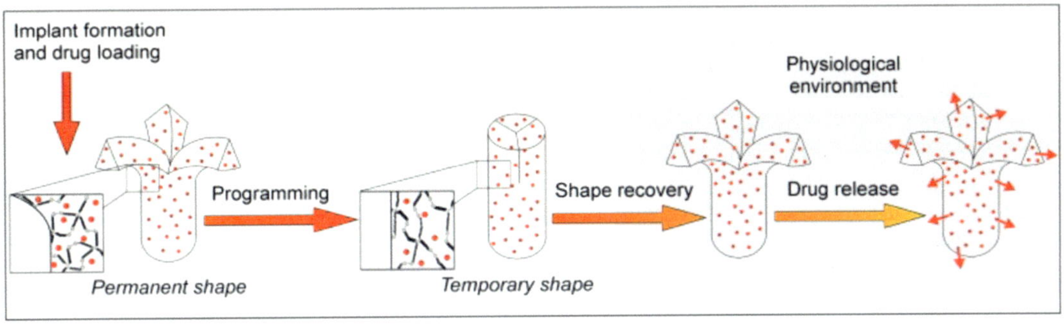

Figure 11: Programming, shape recovery and subsequent drug release of a copolyester-based SMBP system [19].

CONCLUSION

The following table presents a summary of the SMBPs that discussed. Pertinent details have been highlighted to show the type of customisation carried out which is dependent on the respective bio-medical application.

Table 3. Overview of shape-memory biopolymeric systems for biomedical applications.

Biopolymer System	Shape Memory Trigger/Transition	Biomedical Application	References
Hexamethylene diisocyanate cross-linked branched star shaped oligo(ε-caprolactone) polymer networks	Temperature-sensitive shape recovery within a 2°C range (37-39°C)	Prolonged and sustained release of loaded theophylline without initial burst-release at 37 °C.	[49]
Linear, phase-segregated multiblockcopolymers of oligo(ε -caprolactone)diol and oligo(*p*-dioxanone)diol	Temperature-sensitive shape recovery with fiber shrinkage (20-41°C)	Degradable shape-memory suture for wound closure.	[20]
Genipin-crosslinked chitosan membranes and scaffolds; Chitosan/Poly(L-lactide) Composites	Controlled hydration	Biomaterials architectures for tissue regeneration and drug delivery	[6, 7, 28]
Covalently-crosslinked, partially-oxidized alginate modified with RGD peptides	Water uptake	Carrier for transplanting cells and/or release of growth factors for tissue repair	[43]
Bioresorbable terpolymer double layer system based on l-lactide, glycolide and trimethylene carbonate	Elastic recoil	Paclitaxel delivery systems	[50]

Table 3: contd…

Crosslinked star-shaped tetrahydroxy telechelics of oligo[(rac-lactide)-coglycolide] based copolyester urethane network	Thermomechanical shape recovery under physiological conditions	Controlled drug delivery	[19]
Heat-annealed solvent cast films based on Bacterial Poly[(3-hydroxybutyrate)-co-(3-hydroxyvalerate)]	Thermoresponsive shape recovery	Biomedical device design	[51]
Poly(D,L-lactide)/ hydroxyapatite composites	68.0°C and 70.0°C as shape memory deformed and recovered temperature	Application in minimally invasive surgery and bone and tissue repair	[32]
Covalently cross-linked macroporous alginate hydrogel scaffolds	Controlled hydration	Efficient *in vivo* cell seeding and neotissue formation with desired geometries; implantable intervertebral disc	[40, 41]
Hydrogel networks of recombinant telechelic polypeptides with collagen-like transient nodes	Thermomechanical shape recovery	Biomaterial architectures compatible with biological tissues	[18]
Biodegradable cylindrical rod based on chitosan-functionalized PLGA microspheres	High intensity focused ultrasound	Spatiotemporally modulated drug delivery	[35]

This chapter provides a unique perspective on the customisation of SMBP systems in terms of biofunctionality, hydrolytic degradability, and shape memory functionality. Special consideration is placed on naturally sourced biopolymers. Examples of selected benchmark bio-medical systems have been summarised in Table **3**. The cited research provides valuable insight into the design of integrated and innovative biomedical SMBPs. Reported design criteria include the judicious selection of netpoints, and/or the introduction of indirect molecular switches *viz.* units responsive to ultrasound, light, moisture, or pH. Discussed previously, netpoints determine the permanent shape and contribute to the molecular architecture of SMBPs. Careful control of composite ratios and main chain segments can tailor overall properties of the material. Many thermal triggers have been presented in the literature; of which switching segments are recommended to direct the shape memory effect at conditions suitable to physiological systems.

Collective design optimizations involve the use of chemical modifications to: a) improve the intrinsic mechanical characteristics of natural biopolymers using biodegradable nanocomposites or adding synthetic thermoplastic biopolymers, b) modify the stimuli responsive mechanism by using molecular switches more practical for use *in vivo*, and c) otherwise tailor the material to fulfill the requirements for a specific bio-medical application. Consideration should also be given to other biomolecules for the preparation of SMBPs, *e.g.* self-assembling peptides that offer benefits such as: a) spontaneous formation of nanogels, b) conjugation to drug molecules and/or antibodies, and c) sites for chemical functionalization.

Continued research in integrated shape memory and drug release systems is required, and in particular research focusing on: a) evaluation of drug release for different classes of (model) drug molecules, b) assessment of the effect of drug loading levels on the shape memory effect, and c) investigation of the structure activity relationship between polymer degradation and drug release. Experimental studies should also be carried out *in vivo* to fully determine the biocompatibility, cell toxicity, and drug release profiles to promote SMBPs entering bench-to-bedside clinical steps.

Research into the molecular modelling of the shape change memory effect is also required. These methods would use molecular descriptors to identify the most stable shape conformations responsible for a specific drug release and can facilitate design of new systems with tailored properties. Reported modelling of shape memory polymers is largely based on mathematical models that strictly treat the shape memory effect as a function of the stress *versus* strain phenomena. In physiological systems the shape memory effect may change with other variables for example hydration, hydrogen bond formation and pH, and must be accounted for in the modelling approach.

Current research trends point towards the continued customisation of SMBPs for application in bio-medical technologies. To this end, SMBPs customised with drug-release functionality can lead to multifunctional carriers that are a nascent but promising technology for the pharmaceutical sciences. It is envisaged that this research area will contribute to numerous innovative and emerging bio-medical platforms in the near future.

ACKNOWLEDGEMENTS

This work was funded by the National Research Foundation (NRF) of South Africa.

CONFLICT OF INTEREST

The authors confirm that this article content has no conflict of interest.

REFERENCES

[1] Behl M, Lendlein A. Shape-memory polymers. Mater Today 2007; 10(4): 20-8.

[2] Iqbal D, Samiullah MH. Photo-responsive shape-memory and shape-changing liquid-crystal polymer networks. Materials 2013; 6(1): 116-42.

[3] Wu T, Su Y, Chen B. Mechanically adaptive and shape-memory behaviour of chitosan-modified cellulose whisker/elastomer composites in different pH environments. Chem Phys Chem 2014; DOI: 10.1002/cphc.201402157

[4] Zhu Y, Hu J, Luo H, *et al.* Rapidly switchable water-sensitive shape-memory cellulose/elastomer nano-composites. Soft Matter 2012; 8(8): 2509-17.

[5] Draget KI, Smidsrod O, Skjak-Braek G. Alginates from algae. In: Steinbuchel A, Rhee SK, Eds. Polysaccharides and polyamides in the food industry. properties, production and patents. Weinheim, Germany, Wiley-VCH Verlag GmbH & Co 2005; pp. 1-30.

[6] Meng Q, Hu J, Ho K, *et al.* The shape memory properties of biodegradable chitosan/poly(l-lactide) composites. J Polym Environ 2009; 17(3): 212-24.

[7] Correia CO, Mano JF. Chitosan scaffolds with a shape memory effect induced by hydration. J Mater Chem B 2014; 2(21): 3315-23.

[8] Lu D, R, Xiao CM, Xu SJ. Starch-based completely biodegrable polymer materials. eXPRESS Polym Lett 2009; 3(6): 366-75.

[9] Shoulders MD, Raines RT. Collagen structure and stability. Ann Rev Biochem 2009; 78(1): 929-58.

[10] Wang J, Su Y, Jia F, *et al.* Characterization of casein hydrolysates derived from enzymatic hydrolysis. Chem Cent J 2013; 7: 62 (DOI:10.1186/1752-153X-7-62).

[11] Rosato DV, Rosato DV, Rosato MV. 2 - Plastic property. In: Rosato DV, Rosato DV, Rosato MV, Eds. Plastic product material and process selection handbook. Oxford, Elsevier 2004; p. 85.

[12] Capadona JR, Shanmuganathan K, Tyler DJ, *et al.* Stimuli-responsive polymer nanocomposites inspired by the sea cucumber dermis. Science 2008; 319(5868): 1370-4.

[13] Shanmuganathan K, Capadona JR, Rowan SJ, *et al.* Biomimetic mechanically adaptive nanocomposites. Prog Poly Sci 2010; 35(1–2): 212-22.

[14] Biron M. Extrinsic smart polymes. Technical Paper. 2 February 2005. Available from: http://www.specialchem4polymers.com/resources/articles/article.aspx?id=2002&q=smart%20polymer Accessed 5 March 2008.

[15] Pişkin E. Molecularly designed water soluble, intelligent, nanosize polymeric carriers. Int J Pharm 2004; 277(1–2): 105-18.

[16] Schmaljohann D. Thermo- and pH-responsive polymers in drug delivery. Adv Drug Deliv Rev 2006; 58(15): 1655-70.

[17] Lendlein A, Behl M, Hiebl B, *et al.* Shape-memory polymers as a technology platform for biomedical applications. Expert Rev Med Device 2010; 7(3): 357-79.

[18] Skrzeszewska PJ, Jong LN, de Wolf FA, *et al.* Shape-memory effects in biopolymer networks with collagen-like transient nodes. Biomacromolecules 2011; 12(6): 2285-92.

[19] Wischke C, Neffe AT, Steuer S, *et al.* Evaluation of a degradable shape-memory polymer network as matrix for controlled drug release. J Control Rel 2009; 138(3): 243-50.

[20] Lendlein A, Langer R. Biodegradable, elastic shape-memory polymers for potential biomedical applications. Science 2002; 296(5573): 1673-6.

[21] Marco D, Eckhouse S; Biodegradable self-inflating intragastric implants for curbing appetite. US Patent application US20070156248.

[22] Wache HM, Tartakowska DJ, Hentrich A. Development of a polymer stent with shape memory effect as a drug delivery system. J Mater Sci Mater Med 2003; 14(2): 109-12.

[23] Huang WM, Lee CW, Teo HP. Thermomechanical behavior of a polyurethane shape memory polymer foam. J Intell Mater Syst Struct 2006; 17(8-9): 753-60.

[24] Mather PT, Liu C, CJ. B; Shape memory polymer orthodontic appliances, and methods of making and using the same. EP1844097, WO2006071520; (2006)

[25] Yang B, Huang WM, Li C, *et al.* Effects of moisture on the thermomechanical properties of a polyurethane shape memory polymer. Polymer 2006; 47(4): 1348-56.

[26] Lv H, Leng J, Liu Y, *et al.* Shape-memory polymer in response to solution. Adv Eng Mater 2008; 10(6): 592-5.

[27] Mano JF. Viscoelastic properties of chitosan with different hydration degrees as studied by dynamic mechanical analysis. Macromol Biosci 2008; 8(1): 69-76.

[28] Correia CO, Caridade SG, JF. M. Chitosan Membranes Exhibiting Shape Memory Capability by the Action of Controlled Hydration. Polymers 2014; 6(4): 1178-86.

[29] Lu X, Cai W, Zhao L. Study on the shape memory behavior of poly(l-lactide). Mater Sci Forum 475–479(III). PRICM 5: The Fifth Pacific Rim International Conference on Advanced Materials and Processing. 2005; 2399-402.

[30] Lu X, Sun Z, Cai W. Structure and shape memory effects of poly(L-lactide) and its copolymers. Physica Scripta T:T129. Second International Symposium on Functional Materials 2007; 231-5.

[31] Li L, Ding S, Zhou C. Preparation and degradation of PLA/chitosan composite materials. J Appl Polym Sci 2004; 91(1): 274-7.

[32] Zheng X, Zhou S, Li X, *et al.* Shape memory properties of poly(d,l-lactide)/hydroxyapatite composites. Biomaterials 2006; 27(24): 4288-95.

[33] ter Haar G, Coussios C. High intensity focused ultrasound: Physical principles and devices. Int J Hyperthermia 2007; 23(2): 89-104.

[34] Zhang H, Xia H, Wang J, *et al.* High intensity focused ultrasound-responsive release behavior of PLA-b-PEG copolymer micelles. J Control Rel 2009; 139(1): 31-9.

[35] Bao M, Zhou Q, Dong W, *et al.* Ultrasound-modulated shape memory and payload release effects in a biodegradable cylindrical rod made of chitosan-functionalized PLGA microspheres. Biomacromolecules 2013; 14(6): 1971-9.

[36] Wong YS, Xiong Y, Venkatraman SS, *et al.* Shape memory in un-cross-linked biodegradable polymers. J Biomater Sci Polym Ed 2008; 19(2): 175-91.

[37] Hsu L, Weder C, Rowan SJ. Stimuli-responsive, mechanically-adaptive polymer nanocomposites. J Mater Chem 2011; 21(9): 2812-22.

[38] Osman AF, Edwards GA, Schiller TL, *et al.* Structure–property relationships in biomedical thermoplastic polyurethane nanocomposites. Macromolecules 2012; 45(1): 198-210.

[39] Saralegi A, Gonzalez ML, Valea A, *et al.* The role of cellulose nanocrystals in the improvement of the shape-memory properties of castor oil-based segmented thermoplastic polyurethanes. Composit Sci Tech 2014; 92: 27-33.

[40] Guillaume O, Daly A, Lennon K, *et al.* Shape-memory porous alginate scaffolds for regeneration of the annulus fibrosus: Effect of TGF-β3 supplementation and oxygen culture conditions. Acta Biomater 2014; 10(5): 1985-95.

[41] Thornton AJ, Alsberg E, Albertelli M, *et al.* Shape-defining scaffolds for minimally invasive tissue engineering. Transplantation 2004; 77(12): 1798-803.

[42] Hill E, Boontheekul T, Mooney DJ. Regulating activation of transplanted cells controls tissue regeneration. Proc Natl Acad Sci USA 2006; 103(8): 2494-9.

[43] Wang L, Shansky J, Borselli C, *et al.* Design and fabrication of a biodegradable, covalently crosslinked shape-memory alginate scaffold for cell and growth factor delivery. Tissue Eng Part A 2012; 18(19-20): 2000-7.

[44] Wu L, Jin C, Sun X. Synthesis, properties, and light-induced shape memory effect of multiblock polyesterurethanes containing biodegradable segments and pendant cinnamamide groups. Biomacromolecules 2011; 12(1): 235-41.

[45] Rochette JM, Ashby VS. Photoresponsive polyesters for tailorable shape memory biomaterials. Macromolecules 2013; 46(6): 2134-40.

[46] Chaunier L, Lourdin D. The shape memory of starch. Starch/Stärke 2009; 61(2): 116-8.

[47] Véchambre C, Chaunier L, Lourdin D. Novel shape-memory materials based on potato starch. Macromol Mater Eng 2010; 295(2): 115-22.

[48] Beilvert A, Faure F, Meddahi-Pellé A, *et al.* A resorbable shape-memory starch-based stent for the treatment of salivary ducts under sialendoscopic surgery. Laryngoscope 2014; 124(4): 875-81.

[49] Nagahama K, Ueda Y, Ouchi T, *et al.* Biodegradable shape-memory polymers exhibiting sharp thermal transitions and controlled drug release. Biomacromolecules 2009; 10(7): 1789-94.

[50] Musiał-Kulik M, Kasperczyk J, Smola A, *et al.* Double layer paclitaxel delivery systems based on bioresorbable terpolymer with shape memory properties. Int J Pharm 2014; 465(1–2): 291-8.

[51] Kim YB, Chung CW, Kim HW, *et al.* Shape memory effect of bacterial poly[(3-hydroxybutyrate)-co-(3-hydroxyvalerate)]. Macromol Rapid Comm 2005; 26(13): 1070-4.

CHAPTER 6

Biotextiles for Tissue Engineering

Navdeep Grover[1,*] and Lomas K. Tomar[2]

[1]Department of Chemical and Biological Engineering, University of Colorado Boulder, 3415 Colorado Avenue, Boulder, Colorado, USA and [2]University of the Witwatersrand, Faculty of Health Sciences, Department of Pharmacy and Pharmacology, 7 York Road, Parktown 2193, Johannesburg, Gauteng, South Africa

Abstract: The porous textile fabrics as scaffolds combining with extracellular matrix proteins (*e.g.* fibronectin, collagen *etc*), and living cells give a possibility to develop biological tissues for human body repair, such as organ transplants. The flexible biotextiles act as scaffold materials to provide support for cell adhesion, growth and proliferation. A key requirement for an artificial implant is to exactly mimick the biological and mechanical functions of an injured tissue or organ to be replaced without evoking an immune response from the host. Due to their easily tunable properties, the "weak" textiles, such as nonwoven, woven, knitted structures, braided structures *etc*, are promising candidate for tissue engineering applications. These structures provide compliance, porosity and 3D microstructure to induce cell attachment and proliferation to regenerate complex tissues. In this chapter, scaffold designing in tissue engineering using textile structures has been reviewed. The chapter elaborates on 3-D fabric (woven, knitting, nonwoven *etc*.) and their composites for designing scaffolds with required mechanical and biological properties.

Keywords: Biotextile, Tissue Engineering, Scaffolds, Extracellular Matrix (ECM).

1. INTRODUCTION

Textile materials are a promising class of scaffolds for tissue engineering applications. Their properties can be easily modified as per the requirements of the implants, such as high mechanical strength for hip and knee replacements, high degradability for sutures, permeability for skeletal tissues *etc*. Ideal biotextile implants need selective modification of chemical, interfacial, mechanical and biological properties in order to develop the required biocompatible structures [1-3]. In USA alone, 6,439 people were died in 2001 due to unavailability of organs for transplantation [4]. Thus, we can no longer rely on transplantation alone in case of a tissue and organ malfunctioning. Researchers have spent decades to

***Corresponding author Navdeep Grover:** Department of Chemical and Biological Engineering, University of Colorado Boulder, USA; E-mail: navdeepiitd@gmail.com

Viness Pillay, Yahya E. Choonara and Pradeep Kumar (Eds.)

understand the biological interactions of the scaffolds. Attempts have been made to mimick naturally occurring biological processes or properties to develop biologically active materials for tissue engineering [5, 6]. In Europe the share of medical textile is already 10% of the technical textile market, with 100,000 tones of fibres, a growth rate of 3 to 4% per year and a market of US $7 billion. Nonwoven and knitting technologies are making dramatic inroads in this sector and more specifically in some niches having a high specific content [7].

2. BIOTEXTILES: STRUCTURAL AND BIOLOGICAL ASPECTS

Biotextiles are biocompatible and biostable materials comprising of textile fibres, developed with properties specific for their functions in a biological environment [3]. The porosity, its distribution and reproducibility are vital structural parameters of the textile scaffolds, required for the growth and spatial arrangement of the host cells and tissues. It also provides the space for vascularization, neotissue formation and remodeling [8]. The interconnectivity and reproducibility of pores determine the channeling of nutrients and waste and dimensional stability of textile scaffolds respectively, thus influencing the success of tissue engineering [9, 10]. Table **1** compares the various microstructural aspects of foams and textile structures [6].

Table 1. Microstructural aspects of scaffolds [6].

Fabrication	Foam/sponge	Nonwoven	Woven	Braided	Knitted
Pore size (μm)	0.5-500	10-1000	0.5-1000	0.5-1000	50-1000
Porosity (%)	0-90	40-95	30-90	30-90	40-95
Pore distribution	Random to uniform	Random	Uniform	Uniform	Uniform
Reproducibility of porosity	Poor to good	Poor	Excellent	Excellent	Good to excellent
Pore connectivity	Good	Good	Excellent	Excellent	Excellent
Processability	Good	Good	Excellent	Excellent	Good
Other comments	Current techniques are associated with processing undesireable residues such as solvents, salt particles	Equipment cost is high. Control over porosity is always questionable	Shapes are limited	Limited to tubular or uniform cross-sectional shapes	Limited by the low bending properties of current biodegradab-le fibres

The surface chemistry of scaffolds is a vital parameter to determine the biological behavior. The influence of different functional groups on adhesion, spreading and growth of the Chinese hamster ovary (CHO) cells was demonstrated [11]. The surfaces with amino groups were found to be better than the hydroxyl than amide groups for the growth of CHO cells. The positively chargeable character of amino groups in aqueous cell culture medium attracts the negatively charged serum proteins, resulting cell adhesion and proliferation. The surfaces with neutral hydroxyl groups might form hydrogen bonds with the cell surfaces, resulting improved cell adhesion and proliferation than amide groups [11-13]. The influence of different functional groups such as –OH, $-NH_2$, -CFx and –COOH, were also investigated as a function of host tissue response. The plasma discharge induced polymerization was used to generate these functional groups on polypropylene particles and subsequently these particles were implanted subcutaneously in mice. The thick fibrous capsule formation was observed with surfaces which can form hydrogen bonds such as –OH and $-NH_2$. Other surface functional groups (-CFx and –COOH) showed least inflammatory responses [14, 15].

There is a significant effect of biomolecules on cell growth, proliferation, morphology differentiation, organization of the actin cytoskeleton and transcription activity *via* mechanical and chemical stimuli. The biomolecules used for immobilization on the surface of scaffolds are mainly extracellular matrix (ECM) proteins and bioactive peptides (*e.g.* RGD, RGDS *etc*). The ECM is composed of three major classes of biomolecules: structural fibrillar proteins (*e.g.* collagen and elastin), an amorphous interfibrillar matrix composed mainly of proteoglycans or glycosaminoglycans (GAGs) and specialized proteins (*e.g.* fibrillin, fibronectin and laminin). Adsorbed layer of ECM proteins helps in indirect interaction of cells and biomaterial surfaces and binding is mediated by interactions with cell surface receptors [16-23].

3. TEXTILE SCAFFOLDS: STRUCTURE AND DESIGNING

Scaffold serves numerous functions critical for the success of tissue regeneration. It allows cells to attach, grow, proliferate, migrate and differentiate. Scaffolds also enable the efficient transport of nutrients, growth factors, blood vessels and removal of waste materials. To perform these functions, scaffolds need to be structurally and biologically appropriate [24-26]. Langer and Vacanti [24] summarized work done with matrices, cells, *in vitro* bioreactor systems, and the generation of devices with their own vascular supply (Fig. **1**). Systems either open or totally integrated into the recipient, or closed (encapsulated) have been designed to avoid reaction from host's immune system [24-26]. Textile-based scaffolds such as 3-D fabrics (woven, knitting, nonwoven *etc.*) and their composites can be designed with the required mechanical and biological properties [27].

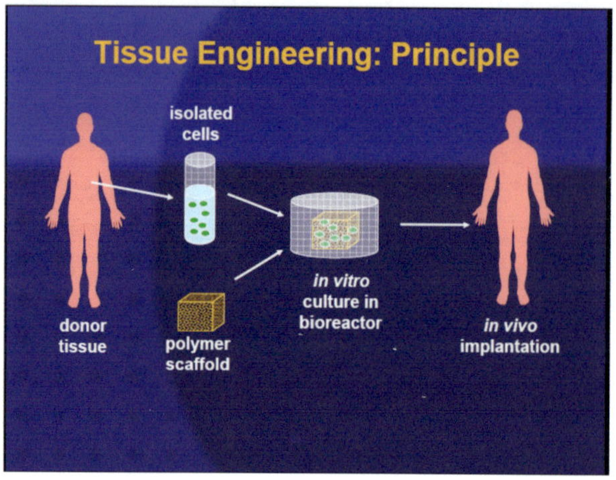

Figure 1: Tissue engineering process [24].

3.1. Nonwovens

Nonwovens are open three dimensional structure, porous enough to allow cell proliferation by continuous supply of nutrients. The surface to volume ratio of the implanted material is very high for fibre based implants. The multifilaments must be converted to staple fibres to produce nonwovens. At present scaffolds are developed for the regeneration of periodontal defects and for the tissue engineering of bladder walls [28]. The spunlaid nonwovens and staple-fibre nonwovens are mainly used as scaffold materials for tissue engineering applications.

Spunlaid nonwovens are formed by using binding techniques such as heat or other consolidation techniques, such as needle punching or hydro entanglement *etc.* These preparations form highly porous structures with decreased pore size which improves biocompatibility and structural stability. Staple-fibre nonwovens are the structures made from short fibres. The fibres in these structures cannot be oriented using drawing process. Thus, there is a problem of implant shrinkage in patients which is a major drawback of these nonwovens. Recently, absorbable nonwoven biotextiles are mainly processed by staple fibre process to create high pore volume and fine fibres without shrinkage. These fibres are needle punched to assure the required strength. When cartilage cells are brought into these nonwoven structures, they attach to the fibres and produce collagen and other components of ECM. Specific articular cartilage can be preferably generated by additional mechanical stimulation [29].

Wu *et al.* [30] fabricated tissue engineered trachea with poly(lactic-glycolic acid) (PLGA) nonwoven mesh enforced by collagen type I. The C-shaped nonwoven mesh was formed by coating and crosslinking of collagen on nonwoven PLGA mesh. The mesh was fabricated into the shape of a human trachea. The PLGA/collagen nonwoven mesh was subsequently seeded *in vitro* with chondrocytes from sheep nasal septum cartilage. The resulted cartilage like tissue was implanted successfully in mice without obvious inflammatory response. "The amount of GAGs in tissue engineered trachea cartilage reached 71.42% of normal value in native cartilage. This study demonstrated that collagen-enforced PLGA nonwoven mesh facilitated the adhesion and proliferation of chondrocytes, it also owned adequate mechanical strength to serve as an ideal scaffold for trachea tissue engineering without internal support" [30]. Gravin *et al.* [31] prepared a bioartificial tissues (BATs) as tendon constructs using nonwoven nylon mesh anchors. Most tissues will require a microvascular network to supply oxygen and nutrients. Therefore, Wu *et al.* [32] created a nonwoven PGA-PLLA microvascular network seeded with $CD34^{+}/CD133^{+}$ endothelial progenitor cells (EPC). The degradation and mechanical properties of electrospun nonwoven membranes composed of poly(glycolide-co-lactide) (GA/LA: 90:10, PLA10GA90) were investigated to design the resorbable scaffolds. The properties of membranes were found to be varied with its degree of crystallinity and orientation. The drawn and annealed membranes had high crystallinity and orientation than annealed only membranes. The annealed only membranes had high degree of crystallinity but no orientation. The degradation rate of crystalline membranes were found to be slower than that as spun amorphous membranes and drawn membranes possessed high tensile strength due to orientation of polymeric domains within the fibres of nonwoven [33]. The degradation also depends upon the blending and copolymerisation. Therefore, one can easily tune the biodegradation rate as well as mechanical properties with different blending and copolymer compositions. The hydrophilicity of membranes could also be a tuned, using blending and copolymerization techniques, which is an important parameter for cell adhesion and proliferation [34-36].

Xu *et al.* [37] prepared electrospun composite fibres by blending of poly(Lactide) (PLA) with nanoparticles prepared from poly(L-lactide)-grafted hydroxyapatite (PLA-g-HAP). The strength of the fiberous mat was found to be higher at low content (4 wt%) of PLA-g-HAP due to the uniform dispersion of the particles within the fibres. The deterioration of mechanical properties was observed with high content of nanoparticles because of the aggregation of nanoparticles within the fibres. The degradation rate of composite mat was also much faster at high content of nanoparticles because of improved wettability which helped nanoparticles to escape from composite fibres. Jeong *et al.* [38] prepared hybrid vascular scaffold by electrospinning of PLGA on freeze-dried porous tubular collagen (Fig. **2**). The average pore diameter of hybrid scaffold was found to be 150±50 μm. The mechanical properties of hybrid scaffolds were improved due the presence of electrospun PLGA fiberous matrix (Table **2**). The mechanical properties of the hybrid scaffold were limited after culturing of smooth muscle cells (SMCs)- and endothelial cells (ECs). A new application of needle punched method for the preparation of three dimensional fiber matrix from regenerated collagen fiber has been reported with high porosity (93%) which concluded the compressive properties to be similar to natural articular cartilage. *In vitro* human bone marrow stromal cells proliferation and cell morphology were evaluated which revealed good cytocompatibility of the matrix indicating its great potential in articular cartilage [39, 40].

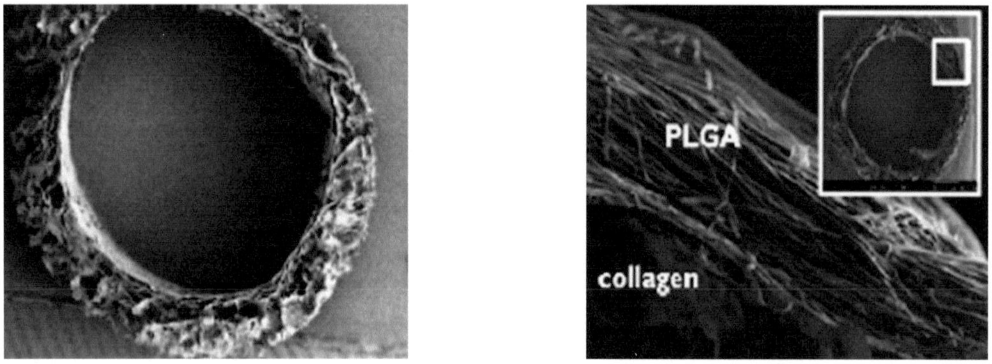

(a) **(b)**

Figure 2: Representative SEM images of fabricated jellyfish collagen/PLGA hybrid scaffolds with bilayered tubular structure: **(a)** cross-section of tubular collagen scaffolds, **(b)** cross-section of collagen/PLGA hybrid scaffolds [38].

Table 2. Mechanical properties of tubular porous scaffolds [47].

Mechanical Properties	PLGA scaffold	Collagen scaffolds	Collagen/PLGA hybrid scaffolds	Tissue Engineered Collagen/PLGA hybrid scaffolds
Tensile Strength (MPa)				
Dry	0.25±0.01	0.12±0.01	0.49±0.04	0.49±0.02
Wet	0.24±0.08	0.06±0.01	0.14±0.02	0.14±0.05
Elongation at break (%)				
Dry	20±1.02	12±0.65	17±0.24	17±0.72
Wet	20±0.61	34±1.40	41±1.89	41±2.01

3.2. Wovens

The first commercial prostheses were woven on two sets of yarn that had a high fabric count [41]. To maintain porosity and interconnectivity of woven structure, the biodegradable coating on woven polyethyleneterephthalate (PET) fabrics were done to design scaffold materials suitable for cell adhesion and proliferation [42, 43]. The coating of collagen and chitosan were shown to have better hepatocyte adhesion and proliferation. The coating of PLGA on one side of PET fabric created a polar structure with improved matrix porosity for tissue engineering of liver. The coating of cyclodextrin (CD) on woven PET vascular grafts showed no toxicity in biological assays [44]. Tofigh *et al.* [45] studied above-knee femoropopliteal bypass grafting with collagen impregnated woven polyester prosthesis. The use of collagen impregnated woven polyester prosthesis was a reasonable alternative in femoropopliteal bypass grafting with acceptable short-term patency rates. The tubular graft having identical mechanical and structural properties with human carotid artery was developed using woven technology with prestretched polyurethane monofilament with low elastic modulus and bulked polyester multifilament yarn with high elastic modulus. The pressure-circumferential stretch ratio (>1.4) of the graft was similar to that of human artery. The compliance and permeability properties of these hybrid grafts were found to be similar with dog's carotid artery (Fig. **3**) [46]. A novel method of constructing multifunctional scaffolds for cartilage tissue engineering was reported by Moutos *et al.*, where they prepared a 3-D woven PCL scaffolds infiltrated with slurry of homogenized cartilage-derived matrix (CDM) of porcine origin, seeded with human adipose-derived stem cells (ASCs), and cultured. Further these constructs were

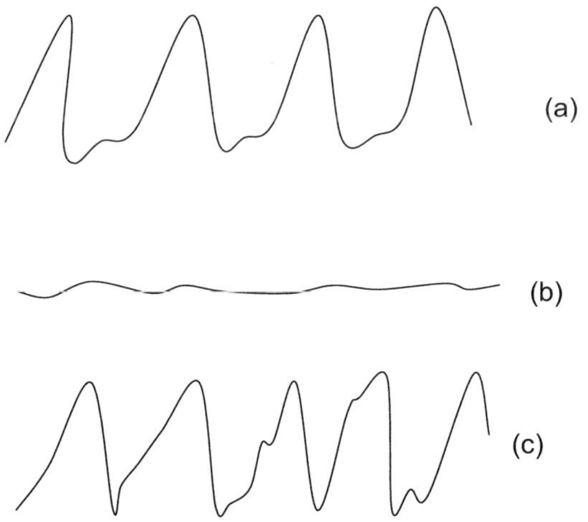

Figure 3: Shape of the pulse waves: **(a)** dog carotid artery, **(b)** commercial graft, and **(c)** hybrid vascular graft HVG-1 [45].

compared to scaffolds derived solely from CDM as well as 3-D woven PCL fabric without CDM. These findings show that fiber-reinforced hybrid scaffolds can be produced with biomimetic mechanical properties as well as the ability to promote ASC differentiation and chondrogenesis *in vitro* [47].

3.3. Knittings

The yarn used is interloped around each other in two directions: warp knit (lengthwise which loosens with time) and weft knit (transverselly) (Fig. **4**) [48]. Various knitting techniques and their application are reviewed by Wang *et al.* [49]. The knitted structure has more pores in the center of loops than the woven fabrics. "The knitted tube is then shrunk by compaction and this makes it possible to alter the water permeability of the knitted prostheses. The knitted prostheses are easier to handle than the ones made with woven fabrics. They can also elongate thereby affecting its dimensional stability" [49]. Velour fabric is also a type of graft that has a knitted construction to improve healing. A high textured yarn is needed rough surface to provide anchor support for thrombotic matrix to reduce blood loss during surgery [41]. The replacement of arterial vessels having diameter ≥6 mm is conducted by using knitted polyester vascular prosthesis. Enhanced ingrowth of tissue was observed due to porous structure of knitting [50].

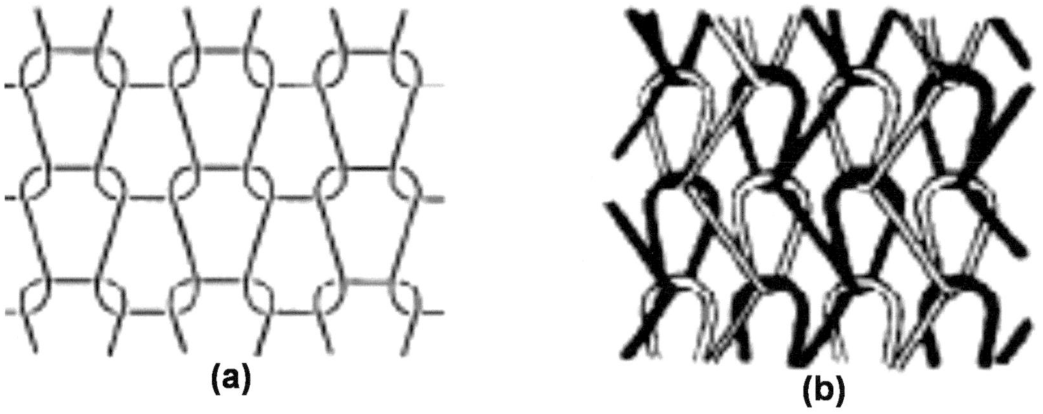

Figure 4: Knitted polyester vascular graft structure **(a)** weft knitting **(b)** wrap knitting [48].

A hybrid scaffold for skin graft were prepared by freeze drying the collagen with knitted poly[(lactic acid)-*co*-(glycolic acid)] (10:90, molar composition) mesh [51]. The graft was found to be cytocompatible against rat dermal fibroblasts cells and when implanted in rat, it evoked minimal host tissue response. Zhang *et al.* [52]

developed a chitosan-based tubular knitted scaffold and examined its feasibility of being applied in tissue engineering of vascular grafts. The knitted chitosan was coated with chitosan and 4:1 ratio solution of chitosan/gelatin and then freeze dried. The graft was observed to have high mechanical strength with an average loss of 18% after *in vitro* degradation. The growth of vascular smooth muscle cells was also studied and it was found that cells were adhered and proliferated very well (Fig. **5**).

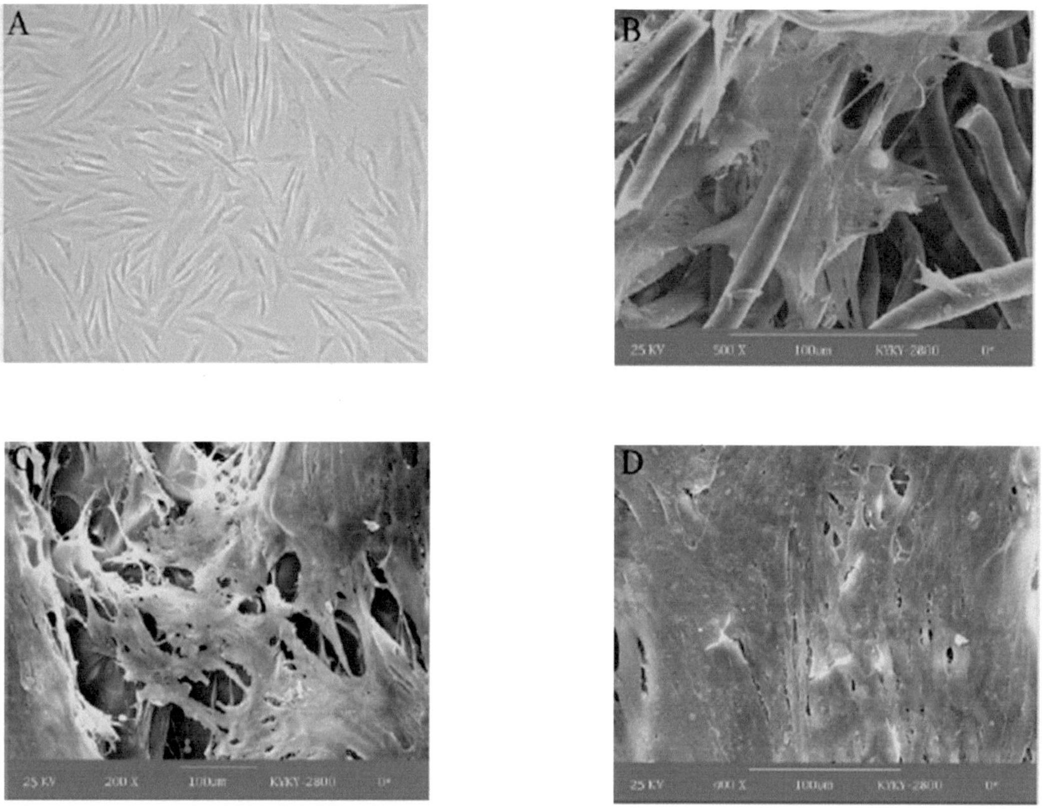

Figure 5: Morphology of cultured rabbit aortic vSMCs on tissue culture flask at 72 h **(A)**, chitosan fibres at 48 h **(B)**, and chitosan–gelatin porous complex at 48 h **(C)**, 72 h **(D)** [52].

"Tissue engineering of the urinary bladder using autologous cells and biodegradable scaffold is a promising method for augmentation" [53]. Abed *et al.* [54] studied the properties of a new hybrid mesh which is composed of a clinical grade polypropylene knitted mesh embedded in a polysaccharide hydrogel. In this study, they evaluated the effect of intramuscular and subcutaneous implantation of a new hybrid mesh in rats. There was improvement in the adverse effect of tissue organization with the fusion of polypropylene knitted mesh with polysaccharide hydrogel as compare to gel alone. In a clinical opinion, Ostergard reviewed the degradation, infection and heat effects on polypropylene mesh for pelvic implantation [55]. Plain two-dimensional constructs of polybutylene succinate (PBS) and raw silk fibers were produced through weft knitting process and cell studies using mouse fibroblasts revealed that the material are good surfaces for adhesion and proliferation. On the basis of initial studies these are potential materials for application in skin, bone or cartilage tissue engineering [56].

3.4. Braiding

Braiding is a complex structural pattern formed by interlacing of strands of textile fibres which have locking angle resulting increase in stiffness of the structure. The intertwining pattern of braid can damage in growing tissue by an opening and closing movement [57]. The tubular braiding is being used as prosthetic implants for the replacement of injured ligaments in joints [58]. The prosthetic implant is formed by tubular woven fabric of polyester yarns with braided structure of polyester monofilaments in which braiding and woven structure act as the reinforcing component and sealing component respectively [59]. Biodegradable braided urethral stents have been in clinical use for more than 10 years. Isotalo *et al.* [60] evaluated the biocompatible properties and degradation time of braided stents, in the rabbit urethra, composed of PLA and barium sulphate blends. The results were also compared with metallic stent. All stents were observed with tissue reactions after 1 week of implantation. At the end of 6 months, chronic inflammatory response and fibrosis were seen on metallic stents whereas inflammatory response was abated in biodegradable specimens. Moreover, the degraded stents promoted the development of epithelial hyperplasia. There is a need of improvement for tissue engineering of anterior cruciate ligament (ACL). The healing of ACL is extremely difficult during extreme rupture condition which is due to deficient vascularization at injured tissues. The twisting braiding was design to overcome this problem using poly(L-lactic acid) (PLLA) yarns [61]. The scaffold could mimic mechanical and biomechanical properties similar to ACL. The stress-strain behavior of these grafts is shown in Fig. (**6**). The twisting of braiding fibres significantly improved its tensile strength and stress-strain behavior. Further in the advancement of ligament tissue engineering. Freeman *et al.*, prepared the PLLA fibers yarn and further braided at 69° to form a scaffold to improve the mechanical strength of the scaffold. They further filled the scaffold with PEGDA hydrogel and concluded that 10% of the PEGDA hydrogel improved the mechanical properties as suitable for ACL replacement [62]. Walters *et al.*, fabricated collagen type I fibers and braided further to prepare a scaffold. For the improvement in the mechanical properties of the scaffold the effect of crosslinking with 1-ethyl-3-(3-dimethylaminopropyl) carbodimide (EDC) and addition of gelatin as well to evaluate the cell activity were premeditated. Addition of the gelatin resulted in lowering the mechanical properties while improved cell activity was observed by increasing the cross linking [63]. The selection of scaffold material is very critical to determine the biological response of the ACL graft. The ideal material should have degradation profile, mechanical properties, biomechanical profile similar to the ACL [64, 65]. The surface chemistry of material determines its biological behavior. The pre-coating of fibronectin (Fn) on materials used for ACL tissue engineering was found to be improved biological response of scaffolds. The idea of using Fn coating is came from the fact that increase production of Fn during ACL healing [65].

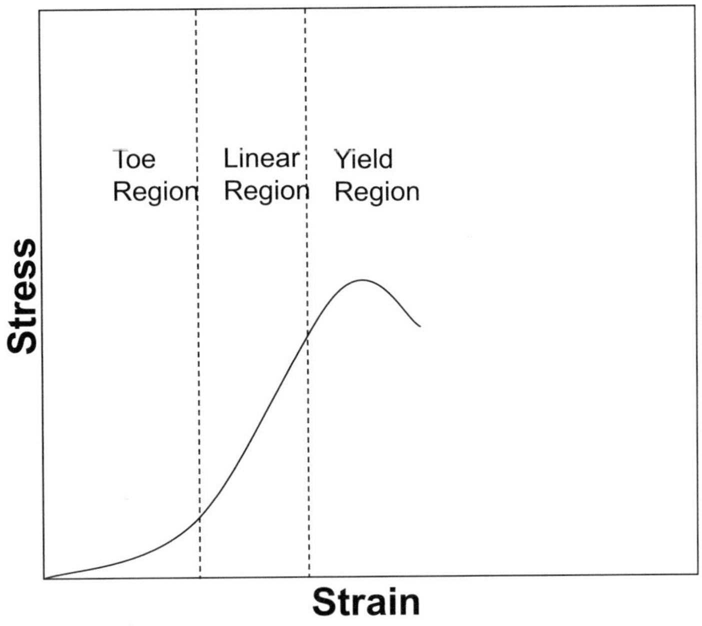

Figure 6: Stress-strain behavior of ligament. The graph displays the 3-staged behavior of ligament (toe region, linear region, and yield region) [61].

3.5. Composite Materials

A textile composite material composed of impregnated matrix, polymeric, ceramic or metallic with textile structure such as a woven, braided or knitted fabric. The reinforcement of textile structures to generate composite materials is generally done by injection molding, liquid molding and film infusion techniques [66-68]. For example, bone plates can be designed by using braided fabrics and ceramic based textile composites [67]. The braided composite materials for dental post were also formed by pultrusion process using carbon fibres as core and glass for sheathing. This unidirectional carbon/epoxy composite post had high stiffness which can avoid risk of root fracture [69]. A knitted polyester reinforced with collagen type I based hybrid composite material had been developed for vascular prostheses. The composite was mixed with a cold solution of bovine smooth muscle cells to control the diameter of tubular scaffold. This hybrid structure showed improved mechanical strength and compliance with native arteries [70, 71]. A 3-D composite scaffold for cartilage tissue engineering were also designed using woven structure reinforced with chrondocyte-hydrogel mixture [69, 72].

Fujihara *et al.* [73] have proposed composite compression bone plates using a braid preform. The change in the thickness of plates and angle of braiding influenced the behavior of the bone plates. The composites plates having thickness of 2.6 mm and 3.6 mm were found to be acceptable for the treatment of forearm and tibia fixation respectively. The static and fatigue bending properties of braided laminated composites were also discussed by Fujihara *et al.* [74] used compression bone plate to repair diapheseal fracture of a bone. From micro-structural viewpoint of braided fabric reinforcement, three different braiding angles (15º, 20º, 25º) were studied. The bending property of materials was found to be better with small braiding angle whereas materials with high braiding angle gave better fracture resistance. Composite living fibers using sodium alginate, gelatin and methacrylic anhydride were prepared and assembled using regular textile processing including weaving, knitting, braiding, winding and embroidering to form a cell encumbered structure and the study concluded by declaring the feasibility of these processes for tailoring functional 3D tissue construct [75].

3.6. Nanofibrous Scaffolds

The most of tissue engineering using biotextile materials is shifting towards electrospinning. The formation of nanofibreous scaffolds using electrospinning could mimic the natural structure of extracellular matrix (ECM). The electrospinning technique is so easy and useful that one can prepare nanofibrous mat virtually from all polymer solutions whether it's aqueous or organic solvents. The continuous structure of nanofibrous mat allows favorable cell-cell and cell-matrix communications. The advantages of a nanofibrous scaffolds are its high surface area, uniform distribution of pores with continuous connectivity and its structural similarity with natural ECM [76]. The differentiation of mesenchymal stem cells towards osteogenic lineage were also promoted by these nanofibrous scaffolds [77-80]. The continuous porous structure of mat also helped their amoeboid movement and transfer of nutrients. A number of other techniques apart from electrospinning were also used to prepare nanofibrous scaffolds such as phase separation and self-assembly with unique properties of fibrous mat [81-83].

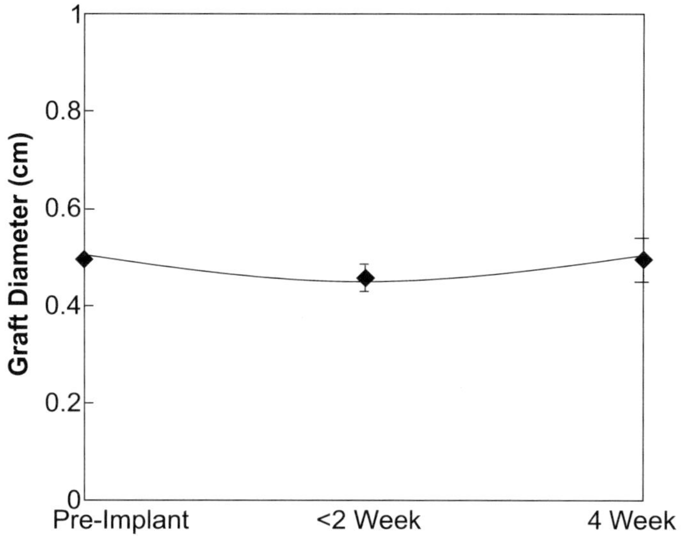

Figure 7: A summary of B-mode ultrasound measurements revealed a stable lumen caliber for the duration of the 1 month study (D) [92].

Although almost every polymer solution can be used for electrospinning, still most of the research was done on the biodegradable poly(α-hydroxyester) polymer family. The common application of these scaffolds is orthopaedic implants [84]. The common examples of this family are poly(ε-caprolactone) and poly(L-lactic acid) which were electrospun for tissue engineering of cartilage and bone tissues. There is another class of bioresorbable polymers known as polyphosphazenes which is being used to prepare nanofibrous scaffolds for various tissue engineering applications. Polyphosphagenes possess excellent biocompatibility and neutral degraded byproducts which make them unique candidate for tissue engineering applications [85-87]. Moreover, various reports in literature indicate that encapsulation improves osteoconductivity and osteointegration [82, 88]. The coating of proteins on polymeric nanofibrous scaffolds were done to improve their cell attachment and growth. The coating of branched and linear peptide amphiphiles (PA) molecules on nanofibrous PGA scaffolds was found to be better as compared to RGDS coating towards bladder smooth muscle cells. The results clearly showed that PA coatings were excellent candidate for bladder patch or implants [89]. Golyglycolic acid scaffolds were prepared and coated with different concentration of poly-L-lactic acid. Equine fibroblast were seeded on the coated and uncoated PGA scaffold to determine the cellularity, cell distribution and extracellular matrix formation of equine fibroblast -like synovium-based meniscal tissue engineering [90]. Tillman *et al.* [91] studied the endothelial cells growth on collagen coated nanofibrous PCL scaffolds and measure patency rate in a rabbit aortoiliac bypass model under physiological conditions. The collagen coating improved adherence and growth of cells and the adherence of platelets abated when endothelialized graft exposed to blood. The scaffold also maintained its structural integrity measured by serial ultrasonography during *in vivo* studies (Fig. **7**). "Further, at retrieval, these scaffolds continued to maintain biomechanical strength that was comparable to native artery (Fig. **8**). This study suggests that electrospun scaffolds combined with vascular cells may become an alternative to prosthetic vascular grafts for vascular reconstruction" [92].

The nanofibreous biotextiles composed of collagen and/or elastin were prepared by electrospinning of respective aqueous solutions. The morphology of the obtained meshes was determined by flow rate, applied electric field, collecting distance and composition of the starting solutions. The diameters of fibres in meshes were increased from 220 to 600 nm with increasing the elastin content.

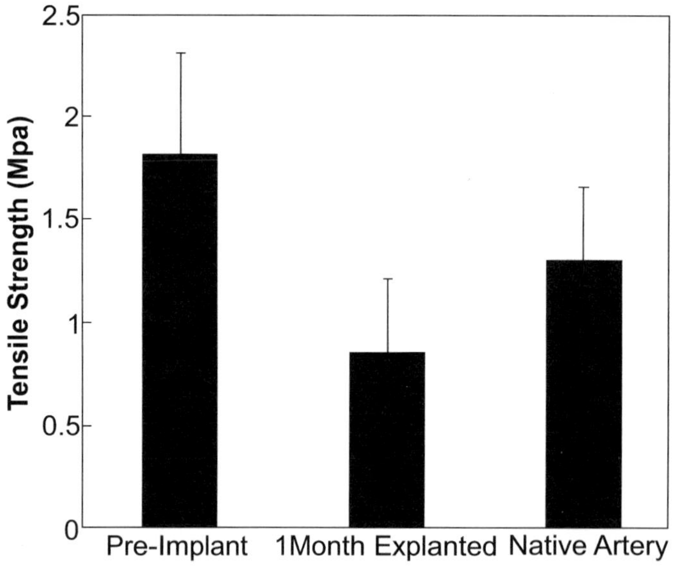

Figure 8: PCL/collagen scaffolds demonstrate comparable biomechanical properties to native aorta at 1 month after implant. Circumferential tensile strength was compared between pre-implant electrospun, 1 month explanted electrospun scaffold and native aortic rings (n = 4) [92].

Collagen/elastin (1:1) meshes were stabilized by crosslinking with *N*-(3-dimethylaminopropyl)-*N*-ethylcarbodiimide hydrochloride (EDC) and *N*-hydroxysuccinimide (NHS). The crosslinking of meshes improved its thermal stability (T_d = 79ºC) without altering their original morphology. Smooth muscle cells grew as a confluent layer on top of the crosslinked meshes after 14 days of culture [92]. Due to its complex neurochemical and neuroanatomical architecture, the nervous system presents a challenge to the field of tissue engineering. However, with the advanced research going on in the field of biomaterial engineering using electrospinning at a molecular level; tissue engineering in the nervous system seems to be possible [93]. Nanofibrous scaffolds diameter also plays a vital role in cell growth and differentiation. The effect of diameter of laminin coated polyethersulfone (PES) naofibres on rat hippocampus-derived adult neral stem cells (NSCs) culture was investigated by Christopherson *et al.* [94]. When cultured under differentiation conditions, rNSCs preferred differentiation into oligodendrocytes and neurons on 283 nm and 749 nm fibres respectively as compared to tissue culture polystyrene surface. The higher degree of spreading and proliferation was observed with decrease in fiber diameter. Biodegradable materials were used for the preparation of the nanofibres by simple freeze-drying method. In this study, different concentrations of the glycogen were used for the preparation of nanofibrous scaffold to assess the structure and porosity of the scaffold. Fibers were further grafted with biodegradable poly(ethyl cyanoacrylate) to make the scaffold water insoluble and scaffolds hydrolytical degradability followed a course of over several months [95]. Novel fibrous gel scaffolds of poly(aspartic acid) were prepared by reactive electrospinning for artificial extracellular matrix [96]. Non-woven scaffolds of polyvinylidene fluoride (PVDF) with different cross section and patterns were prepared to compare the growth pattern of human mesenchymal stromal cells (MSCs) on these scaffolds. Cell adhesion and proliferation was found significantly higher on non-wovens with round fibers than on those with snowflake or trilobal cross sections [97].

CONCLUSION

Applications of textile materials for the replacement of tissue, nerve guideline, arteries and bones as an implant hold a promising future. Due to major developments in biotextile production in the fields of materials as well as technologies in the recent past, these materials find applications in repairing the body whether it is wound closure (sutures) or replacement surgery (vascular grafts, artificial ligaments, nerve grafts, skin grafts *etc*). As transplantation is always not possible to replace defective part of the body, therefore, physicians use an artificial substitute (biomaterials) to transplantation such as biotextiles.

However, there is still much to contribute to the development of scaffold materials for tissue engineering. There is a critical need to develop biocompatible biotextile scaffolds with an ability to interact with cells and extracellular matrix components so as to get desirable cell function and generate functional tissues.

ACKNOWLEDGMENTS

Declared none.

CONFLICT OF INTEREST

The authors confirm that this chapter content has no conflict of interest.

REFERENCES

[1] Butler N, Europe's polyolefin textiles industry is upbeat in Budapest, Technical Textile International. July/August 2004; pp. 21.
[2] Williams DF, Eds. Blood compatibility. Vol 1, CRC Press 1987.
[3] King MW. Designing fabrics for blood vessel replacement. Can Tex J 1991; 108: 24-30.
[4] 2002 Annual report of the US Scientific Registery of Transplant Recipients and the Organ Procurement and Transplantation Network: Transplant Data 1992-2001. Rockville MD, Richmond VA. HHS/HRSA/OSP/DOT and UNOS 2001.
[5] Teo WE, He W, Ramakrishna S. Electrospun scaffold tailored for tissue-specific extracellular matrix. Biotechnol J 2006; 1: 918-29.
[6] Ramakrishna S. Textile scaffolds in tissue engineering in Smart Fibres, Fabrics and Clothing. Tao and Xiaoming, Eds. Wood Head Publishing 2001; pp. 291.
[7] Anand SC, Shah T, Eds. Recent advances in technical textiles. International Conference on Textile and Clothing; 2006; pp. 1, Karachi. Pakistan
[8] Zeltinger J, Sherwood JK, Graham DA, Müeller R, Griffith LG. Effect of pore size and void fraction on cellular adhesion, proliferation, and matrix deposition. Tiss Engg 2001; 7: 557-72.
[9] Lee JJ, Lee SG, Park JC, Yang YI, Kim JK. Investigation on biodegradable PLGA scaffold with various pore size structure for skin tissue engineering.Curr Appl Phys 2007; 7(S1): e37-40.
[10] Evans GRD, Brandt K, Widmer MS, *et al. In vivo* evaluation of poly(L-lactic acid) porous conduits for peripheral nerve regeneration. Biomaterials 1999; 20: 1109-15.
[11] Lee JH, Jung HW, Kang IK, Lee HB. Cell behaviour on polymer surfaces with different functional groups. Biomaterials 1994; 15: 705-11.
[12] Byon BD, Hummert TW, Dean DD, Schwartz Z. Role of material surfaces in regulating bone and cartilage cell response. Biomaterials 1996; 17: 137-46.
[13] Van Wachem PB, Beugling T, Feijen J, Bantjes A, Detmers JP, Van Aken WG. Interaction of cultured human endothelial cells with polymeric surfaces of different wettabilities. Biomaterials 1985; 6: 403-8.
[14] Kamath S, Bhattacharyya D, Padukudru C, Timmons RB, Tang LJ. Surface chemistry influences implant-mediated host tissue responses. Biomed Mater Res 2008; 86A: 617-26.
[15] Chang DT, Jones JA, Meyerson H, *et al.* Lymphocyte/macrophage interactions: biomaterial surface-dependent cytokine, chemokine, and matrix protein production. J Biomed Mater Res 2008; 87A: 676-87.
[16] Hsu S, Kuo CC, Yen HJ, Whu SW, Tsai CL. The effect of two different bioreactors on the neocartilage formation in type II collagen modified polyester scaffolds seeded with chondrocytes. Artif. Organs 2005; 29: 467-74.
[17] Yang XB, Bhatnagar RS, Li S, Oreffo ROC. Biomimetic collagen scaffolds for human bone cell growth and differentiationTiss Engg 2004; 10: 1148-59.

[18] Anand SC. Implantable devices: an overview. In: Anand SC, Kennedy JF, Miraftab M, Rajendran S. Proceedings of MEDTEX03 International Conference and Exhibition on Healthcare and Medical Textiles; Woodhead Publishing Limited; England 2006; pp. 329.
[19] Sarkar S, Chourasia A, Maji S, Sadhukhan S, Kumar S, Adhikari B. Synthesis and characterization of gelatin based polyester urethane scaffold Bull Mater Sci 2006; 29: 475-84.
[20] Sumanasinghe RD, King MW. New trends in the biotextiles - the challenges of tissue engineering. JTATM 2003; 3: 1.
[21] Tissue-engineered vascularised microvessels. In: Materials Today, Research News: Nov 2003; pp. 20.
[22] Couet F, Rajan N, Mantovani D. Macromolecular biomaterials for scaffold-based vascular tissue engineering. Macromol Biosci 2007; 7: 701-18.
[23] Ungaro F, Biondi M, Indolfi L, *et al*. Bioactivated polymer scaffolds for tissue engineering in Topics in Tissue Engineering. Eds. Ashammakhi N, Reis RL. 2005; 2: 1.
[24] Vacanti JP, Langer R. Tissue engineering: the design and fabrication of living replacement devices for surgical reconstruction and transplantation. Lancet 1999; 354: s132-34.
[25] Langer R. Selected advances in drug delivery and tissue engineering. J Cont Rel 1999; 62: 7-11.
[26] Langer R. Biomaterials and biomedical engineering.Chem Engg Sci 1995; 50: 4109-21.
[27] Yeabsley, Karl Mayer Textile Machinery Co Ltd; Uk, Private Communication. April 2005.
[28] Fries T, Ramakers UW, Schedukat N. Tech Text 2002; 45: E128.
[29] Dauner M. Wound Care, In: Textile Asia. Oct 1998; pp. 46.
[30] Wu W, Feng X, Mao T, *et al*. Engineering of human tracheal tissue with collagen-enforced poly-lactic-glycolic acid non-woven mesh: a preliminary study in nude mice. Br J Oral Maxillofac Surg 2007; 45: 272-78.
[31] Grayin J, Qi J, Maloney M, Banes AJ. Novel system for *engineering* bioartificial tendons and application of mechanical load. Tiss Engg 2003; 9: 967-79.
[32] Wu X, Aikawa ER, Guleserian KJ, *et al*. Tissue-engineered microvessels on three-dimensional biodegradable scaffolds using human endothelial progenitor cells. Am J Physiol Heart Circ Physiol 2004; 287: H480-87.
[33] Zong X, Ran S, Fang D, Hsiao BS, Chu B. Structure and morphology changes during *in vitro* degradation of electrospun poly(glycolide-co-lactide) nanofiber membrane. Polymer 2003; 4(2): 416-23.
[34] Kim K, Yu M, Zong X, *et al*. Control of degradation rate and hydrophilicity in electrospun non-woven poly(D,L-lactide) nanofiber scaffolds for biomedical applications. Biomaterials 2003; 24: 4977-85.
[35] Edwards SL, Mitchell W, Matthews JB, Ingham E, Russell SJ. Design of nonwoven scaffold structures for tissue engineering of the anterior cruciate ligament. In: Anand SC, Kennedy JF, Miraftab M, Rajendran S. Proceedings of MEDTEX03 International Conference and Exhibition on Healthcare and Medical Textiles; Woodhead Publishing Limited; England 2006; pp. 355.
[36] Freed LE, Novakovic GV, Biron RJ, Dana BE, Lesnoy DC, Barlow SK. Biodegradable polymer scaffolds for tissue engineering. Biotech 1994; 12: 689.
[37] Xu X, Chen X, Liu A, Hong Z, Xing X. Electrospun poly(L-lactide)-grafted hydroxyapatite/poly(L-lactide) nanocomposite fibers. Euro Polym J 2007; 43: 3187-96.
[38] Jeong SI, Kim SY, Cho SK, *et al*. Tissue-engineered vascular grafts composed of marine collagen and PLGA fibers using pulsatile perfusion bioreactors. Biomaterials 2007; 28: 1115-22.
[39] He N, Ke Q, Huang C, Yang J, Guo Y. Needle-punched nonwoven matrix from regenerated collagen fiber for cartilage tissue engineering. J Appl Polym Sci 2014; 131(13): DOI: 10.1002/app.40404
[40] Vaz CM, Tuijl SV, Bouten CVC, Baaijens FPT. Design of scaffolds for blood vessel tissue engineering using a multi-layering electrospinning technique,Acta Biomat 2005; 1: 575-82.
[41] Chandra P. Blood Compatible Materials and Devices. Technomic Publishing Company, Inc; Lancaster 1991, Pennsylvania.
[42] Risbud MV, Karamuk E, Mayer J. Designing hydrogel coated textile scaffolds for tissue engineering: Effect of casting conditions and degradation behavior studied at microstructure level. J. Mater Sci Lett 2002; 21: 1191-94.
[43] Karamuk E, Mayer J, Wintermantel F, Akaike T. Partially degradable film/fabric composites: textile scaffolds for liver cell culture. Artif. Organs 1999; 23: 881-4.
[44] Blanchemain N, Haulon S, Martel B, Traisnel M, Morcellet M, Hildebrand HF. Vascular PET prostheses surface modification with cyclodextrin coating: development of a new drug delivery system. Eur J Endovasc Surg 2005; 29: 628-32.

[45] Tofigh AM, Wailly DWG, Rhissassi B. Comparing vein with collagen impregnated woven polyester prosthesis in above-knee femoropopliteal bypass grafting. Internat J Surg 2007; 5: 109-13.

[46] Gupta BS, Kasyanov VA. Biomechanics of human common carotid artery and design of novel hybrid textile compliant vascular grafts. J. Biomed Mater Res 1997; 34: 341-9.

[47] Moutos FT, Estes BT, Guilak F. Multifunctional hybrid three-dimensionally woven scaffolds for cartilage tissue engineering. Macromol Biosci 2010; 10: 1355-64.

[48] Wintermantel E, Mayer J, Blum J, Eckert KJ, Luscher P, Mathey M. Tissue engineering scaffolds using superstructures. Bomaterials 1996; 17: 83-91.

[49] Wang X, Han C, Hu X, *et al.* Applications of knitted mesh fabrication techniques to scaffolds for tissue engineering and regenerative medicine. J Mech Behav Biomed Mater 2011; 4: 922-32.

[50] Gupta B, Revagade N, Atthoff B, Hilborn. Radiation-induced graft modification of knitted poly(ethylene terephthalate) fabric for collagen immobilization J. Polym Adv Technol 2007; 18: 281-85.

[51] Woei Ng K, Louis J, Tuan Ho BS, *et al.* Characterization of a novel bioactive poly[(lactic acid)-co-(glycolic acid)] and collagen hybrid matrix for dermal regeneration. Polym Int 2005; 54: 1449-57.

[52] Zhang L, Ao Q, Wang A, *et al.* A sandwich tubular scaffold derived from chitosan for blood vessel tissue engineering. J Biomed Mater Res 2006; 77A: 277-84.

[53] Nakanishi Y, Chen G, Komuro H, *et al.* Tissue-engineered urinary bladder wall using PLGA mesh-collagen hybrid scaffolds: a comparison study of collagen sponge and gel as a scaffold. J Padiat Surg 2003; 38: 1781-84.

[54] Abed A, Deval B, Assoul N, *et al.* A biocompatible polysaccharide hydrogel-embedded polypropylene mesh for enhanced tissue integration in rats. Tiss Eng A 2008; 14(4): 519-27.

[55] Ostergard DR, Degradation, infection and heat effects on polypropylene mesh for pelvic implantation: what was known and when it was known. Int Urogynecol J 2011; 22: 771-74.

[56] Almeida LR, Martins AR, Fernandes EM, *et al.* New biotextiles for tissue engineering: development, characterization and *in vitro* cellular viability. Acta Biomater 2013; 9: 8167-81.

[57] Horan RL, Collette AL, Lee C, Antle K, Chen J, Altman GH. Yarn design for functional tissue engineering. J Biomech 2006; 39: 2232-40.

[58] Irsale S, Adanur S. Design and characterization of polymeric stents. J Indus Text 2006; 35: 189-98.

[59] Ge Z, Yang F, Goh JCH, Ramakrishna S, Lee EH. Biomaterials and scaffolds for ligament tissue engineering J Biomed Mater Res 2006; 77A: 639-52.

[60] Isotalo T, Nuutinen JP, Vaajanen A, *et al.* Biocompatibility and implantation properties of 2 differently braided, biodegradable, self-reinforced polylactic acid urethral stents: an experimental study in the rabbit. J Urology 2005; 174: 2401-4.

[61] Freeman JW, Woods MD, Laurencin CT. Tissue engineering of the anterior cruciate ligament using a braid-twist scaffold design. J Biomech 2007; 40: 2029-36.

[62] Freeman JW, Woods MD, Cromer DA, *et al.* Evaluation of a hydrogel-fiber composite for ACL tissue engineering. J Biomech. J Biomech *2011;* 44: 694-9.

[63] Walters VI, Kwansa AL, Freeman JW. Design and analysis of braid-twist collagen scaffolds. Connect Tiss Res 2012; 53: 255-66.

[64] Lu HH, Cooper Jr. JA, Manuel S, *et al.* Anterior cruciate ligament regeneration using braided biodegradable scaffolds: *in vitro* optimization studies. Biomaterials 2005; 26: 4805-16.

[65] Cooper JA, Lu HH, Ko FK, Freeman JW, Laurencin CT. Fiber-based tissue-engineered scaffold for ligament replacement: design considerations and *in vitro* evaluation. Biomaterials 2005; 26: 1523-32.

[66] Peltola T, Aaritalo V, Haltia AM, *et al.* Urpo. Manufacture and *in vitro* bioactivity of sol-gel derived silica fibre and P(L/DL)LA composite. In: Anand SC, Kennedy JF, Miraftab M, Rajendran S. Proceedings of MEDTEX03 International Conference and Exhibition on Healthcare and Medical Textiles. Woodhead Publishing Limited; England 2006; pp. 329.

[67] Veerabagu S, Fujihara K, Dasari GR, Ramakrishna S. Strain distribution analysis of braided composite bone plates. Comp Sci Tech 2003; 62: 427-35.

[68] Tan P, Tong L, Steven GP. Modelling for predicting the mechanical properties of textile composites-A review. In: Niihara K, Nakano K, Sekino T, Yasuda E. (Eds): High Temperature Ceramic Matrix Composites II. Composites Part A 1997; pp. 903.

[69] Fujihara K, Teo K, Gopal R, *et al.* Fibrous composite materials in dentistry and orthopaedics: review and applications. Comp Sci Tech 2004; 64: 775-88.

[70] Kobashi T, Matsuda T. Fabrication of branched hybrid vascular prostheses. Tiss Eng 1999; 5: 515-24.

[71] Huang ZM, Zhang Y, Ramakrishna S. Modeling of the progressive failure behavior of multilayer knitted fabric-reinforced composite laminates. Comp Sci Tech 2001; 61: 2033-46.

[72] Moutos FT, Freed LE, Guilak F. A biomimetic three-dimensional woven composite scaffold for functional tissue engineering of cartilage. Nat Mat 2007; 6: 162-7.

[73] Fujihara K, Huang ZM, Ramakrishna S, Satknanantham K, Hamada H. Performance study of braided carbon/PEEK composite compression bone plates. Biomaterials 2003; 24: 2661-67.

[74] Fujihara K, Yoshida E, Nakai A, Ramakrishna S, Hamada H. Influence of micro-structures on bending properties of braided laminated composites. Comp Sci Tech 2007; 67: 2191-98.

[75] Akbari M, Tamayol A, Laforte V, *et al*. Composite Living Fibers for Creating Tissue Constructs Using Textile Techniques. Adv Function Mater 2014; 24: 4060-67.

[76] Li WJ, Laurencin CT, Caterson CJ, *et al*. Electrospun nanofibrous structure: a novel scaffold for tissue engineering. J Biomed Mater Res 2002; 60: 613-21.

[77] Li WJ, Tuli R, Huang X, Laquerriere P, Tuan RS. Multilineage differentiation of human mesenchymal stem cells in a three-dimensional nanofibrous scaffold. Biomaterials 2005; 26: 5158-66.

[78] Shin M, Yoshimoto H, Vacanti JP. *In vivo* Bone Tissue Engineering Using Mesenchymal Stem Cells on a Novel Electrospun Nanofibrous Scaffold. Tissue Eng 2004; 10: 33.

[79] Smith LA, Ma PX. Nano-fibrous scaffolds for tissue engineering. Coll Surf B Biointerf 2004; 39: 125-31.

[80] Whitesides GM, Boncheva M. Beyond molecules: Self-assembly of mesoscopic and macroscopic components Proc Natl Acad Sci USA; 2002; 99; pp. 4769-74.

[81] Bhattari N, Edmondson D, Veiseh O, Matsen FA, Zhang M. Electrospun chitosan-based nanofibers and their cellular compatibility. Biomaterials 2005; 26: 6176-84.

[82] Christenson EM, Anseth KS, Van den Beucken JJJP, *et al*. Nanobiomaterial applications in orthopedics. J Orthop Res 2007; 25: 11-22.

[83] Bao X, Teramoto A, Abe Key K. Carboxymethyl chitosan nonwoven scaffold for bone regeneration. Eng Mater 2011; 464: 712-16.

[84] Li WJ, Cooper Jr. JA, Mauk RL, Tuan RS. Fabrication and characterization of six electrospun poly (α-hydroxy ester)-based fibrous scaffolds for tissue engineering applications. Acta Biomater 2006; 2: 377-85.

[85] Nair LS, Bhattacharya S, Bender JD, *et al*. Fabrication and optimization of methylphenoxy substituted polyphosphazene nanofibers for biomedical applications. Biomacromolecules 2004; 5: 2212-20.

[86] Laurencin CT, Nair LS. NATO-ASI Proceedings; Boston/Dordrecht: Kluwer 2004; pp. 281.

[87] Bhattacharya S, Lakshmi S, Bender JD, *et al*. Preparation of poly[bis(carbonylato phenoxy)phosphazene] non-woven nanofiber mats by electrospinning. MRS Fall Meeting Proceedings; 2003; F8: 10.

[88] Bhattacharya S, Lakshmi S, Nair LS, *et al*. Development of biodegradable polyphosphazene-nanohydroxyapatite composite nanofibers *via* electrospinning. Materials Research Society Symposium Proceedings; 2005; 845: 91-96.

[89] Harrington DA, Cheng EY, Guler MO, *et al*. Branched peptide-amphiphiles as self-assembling coatings for tissue engineering scaffolds. J Biomed Mater Res 2006; 78A: 157-67.

[90] Warnock JJ, Fox DB, Stoker AM, *et al*. Culture of equine fibroblast-like synoviocytes on synthetic tissue scaffolds towards meniscal tissue engineering: a preliminary cell-seeding study. 2014; http://dx.doi.org/10.7717/peerj.353

[91] Tillman BW, Yazdani SK, Lee SJ, Geary RL, Atala A, Yoo JJ. The *in vivo* stability of electrospun polycaprolactone-collagen scaffolds in vascular reconstruction. Biomaterials 2009; 30: 583-8.

[92] Buttafoco L, Kolkman NG, Engbers-Buijtenhuijs P, *et al*. Electrospinning of collagen and elastin for tissue engineering applications. Biomaterials 2006; 27: 724-34.

[93] Bellamkonda R, Aebischer P. Review: Tissue engineering in the nervous system. Biotech Bioeng 1994; 43: 543-54.

[94] Christopherson GT, Song H, Mao HQ. The influence of fiber diameter of electrospun substrates on neural stem cell differentiation and proliferation. Biomaterials 2009; 30: 556-64.

[95] Vetrik M, Pradny M, Kobera L, *et al*. Biopolymer-based degradable nanofibres from renewable resources produced by freeze-drying. RSC Adv 2013; 3: 15282-89.

[96] Molnar K, Juriga D, Nagy PM, Sinko K, Jedlovszky-Hajdu A, Zrinyi M. Electrospun poly(aspartic acid) gel scaffolds for artificial extracellular matrix. Polym International 2014; 63: 1608-15.

[97] Schellenberg A, Ross R, Abagnale G, *et al*. 3D Non-Woven Polyvinylidene Fluoride Scaffolds: Fibre Cross Section and Texturizing Patterns Have Impact on Growth of Mesenchymal Stromal Cells. PLOS ONE 2014; 9(4): e94353-62.

CHAPTER 7

Hybrid Semi-Synthetic Polyhydroxy Acid Based-Biomaterials

Swati Jain[1], Sruti Chattopadhyay[2] and Harpal Singh[2,*]

[1]Amity Institute of Nanotechnology, Amity University, Sec. 125, Noida, Uttar Pradesh, India and [2]Indian Institute of Technology Delhi (IIT D), Hauz Khas, New Delhi, India

Abstract: Development of hydrolytically or enzymatically labile bonds is an on-going process in connection to biomaterials. Synthetic biodegradable polymers such as polyhydroxy acids (PHAs), which include homo as well as copolymers of polylactic acid (PLA) and polyglycolic acid (PGA), are inexplicably used in drug delivery and tissue engineering applications. Adjusting the ratio of constituting architecture of block copolymers to manipulate amphiphilic behavior and chemical as well as physical properties is the prime methodology adopted for tailoring PHAs suiting specific requirements. It is critical to develop three dimensional porous scaffolds with enhanced cell adhesion characteristics for improved tissue regeneration while 'modified' release of payload is essential in novel targeted drug delivery systems. Combinatorial polymeric architectures composed of synthetic and naturally derived components aid in attaining these pre-requisites by providing surface functionalities, optimal hydrophobic-hydrophilic ratio and flexibility of support. Synthesis of 'tailored' natural and synthetic biodegradable copolymers is the thrust research area making use of advantageous properties of both individual components for creating a unique structural analogy mimicking biological world. Biomaterials created solely from natural polymers such as polysaccharides and proteins find many uses as drug delivery carriers and 'grafts'. However, they are mechanically weak, their integration with PLA and PGA considerably improve mechanical strength while retaining higher cell adhesion capabilities.

The present chapter details combinatorial approaches for the generation of large arrays of biodegradable polymeric materials entailing multitudes of components to meet their desired properties. Novel synthetic methods generating varied hybrid polymeric products like sponges, meshes, microspheres and nanoparticles are discussed.

Keywords: Alginate, biomaterials, blending, bone growth, cell differentiation, chemical attachment, chitosan, drug delivery, grafting, hyaluronic acid, hybrids, hydrolytic cleavage, nanocarriers, natural polymers, Poly(lactic-co-glycolic) acid, Polyhydroxy acids, polylactic acid, scaffold, surface modification, tissue engineering.

[*]**Corresponding author Harpal Singh:** Room No. 297, Block III, Center for Biomedical Engineering (CBME), Indian Institute of Technology Delhi (IIT D), Hauz Khas, New Delhi-110016, India; Tel: (o): 01126591149; E-mail: harpal2000@yahoo.com

Viness Pillay, Yahya E. Choonara and Pradeep Kumar (Eds.)

1. INTRODUCTION

Biomaterials are critical enabling technology for virtually every biomedical application. Socio-economic developments are endlessly driving considerable scientific and engineering explorations in building medical resources. Specialized materials constituted by metals, ceramics, polymers and combinations of either of them, called as composites are intended for either extracorporeal uses or for permanent or temporary implantations within any body [1, 2]. Polymeric biomaterials have always dominated academician's interests owing to their extraordinary versatility, phenomenal functionalities conferring multidimensional and widespread uses. An array of homo and co-polymeric systems already exist in market, solving problems for generic as well as specialized or customized requirements in biomedical industry [1-4].

Attempts to find innovative solutions for tissue engineering, drug delivery, clinical surgical accessories and implants have consecrated utmost emphasis on biodegradable polymers. These polymers on hydrolytic or enzymatic breakdown are excreted or resorbed from the body without any surgical intervention [5-7]. Various synthetic biodegradable polymers containing hydrolytic susceptible bonds including esters, anhydrides, acetals, carbonates, amides, urethanes, phosphates and hydroxyalkanoates have garnered much attention in recent years. Primarily these families of polymers, differ in their mechanism of erosion including degradation kinetics, rate of action and finally on residues on completion of reactions. Amongst this group, aliphatic polyesters are most studied and researched biodegradable polymers to date as the backbone containing small highly labile esters bonds are bulk erodible, working in the time frame theorized for biomedical applications. The major categories of polymers belonging to this group are polyhydrocarboxylic acids abbreviated as PHAs and their corresponding ester derivatives [8]. PHAs have garnered much attention for the past many years by medical industry ensuing large production of raw materials as well as finished devices.

Omnipresent apprehension exists today regarding preservation of our ecological systems since all forms of pollution have deteriorated our environment with increasing industrialization growing at an alarming rate. This stems from the fact that our reliance on polymers is unmatched and continuously rising. Most of today's synthetic polymers are produced from non-renewable and non-biodegradable petroleum resources having major impact on world resources crisis and plastic litter. Biodegradable polymers are ecologically beneficial as many of them are derived from natural matter [9]. These can be produced by varieties of ecologically friendly bacteria using several waste feed stocks. Fermentation broth is subjected to various generic processes of fermentation isolation, and purification generating renewable matter for polymer production [10].

These polymers form physiologically tolerable non-toxic products on decomposition minimizing interventional revisionary surgeries, controlling the rate of drug release and engineering allograft, cartilage and bone regeneration processes.

However, large absorption of extracellular proteins on PHA based implantable biomaterials is the limiting factor [11]. In order to better address many issues pertaining to medical polymeric materials there is a paradigm shift in aligning degradation parameters according to the requirements of applications. Continual advancements have invoked physical and chemical modifications of PHA and development of semi-synthetic products have taken the center stage.

A whole milieu of natural polymers is integrated with different PHAs creating novel architectures suitably providing desirable unique characteristics specifically pertaining to advancing requirements of biomedical engineering. Natural polymers working as base polymeric materials are also integrated with biodegradable PHAs to improve their mechanical strength and reduce cytotoxicity issues.

We outline the underlying concepts and ideas for preparing semi-synthetic composite materials composed of naturally derived and synthetic polymers in this chapter. Strategic approaches and technologies for developing hybrid biomaterials developed in the last 5-10 years are emphasized in response to their eventual applications. A comprehensive report is formulated to seek future developmental directions in-lieu of progress achieved so far.

2. POLYHYDROXY ACIDS (PHAs)

Polyhydroxy acids are a class of poly(α-ester) polymers that contain aliphatic ester bonds in their backbone. These are mostly derived from α-carboxylic acid such as glycolic, lactic or malic acids or cyclic dimmers and their copolymers such as lactide, glycolide, trimethylene carbonate, ε-caprolactone, δ-valerolactone. Polyesters of β, γ, δ and ε (or 3, 4, 5 and 6) hydroxy acids are clubbed as polyhydroxy alkanoates also abbreviated as PHAs and are categorized on the basis of length of carbon chain as well as the positions of hydroxyl and carboxylic groups. These naturally available biodegradable polymers are seen as an alternative to fossil fuel based products for sustainable future [8, 12]. Amongst this group, polyhydroxy butyric acid (PHB) is used for making scaffolds while another semi-crystalline polyester polycaprolactone

(PCL) shows favorable characteristics as long term implant delivery vehicle. Homopolymer PHB is mechanically weak and is copolymerized with another alkanoate called as polyhydroxy valerate (PHV) creating PHBV for preparing stronger sheets and membranes [12]. The resurgence in PCL is propelled owing to superior viscoelastic and rheological properties attained on low cost production procedures. These promising Food and Drug Administration (FDA) approved materials – PHBs and PCL are chemically manipulated to design implants to match degradation pattern of anatomical site by conjugating with macromolecules for varied biomaterials like allografts, osteogenic grafts, nanofibers for retinal tissues *etc.* [13-16]. The detailed analysis of 2, 3 and 4 HAs or hydroxyl alkanoates and PCL is beyond the scope of this chapter which primarily focuses on α-PHAs only.

PHAs degrade by hydrolytic cleavage of ester linkages into low molecular weight (M. wt.) molecules that can be metabolized or cleared by renal filtration. Mechanistically degradation occur by penetration of water molecules into the amorphous area of polymer disrupting Van der Waals forces of attraction between different layers and hydrogen bonds of the material resulting in lowering of glass transition temperature (Tg). In the second step, covalent bonds are cleaved decreasing the M. wt. of the system which continues constantly resulting in formation of large number of carboxylic acid units. These carboxylic terminating groups auto catalyze enormous cleavage of backbone covalent bonds losing substantial mass and integrity of the whole polymer with the formation of smaller molecules. Finally, the small fragments are further broken down yielding aqueous soluble products such as lactic acid, glycolic acid and eventually carbon as well as oxygen species [17]. In addition to their degradation behavior, these polymers can be processed into different shapes and structures without sacrificing high compressive modulus suitable for applications. Various biomaterials developed from PHAs are mentioned in Table **1**.

Table 1. Different biomaterials formed from PHAs and their applications.

Thrust Area	Applications	Role of Polyhydroxy Acids
Tissue engineering	Soft tissue replacement	PHA form engineered scaffold to provide a temporary pathway for the regeneration of damaged tissue. Biodegradable PHA will degrade either during or after healing; thereby obviating the need of their removal from the body eliminating possible associated side effects. Promote healing in 3-dimensions for appropriate angiogenesis
	Blood vessels	
	Kidney membranes	
	Scaffolds and membranous structures	

Table 1: contd...

Skeletal system	Bone plate for fracture fixation and Bone growth	Allografts and autografts help in augmenting and regeneration of bone working as bone graft materials providing necessary cell anchorage sites, cues, pores growth factors and inhibiting fibrous tissue formation
	Artificial tendons and ligaments	
Cardiovascular devices	Vascular prosthesis	Biodegradable materials create the next generation drug eluting re-absorbable shunts, stents and other cardiovascular implants which can contract and expand according to external conditions.
	Heart valves	
	Cardio-pulmonary bypass	
	Arteriovenous shunts	
Artificial organs	Heart, kidney, liver	
Ophthalmology	Intraocular lenses and biodegradable implants	Drug dispensing contact lenses and ophthalmic implants in both anterior and posterior ocular tissues while maintaining intraocular pressure example Ozurdex™ and Surodex™ from (Allergan Inc., Irvine, CA, United States)
Dental implants		Drug containing tooth extraction sockets
Drug delivery systems	Sustained, controlled, multi-pulsated, targeted, multi-component	Polymers work as biodegradable payload carrier for delivering chemical agents and biomolecules in complex system.
Surgical devices	Sutures, films	PHA are resorbed by the body once healing process is done

In the world of advancing medical fields, explorative new concepts are constantly being established. Regenerative medicine is focused on restoration of pathologically altered tissue architecture owing to damage *via* injury or diseases by transplanting cells in combination with supportive membranous structures and biomolecules. High success rate of PHA for scaffold preparation as compared to biostable polymers is due to their properties which can be tailored for the specific application. These features depend on factors such as anatomical location, severity of trauma and age of the patient as well as the presence of other pathological conditions.

PHAs, comprised of homopolymers - polyglycolic acid (PGA) and polylactic acid (PLA) and their copolymer poly(lactic-co-glycolic) acid (PLGA), are obtained by direct thermal polyesterification methods which work on the principle of condensation of units with the removal of stochiometeric water molecules. The method produces short chain low M. wt. about 2,000-10,000 Da polymer [18, 19].

These do not find much biomedical applications owing to weak mechanical strength. Rapid synthesis technologies progress has enabled the production of relatively high M. wt. polymers by polycondensation reaction by utilizing of chain coupling agents and by removal of water in a highly controlled manner. Fig. (**1**) indicates a general scheme for polymerizing PHAs by thermal condensation mechanism.

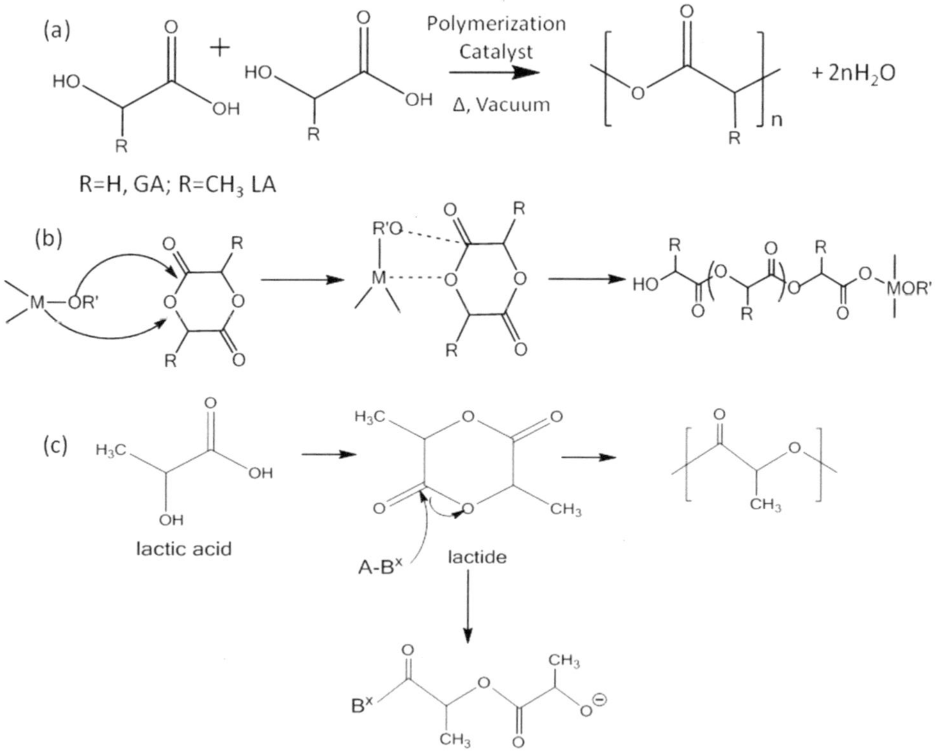

Figure 1: (**a**) Synthesis of PHA by thermal polyesterification mechanism; Formation of polylactide by ring opening polymerization following (**b**) co-ordination insertion mechanism by metal alkoxide (**c**) nucleophilic or electrophilic molecules.

The majority of high M. wt. of the order of 10^5 Mn (number average molecular weight) with substantial high degree of crystallinity and polymerization are formed by ring opening polymerization (ROP) of cyclic monomers which has shorter reaction times, uses milder conditions and does not generate side products as illustrated in Fig. **1** (**b** and **c**). The cyclic di-lactones are prepared from corresponding hydroxyl acids by elimination of water by catalysts. Common catalysts used in ROP include organo tin, antimony or zinc [20]. Many companies produce large quantities of 10^6 Da weight polymers with narrow polydispersities in few hours at 140-220° C commercially in a solvent free environmentally favorable state. Many studies have revealed that the polymerization of lactide is effected by different conditions of reactions including concentration and type of catalyst, monomer purity, and temperature. Fig. (**2**) outlines different chemical approaches for polymerizing PLA to obtain different M wt. products useful for various applications.

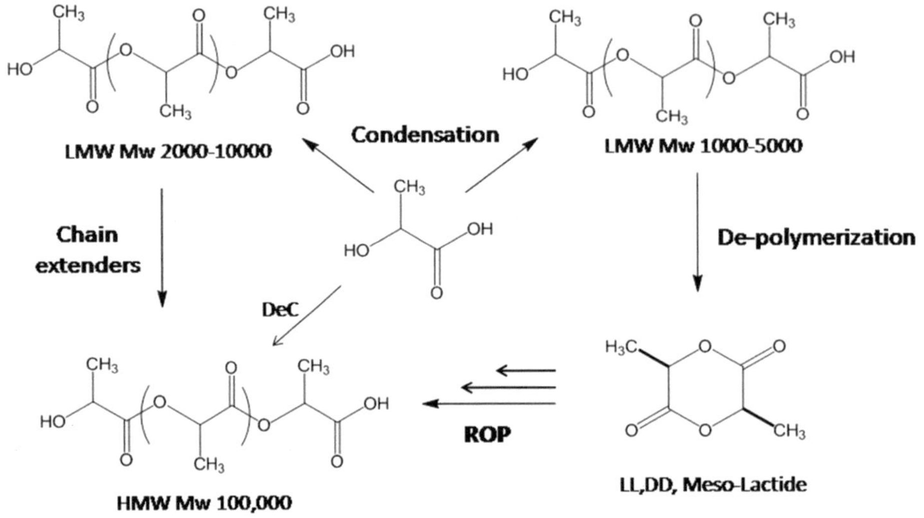

Figure 2: Overview of synthesis of PLA by various methods with different molecular weight (Mw), where L and H denote low and high.

Polymerization catalysts have always been point of focus for chemists in this field. Tin octate and Tin (II)-bis-(2-ethyl hexanoate) are the most famous catalysts for PLA and PGA production respectively. Stannous 2-ethylhexanoate $[CH_3(CH_2)_3CH(C_2H_5)COO]_2Sn$, denoted as $Sn(Oct)_2$, is highly efficient catalyst for cyclic esters which allows complete conversion to final product even at monomer/catalyst ratios of $10^4/1$ [21]. The risk of racemization is low and highly optically pure (99 %) PLLA can be prepared at 150 ^0C, when the reaction time is limited to a few hours. $Sn(Oct)_2$ is soluble in organic solvents and molten lactide; is stable on storage and has been approved by FDA.

2.1. Polygylcolic Acid (PGA), Polylactic Acid (PLA) and their Copolymer

PGA the simplest polyhydroxy acid possessing a moderate degree of crystallinity varying from complete amorphous to a maximum of 52% crystalline state. It is the simplest aliphatic polyester having tight molecular packing and the close approach of ester groups stabilizes the crystal lattice contributing to the high melting point and crystallinity. Its attractiveness lies in the fact that it produces natural metabolite on hydrolytic degradation called as glycolic acid. Importantly, its monofilaments can be degraded *in-vivo* in as little as 2 to 4 weeks due to its hydrophilic nature. The first commercially successful absorbable sutures were developed from PGA as Dexon® and since then many acclaimed biomaterials have been prepared from co-polymeric forms of PGA with other apposite materials.

Polylactic acid (PLA) is the most studied material as evident from the wide array of applications utilizing this polymer. PLA has an edge over other polymeric biomaterials as it can be easily processed by different techniques including film casting, spinning, extrusion and blow molding developing varied designs, shapes and formats. Its higher thermal processability is primarily responsible for PLA wider acceptability in different industries and markets as compared to other biodegradable polymers such as poly(caprolactone) (PCL), poly(ethylene glycols) (PEGs) and poly(hydroxyalkonoates) (PHAs).

Synthesis of PLA through ring opening polymerization (ROP) involves optically active DL, LL, and DD sterioisomers of constitutional unit (2- hydroxy propionic acid) or lactide monomer (Fig. **3**) [22, 23] under the catalytic action of tin octate [24]. The stereochemical composition of lactide monomer governs the final properties of polymer. High purity LL and DD lactide forms are transformed into isotactic hemicrystalline poly(L-lactic acid), PLLA and highly crystalline poly(D-Lactic acid), PDLA respectively which readily crystallize at 175 °C and have glass transition temperature Tg -60 °C. The DL lactide polymerizes to regular chain structure atactic amorphous poly(D,L-lactic acid) PDLLA polymer with Tg – 58 °C. PDLA, PLLA and PDLLA are soluble in common solvents including benzene, chloroform, dioxane, *etc.* and degrade by simple hydrolysis of the ester bond even in the absence of a hydrolases. In comparison to PGA, PLA has different physical and chemical properties owing to presence of pendent methyl group on α-C atom.

L(+) form of lactic acid is produced in humans and other mammals, whereas both D(-) and L(+) enantiomers are produced in bacteria. PLLA is the foremost biodegradable biomaterial, while atactic meso forms are important in different industrial sectors such as automobiles, electronics, and commodity applications including packaging. In fact since large quantities of D (-) LA is harmful to humans, L(+) isomers is the preferred choice for medical and non-medical industries. Chiral monomer lactic acid is produced by chemical as well as biological processes. With growing concerns for ecological pollution and environmental degradation along with the fact that all forms of this polymer are required in large quantities lactic acid is obtained from renewable energy sources. It is produced by fermentation of carbohydrates like corn, potato starch, sugar, cane, sugar beets, woody materials, cheese whey by lactic bacteria belonging to *Lactobacillus* family. Incorporation of carbon dioxide during the production process is also an added benefit as it reduces greenhouse effect rendering it biodegradable, recyclable, and compostable [25, 26].

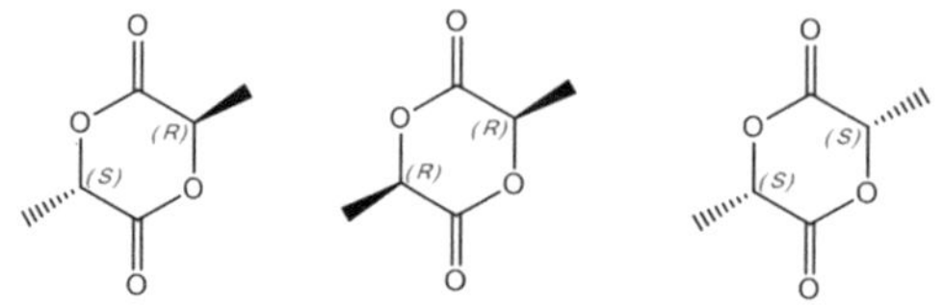

Figure 3: Different stereoisomers of cyclic ring 'lactide' intermediate governing the production of various types of PLA.

The third category of PHA is the copolymer of both PGA and PLA denoted as poly(D,L-lactic-*co*-glycolic) acid or PLGA and D & L- lactic acid forms are in equal ratio and is preferred choice for drug delivery applications in comparison to its constituent homo-polymers [27]. PLGA offers superior control in degradation properties by varying the ratio of the monomers maintaining hydrophilic/ hydrophobic balance, crystallinity, chain structure. Since methyl side groups in PLA render it more hydrophobic than PGA, it has been observed that lactide rich PLGA copolymers absorb low volumes of water owing to lesser hydrophilicity and subsequently degrade more slowly allowing a larger time frame for working with biodegradable materials [28]. Glycolic content is the critical parameter in determining final hydrophilicity and degradation pattern of PLGA structure. Enhanced hydrolytic cleavage with high glycolic percentage has established extensive use of PLGAs.

Solution phase poly-condensation reaction of the two monomers LA and GA above 120 ^{o}C temperature under water removal conditions yield low molecular weight ~10KDa copolymer. Over time many variations of this method have been proposed such as azeotropic solvent and melt/solid condensation for enhancing the physico-chemical properties of the resultant product. However, ROP remains the best synthetic methodology conferring high M wt., mechanically strong, crystalline polymers with random orientation of lactide and glycolide. High M wt. PLGA are fabricated into sutures, bone fixation screws and nails while lower M wt. amorphous materials are extremely useful in biodegradable drug delivery systems for small drug molecules, polypeptides, proteins and other macromolecules in commercial uses and in research activities [17, 29, 30]. The strong ability of PLGA based carriers also stems from the fact that it can be synthesized into micro and nanoparticles either encapsulating or adsorbing the payload. Simultaneously its scaffolding capabilities are also explored with the creation of micro and nano-structured materials *via* series of processing techniques including micropatterning, sintering, porogen leaching, electrospinning *etc*.

2.2. Characteristics of PHAs

Amongst the plethora of available biodegradable polymers the co-polymeric PLGA is extremely popular because of its favorable degradation parameters characterized with long clinical experiences, possibilities of modulation of properties and promising sustained drug delivery through suitable implantable device, at desirable time intervals without any surgical procedures. Comparisons of physical properties of different PHAs is presented in Table **2** to highlight their uses as per demand.

Table 2. Characteristics of different PHAs.

Polymer	Melting Point (°C)	Glass Transition Temperature (°C)	Strength (GPa)	Half-Life in 37°C Normal Saline	Approximate Degradation Time (Months)
PGA	225-230	35-40	7.0	-	6-12
PLLA	180	60-65	2.7	4-6	>24
PDLA	173-180	50-60	-	4-6	-
PDLLA	Amorphous	55-60	1.9	-	12-16
PDLLA/GA 85/15	Amorphous	-	1.4-2.8	-	2-3
75/25	Amorphous	-		-	1-2
50/50	Amorphous	-	1.31	-	1

3. NATURAL POLYMERS USED FOR MODIFYING PHAs

In the coming few decades' bioplastics are expected to complement and gradually replace some of the materials completely or partially derived from fossil oil. Resemblance of naturally occurring polymers with macromolecules of extracellular matrix (ECM) of living organic matter makes them interesting candidates for building implantable biomaterials and drug delivery carriers. Natural polymers include ECM proeteinaceous derivatives, agropolymers (plants and seaweed produce) and animal or human originating molecules. Most of these naturally available polymers and oligomers in their native state are responsible for myriad of functions in cellular organizations. Primarily intercellular membrane based communications and multitudes of interactions are carried by polysaccharides enabling cell movement and sensory functions. Proteins catalyze an array of reactions accounting cellular activities and also work as structure building and restoring components while lipids are the chief energy storing molecules in skin layers.

Profound curiosity in biopolymers stems from their biocompatibility, resounding hydrophillicity and enzymatic/hydrolytic degradability [31]. They are abundantly present in nature while their extraction methods have been optimized yielding commercially viable large quantities. Ample amount of surface functionalities promote multidirectional interactions with cellular and sub cellular structures necessary for the three principal market segments for these materials: wound management products, drug delivery systems (DDS), and tissue engineering. Similarities with cellular components induce them with possibilities to mimic the micro-environments of soft tissues and cartilages extremely significant feature for tissue regeneration applications which is not achievable by synthetic non-biodegradable and biodegradable polymers. As a result, the problems of toxicity and simulation of chronic inflammatory reaction, which are provoked by many synthetic polymers, are suppressed. Apart from working as the matrix completely formed from naturally derived polymers, they find significant use in polymer processing methods.

All of these polymers differ in their compositions originating from various resources and are classified according to their chemical character. Nature can provide a striking selection of polymers, potential candidates for making fibers, adhesives, gels, foams, coatings, films, thermoplastics and thermoset resins. Some of them are actively used in medical field like thermoplastic starch, gums, alginic acids, chitin, chitosan, polyamino acids, cellulosic derivatives, while other remain underutilized [32]. Some of these naturally occurring polymers are listed below in Table **3**.

Table 3. Different naturally derived polymers.

Composition	Origin	Polymer
Polysaccharides	Plants	Starch, cellulose, pectin, konjac, alginate, caragreenan, gums
	Animal/Human	hyluronic acid
	Microorganisms Fungal	pulluan, elsinan, scleroglucan
	Bacterial	chitin, chitosan, levan, xanthan, polygalactosamine, curdlan, gellan, dextran, gum, bacterial cellulose
Proteins	Multiple sources-plants, weeds and animals	Soy, zein, wheat gluten, casein, serum, albumin, collagen/gelatine, silks, resilin, polylysine, polyamino acids, poly(g-glutamic acid), elastin, polyarginyl–polyaspartic acid
Lipids/surfactants	Plants	Acetoglycerides, waxes, surfactants, emulsan
Specialty polymers	Plants	Lignin, shellac, natural rubber

Hydrophilic groups such as carboxyl, hydroxyl and amino are an integral component of most natural polysaccharides imparting charges to them and incur non-covalent bioadhesion or react with functional molecules.

3.1. Chitin and Chitosan

Chitin a nitrogen containing polysaccharide is the most abundant natural polymer after cellulose. These biopolymers are derived from shells of prawns and shell fish sources like crabs and shrimps for commercial use at industrial scale by chemical extraction methods. Chemical extraction of chitin is quite an aggressive process based on demineralization by acid and deproteination by the action of alkali. Further, chitosan is currently obtained by the deacetylation of chitin-(poly-β-(1–4)-N-acetyl-D-glucosamine). It is composed of 1,4 linked 2-amino-2-deoxy- β -D-glucan comprising of glucosamine and N-acetylglucosamine.

Chitosan molecules illustrates several interesting parameters as they are biologically stemmed amino-polysaccharides and exhibit multiple bioactivities, for example, low toxicity, biocompatibility, biodegradable, antimicrobial, and wound-healing properties. Moreover, chitosan elicits minimal foreign body reaction and favors both soft and hard tissue regeneration. Importantly a range of interrelated properties can be obtained for chitosan depending on the degree of acetylation. Chitosan has been investigated as drug delivery device usually as blended material with other natural or synthetic polymers. It has been rendered as a promising biomaterial for vascular surgery, tissue culture and tissue repair as well as a hemostatic agent [33].

3.2. Alginates

They are linear polymer of α-D-mannuronic acid and β-L-guluronic acid linked by a 1–4 glycosidic bonds as well as segments of alternating guluronic and mannuronic acids. It is typically synthesized by brown seaweeds and soil bacteria and finally extracted *via* a basic solution followed by precipitation in acidic conditions to achieve alginic acid while sodium derivative is obtained on algal purification. They are used in many industries as viscosifiers, stabilizers and gel-forming, film-forming or water-binding agents. They have also shown highly promising biomedical applications since they have the capacity to undergo spontaneous gelation on exposure to divalent cations like calcium. Simple and mild reaction conditions for generating hydrogels have propelled the use of alginate as suitable skin care products for wound healing.

3.3. Hyaluronic Acid (HA)

It is linear anionic polysaccharide found naturally in the extracellular matrix of skin, cockscomb, cartilage, vitreous humor, and other body tissue and plays a role in the movement and proliferation of cells acting as structural materials in articular cartilage and skin. It is highly hydrophilic in nature adopting extended random coil conformations which are substantially flexible and viscous. It consist of alternating units of N-acetyl-D-glucosamine and guluronic acid making it a member of the glycosaminoglycan family. The crosslinked form performs lubricating functions in synovial fluid and vitreous humor contributing to viscoelastic properties and serves as a regulator in the lymphatic system. It has been shown to scavenge free radicals, cause bacteriostasis and assist in tissue repair. Crosslinked network of homopolymer has been adapted commercial hydrogels such as HYAFF® (Fidia Farmaceutici, Italy), SYNVISC ONE® (Genzyme, USA) and ORTHOVISC® (Johnson & Johnson, USA) for treating a number of tissue related conditions, including osteoarthritis of the joints, and facial wrinkles and folds of the skin

3.4. Thermoplastic Starch

Starch is an exceptional bio-based polymer because it occurs abundantly in nature as discrete granules and is the principal carbohydrate storage product of plants. Essentially it is composed of linear and branched chains of polymer called amylase and amylopectins whose composition decides the structural properties of the starch. Growth of thermoplastic starch (TPS) is rapid in many industries as its thermal and mechanical properties are controlled and variable (Tg -50°C and 110°C), depending on its composition and additives. It is renewable, biodegradable, inexpensive, raw material and is primarily used as property modifier for medical materials. They are studied in several biomedical applications ranging from bone replacement implants, bone cements to drug delivery systems and tissue scaffolds.

3.5. Dextran

It is a β-D-1,6-glucose-linked glucan with side-chains 1-3 linked to the backbone units with approximately 5%.degree of branching. It can be obtained from fermentation of sucrose-containing media by Leuconostoc mesenteroides B512F. Owing to its reasonable biodegradability, wide availability, ease of medication and high solubility in water irrespective of the pH, it is FDA approved for biomedical applications.

3.6. Collagen and Other Proteins

Insolubility of collagen in body fluids configures its structure building functionalities and it is ubiquitously present in animal world. The triple-helix proteinaceous structure is present as long fibrils conjoined into highly strengthened fibers are found in connective tissues and extracellular matrix. It chiefly provides mechanical stability to ligaments, all types of bones, tendons giving rigidity to skin layers. It forms a significant component of connective tissue such as bone, tendons, ligaments, and skin. They are polypeptide strands bearing triamino acid blocks of different amino acids attached with glycine represents as Gly-X-Y where X and Y are a number of different amino acids, most commonly proline and hydroxyproline. The parallel arrangement of all collagen fibers renders the structure with the tensile strength of a light steel wire. The fibrils show enzymatic degradability by collagenases and metalloproteinases helpful while implanting homopolymer in sponge form. Many protienaceous entities, smaller peptides are in practice to decorate polymeric surfaces rendering them suitable to covalently link biomolecules.

3.7. Specialty Polymers

Lignin and cellulose: A large sub-type of aromatic polymers formed on oxidative combinatorial coupling of 4-hydroxyphenylpropanoids is comprised under generic name lignin. They are important constituent of the cell walls of almost all dry land plant cells proving structural function as primary load-bearing elements along with cellulose.

As lignin is much less hydrophilic than cellulose and hemicelluloses, it prevents the water uptake by plant cell-walls while allowing adequate transportation of water in the vascular tissues. It creates an effective barrier against attack by insects and fungi.

Single component materials example scaffold of plain cellulose, chitosan and alginate were in practice, however off late natural-natural polymer composites are synthesized comprising of cellulose-alginate, cellulose-agarose, cellulose-chitosan, chitosan-alginate and chitosan-agarose. They are prepared as fibers, films, porous sponges, highly crosslinked hydrogels and micro & nanoparticles. The two natural polymers are usually blended in water acting as dispersing medium and solvent since these materials are soluble in water. Blending two or more chemically and physically dissimilar natural polymers has shown potential to overcome the difficulties of processability, and brittleness. Recently attention has been employed on making active biomaterials serving on dual fronts of action and passivation. The captivating duality is particularly useful in fabricating antimicrobial surfaces for wound healing dressings where the inherent characteristics of either the native materials such as alginates or their derivatives like quaternary ammonium chitosan are used.

Many of the natural polymers suffer from inherent limitations of poor processability, brittleness, excessive hydrophilicity, inferior compatibility, and have poor moisture and gas barrier, electrical, thermal and physical properties. They are mechanically weak with low compression and elongation strength. Incorporation plasticizing material, additives or fillers to biopolymers, a common practice to improve of these disadvantageous features of materials prepared solely from naturally derived polymers.

4. COMBINATION OF NATURAL AND SYNTHETIC POLYMERS

Combinatorial chemistries of composite materials offer unique characteristics of individual components strategically useful for specific biomedical applications. Polymeric materials have been produced by combining synthetic & natural, natural & natural and synthetic & synthetic polymers resulting in good mechanical properties, easy processability and lowered production and transformation costs. Furthermore, formation of semi-synthetic hybrid products composed of naturally derived and laboratory synthesized biodegradable polymers provide added features.

Requirement of materials in multi-phase niche applications has triggered the development of unique designs of hybrid products. Skeletal tissue, bone and cartilage, are organized into 3-D structures in body and if a natural or artificial 3-D scaffold is used as a temporary load-bearing device. Ideally it should be able to maintain the requisite load for the pre-determined time without showing symptoms of fatigue or failure. Synthetic biomaterials, on the other hand, facilitate restoration of structure and function of damaged or diseased tissues. In current practices considerable emphasis is placed on biologically active scaffolds which are based on similar analogs of the ECM which can induce generation of cellular structures and finally evolve as functioning tissues and organs. An ideal scaffold for clinical applications should have very high structural and functional mimicking capabilities with native ECM.

Synthetic biomaterials guide resurrection of cellular structure. All three PHAs are integrated with various synthetic and natural biodegradable polymers. Natural polymers are not able to sustain their strength for a long time duration requisite for tissue engineering process or degrade early giving undesirable release of payload in uncontrolled kinetics. As compared to naturally derived materials, synthetic systems often lack appropriate quality and quantity of functionalities which are obligatory in advanced controlled and targeted systems for further incorporation of bioactive elements. Unique chemical configurations and surface characteristics pertaining to functionalities, hydrophilicity, roughness, and topography are time and again required and need to be controlled for customized applications attained by moderating a hybrid system composed of natural as well as synthetic biodegradable polymers. Both naturally derived and synthesized polymers are modified by physical and chemical routes changing bulk as well as surface characteristics through a series of post production or *in situ* synthesis processes. These hybrid materials synergistically combine functionalities of the polymers comprising them.

4.1. Hybrid Designs

There is a considerable shift in the concepts of biomaterial scientist in the last 10 years. This paradigm alter is in the working conditions of investigators independently working on narrow research objects to collaborative teams that facilitate solving greater objectives. Amalgamation of various research disciplines with expertise in chemistry, biology, materials, engineering and clinical practice is forwarding the development of requisite biomaterial.

One of the key modalities to expedite progress is designing polymer tailored for the specific need. The hybridized systems are not exception to the rule and massive numbers of integration processes are developed for modifying PHAs with natural polymers are the other way as represented in Fig. (**4**). Bulk modification is primarily done by blending various components in melt phase and by copolymerizing covalently linked monomers. The first process is highly explored while the latter is difficult needing production of natural macromolecule attached with either glycolic or lactic acid. This reaction is difficult and not many scientists are exploring this field. Various formats of surface modifications are highly sought after following the concept of interfacial interactions between multiphasic materials with biological world seeking enhanced bioactivity.

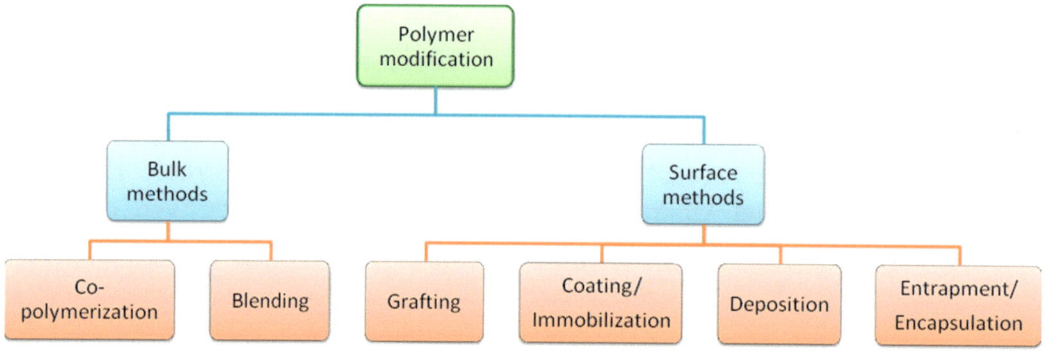

Figure 4: Diverse techniques for constructing hybrid copolymers of naturally derived and synthetic polymers.

5. MODIFICATION OF PHAS WITH NATURAL POLYMERS

5.1. Polylactic acid (PLA) Based Semi-Synthetic Hybrid Polymers

One of the drawbacks of using PLA for tissue engineering scaffold is its relative hydrophobicity due to pendent methyl group and low compatibility. Porous scaffold fabricated from the polymer have low density than water molecules. Due to high porosity scaffolds float along with cell culture medium prior to extensive wetting periods leading to poor distribution of cell during seeding. This also elaborates the number of experimental steps, making the entire system tedious and time consuming. To improve cell affinity of aliphatic polyesters, many efforts are directed to modify surface properties of PLA by adjusting the hydrophilic and hydrophobic ratio. Simple chemical reactions on the surfaces such as developing functional groups on the surfaces have been explored. Rouhi *et al.* (1999) [34] studied self assembled monolayer of PLA with final carboxylic groups and attached fibroblasts cells on it. They concluded that the cell growth and spreading was substantially better on a carboxylic acid terminated hydrophilic surfaces than methyl ended one. Citric acid and lactic acid conjugated blended polymeric system is also studied in detail to evaluate carboxy functionalized PLLA systems [35]. The processing of biomaterials shifted towards developing bioactive intelligent surfaces and functional groups are further used to create linkages with external moieties. Integration of various natural polymers with PLA is presented in the following section.

5.1.1. PLA-Chitosan

Another approach to surface modify PLA is to coat by adsorbing, or depositing various modification species. Typically PLA has been coated with biomimetic apatite, ECM proteins like fibronectin, collagen, vitronectin, thrombospondin, *etc*, peptides and PEG layer [36, 37]. In an innovative work, [38] Zhu *et al.* (2003) have created negative charges on the surface of PLLA scaffold by the ester groups with di-amine. These were electrostatically attached with positive chitosan (CS) molecules in layer-by-layer assembly manner using poly(styrene sulfonate, sodium salt) (PSS) as a negatively charged polyelectrolyte. Such systems respond better in terms of cell attachment, proliferation and cytotoxicity to human endothelial cells. They also postulated that multiple layering of charges created by CS and electrolyte gives scaffold better features than single or doubles layered organization. Owing to its unique properties, CS has been extensively explored with PLLA; many different kinds of hybrids have been prepared with its inclusion either onto the surface or pores of PLLA.

PHA produces acid degradation products – glycolic acid and lactic acid at the implanted site which evoke undesirable reactions leading to local disturbance due to poor vascularization in the surrounding tissues. Localized Excess of LA on degradation of PLA restricts its *in vivo* applications [39]. CS serves not just a functionality generating molecules but also alleviates local toxicity of acid producing biodegradable polymers. Cationic nature of CS aids in minimizing the shifting of pH to acidic conditions resulting from PLLA degradation to minimize foreign body reactions [40] and it acts as buffering agent in the arrangement.

Formation of soft/low density chitosan scaffold into prepared PLLA pores is helpful for biological aptitude to stimulate cell proliferation and hystoarchitecture.

Crosslinked CS directly fills the pores the PLA and resides only in the porous structure maintaining their inter-connectivity. Porous inter-connected networked structure is requisite as these pores facilitate the release of waste and transportation of nutrients to cells by diffusion. Lower density highly hydrated CS molecules serve better in filling the pores while maintaining the 3-D porous morphology of scaffold. This integrated model can have sustained release of drug in CS compartment of the hybrid scaffold [41]. Simple yet effective method for modifying surface requiring no functionalization steps on the host matrix is under evaluation. Molecules accumulate on the surface with changing bulk characteristics of the polymer chains. The concept is to create cell membrane like structure in the external region called as glycolyx which is heavily glycosylated. These molecules dominate the membranes' structure and incur direct specific interactions with certain incoming molecules. They are involved in cell-cell recognition events and contribute to the overall steric repulsion that prevents undesirable non-specific adhesion of other molecules and cells. They impart desirable cell adhesion while repelling non-specific interactions and sometimes internalization of foreign materials. Additionally, 'entrapment' method can be used to generate different morphologies and thicknesses of 3D scaffolds as described in Fig. (**5**), which cannot be achieved by other surface modification methods. The process is based on non-solvent-solvent systems specifically dissolving one component and was first demonstrated by Zhu and his coworkers (2002) [42]. PLLA are soluble in certain solvent while another non-solvent recovers or de-swells the host matrix. Biomacromolecules on the other hand, are soluble in the mixture of polymers and non-solvent. Rapid exposure of PLA with solvent/non-solvent system assists in diffusion of chains of secondary molecule such as biopolymers or other synthetic polymers inside the swollen host matrix. Collapse of PLLA chains on changing to non-solvent condition precede with entanglement of molecules only on the surface of de-swelled or collapsed base polymer. Zhu have entrapped CS, alginate and gelatin using acetic acid/water, calcium chloride and glutaraldehyde solution respectively. PEG, PLA and aspartic acid [43-45] have also been entrapped inside PDLA host material. However, this process is not explored these days with the concept of covalent immobilization gaining more momentum in the last few years. Advancing surface modification technologies including plasma process and aminolysis are better than entrapment since the penetration of biomacromolecules is restricted in this process.

Since scaffold structure is important for guided tissue regeneration, their structure cannot be compromised during hybridization with co-polymer. Both natural polysaccharides and PLA are physically blended together for bioactive film production by casting method. There are certain drawbacks in blending processes;

firstly this way of synthesis is not conceivable for an industrial scale process. Secondly, the initial work carried with melt phase techniques never yielded material with good 'blend' characteristics. Lack of suitable co-solvent for CS and PLA is a major hindrance for making a homogenous molten solution. High glass transition temperature (T_g) of CS sometimes results in possible decomposition of backbone before melting again making their blending process a difficult task. These unfavorable interactions are the cause for large interfacial tension in the melt, poor interfacial adhesion in solid state making it difficult to prepare films with superior mechanical strength. Recently promising method for solid-phase grafting, immersion precipitation phase inversion method with powdered chitosan, solution-blending have been developed for enhancing the properties of blend.

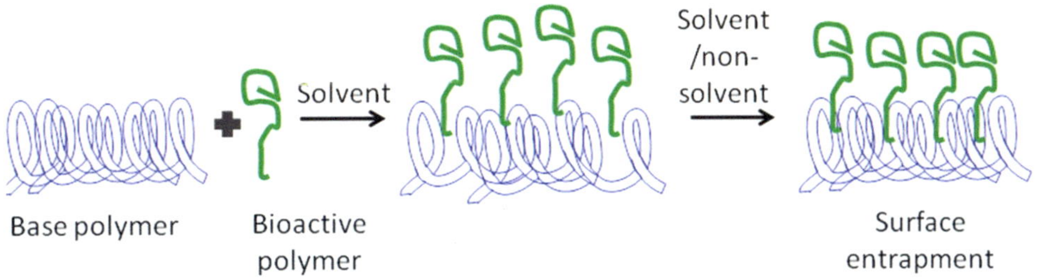

Figure 5: Entrapment of biomacromolecules on the surface of porous polymeric surfaces

Reactive blending involves activation of substrate and reactant during their intensive intermixing by applying external mechanical energy. Better interfacial compatibilization is envisaged with reactive blending ensuring interaction of PLLA chains with chitosan moieties leading to emergence of new crystallite polymorphs ultimately changing the phase content of polymers. The intermixing of solid component with PLLA introduces mechanical stability and improves thermal properties of the basic polymer while the Tg of the polymer is exclusively determined by the CS content. In many of the cases the temperature of mixing is of utmost importance. Low temperature blending allows preservation of mechanical strength by minimizing oxidative degradation of PLLA resulting in partial amorphization of blends. High temperature co-extrusion blending, on other hand, results in low tensile strength of the films owing to aminolysis and alcoholysis reactions of the ester bonds of PLLA by chitosan functional groups. The re-crystallization occurring under the solid state reactive blending (SSRB) conditions [46] proceed in solid state of components eliminating the need of any catalyst and solvent. At higher temperatures a fraction of graft copolymers is formed through amidation of end carboxylic group by amino functionality of CS.

Blending strength can also been increased by mixing powder chitosan in the molten PLLA and authors claim easy production of porous sponge like symmetrical scaffold structure showing improved features than mixing both components in molten state [47].

Fabrication of blended material to microfibers and microbeads will have higher chitosan content with good number of amino groups facilitating interactions with biomolecules, cellular components or even solid materials to form embedded or patterned structure. These hybrid polymers are electrospun into microfibers as matrix having certain degree of antimicrobial properties. More recently substantial improvement in electrospinning has resulted in production of stable nano-dimensional fibers [48]. Different modification reaction occurring on surface of polymers are illustrated in Fig. (**6**) indicating generation of various surface groups which further interact with biomolecules or drugs covalently (Fig. **7**).

Formation of positive charges on CS molecules in acidic medium owing to protonation during spinning was a main cause of worry in making suitable fibers. This can be solved by introduction of negatively charge stabilizers in the systems.

Changing surface characteristics by grafting mechanism is also seen as an important chemical functionalization step achieved by chemical coupling by plasma treatment of the base solid membrane. Lack of reactive functional reactive groups on the surface of PLA hampers its direct chemical interactions with any functionalized macromolecule. So the first step in chemical modifications is the generation of reactive functionalities like COOH, NH_2 or OH with the cleavage of ester bonds by simple surface hydrolysis or aminolysis. Covalent linking can be achieved by the creation of reactive end terminal groups on PLLA, and the attachment with chitosan is facilitated by using bifunctional crosslinkers. In one report Zhu *et al.* [49] used heterobifunctional crosslinker 4-azidobenzoic acid which was bonded to chitosan by free carboxylic acid-amine chemistry. This modified chitosan on solvent casted on PLA films was exposed to UV irradiation resulting in 'grafting to' azide based insertion mechanism of chitosan layer on the surface of PLA film.

In many cases, PLLA membranes are subjected to controlled plasma treatment generating reactive species on the surface capable of attaching macromolecules such as CS, alginates or gelatin on the surfaces. Such surface modified systems are primarily useful for controlling cell morphology and proliferation and enhancing cell adhesion than both homopolymers of natural and synthetic kind [50, 51].

Reactive blending can also result in grafting of molecules on the polymeric chains. In an altogether different scenario, [52] Zhang *et al.* (2012) have successfully fabricated semi-synthetic hybrid by grafting lactic acid onto the amino groups in CS taken as host matrix by a vacuum reaction. Their FTIR analysis revealed the formation of amide I and II suggesting the attachment of PLA side chains onto the main chains of chitosan.

5.1.2. PLA-Collagen System

Collagen distinctly forms an important role in the formation of tissues/organs and is involved in various functional expressions of cells relegating it as significant proteinaceous biomaterial. It is used for administration of antibacterial, anticancer drugs particularly in the field of ophthalmology, wound and burn dressing, tumour treatment, *etc.* However, apart from small intestinal submucosa, materials produced entirely of collagen offer a limited mechanical strength. Though the strength significantly raises after crosslinking its chains with many small homo or heterobifunctional molecules, still the properties do not reach desirable limits. Combination of plastic compressed collagen gels with biodegradable polymers is a simpler way of enhancing tensile strength while retaining structural properties of collagen. Physically coupled collagen--poly (lactic acid-co-ε-caprolactone) (PLAC) hybrids have improved physically relevant parameters and their use as scaffolds for bladder tissue regeneration has been studies by Engelhardt *et al.* [53] 2011. *In vivo* rat model experiments demonstrate the effectiveness of hybrid material with lowered inflammatory response than simple PLAC meshes. Simultaneous enzymatic hydrolysis rate of PLA can also be increased with increasing collagen content improving the hydrophilicity of PLA and PLGA

Covalent linking of targeting species on the surface of Nps is highly recommended for advancing the drug delivery particularly toxic chemotherapeutic agents eventually reducing the side effects. Natural macromolecules on the surface of Nps play multiple roles. Their integration (1) improves the degradation features of the synthetic polymeric vehicle (2) provides surface functionalities to covalently or electrostatically link biomolecules for targeting a particular site of action and (3) may also aid in reducing the steric hindrance by avoiding crowding with the introduction of long chain spacer molecules. Multiple natural molecules can be attached on the surface of PHA.

In another work PLA–chitosan–collagen nanofibrous scaffolds were synthesized by chemical attachment of molecules on PLLA [54]. These matrices displayed sustenance of rat Blood Outgrowth Endothelial Cells (rBOEC) for up to one week. They concluded that the role of collagen was not significant and instead of multiple layers, single deposition of chitosan was sufficed for maintaining and proliferating cells.

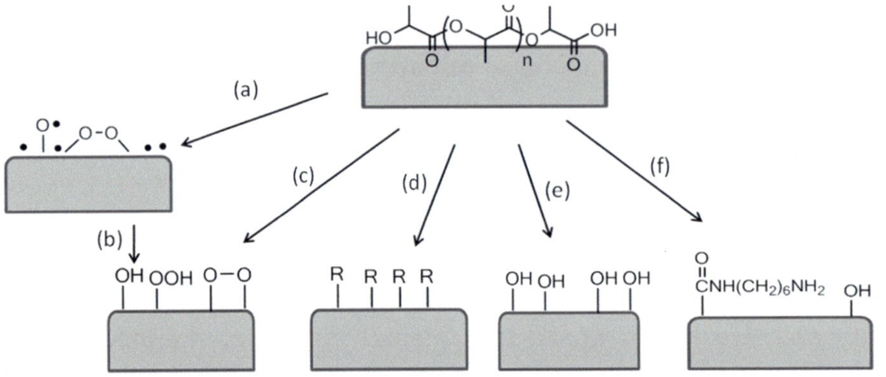

Figure 6: General schemes for surface modification processes for polyhydroxy acid based polymeric scaffolds (**a**) formation of radicals and oxygen species *via* plasma treatment and UV irradiation (**b**) exposure to oxygen generating hydroxyl functionality (**c**) direct oxidation of surface groups (**d**) photoinduced grafting of heterobifunctional molecules like glutaraldehyde, diamines *etc.* (**e**) hydrolysis and (**f**) aminolysis with diamine molecules such as hexamethylene diamine.

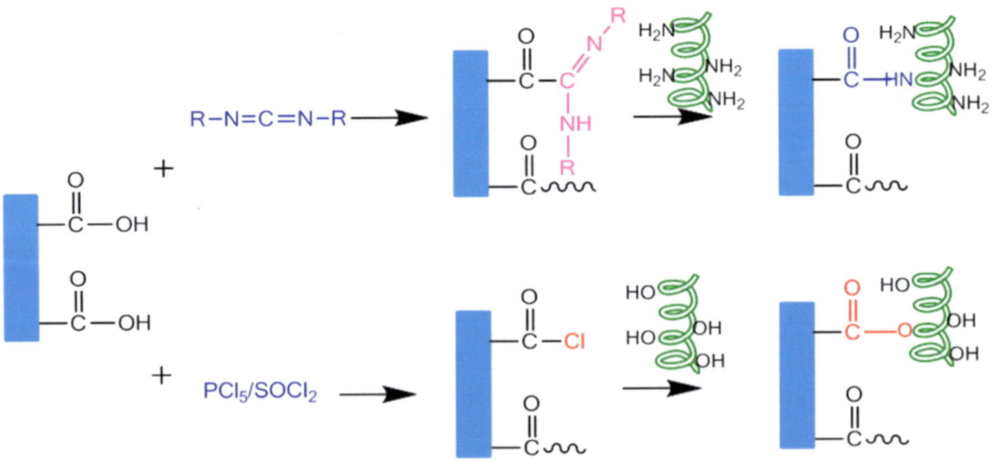

Figure 7: Grafting of biomolecules on carboxy functionalized PLA by activating using water soluble carbodimides and PCl_5 or $SOCl_2$ followed by conjugation with amine or hydroxyl functionalities.

5.1.3. PLA-Alginate Systems

Along with other natural polymers, experiments have been conducted on surface modification of PLA with alginate also. Sodium salt of alginate is suitable porogen to prepare a 3D biodegradable tissue-engineering scaffold encasing the porous structure [55-57].

Novel polymer designs are continuously sought to increase cell adhesion and guided growth of tissues. Recently Liao *et al.* 2014 [58] reported inkjet printing of multiple layers of PLA on alginate dry films. The patterned sites incur preferential adherence on PLA surfaces with higher rigidity.

Spontaneous gelation yielding hydrogels with varied fluid absorption capacity is the biggest draw for alginate in its biomedical applications. The polymer's maximum thrust is in wound care business where absorbent, non-adherent, biodegradable, nonwoven fibers create Alginate dressings. These calcium ions containing materials in contact with sodium-rich solutions of wound drainage impose ion exchange reactions forming soluble sodium - alginate gel. This gel maintains a moist wound bed and supports a therapeutic healing environment.

Alginate coating on other biodegradable materials improves viscoelastic and mechanical properties of the biopolymer while retaining biological activity. PLA sutures, bandages, mucoadhesive surgical barriers and hydrogels are blended, coated, immobilized with sodium alginates to improve wound healing characteristics. In many of these newer modalities, drugs are also loaded for controlled release activity. Recently chitosan-alginate–PLA nano composites have been developed showing multiple features including long time drug delivery properties, antimicrobial properties [59-61].

5.1.4. PLA-Hyaluronic Acid (HA) Systems

HA primarily used for imparting signaling pathways in the carrier or scaffold and is used in several therapeutic applications as well as highly valuable in cosmetic industry. 'Hylation', chemical conjugation of HA to pharmaceutical agents is a common practice to enhance their solubility, stability and retention time. HA in ECM shows high capacity to interact with certain cell receptors primarily called hyaluron binding proteins with cluster of leukocite antigens (clusters of differentiation CD) being the most studied HA receptor. Its non-antigenicity, biodegradability and viscoelastic properties aid in formation of various forms of biomaterials such as gels, foams, tubes, sheets and lightly woven meshes. Their low mechanical strength has propelled the formation of co-polymeric products with a range of polymers are combined with it for preparing wound healing dressings, visco-supplementation materials for joints & dermal fillings, loading chemotherapeutics and for attaching cells.

Bulk incorporation of HA in the matrix of PLA or PLGA is not feasible. Poor compatibility between the two polymers has led to chemical attachment of HA on pre-modified PLA film. Much better dispersion and homogeneity can be obtained by using polymer grafted acrylic acid PLA-g AA instead of planer PLA matrix. Improvement in crystalline structure of AA modified HA-PLA blend is attributed to the chemical interactions between them. Hydroxyl and amine groups from HA react with carboxylic acid functionalities in blending owing to strong interactions and covalent linkages [62].

Nanoparticles of PLA and PLGA are covalently grafted with HA by sequential functionalization steps. Selective reducing end amination on HA and NHS activation of PLA has been employed for making a new co-polymeric system in a carefully controlled set up. The branched molecules are turned into a micelles organization by precipitation/solvent evaporation technique. They are interesting candidate for increasing the solubility of hydrophobic drugs in body fluids by encapsulating them. In another report carboxy ends from HA are activated and attached with amino terminated PLA-PEG di-blocks. These nanostructures are significant drug loaded vehicles decorated with tissue specific entities on the surface carrying drug to targeted site reducing the side effect and simultaneously increasing the solubility of high molecular weight hydrophobic pharmaceutical agents [63].

5.1.5. PLA-Dextran System

Production of dextran in different molecular weights by fractionation and acid hydrolysis along with improved methods for functionalization are contributing for the development of novel biodegradable composite structures for medical uses. Primarily these hybrids find applications in drug delivery systems, carrying necessary payload introducing altered release pattern requisite for improved therapeutics. Nano/microparticles of PLA-Dex can easily form core-shell morphology owing to their varied polarity working as an added benefit in particle based DDS. PLA confers hydrophobicity while dex molecules orient themselves towards aqueous environment carrying hydrophobic drugs. These nanoparticles are reported to stay longer in rat models with higher retention time and efficacy for delivery [64]. These amphiphilic blocks copolymeric structures are designed by chemically conjugating PLA with dextran through specific linkers (Zhao 2014) or by reductive amination of dextran [65, 66].

5.1.6. PLA- Other Biomacromolecules Systems

Application specific surface modifications of PLA are highly noted to tailor its characteristics such as heparinization for improving hemocompatibility. Heparin suppresses platelet adhesion on modified PLA membranes prolongs plasma recalcification time to decrease rate of hemolysis and studied for preparing hemodialysis materials [67, 68].

Few other biopolymers are conjugated with PHAs but they serve primary role in film formation.

Flexural strength and toughness of both synthetic and bio-based polymeric materials is improved by incorporating plasticizers in the polymer mix. This addition also imparts lowering of T_g and easy processing properties to the solution. PLA is compounded with cheap fillers—such as starch—or fibers, such as cellulose and sisal to reduce its cost. In particular, starch increases moisture absorption thereby weakening PLA blends. Though adhesion between the two can be increased using coupling agents, these impose considerable cost and the environmental consequences of their use remain unclear. Combining PLA with starch improves its mechanical strength and important criteria for food packaging materials [69, 70]. Another approach in making renewable resourced bioplastics commercially viable is to use agro-residues and industrial by-products including specialty polymers like lignin, protein isolates and natural fibers. PLA-lignin blends have caught the attention of plastic industry to produce low cost carbon fibers and PLA sheets by regularizing blend characteristics to obtain positive mechanical strength materials.

5.2. Poly(Lactic-Co-Glycolic) Acid (PLGA) Based Semi-Synthetic Hybrid Polymers

Development of innovative biomedical modalities with wide application areas is aided by fine tuning physico-chemical properties of poly(lactic-co-glycolic) acid. They are appealing candidates for creating micro and nano drug conjugate systems for sustained, controlled, targeted and traceable delivery mechanisms Though the underlying exact mechanism of degradation process are not clearly understood and PLGA degradation but a systematic approach has been postulated based on physic-chemical factors and environmental conditions of the drug contributing device. Interplay amongst the three degradation modules – surface diffusion, bulk diffusion and erosion help in releasing the drug molecules while degrading PLGA component of the delivery system. Co-polymeric matrix assist in attaining enhanced pharmacokinetic and biodistribution profile for many drugs, however reports indicate that this profiles follow non-linearity, dose dependency. Accumulation of nanocarriers in renal organ system is common and the clearance is not adequate. PLGA is copolymerized with other polymers to address some of these issues. PLGA/PEG copolymerization forming di-PLGA-PEG and tri-blocks PLGA-PEG-PLGA or PEG-PLGA-PEG reduces the interactions with foreign molecules by steric and hydrated repulsion, giving enhanced shelf stability. Similar motifs are also studies with other polymers like polycaprolactone.

At present patients comfort is considered utmost priority in healthcare procedures; avoidance of invasive surgical interventions or implantations and the focus has consecrated on injectable scaffolds. From clinical perspective, injectable materials are a potential alternative since they reduce the risk of infection, scar formation, and the cost of treatment. FDA approved PLGAs are popular choice for making highly viscous solidifying tissue scaffolds fillings defected site of various sizes and shapes. Currently practiced clinical methodology of *in situ* chemical crosslinking of polymers can be detrimental as they use toxic solidifying agents, destabilize the encapsulated biomolecules, or pose toxicity concerns. Instead electrolytic solidification can overcome these challenges where PLGA colloidal gels with alginate and CS will stiffen the injectable scaffold as they are composed of composed of oppositely-charged Nps at high concentration schematically described in Fig. (**8**). Biodegradability is also simultaneously monitored to evaluate resorption of smaller degraded products in the body. Microspheres are usually prepared by precipitating PLGA in polyelectrolytic solution constituted by CS and alginate resulting in direct coating of both the polymers [71].

Polyelectrolyte multilayers of chitosan and alginates deposited on PLGA nanoparticles (NPs) by layer by layer assembly technique were studied for uptake of the nanostructures by cell lines by Zhou and his coworkers (2010) [72]. HepG2 cell line significantly uptakes PLGA NPs coated with Chi/Alg after the attachment of targeting moiety folic acid.

Implications of nano-fabrication technologies are tremendous for synthesizing nano-carriers of PLGA. Apart from nano precipitation methods hydrophobic drugs are encapsulated by emulsification solvent evaporation technique while double or multiple emulsion method is favored protocol for encapsulating hydrophilic compounds and proteins. They are further advanced into sophisticated pulsed release profile particularly for chemotherapeutic agents and proteins. Newly developed multi-pulsatile drug delivery analogs operate *via* 'switching trigger' mechanism rapidly and ephemerally releasing drug with a pre-denoted on-off regulatory mechanism controlling drug concentration and time. The sequential drug release cycle enable advanced therapeutic regime with multi-drug compartments for treatment and general restoration functions. In a report, a single biodegradable polymeric microchip is constructed with PLGA and/or PLLA in combination with multiple drugs to achieve pulsatile drug delivery over a long period of time [73].

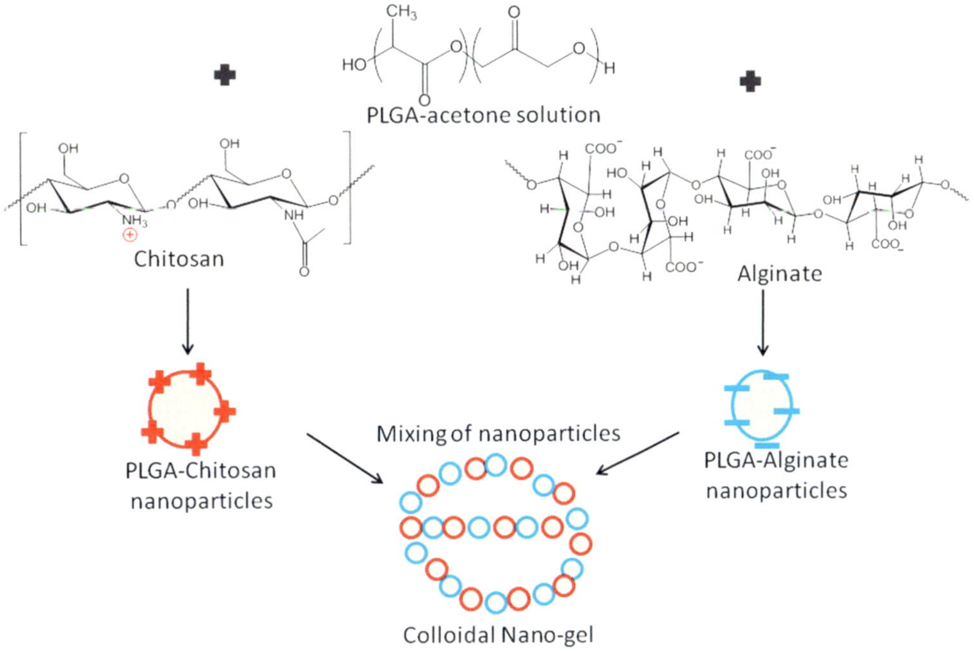

Figure 8: Preparation of cohesive colloidal gel on blending of oppositely charged microspheres. Nanoparticles are formed on precipitating PLGA-chitosan and PLGA-alginate pre-polymer carrying positive and negative charges respectively. Mixing of the two systems produces electrostatically bonded porous stable 3-D networked scaffold.

Simple adsorption of growth factors, proteins, peptides, drug molecules and other payloads into microspheres of natural polymers such as chitosan, alginate allows local burst delivery, but the chronological control over release kinetics is limited. Double layered polymeric matrix grant enhanced controllability on drug release where the host scaffold is endowed with payload containing particles. Wang *et al.* 2013 [74] have developed poly (lactic-co-glycolic) acid and nano-hydroxyapatite scaffolds containing chitosan microspheres with Adrenomedullin (ADM), a bioactive regulatory peptide that affects migration and proliferation of diverse cell types, including endothelial cells, smooth muscle cells, and osteoblast-like cells. Hydrophobic drug loaded CS microparticles can be synthesized by emulsion-ionic cross-linking method in which the two solutions of drug and polymer are mechanically homogenized. Ionic gelation initiated by electrolytic tripoly-phospahate (TPP) nucleates the particles owing to interactions of high positive charges of amino groups of chitosan and negative ones from TPP molecules entrapping the requisite drug molecules with high efficiency. Scaffold preparation is followed by dispersing microspheres in PLGA solution leading to thermally induced phase separation (TIPS) at -20 ^{o}C. It is essential to disperse microspheres homogenously to retain high porosity and maintain interconnectivity between pores suitable for cell and tissue penetration. The two components were proved to coordinately enhance tissue regeneration and extend the release time of growth factors.

Microsphere encapsulation does not affect drug potency. Gainza *et al.* 2013 [75] have demonstrated optimal wound healing capabilities of rhEGF-loaded Poly-Lactic-co-Glycolic-Acid (PLGA)-Alginate microspheres (MS) on rat models which had complete re-epithelization by 11 days. Simultaneously, PLGA has been explored in tissue engineering analysis as biodegradable scaffold. The continued enhancement of biomaterial strategies is known to be highly dependent on the ability to promote rapid and stable vascularization within scaffolds.

Amongst other PHAs, PLGA are most sought after for bone regeneration studies. In spite of high biocompatibility of pure PLGA it is not full developed as substrate for bone regeneration clinically owing to low osteo-conductivity. Its mechanical compliances are also reported to be sub-optimal for making weight bearing devices and to overcome these challenges PLGA is frequently modified with bioceramics natural and synthetic polymers rendering the host biodegradable polymer biomimetic as well as stiff. Chen *et al.* (2013) [76] have prepared a collagen/CS coating on polydopamine-modified PLGA obtaining a suitable material for guided bone regeneration (GBR).

Interestingly apart from conventional naturally derived macromolecules, other extremely specific animal originating polymers have been incorporated with PLGA for its surface functionalization. A new bioactive functionality was introduced from porcine small intestine submucosa (SIS) as natural source for PLGA. The group Lee *et al.* (2004) [77] fabricated SIS-powder-impregnated PLGA (SIS/PLGA) hybrid scaffolds to enhance cell attachment and in-growth.

Their results indicated that the SIS/PLGA scaffolds have osteoconductive effects to allow remodeling and replacement by osseous tissue.

Very recently demineralized bone particles, silk, and small intestine submucosa (SIS) have also been used as modifier for PLGA sponge for engineering vascular endothelial cells and vascular smooth muscle cells (VSMCs) using human adult bone marrow derived mesenchymal stem cells (hMSCs). Natural silk showed the maximum growth effect followed by SIS [78]. Unlike PLA, PLGA-specialty polymer, such as lignin, composites have been explored to get meshes for wound healing purpose. However, their analysis is in a very nascent stage to predict toxicity parameters.

5.3. Polyglycolic Acid Based Hybrid Semi-Synthetic Systems

PGA is highly hydrophilic degrading in diminutive span of time limiting its research for drug delivery systems. Instead most of the work is focused on short-term tissue engineered scaffolds. Easy processability of PGA into various shapes such as porous tubular, fabric, mesh structures renders its use as a carrier for loading nerve, tissue or cartilage repairing [79, 80]. Advancement in electrospinning has particularly benefitted fabrication of PGA fibrous mesh in diameter of about 50 nm smaller than cell size which was not possible with extrusion methods. Macromolecules like alginate loaded or attached PGA meshes improve cell adhesions and these 3-dimensional scaffolds have implications in cell differentiation and tissue development. The bulk modification of polymer constitutes incorporation of natural macromolecules such as chitosan which support chondrogenic activity and allow the expression of cartilage ECM proteins by chondrocytes. Blending is a modified technology utilizing conventional setup and has the advantage of being cost effective while carrying out modifications of polymeric systems.

PGA/chitosan blended artificial nerve grafts reinnervate axons, restore nerve defects and sustain Schwann cell survival [81]. Progressively degrading and absorbable PGA is easily tuned with the regeneration of nerve tissues. Autograft of PGA/CS conduit modified with recombinant neural cell adhesion molecules enhance neurite growth promoting guided regeneration and remyelination as shown by extensive studies to facilitate optic nerve regeneration in rat model [82].

In an important research work carried out at the University hospital infertility center (Taiwan), 12 patients for infertility check ups were applied with polyglycolic acid/chitosan glue into primary endometriotic cells. The results demonstrated this modified glue directly inhibits cell proliferation and induces apoptosis which was further studied for endometriosis therapy [83].

Apart from blending and filling of natural polymer into various forms of PGA, grafting of macromolecules on the surface of PGA is also gradually gaining importance. Films of PGA grafted chitosan have dramatic increase in degradable property of chitosan.

More recently, the focus has shifted to co-polymeric PLGA based hybrid systems from homopolymer of PGA as the degradation of PGA is too fast to be used for longer duration of time. Also the fact that PGA based biomaterials undergo bulk erosion as compared to follow surface eroding mechanism, thereby losing the mechanical integrity.

CONCLUSIONS AND FUTURE DIRECTIONS

Intrinsic properties of biodegradable polymers particularly polyesters have augmented development of specialty biomaterials, surgical instruments and drug delivery systems. However their growth is impeached by hydrophobicity and lack of inherent functionalities. Low mechanical strength and immunogenicity of natural polymers, whereas, is restricting use of biomaterials made from their homo-polymers. The chapter summarizes concepts for combinatorial designs of natural and synthetic biodegradable polymers to overcome some of their inherent limitations. The challenge of modification is manifested by demanding requirements of biomaterial industry for stringent performance and standards required for these applications. The development of innovative biomaterials can be an engine of innovation for new medical treatments and therapies where polymer assisted delivery or growth mechanisms direct enhanced quality (Fig. **9**).

Naturally derived polymers aid in not only creating cell-type conditions on the polymer surface but also solve major crisis of pollution. Biopolymers can successfully replace fossil oil based materials. With the advancing extraction technologies and extortion of raw materials such as lactic acid from bacterial fermentation processes will be responsible for the production of large quantitative amounts of polymers with minimal batch to batch variations. Current trends indicate that the biomaterials' community could largely benefit from a more efficient biomedical materials discovery paradigm conforming to computational designs and algorithmic softwares. Prediction of cellular responses in contact with biomaterials through computers propels high throughput and modeling experimentations accelerating the pace of scientific discovery.

Combination polymers hold great promise for the development of novel materials rendered unique with the introduction of plasticizers, additives, co-monomers and surface modifiers. Natural biopolymeric fillers can retain their bioactivity in a mechanically strengthened environment of synthetic biodegradable synthetic polymers. These systems will become indispensable for building complex implants having multifunctional applications in next few years.

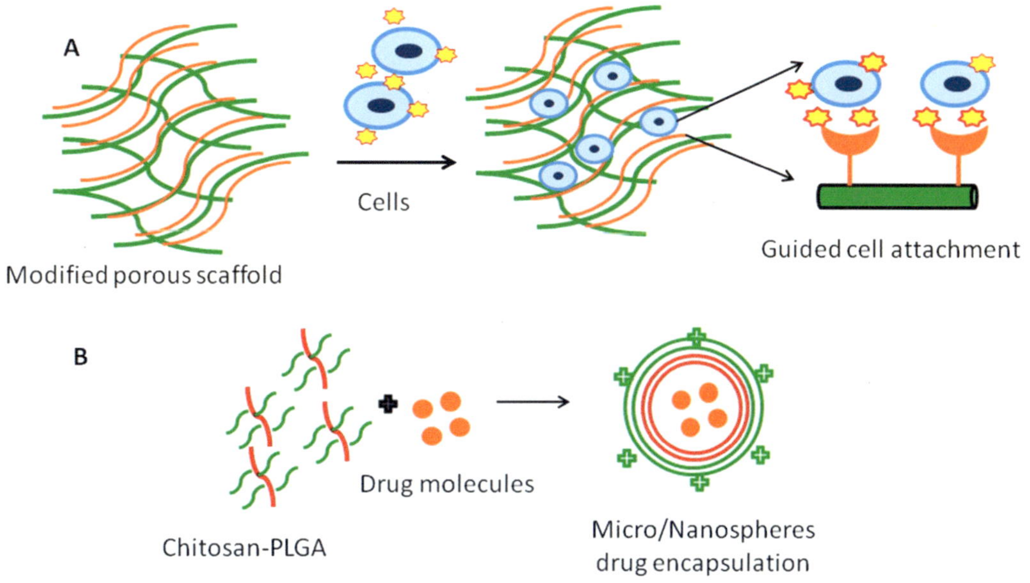

Figure 9: Applications of semi-synthetic polyhydroxy polymers A) attachment of cells is facilitated on functionalized and biodegradable 3-D porous scaffold (B) Encapsulation of bioactive molecules in biodegradable PLGA or PLA polymers while functionalization incurs possibilities for targeted drug delivery

ACKNOWLEDGEMENTS

Dr. Sruti Chattopadhyay gratefully acknowledges Department of Biotechnology (DBT), Govt. of India for the financial support.

CONFLICT OF INTEREST

The authors confirm that this chapter contents have no conflict of interest.

ABBREVIATIONS

Dex	=	dextran
ECM	=	Extra cellular matrix
M wt	=	Molecular weight
Nps	=	Nanoparticles
PGA	=	Polyglycolic acid
PHA	=	Polyhydroxy acids
PHB	=	Poly hydroxyl butyrate
PLA	=	Polylactic acid
PLLA	=	Poly L-Lactic acid
Tg	=	Glass transition temperature

REFERENCES

[1] Hutmacher DW. Scaffolds in tissue engineering bone and cartilage. Biomaterials 2000; 21: 2529-2543.
[2] Chu CC. Biodegradable Polymeric Biomaterials: An Updated Overview In: Park JB and Bronzino JD Eds. Biomaterials: Principles and Applications. 1st ed. Boca Raton FL: CRC Press 2003; pp. 95-115
[3] Shoichet MS. Polymer scaffolds for biomaterials applications. Macromolecules 2010; 43: 581–91.
[4] Ulery BD, Nair SN, and Laurencin CT. Biomedical applications of biodegradable polymers. J Polym Sci B Polym Phys 2011; 49(12): 832–64.
[5] Nair LS, Laurencin CT. Biodegradable polymers as biomaterials. Prog Polym Sci 2007; 32(8–9): 762–98.
[6] Vroman I, Tighzert L. Biodegradable Polymers. Materials. 2009; 2: 307-44.
[7] Langer RS, Vacanti JP. Tissue engineering. Science.1993; 260: 920–26.
[8] Sisson AL, Schroeter M and Lendlein A. Polyesters. In: Lendlein A, Sisson A, Eds. Handbook of biodegradable polymers: synthesis, characterization and applications, Wiley-VCH Verlag GmbH & Co. KGaA 2011; pp. 1-22
[9] Hanggi JU. Requirements on bacterial polyesters as future substitute for conventional plastics for consumer goods. FEMS Microbiol Rev 1995; 16: 213–20.
[10] Kathiraser Y, Aroua MK, Ramachandran KB, Tan IKP. Chemical characterization of medium-chain-length polyhydroxyalkanoates (PHAs) recovered by enzymatic treatment and ultrafiltration. J Chem Tech Biotech 2007; 82: 847–55
[11] Klinge U, Klosterhalfen B, Müller M, Schumpelick V. Foreign body reaction to meshes used for the repair of abdominal wall hernias. Eur J Surg 1999; 165(7): 665-73.
[12] Niaounakis M Biopolymers: Processing and Products eBook ISBN: 9780323279383 2014

[13] Zhang J, Guoping L, Shan GY, *et al.* Monocyte chemoattractant protein-1 released from polycaprolactone/chitosan hybrid membrane to promote angiogenesis *in vivo.* J Bioactive and Compatible Polym 2014; 29(6): 572-88.

[14] Wiens M, Elkhooly TA, Schröder HC, Mohamed TH, Müller WE. Characterization and osteogenic activity of a silicatein/biosilica-coated chitosan-graft-polycaprolactone. Acta Biomater 2014; 10(10): 4456-64.

[15] Chen H, Fan X, Xia J, *et al.* Electrospun chitosan-graft-poly (ε-caprolactone)/poly (ε-caprolactone) nanofibrous scaffolds for retinal tissue engineering. Int J Nanomedicine 2011; 6: 453-61.

[16] Kim MS, Kim G. Three-dimensional electrospun polycaprolactone (PCL)/alginate hybrid composite scaffolds. Carbohydr Polym 2014; 114: 213-21.

[17] Engineer C, Parikh J, Raval A. Review on hydrolytic degradation behavior of biodegradable polymers from controlled drug delivery system. Trends Biomater Artif Organs 2011; 25: 79–85.

[18] Nieuwenhu IS. Synthesis of polylactides, polyglycolides and their copolymers. J Clinical Materials 1992; 10(1-2): 59–67.

[19] Ajioka M, Enomoto K, Suzuki K, Yamaguchi A. Basic properties of polylactic acid produced by the direct condensation polymerization of lactic acid. Bull Chem Soc Japan 1995; 68: 2125–31.

[20] Albertsson AC, Varma IK. Recent developments in ring opening polymerization of lactones for biomedical applications. Biomacromol 2003; 4: 1466–86.

[21] Gupta AP, Kumar V. New emerging trends in synthetic biodegradable polymers – polylactide: a critique. Eur Polym J 2007; 43: 4053–74.

[22] Yoo DK, Kim D. Production of optically pure poly(lactic acid) from lactic acid. Polym Bull 2009; 63(5): 637–51.

[23] Garlotta D. A literature review of poly (lactic acid). J Polym Environ 2001; 9(2): 63–84.

[24] Moon SI, Lee CW, Miyamoto M, Kimura Y. Melt polycondensation of L-lactic acid with Sn(II) catalysts activated by various proton acids: A direct manufacturing route to high molecular weight Poly(L-lactic acid). J Polym Sci Part A: Poly Chem 2000; 3(89): 1673-79.

[25] Mehta R, Kumar V, Bhunia H, Upadhyay SN. Synthesis of poly(lactic acid): a review. J Macromol Sci Polym Rev. 2005; 45: 325–49.

[26] Dorgan JR, Lehermeier HJ, Palade LI, Cicero J. Polylactides: properties and prospects of an environmentally benign plastic from renewable resources. Macromol Symp 2001; 175: 55–66.

[27] Rajeev AJ. The manufacturing techniques of various drug loaded biodegradable poly(lactide-co-glycolide) (PLGA) devices. Biomaterials. 2000; 21: 2475–90.

[28] Middleton JC, Tipton AJ. Synthetic biodegradable polymers as medical devices. Medical Plastics and Biomaterials. 1998; 3-4: 31–38.

[29] Betancourt T, Brown B, Brannon-Peppas L. Doxorubicin-loaded PLGA nanoparticles by nanoprecipitation: preparation, characterization and *in vitro* evaluation. Nanomedicine. 2007; 2(2): 219–32.

[30] Thomas C, Gupta V, Ahsan F. Particle Size Influences the Immune Response Produced by Hepatitis B Vaccine Formulated in Inhalable Particles. Pharmaceutical Res 2010; 27 (5): 905–19.

[31] Ramesh PB, Kevin O'Connor, Ramakrishna S. Prog Biomat 2013; 2: 8-16.

[32] Ige OO, Umoru LE, Aribo S. Natural Products: A Minefield of Biomaterials. International Scholarly Research Network ISRN Mat Sci 2012; (2012): Article ID 983062, 20 pages.

[33] Jin R, Moreira LS, Teixeira PJ, Dijkstra. Injectable chitosan-based hydrogels for cartilage tissue engineering. Biomaterials 2009; 30(13): 2544–51.

[34] Rouhi AM. Contemporary Biomaterials, Understanding surfaces is key to the design of clinically useful materials Chem Eng News 1999; 77(3): 51–63.

[35] Jiang Y, Bai Y, Chen M, Yao F, Zhang H, Yao KD. Modification of poly(L-lactic acid) with L-lactic acid/citric acid oligomers. e-Polymers 2006; 6(1): 841–54.

[36] Chen Y, Mak AFT, Wang M, Li J, Wong MS. PLLA scaffolds with biomimetic apatite coating and biomimetic apatite/collagen composite coating to enhance osteoblast-like cells attachment and activity. Surf Coat Technol 2006; 201: 575–80.

[37] Atthof B, Hilborn J. Protein adsorption onto polyester surfaces: is there a need for surface activation. J Biomed Mater Res Part B Appl Biomater 2007; 80B: 121–30.

[38] Zhu Y, Gao C, He T, Liu X, Shen J. Layer-by-Layer Assembly To Modify Poly(L-lactic acid) Surface toward Improving Its Cytocompatibility to Human Endothelial Cells. Biomacromolecules. 2003; 4(2): 446–52.

[39] Pistner H, Stallforth H, Gutwald R, *et al.* Poly(L-lactide): A long-term degradation study *in vivo*. II: Physico-mechanical behavior of implants. Biomaterials 1994; 15: 439–50.

[40] Mei N, Chen G, Zhou P, *et al.* Biocompatibility of poly(e-caprolactone) scaffold modified by chitosan—The fibroblasts proliferation *in vitro*. J Biomater Appl 2005; 19: 323–39.

[41] Prabaharan M, Rodriguez-Perez MA, de Saja JA, Mano JF. Preparation and Characterization of Poly(L-lactic acid)-Chitosan Hybrid Scaffolds With Drug Release Capability Inc. J Biomed Mater Res Part B: Appl Biomater 2007; 81B: 427–34.

[42] Zhu H, Ji J, Shen J. Surface Engineering of Poly(dl-lactic acid) by Entrapment of Biomacromolecules. Macromol. Rapid Commun. 2002; 3: 819–23.

[43] Quirk RA, Davies MC, Tendler SJB, Shakesheff KM. Surface engineering of poly(lactic acid) by entrapment of modifying species. Macromolecules. 2000; 33(2): 258-60.

[44] Zhang J, Roberts CJ, Shakesheff KM, Davies MC, Tendler SJB. Micro and macrothermal analysis of a bioactive surface-engineered polymer formed by physical entrapment of poly(ethylene glycol) into poly(lactic acid). Macromol 2003; 36: 1215–21.

[45] Rasal RM, Janorkarc AV, Hirta DE. Poly(lactic acid) modifications. Prog Polym Sci 2010; 35: 338–56.

[46] Tatiana A, Tatiana SD, Alexander NS, *et al.* A Novel Approach to Design Chitosan-Polyester Materials for Biomedical Applications. Int J Polym Sci 2012; Vol. 2012: Article ID 827967, 10 pages.

[47] Zuo DY, Li HJ. Structure and Properties of Chitosan/Poly-Lactic Acid Blend Scaffold and Chitosan Powder/Poly-Lactic Acid Blend Scaffold. Adv Mat Res 2013; 647: 9-15.

[48] Shan X, Li F, Liu C, Qun Gao Q. Electrospinning of chitosan/poly(lactic acid) nanofibers: The favorable effect of nonionic surfactant. J Appl Polym Sci 2014; 131(22): 41098-1102.

[49] Zhu A, Zhang M, Wu J, Shen J. Covalent immobilization of chitosan/heparin complex with a photosensitive heterobifunctional crosslinking reagent on PLA surface. Biomaterials. 2002; 23: 4657–65.

[50] Ding Z, Chen J, Gao S, Chang J, Zhang J, Kang ET. Immobilization of chitosan onto poly-L-lactic acid film surface by plasma graft polymerization to control the morphology of fibroblast and liver cells. Biomaterials 2004; 25(6): 1059-67.

[51] Ma Z, Gao C, Gong Y, Ji J and Shen J. Immobilization of natural macromolecules on poly-L-lactic acid membrane surface in order to improve its cytocompatibility J Biomed Mat Res 2002; 63(6): 838–47.

[52] Zhang Z, and Cui H. Biodegradability and Biocompatibility Study of Poly(Chitosan-g-lactic Acid) Scaffolds. Molecule 2012; 17: 3243-58.

[53] Engelhardt EM, Micol LA, Houis S, *et al.* A collagen poly(lactic acid-co-ε-caprolactone) hybrid scaffold for bladder tissue regeneration. Biomaterials 2011; 32(16): 3969-76.

[54] Swarnalatha B, Nair SL, Shalumon KT, *et al.* Poly (lactic acid)–chitosan–collagen composite nanofibers as substrates for blood outgrowth endothelial cells. Int J Biol Macromol 2013; 58: 220–24.

[55] Yu G, Fan YJ. Preparation of poly(D,L-lactic acid) scaffolds using alginate particles. Biomater Sci Polym Ed 2008; 19(1): 87-98.

[56] Wayne JS, McDowell CL, Shields KJ, Tuan RS. *In vivo* response of polylactic acid-alginate scaffolds and bone marrow-derived cells for cartilage tissue engineering. Tissue Eng 2005; 11(5-6): 953-63.

[57] Nanda PK et.al. Use of Polymeric Scaffold for *In vitro* Growth of Fibroblast–Like Cells of Indian Major Carp, Cirrhinus Mrigala. Adv Anim Vet Sci 2014; 2(3): 177–82.

[58] Liao Y-C, Ma Y-T, Huang C-H, Yu J, Hsiao H-M. Rigidity Guided Cell Attachment on Inkjet-Printed Patterns. *ACS Appl Mater Interfaces* 2012; 4(7): 3335–39.

[59] Shanmugasundaram OL, Gowda RVM. Development and characterization of polylactic acid bandage coated with biopolymers and drugs for wound healing. J Text Inst 2012; 1(5): 103.

[60] Shanmugasundaram OL, Gowda RVM, Saravanan D. Drug release and antimicrobial studies on polylactic acid suture. Int J Biotechnol Mol Biol Res 2011; 2(5): 80-89.

[61] Prabhakara MN, Rao US, Babu PK, Subha MCS, Rao KC. Interpenetrating Polymer Network Hydrogel Membranes of PLA and SA for Control Release of Penicillamine Drug. Indian J Adv Chem Sci 2013; 1(4): 240-49.

[62] Wu CS and Liao HT. A new biodegradable blends prepared from polylactide and hyaluronic acid. Polym 2005; 46: 10017–26.

[63] Liu CC, Chang KY, Wang YJ. A novel biodegradable amphiphilic diblock copolymers based on poly(lactic acid) and hyaluronic acid as biomaterials for drug delivery. J Polym Res 2010; 17: 459–69.

[64] Verma MS, Liu S, Chen YY, Meerasa A, Gu FX. Size-tunable nanoparticles composed of dextran-*b*-poly(*D,L*-lactide) for drug delivery applications Nano Res 2012; 5(1): 49-61.

[65] Zhao Z, Yao X, Zhang Z, Chen L, He C, Chen X. Boronic Acid Shell-Crosslinked Dextran-b-PLA Micelles for Acid-Responsive Drug Delivery. Macromol Biosci 2014 21; 14(11): 1609-18.

[66] Verma MS, Liu S, Jones L, Gu FX. Dextran-Based Nanoparticles for Sustained Drug Delivery: Applications in Dry Eye Treatment and Cancer Therapy. In: Figgs GP, Eds. Dextran: Chemical Structure, Applications and Potential Side Effects. Nova Science Publishers 2014; pp. 31-62

[67] Jee KS, Park HD, Park KD, Kim YH, Shin JW. Heparin conjugated polylactide as a blood compatible material. Biomacromolecules. 2004; 5(5): 1877-81.

[68] Gao A, Liu F, Xue L. Preparation and evaluation of heparin-immobilized poly (lactic acid) (PLA) membrane for hemodialysis J Membr Sci 2014; 452: 390–99.

[69] Ke T, Sun SX, Seib P. Blending of poly(lactic acid) and starches containing varying amylose content. J Appl Polym Sci 2003; 89: 3639 46.

[70] Yua L, Deana K, Li L. Polymer blends and composites from renewable resources. Prog Polym Sci 2006; 31: 576–602.

[71] Wang Q, Jamal S, Detamore MS, Berkland C. PLGA-chitosan/PLGA-alginate Nanoparticle Blends as Biodegradable Colloidal Gels for Seeding Human Umbilical Cord Mesenchymal Stem Cells. J Biomed Mater Res A. 2011; 96(3): 520–27.

[72] Zhou J, Romero G, Rojas E, Ma L, Moya S, Gao C. Layer by layer chitosan/alginate coatings on poly(lactide-co-glycolide) nanoparticles for antifouling protection and Folic acid binding to achieve selective cell targeting. J Colloid Interface Sci 2010; 345(2): 241-47.

[73] Yadav AK, Mishra P, Mishra AK, Mishra P, Jain S, Agrawal GP. Development and characterization of hyaluronic acid–anchored PLGA nanoparticulate carriers of doxorubicin. Nanomedicine: Nanotechnol, Biol and Med 2007; 3: 246–57.

[74] Wang L, Li C, Chen Y, Dong S, Chen X, Zhou Y. Poly(lactic-co-glycolic) Acid/Nanohydroxyapatite Scaffold Containing Chitosan Microspheres with Adrenomedullin Delivery for Modulation Activity of Osteoblasts and Vascular Endothelial Cells. BioMed Res Int 2013; 2013: Article ID 530712, 13 pages.

[75] Gainza G, Aguirre JJ, Pedraz JL, Hernández RM, Igartua M. rhEGF-loaded PLGA-Alginate microspheres enhance the healing of full-thickness excisional wounds in diabetised Wistar rats. Eur J Pharm Sci 2013; 50(3-4): 243-52.

[76] Chen G, Xia Y, Lu XL, Zhou XF, Zhang FM, Gu N. Effects of surface functionalization of PLGA membranes for guided bone regeneration on proliferation and behavior of osteoblasts. J Biomed Mater Res Part A 2013; 101A: 44–53.

[77] Lee SJ, Lee IW, Lee YM, Lee HB, Khang G. Macroporous biodegradable natural/synthetic hybrid scaffolds as small intestine submucosa impregnated poly(D,L-lactide-co-glycolide) for tissue-engineered bone. J Biomater Sci Polym Ed 2004; 15(8): 1003-17.

[78] Shim JB, Ankeny RF, Kim H, Nerem RM, Khang G, Shim JB. A study of a three-dimensional PLGA sponge containing natural polymers co-cultured with endothelial and mesenchymal stem cells as a tissue engineering scaffold. Biomed Mater 2014; 9(4): 045015.

[79] Shahin K, Doran PM. Improved seeding of chondrocytes into polyglycolic acid scaffolds using semi-static and alginate loading methods. Biotechnol Prog 2011; 27(1): 191-200.

[80] Hashimoto A, Kuwabara M, Hirasaki Y, *et al.* Reduction of air leaks in a canine model of pulmonary resection with a new staple-line buttress. J Thorac Cardiovasc Surg 2011; 142(2): 366–71.

[81] Jiao HI, Yao J, Yang Y, *et al.* Chitosan/polyglycolic acid nerve grafts for axon regeneration from prolonged axotomized neurons to chronically denervated segments. Biomaterials. 2009; 28: 5004-18.

[82] Xu GI, Nie DY, Wang WZ, *et al*. Optic nerve regeneration in polyglycolic acid-chitosan conduits coated with recombinant L1-Fc. Neuroreport 2004; 15(14): 2167-72.
[83] Wang YC, Fu RH, Hsieh HJ, Chao HT, Kao SH. Polyglycolic acid/chitosan glue and apoptosis of endometriotic cells. Fertil Steril 2005; 84(1): 75-81.

CHAPTER 8

Microbial Exopolysaccharides for Biomedical Applications

Silvia Vasiliu[1,*], Stefania Racovita[1], Maria-Andreea Lungan[2], Jacques Desbrieres[3] and Marcel Popa[2]

[1] "Petru Poni" Institute of Macromolecular Chemistry, Gr. Ghica Voda Alley No. 41A, 700487 Iasi, Romania; [2] "Gheorghe Asachi" Technical University of Iasi, Faculty of Chemical Engineering and Environmental Protection, Department of Natural and Synthetic Polymers, Prof. Dr. Docent Dimitrie Mangeron Street, No. 73, 700050, Iasi, Romania and [3] Pau and Pays de l'Adour University (UPPA), Institut des Sciences Analytiques et de Physico-Chimie pour l'Environnement et les Materiaux (IPREM/EPCP), UMR CNRS 5254, Helioparc Pau Pyrenees, 2, av. President Angot, 64053 Pau Cedex 09, France

Abstract: The environment represents the major source of prokaryotic (*Eubacteria* and *Archaebacteria*) and eukaryotic (fungi, algae and phytoplankton) microorganisms that can posses the ability to synthesize various microbial exopolysaccharides. Since the mid-19 th century, when the dextran was found in the wine, the microbial exopolysaccharides have received a growing attention due to their structural diversity that can supply a broad range of physico-chemical, rheological and biological properties. Exopolysaccharides, known as extracellular polysaccharides because they are secreted outside of the cells, are high molecular weight polymers, that can be classified depending on the monomeric composition into two groups: homopolysaccharides composed of a single type of monosaccharides (dextran, levan) or heteropolysaccharides formed from two or more repeating units of monosaccharides (xanthan, gellan, alginate and hyaluronic acid). Also, microbial exopolysaccharides are produced in two basic forms: capsules and slime polysaccharides. Industrial and technological advancement has led to the use of the exopolysaccharides in various fields such as, food industry, waste water treatment, bioremediation as well as in the pharmaceutical and biomedical fields. Antiulcer, antitumor, anticoagulant and antiviral activities, anti-reflux therapies, cholesterol lowering agents, surgical and ophtalmic applications, treatment of rheumatoid arthritis are some of the health benefits of exopolysaccharides. The objective of this chapter is to make in introduction a concise presentation of the structure and properties of the most popular exopolysaccharides such as, dextran, curdlan, pullulan, xanthan gum, gellan gum, hyaluronic acid and alginate followed by a discussion about their applications for treatment and prevention of some diseases.

Keywords: Alginate, curdlan, dextran, exopolysaccharides, gellan gum, hyaluronan, microbial polymers, pullulan, tissue engineering, xanthan gum.

***Corresponding author Silvia Vasiliu:** "Petru Poni" Institute of Macromolecular Chemistry, Gr. Ghica Voda Alley no. 41A, 700487, Iasi, Romania; Tel: +40 232 217454; Fax: +40 232 211299;
E-mail: msilvia@icmpp.ro; vasisilv@yahoo.com

Viness Pillay, Yahya E. Choonara and Pradeep Kumar (Eds.)

INTRODUCTION

Advances in industrial and technological fields as well as a growing requirement for natural polymers to be used in various fields have led to the discovery and exploitation of different environmental sources such as, plants, animals, fungi, algae and Gram negative or Gram positive bacteria [1-3].

Many bacteria species are the foundation for the production of several types of polymers with various biological functions. In the last decades natural polysaccharides have presented a considerably increased interest to the scientific community due to their great diversity in structural and functional properties. The first nominalization of bacterial polymers began in the mid nineteenth century when Louis Pasteur highlighted the role of microorganisms in nature and laid the foundation of modern microbiology [4, 5]. In the same period two exopolysaccharides (dextran and bacterial cellulose) have been discovered [6]. Then, until the mid twentieth century a wide range of bacterial polymers (polyamide cyanophycin, polyester polyhydroxybutyrate, alginate, xanthan, poly-γ-glutamate and polyphosphate) were developed [7-10].

In literature are known four major classes of bacterial polymers, such as inorganic polyanhydrides, polyesters, polyamide and polysaccharides [7]. Over the last decades a growing interest has been awarded for isolation, biosynthesis and identification of other microbial polysaccharides, that can be used in various applications such as, emulsifiers, adjuvants in food industry, depollution agents or in medical and pharmaceutical fields [11-14].

According to the location in the cell, the microbial polysaccharides can be classified into three groups as presented in Fig. (**1**).

Extracellular polysaccharides, known as exopolysaccharides (EPS) are high molecular weight polymers, which can be produced in two distinct forms:

- capsular polysaccharides, where EPS are covalently linked to phospholipid or lipid A molecules situated on the cell surface [15].
- slime or unbond polysaccharides, where EPS remain unattached to the cell surface [16].

EPS represent the polymers with high molecular weight, which can be produced by certain types of microorganisms:

- Mesophilic bacteria: Baccilus spp., Lactobacillus Bulgaricus, Lactobacillus brevis, Leuconostoc mesenteroides, Streptococcus spp., Pseudomonas spp., Acetobacter spp., Aureobasidium spp., Escherichia spp. [17, 18];
- Thermophilic microorganisms: Thermoccocus, Sulfolobus [19];
- Halophilic groups: Haloferax, Halobacterium, Halococcus, Halomonas [20].

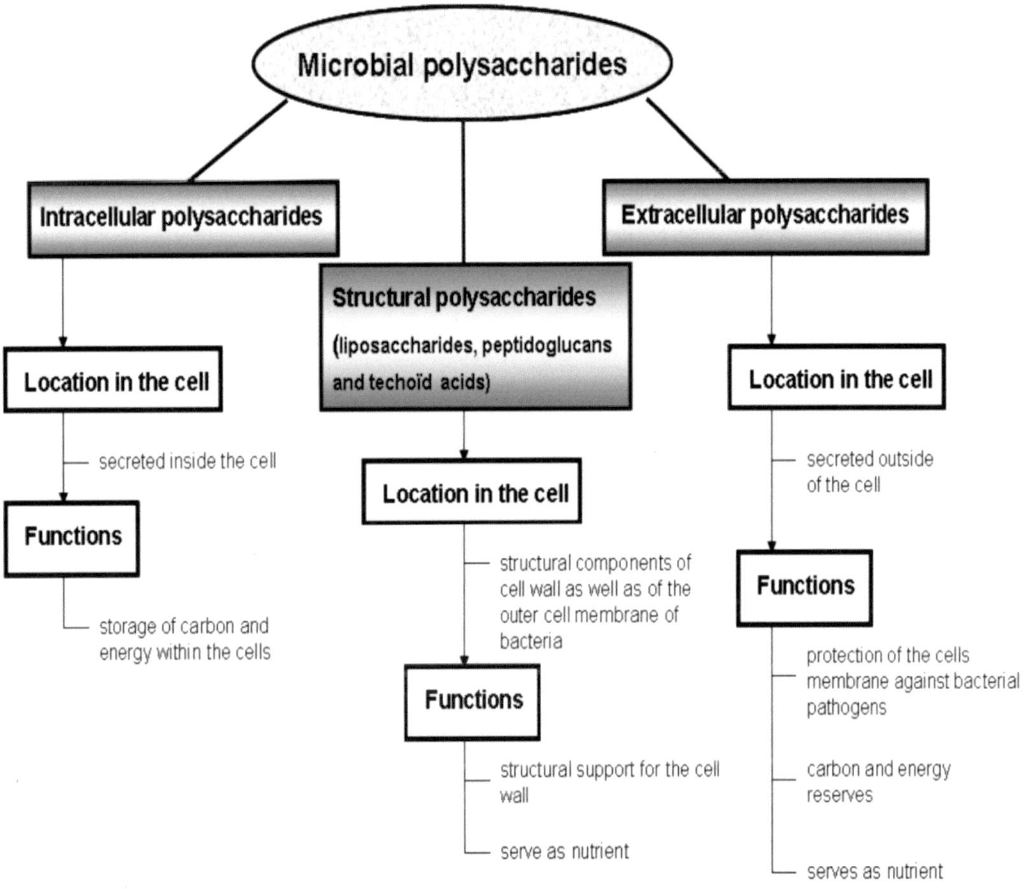

Figure 1: Classification of the microbial polysaccharides.

The history of microbial polysaccharides is presented in Tables **1** and **2**.

Table 1. History of microbial homopolysaccharides.

Name of Homopolysaccharides	History of Microbial Homopolysaccharides	Refs.
Dextran (DEX)	1861 - studies on viscosity development in wine showed that the formation of slime is carried out by microbial action	[21]
	1874 – the carbohydrate with empirical formula $C_6H_{10}O_5$ was named dextran	[22]
	1878 – isolation of bacterium *Leuconostoc mesenteroids* responsible for gelification of sugar cane and beet sugar juice	[23]
	1931 – establishing the conditions for dextran synthesis	[24]
	1941-1942 – dextran production by growing of few strains on media containing sucrose and raffinose	[25]
	1943 – the enzyme responsible for dextran production was named dextransucrase	[26]
	1944 – the first clinical use of dextran in Sweden	[27]
	1953 – industrial production of DEX by fermentation process	[28]
	1954 – 96 bacterial strains were the basis for the characterization and classification of dextran; elucidation of approximate structure of several types of dextran	[29]
	1956 – low molecular weight DEX production by *Streptococcus* strain	[30]
	1962-1963 structural studies on dextran	[31]
	1980 – optimization of fermentation process for improved production of different types of dextrans with various sizes and structures	[32-34]
Curdlan	1964 – discovery of curdlan	[35]
	1967 – gelation of curdlan above 55^0C in aqueous and dilute alkaline solutions was observed	[36]
	1968 – the helical structure of curdlan has been established	[37]
	1986 – curdlan was used in food industry in Korea, Taiwan and Japan	[38]
	1989 – Takeda Chemical Ind. Ltd. (Osaka Japan) was commercialized curdlan for food industry	[38]
	1996 – curdlan was introduced in United States food industry under the name Pureglucan™ as stabilizer, tickener and texture modifier	[39]
Pullulan	1938 – first observations about extracellular polymer production by *Aureobasidium pullulans* (*A. pullulans*)	[40]
	1958 – first isolation and characterization of new exopolysaccharide from culture broth of *A. pullulans*	[41]
	1959 – the new exopolysaccharide has received the name pullulan	[42]
	1962-1965 – studies for elucidating the pullulan structure	[43, 44]
	1976 – commercial production of pullulan	[45]
	1982 – fabrication of pullulan films by Hayashibara Company	[45]

Table 2. History of microbial heteropolysaccharides.

Name of Heteropolysaccharides	History of Microbial Heteropolysaccharides	Refs.
Xanthan gum (XAN)	1950 – discovery of xanthan gum by scientists from the US Department of Agriculture	[46]
	1960 – first industrial production of xanthan gum	
	1964 – xanthan becames available commercially	
	1978 – xanthan gum was authorized for food use in US	
	1982 – xanthan gum was authorized for food use in Japan	[46]
Gellan gum (GLL)	1978 – discovery of gellan that was isolated from *Pseudomonas elodea*	[47]
	1988 – gellan was authorized for use in food industry in Japan	[48]
	1992 – gellan was authorized for use in food industry in US	[49]
	1994 – discovery of gellan isolated from *Sphingomonas paucimobilis*	[50]
Hyaluronic acid or hyaluronan (HA)	1880 – first works on HA	[51]
	1934 – first isolation of HA from the bovine vitreous humor	[52]
	1931-1940 – HA isolation from vitreous body, synovial fluids, umbilical cord, skin and rooster comb as well as from *Streptococci*	[53]
	1942 – first commercial uses of HA	
	1950 – discovery of various techniques for measurement of both molar mass and the polydispersity of HA	[54]
	1954 – the chemical structure of HA was solved	[55]
	1950-1960 – studies about physico-chemical characterization of HA	[51]
	1986 – introduction of term „hyaluronan" to both acid form (hyaluronic acid) and salt form (sodium hyaluronate)	[56]
Alginate (ALG)	1881 – discovery of alginate in brown seaweed	[57]
	1926 – some researchers have established that the main constituent of alginate was uronic acid	[58]
	1929 – industrial production of alginate in California	
	1929-1930 – studies about the nature of uronic acids	[59]
	1939 – it was determined that the β 1,4 is the connection between uronic acid residues from alginate backbone	[60]
	1955 – Fischer and Dorfel have identified α-guluronic acid in the structure of alginate	[61]
	1965 – examination of the structure of alginate produced from different sources as well as the development of the new methods for chemical fractionation of alginate	[62]
	1966 – discovery of alginate isolated from *Azotobacter vineladii* and *Pseudomonas*	[63]
	1975 – first bacterial alginate biosynthesis	[64]

Part of the polysaccharides family, microbial polysaccharides and more precisely polymeric compounds produced by the exterior matrix of cells are favored over other natural polymers obtained from plant and marine source in many fields of activity because of their great characteristics. Some of these extraordinary characteristics are:

- Superior ability to modify the properties of aqueous solutions, thus being often used as thickeners and suspension or emulsion stabilizers. Microbial polysaccharides also have the capacity to flocculate and swell to form gels, films, membranes and colloidal suspensions;

- Wide renewable sources being produced from abundant, well known and cheap raw materials;

- Good resistance to hydrolysis in a wide range of temperature and pH;

- Peculiar rheological properties such as high viscosity and high elasticity even at low shear rates.

In this context, this chapter highlights the structure, properties and applications of some of the most important microbial exopolysaccharides, namely: curdlan, dextran, pullulan, alginate, gellan gum, hyaluronic acid and not least xanthan gum.

Thus, several biomedical applications are explored throughout the chapter starting with drug delivery formulations and continuing with summarizing their wound healing and tissue engineering utility and nonetheless, post-operative adhesion prevention and anti-arthritic benefits.

STRUCTURE AND PROPERTIES OF MICROBIAL EXOPOLYSACCHARIDES

Microbial exopolysaccharides are ionic and non-ionic polymers, structurally comprised of repeating monosaccharide units (pentoses, hexoses, amino sugars or uronic acids) that are linked by glycosidic bonds.

Depending on the monomeric composition EPS can be grouped in two main classes:

- Homopolysaccharides, that are formed from a single type of monosaccharides such as D- glucose or α-fructose and occur in both linear chain (pullulan, curdlan, levan) or ramified chain (dextran) forms. The homopolysaccharides can be also classified as follows:

a. α-D-glucans, produced by *Leuconostoc mesenteroides*;

b. β-D-glucans, produced by *Pedicoccus spp.* and *Streptococcus spp.*;

c. fructans, synthesized by *Streptococcus salivarius*;

d. polygalactans [65].

- Heteropolysaccharides (xanthan gum, gellan gum, hyaluronic acid, alginate), that are formed from two or more repeating units of monosaccharides such as, D-glucose, D-galactose, L-rhamnose, L-fructose and sometime from N-acetylglucosamine, N-acetylgalactosamine or glucuronic acid [66].

The differences between homo and heteropolysaccharides can be given by the chemical structure and the binding mode as well as the site of synthesis and the types of enzymes used [66].

Dextran, a member of α-D-glucans family is a microbial fermentative hydrocolloid produced by dextransucrase, glucosyltransferase secreted mainly by Lactic acid bacteria (LAB). The family of LAB is comprised by the following genus of Gram positive bacteria: *Leuconostoc*, *Lactobacillus*, *Lactococcus* and *Streptoccocus* [67, 68].

DEX represents the first industrial microbial exopolysaccharides produced by LAB that was approved to be used in food industry.

Structurally, DEX is a homopolysaccharide composed of D-glucose units linked by (1,6) α-glycosidic bonds (95%) and different amounts of α-1,2, α-1,3 and α-1,4 branched glycosidic linkages (Table **3**) [32].

Structure and physicochemical properties (molecular weight, optical activity, solubility and physiological action) of DEX are dependent on the conformation of the enzyme synthetized by the strain [69].

Therefore, depending on the microbial strains various types of dextran can be obtained as follows:

- *Leuconostoc and Streptococcus* produce dextransucrase that converts sucrose in DEX which has 50% or more (1,6) α-glycosidic bonds [70];

- *Leuconostoc mesenteroides NRLL B-512F* gives soluble linear DEX containing 95% of (1,6) α-glycosidic bonds [71];

Table 3. Chemical structure of microbial homopolysaccharides.

Name of EPS	Chemical Structures
DEX	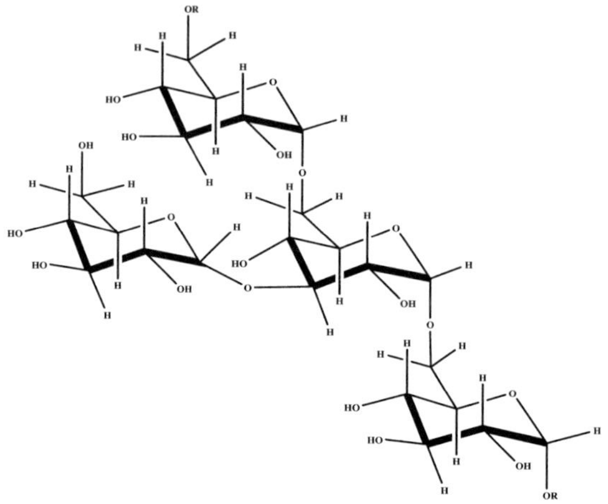
Curdlan	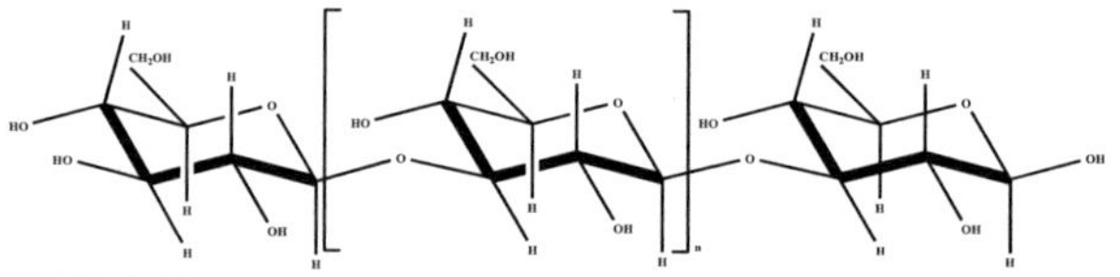
Pullulan	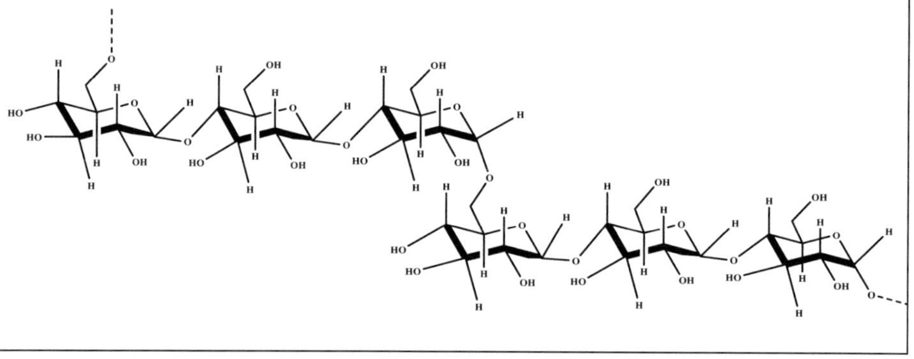

- *Leuconostoc mesenteroides NRRL B-640* produces higly soluble DEX which contains 97% of (1,6) α-glycosidic bonds [72];

- *Leuconostoc dextranicum NRRL B-1146* forms DEX which contains a linear backbone of (1,6) α-glycosidic bonds with (1,4) α-branch linkages and (1,6) α-linkages, respectively;

- *Leuconostoc mesenteroides B-1299* forms intracellular as well as extracellular dextransucrases. This extracellular glucosyltransferase produces two types of DEX, namely fraction L characterized by 27% of (1,2) α-branch linkages and 1% of (1,3) α-branch linkages, respectively and fraction S [35% of (1,2) α-branch linkages] [73, 74];

- *Leuconostoc mesenteroides B-742* produces: fraction L [50% of (1,6) α-linkages, 14% of (1,4) α-branch linkages and 1% of (1,3) α-branch linkages] and fraction S [50% of (1,6) α-linkages and 50% of (1,3) α-branch linkages] [75];

- *Leuconostoc mesenteroides NRRL B-1355* forms two different types of DEX: fraction L [soluble linear DEX which contains 95% of (1,6) α-glycosidic bonds] and fraction S [53% altered sequence of (1,6) α-linkages and (1,3) α-linkages]. The fraction S has been named alternan [76];

- *Leuconostoc mesenteroides PCSIR-4 and PCSIR-9* give dextran with low and high molecular weights [77];

- *Acetobacter capsulatus NCTC 4943* known as *Gluconobacter oxydans 11894* and *Acetobacter viscous NCTC 7216* named *G. Oxydans ATCC 11895* form dextran dextrinase that converts dextrins to DEX [78];

- *Streptococcus strains* form dextransucrase that synthetizes two types of DEX: one water soluble consisting of (1,6) α-bonds and the other insoluble that possesses more than 50% of (1,3) α-bonds with small quantity of (1,6) α-bonds and (1,4) α-linkages [30];

- *Streptococcus mutans strain OMZ 176* produces water insoluble DEX that contains 90% of (1,3) α -glycosidic linkages [79];

- *Streptococcus mutans 6715* produces DEX with 35% (1,3) α-branch linkages [80];

- *Streptococcus sanguis strain 804* forms gelatinous DEX that contains (1,3) α- and (1,6) α-glycosidic linkages in equal amounts [81].

DEX may have various structures, such as comb-like, laminated and ramified structures [82]. The type and degree of branching can be affected by the temperature of the fabrication process of DEX. It was reported that the lower temperature (3-20ºC) can lead to the formation of DEX with expanded conformation while at higher temperature (30ºC) DEX displayed a more compact structure [83]. Also, increasing the temperature from 4ºC to 45ºC the branching increases from 4.8% to 14.7%, respectively [67].

Infrared spectroscopy, catalytic oxidation, optical rotation, capillary gas-liquid chromatography, mass spectrometry, periodate oxidation and methylation reactions were the methods used for the determination of the DEX structure [84, 85]. The native DEX is a polymer with high molecular weight and high polydispersity. DEX with the (1,6) α-glycosidic linkages is soluble in various solvents such as water, formamide, ethyleneglycol, dimethyl sulfoxide and glycerol, the kinetics of dissolution being influenced by the molecular weight of the polysaccharides [86]. DEX is a biocompatible and biodegradable polymer with non-antigenic or non-immunogenic character. The degradation of DEX can be realized by the action of α-1-glycosidases (dextranases) that is localized in various parts of the human body such as, lower part of gastrointestinal tract (GIT), spleen, liver and kidney.

The (1,3) β-D-glucans family found in prokariotic and eukariotic microorganisms includes three types of microbial polymers:

- linear glucans;
- (1,3)(1,6) β-D-glucans with branch-on-branch or cyclic structures;
- Side-chain-branched (1,3)(1,2) β-D-glucans [87].

Curdlan, a microbial exopolysaccharides detected in *Alcaligenes faecalis* var. *myxogenes strain 10C3.* This polymer belongs to the first type of glucans family, namely (1,3) β-D-glucan whose linear structure has been evidenced in 1968 by using the methylation and periodate methods [88]. Structurally, curdlan is formed by repeating glucose units joined by β-bonds between the first and third carbon of the glucose molecule (Table **3**) [89].

Curdlan is soluble in alcaline solutions, dimethyl sulfoxide, formic acid and some aprotic solvents but insoluble in water, alcohols and the most organic solvents. Curdlan forms gels with different characteristics in function of some parameters as heating temperature, heating time, dispersing method, heating rate and concentration of homopolysaccharide solution. The gels can be obtained by two methods: heating the aqueous suspension of curdlan and neutralization from alkaline solution [90]. By means of the first method two types of gels can be obtained:

- high-set thermo-irreversible gels, when the aqueous suspension of curdlan is heated to temperature above 80ºC;

- low-set thermo-reversible gels, by heating the aqueous suspension of curdlan in the range of 55-60ºC and then cooling below 40ºC. Also the low-set thermo-reversible gels can be obtained by addition of cations (Ca^{2+} or Mg^{2+}) [91].

This behavior can be explained by the modification of the curdlan's structure from a mixture of single helices at room temperature to a condensed rod-like triple helices at higher temperature [92]. When the second method is used a low-set gel is formed by neutralization from alkaline solutions without heating while the high-set irreversible gels can be obtained by heating the neutralized gels to above 80ºC [87].

Another homopolysaccharides, namely pullulan, obtained from the fermentation medium of *Aureobasidium pullulans* (black yeast) is comprised of maltotriose units connected by (1,6) α-D-glycosidic linkages [93]. Also, the internal glucose units are joined by (1,4) α-glycosidic bonds (Table **3**) [94].

At the beginning of 1960s, Wallenfels *et al.* have established that the ratio between (1,6) and (1,4) glycosidic linkages is 2:1 [44]. Due to the presence of (1,4) and (1,6) α-glycosidic linkages respecively, the structure of pullulan can be situated between amylose and DEX structures [95]. In addition, it was found that the linear chain of pullulan presents maltotetraose subunits that are randomly distributed throughout the molecule [96].

Alongside the *Aureobasidium pullulans* pullulan can be isolated from other microbial source such as:

- saprophytic fungus *Tremella mesenterica* [97];

- parasitic fungi *Cytaria harioti* and *C. darwinii* [98];

- pathogenic fungus *Cryphonectria parasitica* [99];

- lichenised ascomycete *Teloschistes flavicans* [100];

- *Rhodontotula bacarum* [101].

The alternation of (1,4) and (1,6) α-linkages can lead to a good flexibility of the polysaccharide molecules as well as the increase of pullulan solubility [102].

Pullulan is characterized by a higher solubility in water and dilute alkaline solutions while in alcohol and other organic solvents is insoluble. From medical point of view, pullulan is non-toxic, non-mutagenic, non-carcinogenic, tastless and odorless. Chemical modification (etherification, esterification, chlorination, sulfation, oxidation, CMP/amidification, copolymerization) of pullulan can lead to the formation of the new polymeric materials insoluble or with low solubility that can be used in various biomedical applications [103, 104].

HA is an extracellular polysaccharide from the glycoaminoglycan family, being the only member of this class that does not contain any sulfate groups within its backbone or any covalent bonds between its structure and a proteoglycan core protein [105]. Also, known in literature as hyaluronate or hyaluronan, this natural polymer is a linear heteropolysaccharide found in most connective, epithelial and neural tissues, being the major components of the extracellular matrix of most living organisms (mammals, Ascaris worms, *Pseudomonas* slime, *etc*).

HA can be present in the pericellular wall of organisms, intracellulary and in various tissues and organs like: skin, cartilage, brain, synovial liquid, aeorta, *etc* [106, 107].

First purified from the vitreous body in 1934 by Meyer and Palmer, nowadays HA can be isolated from various sources including joint fluid, rooster combs and several strains of streptococci (ex. *Streptococcus zooepidemians*, *Streptococcus pyogenes, Streptococcus equi* and *Streptococcus uberis*) [105, 108].

Structurally HA is a linear polyelectrolyte composed by alternately repeating unit of (1,4) β-D-glucuronic acid and (1,3) β-D-glucosamine (Table **4**) [109, 110].

Despite its relative simple structure this carbohydrate possesses many interesting properties contributing to the tissue remodeling processes and the mediation of cell behavior during morphogenesis [111].

Several studies have also determined that HA can be an important tool in the progression of some malignant tumors, where it is found in very high dosage, or that it can be a significant signal for activating kinase pathways and regulating angiogenesis [112].

Table 4. Chemical structure of microbial heteropolysaccharides.

Name of EPS	Chemical Structures
HA	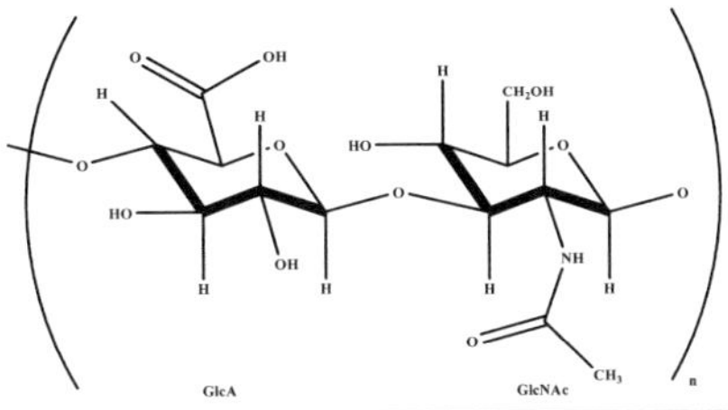
XAN	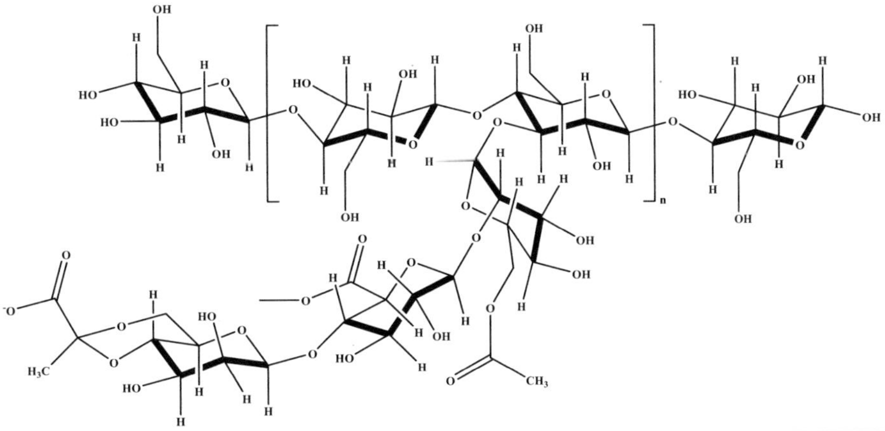
GLL	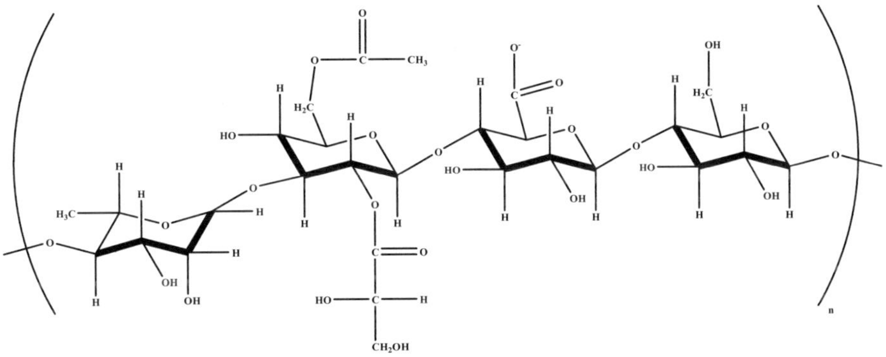

Another essential role played by HA is in inflammation. Recent investigations have shown that the inflammatory activity of HA is in strong correlation with the molecular weight of the polysaccharide. Thus, polymer with low molecular weight can accumulate at the sites of inflammation and stimulate the production of various cytokines, while high molecular weight structures are known to present an anti-inflammatory activity [113]. As it is well known in literature, the molecular weight represents one of the most important properties of a polymer. The value of the molar mass of a certain structure is responsible for its rheological properties, more precisely for its viscous and elastic behavior when exposed to different shear rates.

It is very difficult to ascertain the absolute molecular weight of HA because of possible aggregation phenomena that may occur in solution. An alternative to this problem consists in the determination of the molecular weight distribution and the average molecular weights, namely the number-average molecular weight (M_n) and the weight-average molecular weight (M_w). One of the most used methods to characterize HA from the molecular weight point of view is steric exclusion chromatography (SEC). This method offers a complete analysis of the polymeric solutions of HA as well as a detailed examination of their purity [114].

Depending on the average molecular mass, there are four types of HA molecules with very different characteristics, as follows:

- small HA oligomers with M_w = 400 – 4000 Da can be anti-apoptotic and inducers of heat shock proteins;
- oligosaccharides with an average molecular weight between 6×10^3 and 2×10^4 Da are immune-stimulatory, angiogenic and inflammatory;
- medium size HA chains with M_w between 2×10^4 and 10^5 Da can influence ovulation, embryogenesis and wound repair processes;
- high molecular weight HA ($M > 5 \times 10^5$ Da) can be used as space fillers, immunosuppressive or antiangiomics [115].

For latter applications of HA materials, a crucial characteristic is represented by its highly solubility in water, more than 10 g L^{-1}. Even in the case of high molecular weight sorts, HA is presented as a highly hydrophilic polymer with a capacity of sorption up to 1000 times its solid volume. This capacity to retain great amounts of water conducts in the formations of very loose hydrated networks with interesting hygroscopic and homeostatic properties [51, 116].

In physiological solutions HA molecules are presented as stiffened random coils due to intra- and intermolecular associations, especially hydrogen bonds. Nevertheless, the stiffened coil structure is also a consequence of other associations such as the interactions established between the molecules and the solvent. A twisting ribbon structure is formed with the help of two mechanisms. In the first place the axial hydrogen atoms interact to form a non-polar, slightly hydrophobic fragment while the equatorial side chains interact to form a more polar and hydrophilic fragment [117]. The structure obtained exhibits a very unusual rheological behavior. Thereby, even at low concentration, HA solutions are very viscous and this viscosity only increases with further increasing of the concentration and the chain length of the polymer [51]. In the case of high concentration solutions, the HA molecules reflect a molecular entanglement although there is no evidence of chain-chain associations. Studies of molecular diffusion and molecular dynamics have proven to be very helpful in elucidating these structural characteristics. Thus, the analyses carried out on HA solutions of different concentrations have illustrated very mobile HA chains that are ordered and stiffened due to hydrogen bonds formed between fragments which also restrict the glycosidic bonds rotation of adjacent saccharide units.

HA is often referred in literature as a pseudo-plastic material due to its visco-elastic behavior and it can easily be used in many applications as lubricant or emulsifier [118]. In solid state hyaluronan has been investigated by electron microscopy and nuclear magnetic resonance spectroscopy. Electron microscopy has shown that HA has the arrangement of rope while NMR spectras suggested a stable tertiary structure for the polysaccharide molecule. Still, this behavior was not seen in the case of HA solutions. For dilute solutions and even for more concentrated ones, hyaluronan molecules present very mobile chains that do not undergo any transition to a gel-like state [119]. In physiological conditions HA is a highly charged polyelectrolyte and it is often found in solution in association with different counterions, such as Na^+, K^+, Ca^{2+}, or Mg^{2+}.

Because HA is a natural polymer, it has non-allergenic properties that make it suitable in medical usage [108].

Also, HA has an antioxidant character and it can be used to regulate the inflammatory response and to stabilize the granulation tissue matrices [120].

Xanthan gum is one of the most significant microbial extracellular polysaccharide being obtained by fermentation process of the plant-pathogenic bacterium *Xanthomonas campestris*. Alongside *Xanthomonas campestris*, several other bacterium can produce this heteropolysaccharide, namely: *Xanthomonas phaseoli*, *X. malvacearum*, *X. carotae*, *X. citrumelo* or *X. juglandis* [121]. First commercialized in the 1960's it was believed that XAN was a heteropolysaccharide consisting in D-glucose, D-mannose, D-glucuronic acid acetal linked pyruvic acid and O-acetyl structures [122]. Nowadays, after thorough investigation, the accepted structure of XAN is illustrated in Table **4**.

Thus, the primary structure of XAN consists in a pentasaccharide repeating units of D-glucopyranosyl, D-mannopyranosyl and D-glucopyranosyl glucuronic acid in a molar ratio of 2:2:1.

Presenting a similar conformation to cellulose, XAN's backbone is formed by linking the D-glucopyranosyl fragments in the (1,4) β positions of adjacent units. To alternate D-glucosyl residues at the O-3 position, a trisaccharide side chain of D-glucuronosyl fragments is introduced between two D-mannosyl residues. Further more, the terminal β-D-mannopyranosyl unit is linked by glycosidic bound to the O-4 position of the β-D-glucopyranosyluronic acid which in return is linked to the α-D-mannopyranosyl residue in the O-2 position [123, 124]. Approximately one half of the terminal mannose units present as side chains pyruvic acid moieties, while the internal mannose units present O-acetyl groups as substituents. Structural studies have shown that in some cases the XAN molecule has on the external mannose residue another O-acetyl group linked in the C_6 position [125].

The amount and type of substituents attached on the backbone depend mostly on the bacterial growth conditions. These characteristics further influence the properties of XAN sorts and their applications. The secondary structure of XAN depends on the conditions under which the material is characterized. Thus, depending on some parameters (temperature and ionic strength) of the exopolysaccharide solution, the macromolecular chains can be found in an ordered or a disordered conformation. Stokke *et al.* performed a very detailed analysis regarding these characteristics [126].

Another technique employed to elucidate the XAN structure is X-ray scattering. With the help of X-ray analyses, scientists demonstrated that under ordered conformation, polysaccharide molecule exists as a right-hand helix with five-fold symmetry with a diameter of 1.9 nm and a pich of 4.7 nm [127].

Nowadays, most authors sustain the idea of a double helix conformation as the secundary structure of XAN. The stabilization of this conformation is assured by hydrogen bonds, electrostatic interactions and steric effects.

In solid state XAN is found as a cream colored powder with high solubility in water, the polysaccharide forming high viscous solutions even at low concentrations. This unusual rheological behavior, similar to HA and gellan gum, makes XAN usable as thickener and stabilizer in many applications [128]. XAN is a highly stable polymer that it is not easily degradated by most microorganisms. Under certain conditions and in contact with different microorganisms, studies have shown that XAN preserves its intact structure even after long periods of time that can range from one month up to one year in storage [125].

In solution, XAN shows a pseudoplastic behavior and the viscosity depends on the shear rate. The structure conformation of XAN may suffer modifications under the influence of ionic strength and temperature. When the temperature is changed, the transition from an ordered to a disordered structure is usually associated with a complete or partial separation of the double helix. Nuclear magnetic resonance spectroscopy (NMR spectroscopy) has demonstrated that XAN exhibits an ordered conformation below 55°C and a disordered random coil at temperatures higher than 55°C. The order-disorder transition was also investigated by circular dichroism spectroscopy, FT-IR spectroscopy and optical rotation measurements. Under specific conditions, precisely at temperatures below the transition point and in high salt concentration medium, a renaturation process can happen.

In the presence of electrolytes, depending on the solution concentration two phenomena may appear:

- for low concentration solutions, lower than 0.15%, the presence of the electrolytes (NaCl, KCl) reduces the viscosity of the system;

- for high concentration solutions the addition of electrolytes leads to an opposite effect.

The salt concentration also affects the rheological behavior of the solutions. Thereby, a peak viscosity is obtained at a salt concentration of 0.02 – 0.07% above this level any addition of electrolytes being practically irrelevant to the visco-elastic modulus [128, 129]. With respect to the molecular weight, XAN is a polymer with high molar mass. It is very difficult to determine the accurate weight average molecular weight of this polymer. Nevertheless, with the help of gel-permeation chromatography and asymmetrical flow field fragmentation combined with multi-angle laser light scattering, a molecular mass between 4 and 6×10^6 g mol^{-1} has been estimated [126,130]. Using sedimentation analysis, several sorts of XAN were investigated and a mass from 4 to 62 million g·mol^{-1} has been determined [128]. Finally, although XAN is a completely biodegradable polysaccharide, it is very resistant to enzymatic degradation by many enzymes, such as protease, pectinase, amylase, cellulase and hemicellulase. The excellent properties alongside its biocompatibility and bioadhesion make this microbial exopolysaccharide very useful in many areas from cosmetics and food industry to pharmaceutical and medical applications [131].

Gellan gum is a microbial heteropolysaccharide obtained by aerobic submerged fermentation of bacterium *Sphingomonas elodea* also known as *Pseudomonas elodea* [132, 133].

Discovered in 1978, GLL is one of the most extensively investigated exopolysaccharide, in recent years being used in a various fields of applications [134].

Structurally, GLL is a linear tetrasaccharide constituted by repeating units of α-L-rahmnopyranosyl, β-D-glucopyranosyl and β-D-glucuronopyranosyl in a molar ratio of 1:2:1. In native state, on the 3-linked glucose residue, the polymer has two different substituents, namely a L-glyceryl group at the O-2 position and an acetyl group at the O-6 position [135]. The primary structure of GLL is represented in Table **4**.

To elucide the secundary structure of GLL, various methods of characterization have been employed. Among them, osmotic pressure and light scattering techniques helped to determine the 3-fold double helix structure of GLL. The double helix conformation is formed by two left-handed polymeric chains. On the periphery of the macromolecular chains, the double helix has various acetate residues, while glyceryl groups are attached internaly. These glyceryl groups alongside the hydrogen bonds form interchain associations with the role to stabilize the structure of the polysaccharide [136]. As in previous cases, the molecular weight of the polymer is a very important characteristic that influences the latter applications of the polysaccharide. Through light scattering measurements it was revealed that GLL possesses an average molecular weight situated between 23.8 and 43.4 x 10^4 g mol^{-1} [188]. Another important characteristic of GLL is its distinctive gelation temperature which leads to the formation of transparent gels [137]. In water, GLL is completly soluble above the temperature of 90°C, while cooled at around 30-35°C it forms strong hydrogels. Also, thorough investigations have demonstrated that at room temperature GLL has a double helix conformation that upon heating transforms into two single macromolecular chains.

The gelation process usually occurs in a two step mechanism over a period of 24 hours, as follows:

- in the first step a transition from a random coil to a double helix structure takes place;
- in the second step the double helix molecule is transformed into a helical aggregate [138].

At high temperature values, the GLL solutions are found in a sol state while at low temperatures a gel is formed. Alongside temperature, various other factors influence the gelation process and the strength of the final networks [139]. Among them, some of the most important parameters are: acetyl content, type and concentration of cations, pH, presence of hydrophilic ingredients and temperature.

In the case of the acetylation degree, studies have detemined that an acetylated GLL provides a soft, elastic and flexible thermoreversible gel, while deacetylated GLL molecules form more brittle and stiffened thermoreversible gels. These characteristics may be attributed to the fact that the presence of acetyl groups on the polymeric chains hinders the formation of double helix and thus the formation of more compact hydrogels [132].

Since GLL is a polyelectrolyte, gel properties are influenced by the presence of different salts in the medium [133]. By introducing various counterions in GLL solutions it was observed an increase in the elastic modulus that resulted in the end in an increase in the gel strength and the brittelness of the final materials [140].

More detailed analyses highlighted that the most efficient cations are the divalent ones, for example Ca^{2+} and Mg^{2+} cations.

Studies elaborated by Sanderson and Clark revealed that gel strength can be enhanced by increasing the pH from 3.5 to 8. Although a change in pH does not affect the setting point of the gel, it can modify the melting temperature of a certain hydrogel. As an example, the melting point of a specific gel can be increased by simply decreasing the pH from a neutral value to a value of 3.4, while maintaning all the other parameters constant [141].

Kasapis *et al.* examinated the influence of different sugars on the gelation process through transmission electron microscopy. It was observed that by adding such type of hydrophilic ingredients to a system the ion concentration needed to obtain an optimal gel strength had significantly decreased [142].

GLL gel is stable at high temperatures up to 90ºC. The melting point of this anionic exopolysaccharide depends strongly on the concentration and type of cations present in the medium. Thus, depending on the final application, the melting temperature of a GLL based material can be modified by simply adjusting the counterion composition and concetration.

The reological behavior of GLL structures as well as the gel formation mechanism were investigated with the help of multiple methods of characterization. Among then we recall: rheological measurements, tensile testing, atomic force microscopy, differential scanning calorimetry, NMR spectroscopy, texture analysis and not least confocal laser scanning microscopy [143-146].

Alginates, a polyanionic heteropolysaccharides family, are found either as structural components in various marine brown algae (*Laminaria hyperborea*, *Macrocystis pyrifera*, *Laminaria digitata*, *Ascophyllum nodosum*, *Laminaria japonica*, *Eclonia maxima*, *Lessonia nigrescens*, *Durvillea antartica* and *Sargassum species*) or as capsular polysaccharides in two bacterial genera: *Pseudomonas* (*Pseudomonas aeruginosa*, *Pseudomonas fluorescens*, *Pseudomonas mendocina*, *Pseudomonas putida* and *Pseudomonas syringae pathovar glycina*) and *Azotobacter* (*Azotobacter vinelandii*) [147, 148].

The composition and sequential structure of ALG depend on the type of algae used as well as on the seasonal and growth conditions [149].

Structurally, ALG are linear polymers composed of variable proportions of β-D-mannuronic acid (M) and α-L-guluronic acid (G) residues connected by (1,4) glycosidic bonds (Scheme **1**).

Haug *et al.* [150] have provided the first information about the structure of alginate, when by partial acid hydrolysis and fractionation methods they have separated the ALG in three fractions with different composition as follows:

- two fractions that are formed by two types of monomers: β-D-mannuronic acid and α-L-guluronic acid, respectively;
- one fraction that contains approximatively equal proportions of both monomers and a large number of MG dimer units.

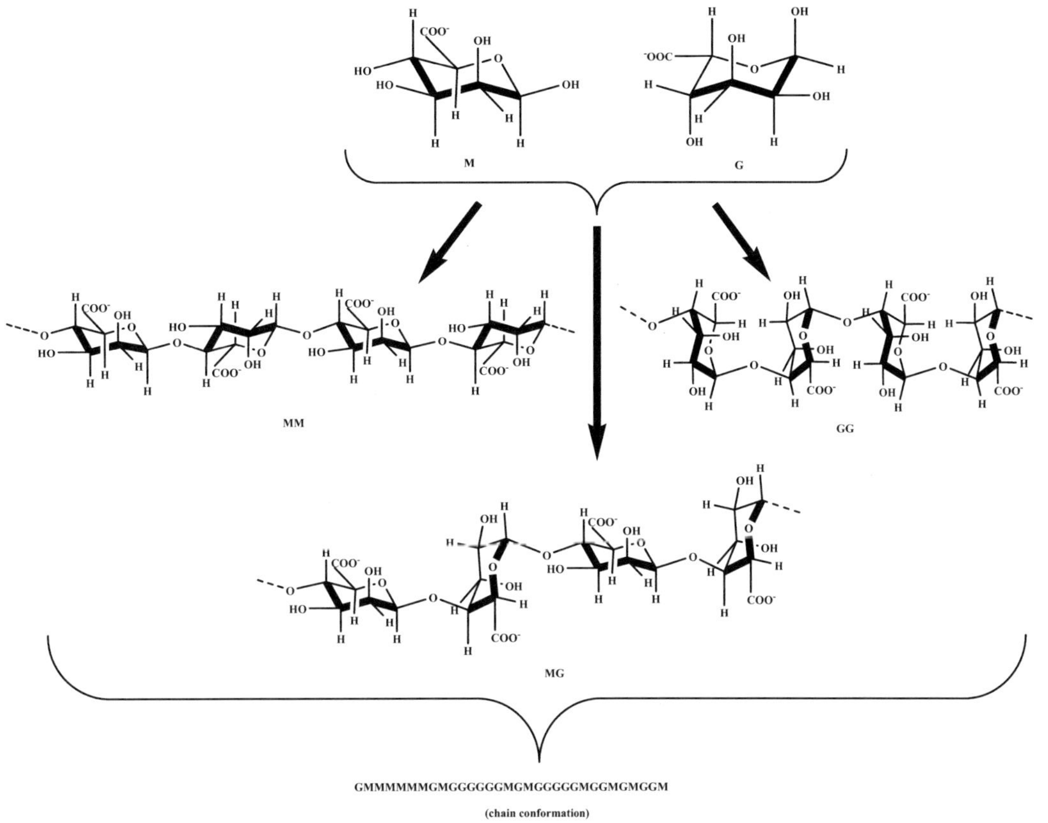

Scheme 1: Chemical structure of alginate.

The conclusion was that ALG is a block copolymer composed of M, G and MG blocks [150]. Using X-ray diffraction method it has been demonstrated that both types of regular blocks (M and G blocks) present the following conformations: two-fold helix (mannuronic acid blocks); three-fold helix in presence of sodium salt or two-fold helix in presence of calcium salt (guluronic acid blocks) [151].

The main property of ALG is their capacity to form gels in the presence of various di- and multivalent cations, the affinity of the ALG increasing in the following order: Mg<< Fe< Mn< Zn< Ni< Co< Ca< Cd< Sr< Cu< Pb< Ba [152]. The gelation process in the presence of Ca^{2+} ions can lead to the formation of strong interactions between Ca^{2+} ions and guluronate blocks of ALG that allows a high degree of coordination in accordance with the "egg-box" model [153].

The mechanical properties of ALG gels can be improved by increasing the molecular mass as well as the length of G blocks.

BIOMEDICAL APPLICATIONS OF MICROBIAL EXOPOLYSACCHARIDES

The discoveries regarding different types of exopolysaccharides with unique properties as well as the beneficial effects of bacteria to the human health have contributed to the creation of new advanced materials that could be used in biomedical field for the elaboration of various modern therapeutic strategies. Some of the health benefits of exopolysaccharides are presented in Table **5**.

Table 5. Some biomedical applications of exopolysaccharides studied in this chapter.

<table>
<tr><th>Name of EPS</th><th>Biomedical Properties</th><th>Biomedical Applications</th><th>References</th></tr>
<tr><td>DEX
Pullulan</td><td rowspan="6">Nontoxic
Non-immunogenic
Biodegradable
Biocompatible
Antitumor, anticoagulant and antiviral activity</td><td>Blood plasma substitutes</td><td>[154, 155]</td></tr>
<tr><td>DEX, pullulan, curdlan, XAN, GLL, HA, ALG</td><td>Drug delivery systems</td><td>[156-161]</td></tr>
<tr><td>HA, XAN, GLL</td><td>Ophthalmic applications</td><td>[162-164]</td></tr>
<tr><td>HA, XAN, GLL and ALG</td><td>Wound healing</td><td>[165, 166]</td></tr>
<tr><td>HA</td><td>Adhesion prevention</td><td>[167]</td></tr>
<tr><td>HA, GLL, ALG</td><td>Tissue engineering</td><td>[168-172]</td></tr>
</table>

Drug Delivery Systems (DDS)

The low efficiency of small molecular weight drugs in treating severe diseases and the series of disadvantages presented by the administration of this type of drug (the necessity of multiple drug doses in a short period of time in order to assure the proper therapeutic dosage; the possible fluctuation of the drug concentration in the blood flow which may lead to low or toxic dosage and the intermittently and/or incomplete sorption of drug) have conducted to the necessity of finding new strategies in elaborating improved controlled drug delivery systems or target systems.

The drug administration can be performed by several routes: oral, sublingual, pulmonary, ocular, nasal, anal, vaginal, transdermal and parenteral.

Oral Administration of Drug

The hydrogel based on DEX, prepared by radical polymerization of N-isopropylacrylamide and DEX macromer containing oligolactate-2 hydroxyethyl methacrylate units was loaded with two hydrophilic model drugs such as, methylene blue and bovine serum albumine. The drug release kinetics were influenced by several factors like: molecular size of the drugs, temperature, swelling degree and degradation stages of the hydrogel [173].

5-Aminosalicylic acid and nalixidic acid were used for the preparation of the drug-dextran conjugates in order to treat the inflammatory bowel disease [174].

1,3 β-D-glucans have potential in biomedical fields due to their properties such as helical structure and gel forming ability. Dry tablets based on curdlan encapsulating theophylline with good pharmacokinetics were obtained by spray-drying technique.

Another exopolysaccharide that has found applications in drug delivery is pullulan [175]. In order to obtain the controlled drug delivery systems, the pH and temperature sensitive material was prepared by grafting poly(N-isopropylacrylamide-*co*-acrylamide) onto pullulan microparticles [176].

Pullulan presents a high affinity to the liver, this property being exploited for targeted drug delivery to this organ by introducing a chelating residue (diethylenetriaminepentaacetic acid, DTPA) on pullulan. After that the interferon was conjugated by coordination with DTPA-pullulan in the presence of zinc ions. The results showed that this conjugate may be used for the treatment of patients with chronic liver disease in order to eliminate the hepatitis C virus [177]. Also, hydrophobized pullulan conjugates were used as macromolecular supports for various bioactive substances like: metronidazole, nicotinic acid, sulfathiazole, mitoxantrone, epirubicin and artificial vaccines [178, 179].

To prepare a proper DDS, depending on the route of administration and the place of action, HA can be used alone in a native state or it can be modified through grafting or crosslinking reactions. Several drugs have been loaded into these vehicles among them being pilocarpine hydrochloride, a parasympathomimetic alkaloid for dry mouth and glaucoma treatment [159].

Core-shell microparticles made of two biocompatible natural polymers (chitosan and HA) were obtained by means of the simple coacervation method [180]. The retention and the release processes of two antibiotics such as, chloramphenicol succinate sodium salt and cefotaxime sodium salt have led to the idea that chitosan-HA complex microparticles have potential applicability in oral administration of antibiotics [181].

XAN has a wide range of industrial applications due to its special rheological properties such as high viscosity even at low concentrations or high elasticity at low shear rates. It can be easily used as stabilizer for solid suspensions and foams, as thickener for solutions or as stabilizer for emulsion systems in cosmetics and food industry. More recently XAN has been used with great success in biomedical applications.

Furthermore, associated with other natural polymers such as glucomannan, carragenan, ALG and gelatin, XAN has exhibited excellent results both in pharmaceutical and medical fields. For example in combination with galactomannan, two types of hydrogels can be obtained, one at approximately 20ºC when XAN is found in a random coiled conformation and a second one at around 60ºC when the exopolysaccharide presents a helical conformation.

Being relative stable at a wide range of temperatures and pH, XAN can easily be manipulated under the form of nano- and microparticles, films, hydrogels or 3D matrices with the purpose to retain and further release various types of biomolecules.

Thus, for oral administration, in the case of drug release from tablets formulations, the presence of XAN in the systems has always lead to a slow solubilization of drug and a more sustained release of the biomolecule.

For example, to treat oral candidiasis, Ruissen *et al.* proposed the preparation of XAN based formulations obtained either by physical blending or by covalent coupling of the active biomolecule and the exopolysaccharide. Precisely, their study has been concentrated on determining the activity against *Candida albicans* of synthetic cationic peptide dhvar 1, based on the human fungicidal salivary peptide histatin 5. In both cases the results indicated that the presence of XAN in the system significantly increased the LC_{50} of the peptide from 2.6 (S.D. = 0.6) to 5.8 (S.D. = 4.0) for the physical blends and to 18.4 (S.D. = 6.7) for the covalent coupling. The presence of the polysaccharide led to a slower degradation process being concluded that XAN can constitute an appropriate carrier for antimicrobial peptides when a sustained delivery system is desired [182].

XAN pellets prepared by extrusion-spheronisation method were evaluated for the release of diclofenac sodium and ibuprofen [160].

The release profiles for the two model drugs revealed very different behaviors as follows:

- tablets containing ibuprofen released the drug in a bimodal manner and the release profile was characterized as a class II transport mechanism;

- tablets obtained from XAN pellets and diclofenac sodium exhibited an anomalous release pattern. In this case drug release was controlled by both diffusion and erosion mechanisms.

Diclofenac sodium release was also investigated from XAN matrices [183]. Gastrointestinal transit of the formulation and *in vivo* dissolution of tablets were evaluated both under fed and fasted conditions. Results gathered stated that a faster dissolution rate and a better release profile were obtained in the first case in comparison with release profiles for patients deprived of any nutritious source. It was determined that for the fasted state more than 60% of the loaded drug was delivered to the colon.

Talukdar *et al.* compared the effect that model drugs could exert on the rheological properties of natural materials namely XAN and cellulose. As model drugs they opted for three types of hydrophilic biomolecules precisely caffeine, indomethacin and sodium indomethacin. In the case of XAN, studies have shown that among the three drugs only sodium indomethacin seemed to influence the rheological properties of the solutions. This behavior is probably due to the modification of ionic strength in the system, being known that XAN solutions are susceptible to this kind of changes. Being an anionic molecule with high solubility in aqueous solutions, sodium indomethacin seemed to have a pronounced influence upon rheological properties but only in the case of dilute XAN solutions. For more concentrated systems the presence of the drug appeared to have no major effect and in some cases being practically negligible. The ascertain differences in rheological behavior also exerts a major impact on the release mechanism for the biomolecules [184].

Thus, for tablet formulations with high contents of XAN it seems that the general conduct consists in higher rates of release owed to higher swelling degrees of systems.

For controlled drug release of antihypertensive drug acrylamide-grafted-XAN copolymers were investigated. Copolymers obtained *via* free radical polymerization were used as potential carriers for the oral administration of atenolol and carvedilol. In both cases drug release kinetics and transport mechanisms followed a non-Fickian trend, diffusion through the matrix being in strong relation with polymer chain relaxation [129].

To improve characteristics, composite beads consisting in ALG and XAN have been prepared *via* ionotropic gelation technique. To evaluate their capacity as drug delivery systems diclofenac sodium was elected as hydrophilic molecule. Investigations performed on the obtained complexes revealed the essential role played by XAN in release profiles. Thus, due to erosion process of small aggregates of XAN situated on surface beads, a higher release rate for diclofenac could be determined. This behavior suggested the presence of XAN in the system could modulate physicochemical properties of structures and in the consequence drug release profiles [185].

XAN hydrogels crosslinked with epichlorohydrin were successfully used as carriers for several drugs such as theophylline, isosorbate dinitrate (isoket) and medroxyprogesteron. The loading mechanism in all cases was diffusion process. Studies have determined an affinity displayed by the hydrogels towards isoket, with maximum values of inserted drugs being as follows: 28 mg/g for metroxyprogesteron, 80 mg/g for theophylline and 150 mg/g for isosorbate respectively [186].

The complex with chitosan under the name of ChitoXan has also been successfully used as vehicle for controlled drug delivery. Using fibroblasts as cell culture, the cytotoxic effects of both the matrix and the resulting fragments from degradation process were assessed. *In vivo* tests monitored by light-microscopic observations have indicated a good tolerance for the hydrogel with no significant cell culture damage. On the contrary, the elevated hydrophilicity and high porosity allowed the penetration of phagocytic cells into the matrix followed by the resorption and clearance from the body [187].

Polyelectrolyte complex hydrogels based on XAN and chitosan were also studied from a biodegradation point of view [188]. For this purpose, microparticles based on the two polysaccharides were prepared *via* complexation process and their characteristics were evaluated. Investigations upon degradability were performed under physiological conditions. Based on the results, microspheres proved to have potential for latter use as DDS in the gastrointestinal tract.

Some pharmaceutical formulations have been obtained from XAN and galactomannan blends. Highly substituted galactomannan procured from *Mimosa scabrella* mixed with XAN was investigated as possible carrier for diclofenac sodium. Both as tablets and capsules, the formulated systems indicated a drug release profile respecting a zero-order kinetic with erosion being the dominant mechanism [189].

Polyelectrolyte complexes between an ion exchange resin and XAN or GLL, in form of microparticles were obtained for inclusion of cefotaxime sodium salt. The results obtained showed that the microparticles present a higher adsorption capacity and sustained release profile [190].

In our research, theophylline was chosen as model drug for encapsulation in microparticles based on synthetic polymers [glycidyl methacrylate (GMA) and dimethacrylic esters] and XAN. Depending on the nature of dimethacrylic esters [ethylene glycol dimethacrylate (EGDMA), diethylene glycol dimethacrylate (DEGDMA) and triethylene glycol dimethacrylate (TEGDMA)] that have the role of crosslinking monomers, the microparticles without drug showed different structures (Fig. **2**) [191].

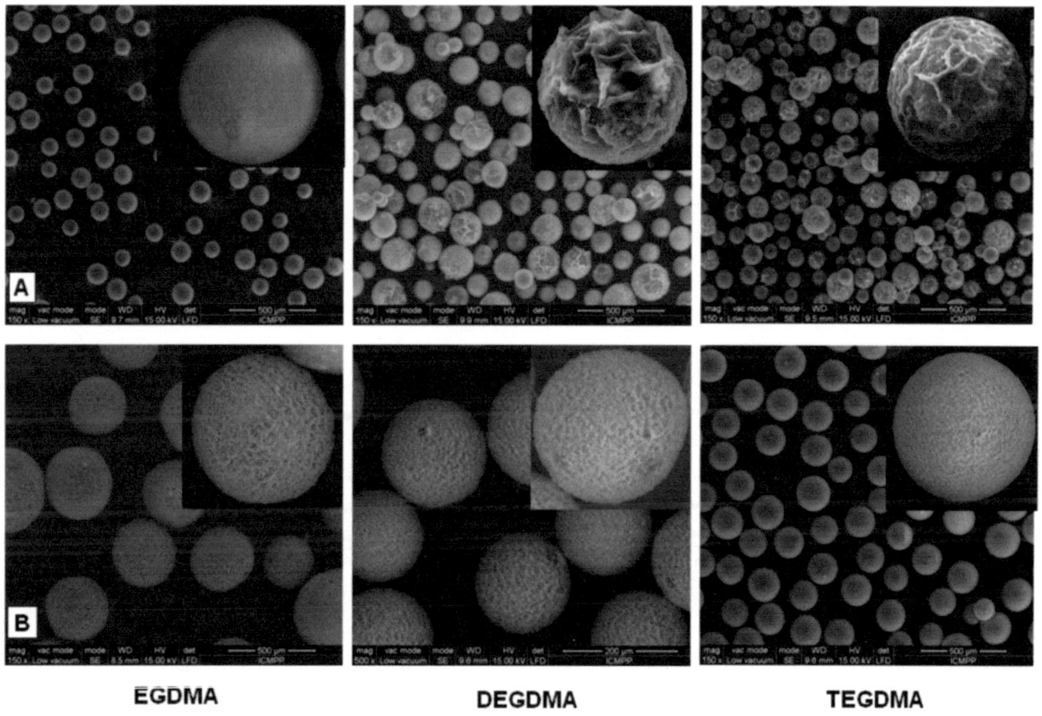

Figure 2: SEM micrographs of microparticles: **A**) GMA-XAN crosslinked microparticles; **B**) GMA crosslinked microparticles.

It was found that XAN was linked at the surface of microparticles forming a layer surrounding the crosslinked microparticles. The grafting mode of XAN on synthetic polymer matrix is shown in Scheme **2**.

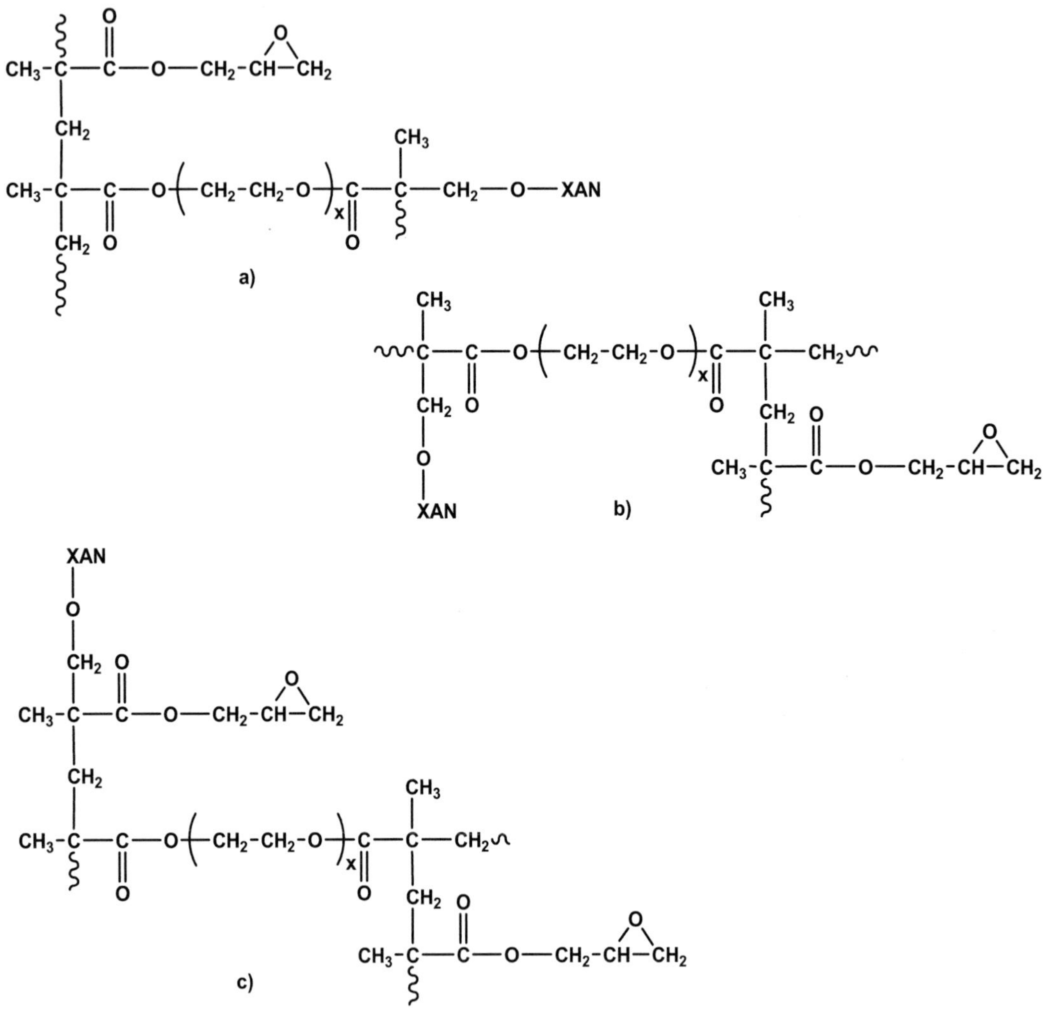

Scheme 2: Grafting mode of XAN on synthetic polymer matrix.

The presence of XAN in the structure of microparticles increases the biocompatibility of the system as well as the hydrophylicity and the specific surface area. By theophylline encapsulation the spherical shape and the porous structure of microparticles were preserved. The kinetic release studies demonstrated that microparticles based on synthetic polymers and XAN showed a sustained release of theophylline.

Being one of the most extensive studied bacterial polysaccharides, GLL is characterized through a number of excellent properties such as biocompatibility, biodegradability, non-toxicity, widespread availability and low production costs which leads to the application of this natural gum in various areas of our life.

Mainly used in food industry and cosmetics as stabilizer or thickener, GLL has been used in recent years to formulate novel DDS for biomedical applications.

Being an anionic polysaccharide, GLL can undergo crosslinking processes in the presence of suitable counterions, especially in the presence of divalent cations. This major characteristic possessed by GLL makes it attractive for the incorporation of many biological compounds such as active substances, enzymes or proteins.

For oral administration, formulations based on GLL were thorough investigated especially for the treatment of various gastrointestinal disorders. The major advantage presented by GLL is its capacity to form gels *in situ* in the presence of suitable counterions and due to the acidic pH found in stomach, the polysaccharide demonstrated to be adequate for preparing DDS in the form of tablets.

To evaluate the performance of the polysaccharide as drug carrier, Miyazaki *et al.* prepared several aqueous solutions containing 1% (wt/v) GLL, calcium ions in a complex form and theophylline as model drug. The prepared solutions were further administrated orally in both rats and rabbits and the release rate of theophylline was studied. Once reached the stomach the GLL solutions underwent a gelation process retaining the active substance in the polymer matrix mesh and thus assuring a sustained drug release rate over a longer period of time. The studies performed indicated a better release profile for the newly prepared systems in comparison with some common commercial theophylline formulations [192].

Interpenetrating networks based on GLL and poly(vinyl alcohol) was obtained *via* emulsion crosslinking method [161]. The prepared microspheres were then loaded with carvedilol, a common antihypertensive drug, with an encapsulation efficiency of more than 87%. Studies regarding the release profile of carvedilol from the polymeric networks have indicated that both the mechanisms and the rate of release are in correlation with the crosslinking density of the interpenetrating networks and with the molar mass between crosslinks. As predicted, by increasing the density of crosslinking the swelling degree of the matrices has decreased significantly leading in the end to a lower rate of release for the model drug.

Literature also mentions GLL/chitosan complexes prepared by self-assembly at an aqueous solution interface. In this case, spherical capsules were formed by electrostatic interactions between the two polysaccharides *via* a polyanion complex formation. The newly synthesized network was investigated regarding its release capacity of both highly molecular weight proteins and low molecular weight biomolecules [193].

The results obtained in both cases indicated that chitosan/GLL capsules could be a useful alternative for the preparation of DDS for controlled drug release.

Materials derived from natural sources like alginates are abundant, cheap and present interesting properties. ALG are characterized by two main properties:

- the presence of carboxylic groups on both M blocks and G blocks, making them suitable for chemical modifications;
- ability of G blocks to form gels in the presence of some divalent ions.

The drug release process can be modulated by controlling the rate of delivery, which can be achieved by several ways such as, physical crosslinking with divalent ions, preparation of the semi-interpenetrated network polycomplexes and functionalized-ALG/chemically crosslinked ALG and application of drying and freeze-thawing methods.

ALG gels have been used as controlled DDS for delivery of some types of drugs: flurbiprofen [194], antineoplastic agents like mitoxantrone and doxorubicin [195], diclofenac sodium [185], theophylline [196], albendazole [197], metronidazole [198], amoxicillin [199], verapamil hydrochloride [200], prednisolone [201] and pindolol [202]. Floating drug delivery system incorporating riboflavine as drug model was developed by Stops *et al.* [203]. The riboflavine release was slow in acidic media and more rapid in alkali media. Taha *et al.* have investigated a new opportunity to prepare ALG matrices by incorporation of sodium laurylsulfate or sodium dodecylsulfate in ALG-drug (chlorpheniramine maleate) solution prior to crosslinking with divalent ions (Ca^{2+} or Zn^{2+}) in order to improve the amphiphilic drug loading into ionically crosslinked ALG networks [204]. Chitosan-ALG microparticles have been studied as colon-specific delivery systems of 5-aminosalicylic acid [205]. *In vitro* drug release studies have demonstrated that the drug is released in a controlled manner, the exponent of diffusion coefficient indicating an anomalous transport of drug characterized by chain relaxation, erosion and degradation of polymer networks.

DDS based on ALG and its derivatives have been developed either for neutral regeneration and neuroprotection [206] or for use as mucoadhesive drug delivery systems [207, 208].

Cancer, one of the most aggressive diseases of this century represents the main cause of death of millions of people around the world. For this reason, the researchers from whole world work to find the cause and the adequate strategies for elucidating the mystery of this disease.

Dexamethasone and methylprednisolone were covalently attached to DEX using as a spacer the succinic anhydride and succinic acid/glutaric acid, respectively [209, 210]. Methotrexate-DEX, sodium phenylacetate-DEX and mitomycin-DEX conjugates are another exemples of DDS based on DEX that have proved their utility in the treatment of cancer and in the inhibition of tumoral cells [211, 212].

Pullulan was used by Scomparin *et al.* to obtain two new anticancer polymers (doxorubicin-derivatized pullulan and folic acid-functionalized bioconjugate) that were designed for tumor cell targeting [158]. The results obtained suggested that:

- both bioconjugates displayed similar drug release features in pH= 7.4 or in plasma, when ~ 20% of doxorubicin was released within 3 days;
- a complete release of drug in about 40 h is observed at pH = 5.5;
- pullulan is a suitable macromolecular support for anticancer drug delivery

The potential applications as DDS are in direct and strong relation with the number of receptors in the body with specific affinity for HA. The most common receptors capable to be targeted in hyaluronan are CD44 and RHAMM [13]. Both receptors have demonstrated affinity to binding cancer cells. Thus, several bioconjugates based on HA have been investigated in tumor targeting, one example being HA-carborane conjugates [213].

Tumor-target systems were also prepared from self-assembled HA matrices grafted with poly(L-histidine). The pH responsive conjugate showed favorable results as anti-cancer drug carrier, more precisely for target release of doxorubicin [214].

Doxorubicin encapsulated into HA-ceramide coated hybrid liposome was used for cancer diagnosis and therapy. *In vitro* release profiles for doxorubicin and *in vivo* studies over target capacity evidenced a better behavior for hybrid liposome compared with some conventional ones. The results determined that HA-ceramide derivatives can be useful both in cancer imaging and drug delivery [215].

Currently a novel drug delivery labeled ONCOFID^{TM}-P is used with encouraging results in phase II clinical studies for the treatment of refractory bladder cancer. Also, HA-irinotecan conjugates are being used to treat patients with metastasis cancer in phase II and phase III clinical trials [216].

Magnetic nanoparticles based on modified HA and iron oxide were prepared by one-pot hydrothermal approach. The obtained nanoparticles were characterized as stable, haemocompatible and cytocompatible and the properties assured their applicability in tumor target magnetic resonance imaging (MR imaging). For their usage in MR imaging, the nanoparticles were previously modified with fluorescein isothiocyanate [217].

Nanostructured lipid carriers coated with HA were prepared by melt emulsion technique. Loaded *via* electrostatic attraction with paclitaxel, the obtained system exhibited good results in tumor targeting and cancer treatment [218].

Interpenetrated networks based on HA also showed that these types of scaffolds can provide a desirable environment for the adhesion and chondrogenic differentiation of human adipose derived stem cells. The results so far suggested the 3D networks can be applied successfully as drug carries [219].

In pulmonary applications, peptide-HA nanoparticles showed good results in the treatment of asthma. Dexamethasone was the drug of choice and the conjugate obtained exhibited positive therapeutic effects in both neutrophilic asthma and eosinophilic asthma in mice [220].

Ophthalmic Applications

Being a major component of the vitreous humor HA is one of the most used biomaterial in ophthalmic applications. Some of the first works made date since the 1960's and 1970's when Balazs and a team of surgeons successfully carried out a series of clinical trials using HA as biomaterial [162]. In the same period of time the first device based on HA alone became available on the market under the name of Healon®.

Since then many other preparations based of various copolymers have been developed with the purpose to improve efficiency and patient compliance. Some of the commercial brands available today are: Viscoat®, Discovic® or VISC26 [221].

The growing interest showed not only in ophthalmology, but also in other medical fields is attributed to the unusual rheological properties of HA which are in strong correlation with the average molecular weight of the polymer [111].

Depending on its molar mass HA based materials for ophthalmic applications can be classified in three categories as follows [162, 222]:

- viscoadhesive ophthalmic viscoelastic devices – materials with very high viscosity and elasticity even at low shear rates;
- viscodispersive ophthalmic viscoelastic devices – materials with low viscosity and low elasticity at different shear rates;
- viscoadaptive ophthalmic viscoelastic devices – elastic and pseudoplastic materials.

As a viscoelastic material, HA can often be used in cataract surgery. It has been shown that even a 1% (wt/v) solution of HA can significantly enhance the repair process after corneal injury [223]. In surgery, HA solution can also be used for corneal transplantation and replacement of vitreous humor as well as in glaucoma surgery or to manipulate detached retina [162].

More recent research determinates that HA hydrogels can be easily used for corneal blindness [224]. Corneal blindness is a worldwide disease caused by limbal stem cell deficiency (LSCD). To overcome limitations of classical therapy for LSCD, hydrogels based on natural polymers represent a very attractive alternative. Taking this into consideration, hydrogels prepared from HA or HA derivatives demonstrated to be suitable scaffolds for cell culture and proliferation and ultimately for corneal regeneration.

Due to the viscoelastic properties, this natural polymer can be used for the protection of eye surface. It is well known that HA solutions do not flow as easily as physiological salt solutions do, this characteristic making them usable as humidifiers for patients with dry eye syndrome or for patients that use eye lens [225].

Nonetheless, because of its high residence time on eye surface, HA is often used as carrier for many topical ophthalmic drugs such as: adrecaidine propargyl ester, pilocarpine, gentamicin, timolol and tropicamide [226-230].

In ophthalmic applications XAN seems to exert very good therapeutic effect on corneal epithelial disorder. Also, in the case of corneal epithelial protecting effect the exopolysaccharide showed positive results being more and more considered as a potential pathway to follow for many ophthalmic applications [231].

XAN dispersions were prepared in an ophthalmic dosage and their interaction with mucin in the corneal area was investigated [163]. It is well known that XAN solutions are characterized by very high viscosity and thus patient acceptability is very limited for these types of preparations. To overcome the obstacle, a new method of synthesis was suggested by the authors consisting in dispersion of XAN powder in a phosphate buffer solution, pH 7.4, at 70ºC. To ensure a complete hydration of the polymer, the dispersion was further maintained at 6ºC for a total of 12 hours. To investigate the interaction between XAN and mucin, equal amounts of both components were mixed together and rheological determinations were elaborated. Studies effectuated have revealed that the mucoadhesive capacity of XAN is in strong correlation with the average molecular weight and the structural conformation of molecules. Taking this into consideration XAN solutions could be used with much success in corneal epithelial disorder as protecting agent.

Moreno *et al.* have been investigated XAN solutions as possible treatment for dry eye syndrome [232]. Their research laboratory performed a comparison between two types of ophthalmic solutions in terms of safety and efficiency. Thus, two solutions were prepared, one based on chondroitin sulfate and XAN and the other based on poly(ethylene glycol), propylene glycol and hydroxypropyl guar. Undergone analyses have showed a better performance for chondroitin sulfate/XAN solutions in dry eye treatment. These results may be explained by the high viscosity of XAN which acts as a stabilizer for the solution and leads to a longer residence time on eye surface.

In combination with HA, XAN exhibited good results as re-epithelializing agent [233]. It is well known that epithelial cells such as corneal cells can undergo severe damage when they come in contact with foreign objects. Some of these injuries may consist in postoperative ulcers or cuts and wounds due to accidents and surgical interventions. To prevent or treat such lesions, new conjugates needed to be formulated and XAN and HA have shown that they could represent an attractive alternative. Under investigations, films based on these biomaterials have shown good capability to accelerate the formation of new epithelial cells being thus suitable for ophthalmic utility.

To investigate effects in traumatic corneal abrasions, Faraldi *et al.* elaborated hydrogels containing HA and XAN. To this mixture a model drug was added in the presence of netilmicin antibiotic [234].

As previously mentioned, corneal abrasion is a defect of the corneal epithelial surface characterized by acute ocular pain and patient discomfort. Traditional treatments consist in using eye patches, topical antibiotics or oral analgesics.

A possible alternative to classical therapy may be presented by usage of eye gels capable of *in situ* crosslinking in the presence of ions from the lacrimal fluid. Both HA and XAN have the capacity to promote wound healing and cell proliferation due to their extraordinary characteristics thus the combination of the two extracellular polysaccharides alongside an antibiotic may present a solution in corneal abrasion treatment and in tissue protection against bacterial infections.

In situ gelling vehicles were synthesized by utilizing poloxamer-407 in combination with XAN and guar gum. The influence of the two natural polymers over the physicochemical characteristics and biological performance of the material was investigated. *In vitro* results have showed a slower rate of drug release for the systems containing the polysaccharides while *in vivo* tests revealed an increase in bioavailability due to longer corneal residence time. Decreasing the frequency of dosage administration the new proposed systems can lead to a better patient compliance being an attractive alternative to conventional ophthalmic therapy [235].

Traditional ophthalmic therapy consists in the administration of solutions or suspensions containing an active substance or biomolecule in the ocular globe. The major disadvantage for these classic methods of treatment is the short residence time on the eye surface. Thus, shortly after its administration, dilution takes place due to blinking and the active substance is eliminated from the corneal surface. In this case, to ensure a proper therapeutic response a new dosage is required which can cause some discomfort for patients.

To prevent this situation, an attractive alternative is presented by the use of new *in situ* gelling systems with sustained drug release [236].

One of the most common biomaterials adequate for this type of applications is GLL. Capable to form *in situ* gels, this extraordinary exopolysaccharide has shown great interest in the scientific community regarding its use in ophthalmic therapy.

In this context GLL solutions with *in situ* gelling capacity were used with great success by Balasubramamian *et al.* for the sustained release of indomethacin [237].

Presented in a complex ocular formulation containing gatifloxacin, the extracellular polysaccharide exhibited good mucoadhesive properties and a sustained release rate of the drug [164]. For a more thorough evaluation of the capacity to treat ocular disorders, three types of polymeric vehicles were prepared as follows: a first set was obtained from unmodified GLL, a second set was based on GLL combined with sodium carboxymethylcellulose and finally a third type was prepared from GLL and ALG, respectively. The three systems where further investigated in terms of gelation ability, rheological behavior, *in vitro* release, muchoadhesive force and antibacterial activity with positive results in all aspects. Due to their ability to enhance bioavailability prolonging the precorneal residence time and due to the ability to assure a sustain release rate of the drug, the newly designed systems could represent a viable alternative for ophthalmic applications.

Because of its ability for *in situ* gelation, low-acyl GLL has been investigated by Rozier *et el.* with positive results regarding the synthesis of novel ophthalmic formulations [238].

Furthermore, Sanzgiri *et al.* have realized a comparison between three types of formulations based on GLL and evaluated their capacity for drug delivery to the eye surface. The authors concentrated their work on evaluating the properties of the newly designed systems and the release profiles of methylprednisolone (MP) from GLL-MP films, physical blends of GLL films and MP, and GLL dispersions in aqueous solutions containing suspended MP respectively. *In vitro* experiments have indicated that in the case of covalently linked films the drug release mechanism followed a zero-order kinetic while the physical blends and water dispersions followed a square root of time relationship and an anomalous pattern [239].

Aqueous solutions of GLL were also employed for the treatment of glaucoma, encountering positive response [240].

Tayel *et al.* prepared novel ocular formulations under the form of drug-loaded nanoemulsions. The aim of their work was to elaborate new polymeric carriers for the sustained release of terbinafine hydrochloride [241]. *In vitro* drug release tests were performed for the synthesized structures both in aqueous solutions and in the presence of mucin. The conducted experiments concluded a more retarded, zero-order drug release rate for the systems immersed in solution alongside mucin. *In vivo* tests effectuated on rabbit aqueous humor revealed a good biocompatibility and an elevated bioavailability for the drug-loaded systems suggesting these novel materials could be suitable for the treatment of ocular irritation.

Nasal Administration

Sephadex, a crosslinked DEX gel in bead form, possesses the ability for delivery of the protein and peptide drugs across the mucosal membrane.

For this purpose, Sajadi Tabassi *et al.* have used Sephadex containing tetanus tanoid as a proper mucosa antigen delivery system for nasal administration [242]. Since its discovery XAN has been widely investigated as tablet excipient with the role of increasing drug release rate in oral administration, but about the nasal applications have not been many reports mentioned in literature. In this context, Bertram and Bodmeier proposed novel sponge-like *in situ* gelling materials prepared for the delivery of proteins and peptides in the nasal cavity [243]. Studies elaborated determined that by using high molecular weight XAN sorts a better bioadhesion to the nasal mucosa was revealed and thus a more extended drug release profile was obtained. Although it is very unlikely for the biomaterials to remain in the nasal cavity for a long period of time, the sponge-like structures displayed a release profile over an 8 hour period suggesting the structures could be used as potential carriers for sustained drug release.

Anal Administration

Curdlan has been used to prepare suppositories loaded with three types of drugs: indomethacin, prednisolone and salbutamol sulfate for rectal administration [157]. The preparation conditions (heating time, concentration of curdlan solution and the toxicity of the medium) of the suppositories based on curdlan have little effect on the drug release profile. The sustained drug release mechanism was influenced by the heating temperature during gel preparation, the drug amount in the pharmaceutical formulation and the type of release media [157].

Subcutaneous Administration

Literature mentions drug delivery systems prepared in forms of microspheres *via* a simple spray-drying technique. The microparticles were further used as bone cements for the sustained release of erythrosin B [244]. Both high molecular weight and low molecular weight HA-modified docetaxel-loaded liposomes were prepared through electrostatic attraction to evaluate the lymphatic targeting after subcutaneous administration. *In vivo* evaluation of the novel systems suggested that high molecular weight materials promote a better targeting by enhancing lymph node and lymphatic drainage making them more suitable for diagnostics and treatment of lymphatic disease [245].

Transdermal Administration

Liu *et al.* have reported the use of HA to prepare novel self-dissolving microneedle arrays for transdermal drug delivery. As a model drug the authors have chosen a relative high molecular weight molecule in the presence of fluorescein isothiocyanate-labeled DEX. The analyses effectuated revealed that the DDS have lead to an increasing in drug penetration and accumulation at the place of action and that it could be an attractive pathway to improve transdermal delivery [246].

Implant

An interesting application of the polysaccharide consists in the preparation of GLL based films for the incorporation of insulin. Implanted in rats, the obtained devices exhibited a sustained release rate of insulin for the treatment of diabetes. *In vivo* experiments performed on several batches of rats have shown that the release effect of insulin has lasted for one week concluding that GLL scaffolds could be appropriate materials for the preparation of sustained release systems in diabetes [247].

Arthritic Diseases

Rheumatoid arthritis is an autoimmune disease, in which the immune system attacks healthy tissues (joints), being characterized by chronic progressive inflammation and gradual joint and bone distruction [248]. This systemic inflammatory disorder affects the lining of the joints causing swelling, stiffness and pain. The therapeutic agents available on the market (etanercept, infliximab and adalimumab) present some disadvantages such as high clearance rates and low bioavailability [249]. The challenge for the scientific world is to develop the new carriers for proteins that are able to prolong their therapeutic effect with reduced side effects [250]. A new system based on temperature sensitive copolymer [succinylated pullulan-g-oligo(L-lactide)] and therapeutic protein (etanercept) was prepared and it was demonstrated that this formulation improve the long-term stability of etanercept and exhibited the therapeutic effect in animal model with rheumatoid arthritis [251].

HA presents a series of interesting characteristics that make it applicable in many arthritic diseases, especially in osteoarthritis. Being a major component in joints, specifically in synovial fluid, HA fulfils the role of lubricant and shock absorber in many cases. In addition, HA has the property to restrict the entry of plasma proteins into synovial fluid while facilitating the transport of water and small solutes between synovial capillaries, cartilage and other joint components [118].

In arthritic disease it has been determined by thorough examination that both the average molar mass and the concentration of HA are severely decreased. Thus, due to inflammation of joints in patients, the molecular weight of the polysaccharide can drop from 4 – 5 x 10^6 g mol^{-1} in a healthy tissue to 0.5 x 10^6 g mol^{-1} in a sick one [114].

To counteract this process, intra-articular injections of HA can be performed. Studies performed on therapeutic effects of HA injections have indicated a significant improvement in articular functionality followed by pain relief in patients [252].

Yu *et al.* performed several surveys on treatment of osteoarthritis using injectable *in situ* crosslinkable hydrogels [253]. Also, HA injections have demonstrated to possess chondroprotective properties being useful in clinical applications [254].

Smaller benefits, but not of statistical significance, have been determined in temporomandibular joint arthritis and in adhesive capsulitis of the shoulder [255].

In recent years, more and more products based on HA with usage in osteoarthritis are available on the market, among them some of the most common ones being: Arthrum®, Suplasyn®, Hyalgan®, Durolane®, Hylan®, Synvisc®, Orthovisic® or Viscorneal® [256].

To investigate the influence of molar mass upon the efficiency of HA materials in osteoarthritis both *in vitro* and *in vivo* tests were undergone. Thereby, although in the *in vitro* experiments high molecular weight HA exhibited a better efficiency in the curing process, this behavior was not observed in the *in vivo* experiments. Thus, while materials with a molecular weight of approximately 2 x 10^6 g mol^{-1} have proven to be suitable for *in vitro* investigations, for *in vivo* tests the HA with lower molecular weight displayed a better result. This discrepancy in the results could be explained by the fact that smaller length polymeric chains can easily penetrate into damaged tissues and accumulate in higher quantities at the desired place of action [257].

More recently attention has been focused on the role of HA in rheumatoid arthritis. Elevated joint levels of HA have been associated with increase of the serum levels and some abnormal immune response and thus, in this case, HA is rather seen as the culprit for the disease and not the remedy [255].

Wound Healing

Wound healing is a dynamic and complex process in the human body that can be divided into 4 distinct phases: the hemostasis phase, the inflammatory phase, the proliferative phase and the remodeling phase [258].

One of the first therapeutic applications of HA dates from the 1960's when it was used in wound healing and burn injury. Nowadays, high molecular weight HA based materials like Bionet and Hyalomatrix are often employed as skin bandages or as patches for healing diabetic foot lesions, nasal mucosal and leg ulcers [118, 165]. Hylaff scaffolds have been widely used in burn care and post-traumatic or complicated surgical wounds while more recently, copolymers based on HA-agarose and HA-fibronectin have attracted much attention due to their properties as potential wound healing materials [259].

As mucoadhesive materials, there are several studies made on injectable autocrosslinked sponges [112, 260]. Results obtained showed that HA hydrogels have the capacity to modulate wound healing and to restore the gliding function of the tendon and pheripheral nerve structure. To further improve its characteristics the polysaccharide has been mixed with different proteins such as gelatin or collagen. The collagen addition to the sponge has demonstrated to elevate wound healing capacity for *in vivo* experiments [261].

In wound healing GLL materials were designed to inhibit postsurgical adhesion and to prevent scar formation. Water insoluble films formed from low-acyl GLL crossliked with 1-ethyl-3-(3-dimethylaminopropyl) carbodiimide were prepared to accelerate wound repair processes. Displaying *in vitro* biocompatibility to both L929 fibroblast cells and blood the newly synthesized films were further investigated for *in vivo* tests. Thus, after subcutaneous implantation in rats, GLL films exhibited good biocompatibility with high potential use in clinical applications [166].

Cencetti et al. elaborated several investigations on biocompatible materials composed of nonwoven GLL and Hyaff®. In order to enhance the entrapment of silver in the dressing material and also to modulate its release, the nonwoven matrix was treated with a poly (vinyl alcohol)/borax system. Studies performed showed that the new hydrophilic material exhibited a strong antimicrobial activity against two model bacterium, namely *Pseudomonas aeruginosa* and *Staphylococcus aureus*. As in previous cases, the results obtained concluded that GLL based biomaterials are an attractive alternative for wound healing and prevention of various infections [262].

ALG-based wound dressings offer numerous advantages such as maintenance of physiologically moist microenvironment, facilitation for wound healing, minimization of bacterial infection at wound site and promotion of rapid epithelization [263]. Some types of alginate dressings are commercially available:

- Algicell^{TM} (Derma Science) - antimicrobial silver dressing in absorbent/gelling format as well as calcium ALG dressing;
- Algisite M^{TM} (Smith & Nephew) – calcium ALG dressing in soft gel format;
- Comfeel Plus^{TM} (Coloplast) – hydrocolloid dressing with calcium ALG;
- Kaltostat^{TM} (ConvaTec) – calcium/sodium ALG dressing produced as absorbent fibrous fleece;
- Sorbsan^{TM} (UDL Laboratoires) – calcium ALG dressing prepared as textile fibers;
- Tegagen^{TM} (3M Healthcare) – ALG dressing;
- Melgisorb^{R} (Molnlycke Health Care) – silver ALG dressing.

All ALG-based wound dressing are indicated in treatment of skin abrasions, minor lacerations, superficial cuts and burns, diabetic foot ulcer, leg ulcers, pressure ulcers, traumatic and surgical wounds. The presence of silver ions into ALG dressing leads to the increase of antimicrobial activity as well as to enhance the antioxidant capacity. Also, the presence of zinc ions in ALG dressing may enhance keratinocyte migration as well as it can generate immunomodulatory and antimicrobial effects [264].

Adhesion Prevention Applications

It is well known that adhesion phenomenons are very common in postsurgical situations [265]. Surgeons usually use certain materials or films derived from natural polymers with the purpose to protect injured tissues or to prevent the occurrence of these types of processes. One example of coating material with excellent results in post-surgical adhesion prevention consists in hydrogel structures based on N, O-carboxymethyl chitosan and aldehyde HA [266].

HA is a polysaccharide with anti-adhesive properties often used as coating material for many devices such as sensors, catheters or guidewires. Alongside preventing fouling and tissue abrasion, the coating procedure can also improve the biocompatibility and lubricity of these devices [167].

Commercially, some of the best known products based on HA or HA copolymers are ACP^{TM}, a gel type material obtained through autocrosslinkable process and Seprafilm composed of chemically modified HA and carboxymethylcellulose.

As protective devices, HA materials have been manipulated with great success in pelvic operation, ear and sinus surgery and more recently in tendon and nerve surgery [118].

Tissue Engineering

Throughout the world, millions of patients suffer the loss or damage of some organs or tissues as a result of various diseases or accidents [267]. Tissue engineering covers a wide range of applications and aims to replace, repair or regenerate tissue/organ function in order to be used in transplantation [168]. Today, tissue engineering is focused on three main directions:

- the use of cell substitutes and isolated cells to replace damaged cells that supply certain functions in the body;
- the delivery of growth factors at targeted locations;
- directing the cells growth in a three-dimensional architecture [268].

In the last direction the scaffolds play an important role because they provide macromolecular support for adhesion, migration and proliferation of the cells. Three dimensional hydrogel scaffolds based on DEX and pullulan have been successfully used in tissue engineering allowing a better control of stem cell differentiation [269].

HA is one of the most used EPS for cell proliferation and differentiation, cell migration and cell signaling [169]. Literature mentions the use of HA networks for proliferation of various biomolecules such as chondrocytes, hepatocytes, fibroblasts, neurons or keratinocytes [165, 270]. Keratinocytes and fibroblasts were co-cultured on scaffolds prepared by freezing and lyophilizing methods to help construct artificial bilayer skin *in vitro*.

By activating the CD44 receptor, HA matrices have been used in cell signaling to promote keratinocyte activity. The receptor activation apparently stimulates cell activity and improves the abnormal epidermal functions, leading in the end to anti-aging and skin treatment [271].

Alongside wound healing, materials based on HA show great interest in tissue engineering. Because HA is highly susceptible to degradation, a modification process is usually needed for the preparation of suitable scaffolds for tissue engineering applications. The chemical modification steps consist in grafting reactions or crosslinking processes [169].

Three-dimensional networks obtained from HA and poly(D-lysine) were used for brain tissue engineering, while matrices based on HA and collagen were employed for cartilage tissue engineering [261].

Recent research made by Skaalure *et al.* demonstrated that semi-interpenetrated networks formed from biodegradable poly (ethylene glycol) hydrogels and HA gels have promising results in tissue engineering [272]. To evaluate their performance, mature chondrocytes were encapsulated into the networks and the long-term culture effects were determined. The investigations elaborated highlighted that the use of hydrogels leads to an increase production of collagen and sulfated glycosaminoglycan (sGAG) in the damaged tissue. These findings supported the idea that HA improves bioactivity and chondrocyte anabolism long term and that materials based on this polysaccharide may be a proper strategy to improve cartilage tissue engineering.

Composite hydrogels based on HA and poly (ethylene glycol) were synthesized for intervertebral disc regeneration or replacement of them [170]. These artificial networks played the role of supports for nucleus pulpous and annulus fibrous cell culture demonstrating to be a suitable environment for cell proliferation and sGAG production.

Polyelectrolyte complexes based on HA fibers coated with chitosan were mentioned as an attractive alternative for cartilage tissue engineering [273].

Mixed with DNA, HA hydrogels exhibited good results both in tissue engineering and gene delivery while blended with inorganic materials such as hydroxyapatite they showed a good mechanical resistance for bone implant [274].

With interest in regenerative medicine applications, Lam *et al.* prepared HA hydrogel scaffolds [275]. Thermoreversible hydrogel materials obtained from crosslinked HA and poly (N-isopropylacrylamide) were investigated for possible intervertebral disc regeneration [276]. Injected at the place of action and crosslinked *in situ* the network proved to be an excellent environment for cell culture and differentiation of human messenchymal stem cells.

To develop new repair approaches for non-vascular meniscus, Freyman *et al.* dedicated their work to synthesize novel materials based on HA [277].

In central nerve tissue engineering, Wang *et al.* elaborated a novel crosslinked thiol-modified heparin, HA-gelatin scaffold. Undertaken investigation on the material has shown an improvement in cell survival and a significant decreasing in glial scar formation and local inflammation.

Biomimetic matrices based on HA have demonstrated excellent influence on regeneration of motor functions being thus suitable for possible nerve and brain tissue repair [259].

Recent research has indicated a possible utility of GLL in tissue engineering and wound healing even though its applicability in cartilage reconstruction after degradation or mechanical damage is still limited. This limitation is due to the relative poor mechanical strength of GLL. Thus, a solution to overcome this problem can be represented by the chemical modification of the polysaccharide through grafting, crosslinking, *etc*.

Novel biomaterials based on GLL have been exploited as injectable carriers for various autologous cells such as: chondrocytes and articular chondrocytes, bone marrow cells and chondrogenic and non-chondrogenic adipose stem cells.

Thus, low-acyl GLL dissolved in deionized water at high temperature led to the formation of 1% (wt/wt) solution which served as a matrix for bone marrow cell culture [278]. Investigations that employed a Live/Dead Viability/Cytotoxicity Kit showed promising results in cell proliferation and differentiation.

Oliveira et *al.* suggested that GLL could be used as a minimally invasive injectable system capable of *in situ* gelation under physiological conditions [171].

Used as culture environment for human articular chondrocytes, the hydrogels exhibited good bioavailability when implanted subcutaneously in mice.

Further work was elaborated by encapsulating human nasal chondrocytes in hydrogels based on GLL and agarose followed by cell viability tests and histological analysis, and by encapsulating autologous cells in GLL hydrogels for the treatment of rabbit articular defects. Both cases displayed positive results in cell viability during storage with good cell proliferation and differentiation.

Also, thorough *in vivo* investigation after implantation confirmed GLL's great biocompatibility suggesting a slight tendency of the polymeric matrix to integrate with the environment. By studying the mechanical properties of the networks before and after implantation a small decrease in the molecular weight was observed. Although the mechanical characteristics were similar for both cases, this tendency could be explained by a possible resorption of the matrix and a migration process of the cells to the surrounding tissues. Showing capacity for tissue regeneration, injectable hydrogels could constitute a proper alternative for the treatment of various cartilage defects.

Because GLL is a natural polymer with poor mechanical properties, it is highly recommended to combine the polysaccharide with other polymers either in physical mixtures or covalent coupling.

Hydrogel structures showed improvement in mechanical strength, low cytotoxicity and good capacity for chondrocyte proliferation when the GLL has reacted with carboxymethyl chitosan. [279].

Literature mentions several other methods for the improvement of mechanical parameters for GLL [280]. Thus, Cautinho *et al.* proposed the preparation of GLL hydrogels crosslinked by both physical and chemical reactions. Used as scaffold for NIH-3T3 fibroblasts culture, GLL exhibited *in vitro* biocompatibility while presenting a high cell survival rate making it useful for tissue regeneration.

Studies effectuated on photo-crosslinked methacrylated GLL indicated a significant increase in structure density due to its modification [281]. *In vitro* results stated that the hydrogels maintained their non-toxicity and their ability to proliferate various cells being promising materials for intervertebral disc regeneration.

Over the past several decades the ALG gels have been explored as scaffolds in regeneration of different tissues and organs as follows:

- new blood vessel formation to treat patients with coronary and peripheral arterial diseases [172];
- bone and cartilage regeneration [282];
- skeletal muscle regeneration by cell transplantation, growth factor delivery or a combination of both methods;
- in neurology and neurosurgery for repair of the central and peripheral nerve system [283];
- to provide hepatic tissue and for transplantation of encapsulated pancreatic islet allografts and xenografts [284].

Other Biomedical Applications

In case of the *in vitro* fertilization, the replacement of albumin protein with HA has demonstrated to have positive results. Because albumin may suffer contamination with viruses or prions, it is essential to remove the protein from blood before fertilization. Due to its high viscosity and anti-adhesive property HA represents one of the best alternatives as cell culture environment for further *in vitro* fertilization.

In this context, recent studies have determined that by introducing HA in the culture medium, after a frozen-thawing process, the embryos do not undergo any transformation [285]. Furthermore, it seems that HA scaffolds can lead to a higher quality of the embryos and finally a better pregnancy [286]. HA can be used with great success in laryngology and urology. Being produced by fibroblasts and macrophages HA is one of the most important components of the extracellular matrix in vocal cords. Thus, autocrosslinked or modified with other materials, HA can be suitable as space filler, shock absorber or wound healer in tissue damaging [287]. As space filler, it can be used as sphincter repairing to treat urinary incontinence, to maintain alveolar space or for reconstructive surgery [259].

Recent applications are being developed in face lifting while in cosmetics HA based products are being used to enhance skin hydration or as antioxidant [288].

Because the level of HA increases in liver diseases, devices prepared form the polysaccharide have proven to be useful as landmarks for liver failure diagnostic [289].

FUTURE PERSPECTIVES

As research continues to move forward, more and more intelligent and cleaner systems based on natural polymers are being prepared in order to satisfy the continue growing demand of human kind necessities. The discovery of new EPS from microbial sources can provide multiple opportunities for their biomedical applications. In combination with suitable cells, biomolecules and drugs, the EPS can be promising materials for cardiac tissue engineering. Also, the synthesis of new hybrid materials based on EPS and magnetic nanoparticles loaded with drugs may be used for medical imaging or for preparation of target DDS. Another challenge in biomedical applications of EPS is the design of new materials in order to use them in the treatment of posterior ocular segment diseases.

CONCLUSION

Microbial exopolysaccharides represent a uniquc family of polysaccharides that have received a growing attention from the scientific world due to their excellent properties. The long way of EPS from their discovery to applications involves a number of steps starting with screening and isolation of bacterial sources followed by extraction and optimization of EPS production, determination and validation of the chemical structure, establishment of physico-chemical, rheological and biological properties and finally the biomedical applications and clinical testing.

The structural diversity of EPS has led to the use of different techniques (methylation analyses, NMR spectroscopy, FT-IR spectroscopy, atomic force microscopy, scanning electron microscopy, confocal laser scanning microscopy, X-ray scattering, *etc*) for elucidation and understating of their structures. In the last years many types of EPS were discovered, but only some of them are used in biomedical applications.

From this point of view, EPS represent important tools in the treatment of many lethal diseases (cancer and hepatitis) that kill millions of people every year. Also, antiulcer, anticoagulant and antiviral activities, anti-reflux therapies, cholesterol lowering agents, surgical and ophtalmic applications, treatment of rheumatoid arthritis, wound healing and tissue engineering represent the relevant solutions to enhance the human health.

AKNOWLEDGEMENTS

Paper dedicated to the 65th anniversary of “Petru Poni” Institute of Macromolecular Chemistry of Romania Academy, Iasi, Romania. This chapter was supported by the project PNII-PT-PCCA-2013-4-1570, financed by the Romanian National Authority for Scientific Research, CNCS-UEFISCDI.

CONFLICT OF INTEREST

The authors confirm that this chapter content has no conflict of interest.

REFERENCES

[1] Kumar AS, Mody K, Jha B. Bacterial exopolysaccharides - a perception. J Basic Microbiol 2007; 47: 103-17.

[2] Freitas F, Alves VD, Reis MAM. Advances in bacterial exopolysaccharides: from production to biotechnological applications. Trends Biotechnol 2011; 29: 388-98.

[3] Donot F, Fontana A, Baccou JC, Schorr-Galindo S. Microbial exopolysaccharides: Main exemples of synthesis, excretion, genetics and extraction. Carbohydr Polym 2012; 87: 951-62.

[4] Bornsheuer UT, Buchholz K. Highlights in biocatalysis – historical landmarks and current trends. Eng Life Sci 2005; 5: 309-23.

[5] Fruton JS. Early theories of protein structure. Ann N Y Acad Sci 1979; 325: 1-20.

[6] Brown AJ. An acetic ferment which forms cellulose. J Chem Soc 1886; 49: 432-39.

[7] Rehm BHA. Bacterial polymers: biosynthesis, modifications and applications. Nat Rev Microbiol 2010; 8: 578-92.

[8] Rehm BHA, Valla S. Bacterial alginates: biosynthesis and applications. Appl Microbiol Biotechnol 1997; 48: 281-88.

[9] Vandamme EJ, Bruggeman G, DeBaets S, *et al.* Useful polymers of microbial origin. Agro Food Ind Hi Tech 1996; 7: 21-25.

[10] Berg H, Ziegler K, Piotukh K, *et al.* Biosynthesis of the cyanobacterial reserve polymer multi L-arginyl-poly-L-aspartic acid (cyanophycin): mechanism of the cyanophycin synthetase reaction studied with synthetic primers. Eur J Biochem 2000; 267: 5561-70.

[11] Sutherland IW. Novel and established applications of microbal polysaccharides. Trends Biotechnol 1998; 16: 41-46.

[12] Iyer A, Mody K, Jha B. Emulsifying properties of a marine bacterial exopolysaccharide. Enzyme Microbiol Technol 2006; 38: 220-22.

[13] Okutni K. Antitumor and immunostimulant activities of polysaccharides produced by a marine bacterium of the genus *Vibrio*. Bull Jpn Soc Sci Fish 1984; 50: 1035-37.

[14] Colliec Jouault S, Chevolot L, Helley D, *et al.* Characterization, chemical modificiations and *in vitro* anticoagulant properties of an exopolysaccharide produced by *Alteromonas infernus*. Biochim Biophys Acta 2001; 1528: 141-51.

[15] Deng L, Kasper DL, Krick TP, *et al.* Characterization of the linkage between the type III capsular polysaccharide and the bacterial cell wall of group B*Streptococcus*. J Biol Chem 2000; 275: 7497-04.

[16] Sandford PA. Exocellular, microbial polysaccharides. Adv Carbohydr Chem Biochem 1979; 36: 265-13.

[17] Lotfipour F, Mirzaeei S, Maghsoodi M. Evaluation of the effect of $CaCl_2$ and alginate concentrations and hardening time on the characteristics of Lactobacillus acidophilus loaded alginate beads using response surface analysis. Adv Pharm Bull 2012; 2: 71-78.

[18] Marvasi M, Visscher PT, Martinez LC. Exopolymeric substances (EPS) from *Bacillus subtilis*: polymers and genes encoding their synthesis. FEMS Microbiol Lett 2010; 313: 1-9.

[19] Rinker KD, Kelly RM. Growth physiology of the hyperthermophilic archaeon Thermococcus litoralis: development of a sulfur-free defined medium, characterization of an exopolysaccharide and evidence of biofilm formation. Appl Environ Microbiol 1996; 62: 4478-85.

[20] Nicolaus B., Lama L, Esposito E, *et al.* *Haloarcula* spp able to biosynthesize exo- and endopolymers. J Ind Microbiol Biotechnol 1999; 23: 489-96.

[21] Pasteur L. On the viscous fermentation and the butyrous fermentation. Bull Soc Chim Paris 1861; 11: 30-31.

[22] Scheibler C. Investigation on the nature of the gelatinous excretion (so-called frog's spawn) which is observed in production of beet-sugar juices. Z Dtsch Zucker Ind 1874; 24: 309-35.

[23] Van Tieghem P. On sugar-mill-gum. Ann Sci Nat Bot Biol Veg 1878; 7: 180-03.

[24] Tarr HLA, Hibbert H. Studies on reactions relating to carbohydrates and polysaccharides XXXVII. The formation of dextran by *Leuconostoc mesenteroides*. Canad J Res 1931; 5: 414-27.

[25] Hehre EJ, Sugg JY. Serologically reactive polysaccharides produced through the action of bacterial enzymes: I. Dextran of Leuconostoc mesenteroides from sucrose. J Exp Med 1942; 75: 339-53.

[26] Hestrin S, Avineri-Shapiro S, Aschner M. The enzymic production of levan. Biochem J 1943; 37: 450-56.

[27] Leathers TD. Dextran. In: Vandamme EJ, DeBaets S, Steinbuchel A, Eds. Biopolymers vol 5 Polysaccharides I: Polysaccharides from prokaryotes. Weinheim, Wiley-VCH 2002; pp. 299-321.

[28] Bixler GH, Hines GE, McGhee RM, *et al.* Dextran. Ind Eng Chem 1953; 45: 692-05.

[29] Jeans A, Haynes WC, Wilham CA, *et al.* Characterization and classification of dextrans from ninety-six strains of bacteria. J Am Chem Soc 1954; 76: 5041-52.

[30] Hehre EJ, Senti FR, Hellman NN. Natural synthesis of low molecular weight (clinical type) dextran by a *Streptococcus strain*. J Biol Chem 1956; 222: 739-50.

[31] Bourne EJ, Hutson DH, Weigel H. Studies on dextrans and dextranases. 3 Structures of oligosaccharides from *Leuconostoc mesenteroides* (Birmingham) dextran. Biochem J 1963; 86: 555-62.

[32] Seymour FR, Knapp RD. Structural analysis of dextrans, from strains of leuconostoc and related genera that contain 3-O-α-D-glucosylated α-D-glucopyranosyl residues at the branch points, or in consecutine, linear position. Carbohydr Res 1980; 81: 105-29.

[33] Purama RK, Goswami P, Khan AT, *et al.* Structural analysis and properties of dextran produced by *Leuconostoc mesenteroides* NRRL B-640. Carbohydr Polym 2009; 76: 30-35.

[34] Patel S, Kasoju N, Bora U, *et al.* Structural analysis and biomedical applications of dextran produced by a new isolate *Pediococcus pentosaceus* screened from biodiversity hot spot Assam. Biores Technol 2010; 101: 6852-55.

[35] Harada T, Yoshimura T. Production of a new acidic polysaccharide containing succinic acid by soil bacterium. Biochem Biophys Acta 1964; 83: 374-76.

[36] Maeda I, Saito H, Masada M, *et al.* Properties of gels formed by heat treatment of curdlan, a bacterial β-1,3-glucan. Agric Biol Chem 1967; 31: 1184-88.

[37] Sundaralingam M. Some aspects of stereochemistry and hydrogen bonding of carbohydrates related to polysaccharide conformations. Biopolymers 1968; 6: 189-13.

[38] Miwa M, Nakao Y, Nara K. Food applications of curdlan. Food Hydrocol 1993; 119-24.

[39] Spicer EJF, Goldenthal EI, Ikeda T. A toxicological assessment of curdlan. Food Chem Toxicol 1999; 37: 455-79.

[40] Bauer R. Physiology of *Dematium pullulans* (de Bary). Zentrabl Bakteriol Parasit Infekt Hyg 1938; 98: 133-67.

[41] Bernier B. The production of polysaccharide by fungi active in the decomposition of wood and forest litter. Can J Microbiol 1958; 4: 195-04.

[42] Bender H, Lehmann J, Wallenfels K. Pullulan, ein extracellulares Glucan von *Pullularia pullulans*. Biochim Biophys Acta 1959; 36: 309-16.

[43] Bouveng HO, Kiessling H, Lindberg B, *et al.* Polysaccharides claborated by *Pullularia pullulans*. Part I. The neutral glucan synthesized from sucrose solutions. Acta Chem Scand 1962; 16: 615-22.

[44] Wallenfels K, Keilich G, Bechtler G, *et al.* Investigations on pullulan. IV Resolution of structural problems using physical, chemical and enzymatic methods. Biochem Z 1965; 341: 433-50.

[45] Prajapati VD, Jani GK, Khanda SM. Pullulan: An exopolysaccharide and its various applications. Carbohydr Polym 2013; 95: 540-49.

[46] Garcia-Ochoa F, Santosa VE, Casasb JA, *et al.* Xanthan gum: production, recovery, and properties. Biotechnol Adv 2000; 18: 549-79.

[47] Prajapati VD, Jani GK, Zola BS, *et al.* An insight into the emerging exopolysaccharide gellan gum as a novel polymer. Carbohydr Polym 2013; 93: 670-78.

[48] Bajaj IB, Survase SA, Saudagar PS, *et al.* Gellan gum: Fermentative production, downstream processing and aplications. Food Technol Biotech 2007; 45: 341-54.

[49] Pszczola DE. Gellan gum wins IFT's food technology industrial achievement awards. Food Technol 1993; 47: 94-96.

[50] Takeuchi M, Sawada H, Oyaizu H, *et al.* Phylogenetic evidence for *Sphingomonas* and *Rhizomonas* as nonphotosynthetic members of α-4 subclass of the *Proteobacteria.* Int J Syst Bacteriol 1994; 44: 308-14.

[51] Deschrevel B. Hyaluronan: Its unique properties make it an appliation-rich medical system. In: Popa M, Ottembrite RM, Uglea CV, Eds. Medical applications of polymers vol1. California, American Scientific Publishers 2011; pp. 151-178.

[52] Meyer K, Palmer JW. The polysaccharide of the vitreous humor. J Biol Chem 1934; 107: 629-34.

[53] Meyer K. The biological significance of hyaluronic acid and hyaluronidase. Physiol Rev 1947; 27: 335-59.

[54] Laurent TC, Gergely J. Light-scattering studies on hyaluronic acid. J Biol Chem 1955; 212: 325-33.

[55] Liu L, Liu Y, Li J, *et al.* Microbial production of hyaluronic acid: current state, challenges and perspectives. Microbial Cell Fact 2011; 10: 99-08.

[56] Balazs EA, Laurent TC, Jeanloz RW. Nomenclature of hyaluronic acid. Biochem J 1986; 235: 903.

[57] Stanford ECC. New substance obtained from some of the commonomer species of marine algae: Algin. Chem New 1883; 47: 254-57.

[58] Schmidt E, Vocke F. Zur kenntnis der Poloyglykuronsauren (I.). Chem Ber 1926; 59: 1585-88.

[59] Miawa T. Alginic acid. J Chem Soc Jpn 1930; 51: 738-45.

[60] Hirst EL, Jones JKN, Jones WO. The structure of alginic acid. Part I. J Chem Soc 1939; 1880-85.

[61] Fischer FG, Dorfel H. Die Polyuronsauren der Braunalgen (Kohlenhydrate der Algen). Z Physiol Chem 1955; 302: 186-03.

[62] Haug A, Smidsrod O. Fractionation of alginates by precipitation with calcium and magnesium ions. Acta Chem Scand 1965; 19: 1221-26.

[63] Gorin PA, Spencer JFT. Exocellular alginic acid from *Azotobacter vinelandii.* Can J Chem 1966; 44: 993-98.

[64] Pindar DF, Bucke C. The biosynthesis of alginic acid by *Azotobacter vinelandii.* Biochem J 1975; 152: 617-22.

[65] Stewart-Tull DES. The immunological activities of bacterial peptidoglycans. Annu Rev Microbiol 1980; 34: 311-40.

[66] Nwodo UU, Green E, Okoh AI. Bacterial exopolysaccharides: Functionality and prospects. Int J Mol Sci 2012; 13: 14002-15.

[67] Kim D, Robyt JF, Lee SY, *et al.* Dextran molecular size and degree of branching as a function of sucrose concentration, pH and temperature of reaction of *Leuconostoc mesenteroides* B-512FMCM dextransucrase. Carbohydr Res 2003; 338: 1183-89.

[68] Robyt JF, Yoon SH, Mukerjea R. Dextransucrase and the mechanism for dextran biosynthesis. Carbohydr Res 2008; 343: 3039-48.

[69] Abbot D, Bourne EJ, Weigel H. Studies on dextrans and dextranases. Part VIII. Size and distribution of branches in some dextrans. J Chem Soc C 1966; 827-31.

[70] Monsan P, Bozonnet S, Albenne C, *et al.* Homopolysaccharides from lactic acid bacteria. Int. Dairy J 2001; 11: 675-85.

[71] Kim D, Robyt JF. Properties of *Leuconostoc mesenteriodes* B512 FMC constitutive dextransucrase. Enzyme Microbiol Technol 1994; 16: 1010-15.

[72] Majumuder A, Bhandari S, Purama RK *et al.* Enhanced production of a novel dextran from *Leuconostoc mesenteroides* NRRL B-640 by response surface methodology. Ann Microbiol 2009; 59: 309-15.

[73] Dols M, Remaud-Simeon M, Monsan PF, Dextransucrase production by *Leuconostoc mesenteroides* NRRL B 1299. Comparison with *L. mesenteroides* NRRL B-512F. Enzyme Microb Technol 1997; 20: 523-30.

[74] Kim D, Robyt JF. Dextransucrase constitutive mutants of *Leuconostoc mesenteroides* B-1299. Enzyme Microb Technol 1995; 17: 1050-56.

[75] Khalikova E, Susi P, Korpela T. Microbial dextran-hydrolyzing enzymes: Fundamentals and applications. Microbiol Mol Biol Rev 2005; 69: 306-25.

[76] Cote GL, Robyt JF. Isolation and partial characterization of an extracellular glucansucrase from *Leuconostoc mesenteroides* NRRL B-1355 that synthesizes an alternating (1-6), (1-3)-α-D-glucan. Carbohydr Res 1982; 101: 57-74.

[77] Eui-Qader SA, Iqbal L, Aman A, *et al.* Production of dextran by newly isolated strains of *Leuconostoc mesenteroides* through batch fermentation. Turk J Biochem 2006; 31: 21-26.

[78] Mao X, Wang S, Kan F, *et al.* A novel dextran dextrinase from *Gluconobacter oxydans* DSM-2003: Purification and properties. Appl Biochem Biotechnol 2012; 168: 1256-64.

[79] Figures WR, Edwards JR. D-Glucosyltransferase of *Streptococcus mutans*: isolation of two forms of the enzyme that bind to insoluble dextran. Carbohydr Res 1981; 88: 107-17.

[80] Mukasa H, Shimamura A, Tsumoni H, Purification and characterization of basic glucosyltransferase from *Streptococcus mutans* serotype *c.* Biochim Biophys Acta 1982; 719: 81-89.

[81] Grahame DA, Mayer RM. Purification and comparison of two forms of dextransucrase from *Streptococcus sanguis*. Carbohydr Res 1985; 142: 285-98.

[82] Sidebotham RL. Dextrans. Adv Carbohydr Chem Biochem 1974; 30: 371-44.

[83] Sabatie J, Choplin L, Doublier JL, *et al.* Rheology of native dextrans in relation to their primary structure. Carbohydr Polym 1988; 9: 287-99.

[84] Cheetham NWH, Fiala-Beer E, Walker GJ. Dextran structural details from high-field proton NMR spectroscopy. Carbohydr Polym 1991; 14: 149-58.

[85] Taylor C, Cheetham NWH, Walker GJ. Application of high-performance liquid cromatography to a study of branching in dextrans. Carbohydr Res 1985; 137: 1-12.

[86] Heinze T, Liebert T, Heublein B, *et al.* Functional polymers based on dextran. Adv Polym Sci 2006; 205: 199-91.

[87] Mcintosh M, Stone BA, Stanisich VA. Curdlan and other bacterial (1,3)β-D-glucans. Appl Microbiol Biotechnol 2005; 68: 163-73.

[88] Nakanishi I, Kimura K, Kanamaru T. Studies on curdlan-type polysaccharide.I. Industrial production of curdlan- type polysaccharide. J Takeda res Lab 1992; 51: 99-08

[89] Moscovici M, Ionescu C, Caraiani T, *et al.* Curdlan- type polysaccharide obtained using a strain of *Agrobacterium rhizogenes*. Rom Biotechnol Lett 2009; 14: 4530-37.

[90] Zhang H, Nishinari K, Williiams MAK, *et al.* A molecular description of the gelation mechanism of curdlan. Int J Biol Macromol 2002; 30: 7-16.

[91] Manjanna KM, Shivakumar B, Pramodkumar TM. Natural exopolysaccharides as novel excipients in drug delivery: A review. Arch Apl Sci Res 2009; 1: 230-53.

[92] Zhang R, Edgar KJ. Properties, chemistry and applications of the bioactive polysaccharide curdlan. Biomacromolecules 2014; 15: 1079-96.

[93] Bishwambhar M, Suneetha V, Ramalingam C. An overview of mechanistic characterization and optimization of pullulan producing microorganism. South Asian J Experim Biol 2011; 1: 147-51.

[94] Cheng KC, Demirci A, Catchmark JM. Pullulan: biosynthesis, production and applications. Appl Microbiol Biotechnol 2011; 92: 29-44.

[95] Singh RM, Saini GK, Kennedy JF. Pullulan: Microbial sources, production and applications. Carbohydr Polym 2008; 73: 515-31.

[96] Carolan G, Catley BJ, McDougal FJ. The location of tetrasaccharide units in pullulan. Carbohydr Res 1983; 114: 237-43.

[97] Fraser CG, Jennings HJ. A glucan from *Tremella mesenterica* NRRL-Y6158. Can J Chem 1971; 49: 1804-07.

[98] Waksman N, De Lederkremer RM, Cerezo AS. The structure of an α-D-glucan from *Cyttaria harioti* Fischer. Carbohydr Res 1977; 59: 505-15.

[99] Corsaro MM, De Castro C, Evidente A, *et al.* Phytotoxic extracellular polysaccharide fractions from *Cryphonectria parasitica* (Murr.) Barr strains. Carbohydr Polym 1998; 37: 167-72.

[100] Reis RA, Tischer CA, Gorin PAJ, *et al.* A new pullulan and a branched (1→3)-(1→6)-linked β-glucan from the lichenised ascomycete *Teloschistes flavicans*. FEMS Microbiol Lett 2002; 210: 1-5.

[101] Chi Z, Zhao S. Optimization of medium and cultivation conditions for pululan production by a new pullulan-producing yeast strain. Enzym Microb Technol 2003; 33: 206-11.

[102] Oguzhan P, Yangilar F. Pullulan: Production and usage in food industry. African J Food Sci Technol 2013; 4: 57-63.

[103] Mocanu G, Mihai D, Le Clerf D, *et al.* Synthesis of new associative gel microspheres from carboxymethyl pullulan and their interactions with lyzozyme. Eur Polym J 2004; 40: 283-89.

[104] Mahner C, Lechner MD, Nordmeier E. Synthesis and characterisation of dextran and pullulan sulphate. Carbohydr Res 2001; 331: 203-08.

[105] Boeriu CG, Springer J, Kooy FK, *et al.* Production methods for hyaluronan. Int J Carbohydr Chem 2013; 2013: 1-14.

[106] Evanko SP, Wight TN. Intracellular localization of hyaluronan in proliferating cells. J Histochem Cytochem 1999; 47: 1331-41.

[107] Kan FWK. High-resolution localization of hyaluronic acid in the golden hamster oocyte–cumulus complex by use of a hyaluronidase–gold complex. Anat Rec 1990; 228: 370-82.

[108] Shen B, Wei A, Bhargav D, *et al.* Hyaluronan: its potential application in intervertebral disc regeneration. Orthopedic Res Rev 2010; 2: 17-26.

[109] Lequeux I, Ducasse E, Jouenne T, *et al.* Addition of antimicrobial properties to hyaluronic acid by grafting of antimicrobial peptide. Eur Polym J 2014; 51: 182-90.

[110] Kablik J, Monheit GD, Yu L, *et al.* Comparative physical properties of hyaluronic acid dermal fillers. Dermatol Surg 2009; 35: 302-12.

[111] Garg HG, Hales CA, Eds. Chemistry and biology of hyaluronan. Elsevier Ltd 2004.

[112] Leonelli F, La Bella A, Migneco LM, *et al.* Design, synthesis and applications of hyaluronic acid-paclitaxel bioconjugates. Molecules 2008; 13: 360-78.

[113] Cantor JO. Potential therapeutic applications of hyaluronan in the lung. Int J Chron Obstruct Pulmon Dis 2007; 2: 283–88.

[114] Balazs EA. Therapeutic use of hyaluronan. Struct Chem 2009; 20: 341-49.

[115] Milas M, Rinaudo M. Characterization and properties of hyaluronic acid (Hyaluronan). In: Severian D, Ed. Polysaccharides structural diversity and functional versatility. New York, Marcel Dekker 2005; pp. 535-49.

[116] Mero A, Campisi M. Hyaluronic acid bioconjugates for the delivery of bioactive molecules. Polymers 2014; 6: 346-69.

[117] Necas J, Bartosikova L, Brauner P, *et al.* Hyaluronic acid (hyaluronan): a review. Vet Med CZECH 2008; 53: 397-11.

[118] Mori S, Naito M, Moriyama S. Highly viscous sodium hyaluronate and joint lubrication. Int Orthop 2002; 26: 116-21.

[119] Aravamudhan A, Ramos DM, Nada AA, *et al.* Natural polymers: polysaccharides and their derivatives for biomedical applications. In: Kumbar SK, Laurencin CT, Deng M, Eds. Natural and synthetic biomedical polymers. San Diego, Elsevier Inc 2014; pp. 69-71.

[120] Dahiya P, Kamal R. Hyaluronic acid: A boon in periodontal therapy. N Am J Med Sci 2013; 5: 309-15.

[121] Leela JK, Shama G. Studies on xanthan production from *Xanthomonas campestris*. Bioprocess Eng 2000; 23: 687-89.

[122] Hamcerencu M, Desbrieres J, Popa M, *et al.* New unsaturated derivatives of xanthan gum: synthesis and characterization. Polymer 2007; 48: 1921-29.

[123] Sharma BR, Naresh L, Dhuldhoya NC, *et al.* Xanthan gum – a boon to food industry. Food Promotion Chronicle 2006; 1: 27-30.

[124] Kassim MBI. Production and characterization of the polysaccharide “xanthan gum” by a local isolate of the bacterium *Xanthomonas campestris.* Afr J Biotechnol 2011; 10: 16924-28.

[125] Cadmus MC, Jackson LK, Burton KA, *et al.* Biodegradation of xanthan gum by *Bacillus sp.* Appl Environ Microbiol 1982; 44: 5-11.

[126] Born K, Langendorff V, Boulenguer P. Xanthan. In: Steinbüchel A, Ed. Biopolymers online. Wiley-VCH Verlag GmbH & Co. KGaA 2005; pp. 261-81.

[127] Lawson CJ, Sutherland W. Polysaccharides. In: Rose AH, Ed. Microbiology: primary products of metabolism. London, Academic Press Inc 1978; pp. 349-58.

[128] Lachke A. Xanthan – a versatile gum. Resonance 2004; 9: 25-33.

[129] Mundargi RC, Patil SA, Aminabhavi TM. Evaluation of acrylamide-grafted-xanthan gum copolymer matrix tablets for oral controlled delivery of antihypertensive drugs. Carbohydr Polym 2007; 69: 130-41.

[130] Sworn G, Montsanto. Xanthan gum. In: Phillips GO, Williams PA, Eds. Handbook of hydrocolloids, Second Edition. Cambridge, Woodhead Publishing 2009; pp. 105-15.
[131] Desbrieres J, Hamcerencu M, Popa M. Xanthan and gellan as polysaccharides for biomedical applications. In: Popa M, Ottenbrite RM, Uglea CV, Eds. Medical applications of polymers. California, American Scientific Publishers 2011; pp. 117-50.
[132] Coviello T, Matricardi P, Marianecci C, *et al.* Polysaccharide hydrogels for modified release formulations. J Control Release 2007; 119: 5-24.
[133] Nitta Y, Nishinari K. Gelation and gel properties of polysaccharides gellan gum and tamarind xyloglucan. J Biol Macromol 2005; 5: 47-52.
[134] Morris ER, Nishinari K, Rinaudo M. Gelation of gellan – a review. Food Hydrocolloid 2012; 28: 373-11.
[135] Nakamura K, Tanaka Y, Sakurai M. Dynamic mechanical properties of aqueous gellan solutions in the sol-gel transition region. Carbohydr Polym 1996; 30: 101-08.
[136] Shah JH, Jani GK, Parikh JR. Gellan gum and its applications – a review. Latest Reviews 2007; 5: 1-14.
[137] Laaman TR, Ed. Hydrocolloids in food processing. Iowa: Blackwell Publishing 2011.
[138] Hellriegel J, Günther S, Kampen I, *et al.* A biomimetic gellan-based hydrogel as a physicochemical biofilm model. J Biomater Nanobiotechnol 2014; 5: 83-97.
[139] Ding K, Hogbin Z, Katsuyoshi N. Gellan gum and modified gellan gum hydrogels as biomedical materials. Prog Chem 2014; 26: 1172-89.
[140] Maiti S, Ranjit S, Mondol R, *et al.* Al^{+3} ion cross-linked and acetalated gellan hydrogel network beads for prolonged release of glipizide. Carbohydr Polym 2011; 85: 164-72.
[141] Sanderson GR, Clark RC. Gellan gum, a new gelling polysaccharide. In: Phillips GO, Wedlock DJ, Williams PA, Eds. Gums and stabilisers for the food industry, Volume 2. UK, Oxford University Press 1984; pp. 201–09.
[142] Kasapis S, Abeysekera R, Atkin N, *et al.* Tangible evidence of the transformation from enthalpic to entropic gellan networks at high levels of co-solute. Carbohydr Polym 2002; 50: 259-62.
[143] Flores-Huicoche E, Rodríguez-Hernández AI, Espinosa-Solares T, *et al.* Sol–gel transition temperatures of high acyl gellan with monovalent and divalent cations from rheological measurements. Food Hydrocolloid 2013; 31: 299-05.
[144] Picone CSF, Cunha RL. Influence of pH on formation and properties of gellan gels. Carbohydr Polym 2011; 84: 662-68.
[145] Pérez-Campos SJ, Chavarría-Hernández N, Tecante A, *et al.* Gelation and microstructure of dilute gellan solutions with calcium ions. Food Hydrocolloid 2012; 28: 291-00.
[146] Shimizu M, Brenner T, Liao R, *et al.* Diffusion of probe polymer in gellan gum solutions during gelation process studied by gradient NMR. Food Hydrocolloid 2012; 26: 28-32.
[147] Smirdsrod O, Skjakbrlk G. Alginate as immobilization matrix for cells. Trends Biotechnol 1990; 8: 71-78.
[148] Fett WF, Osman SF, Fishman ML, *et al.* Alginate production by plant- pathogenic Pseudomonads. Appl Environ Microbiol 1986; 52: 466-73.
[149] Haug A, Larsen B, Smidsrod O. Uronic acid sequence in alginate from different sources. Carbohydr Res 1974; 32: 217-25.
[150] Haug A, Larsen B, Smidsrod O. A study of the constitution of alginic acid by partial hydrolysis. Aca Chem Scand 1966, 20: 183-90.
[151] Atkins EDT, Isaac DH, Nieduszynski IA, *et al.* The polyuronides: their molecular architecture. Polymer 1974; 15: 263-71.
[152] Agulhon P, Markova V, Robitzer M, *et al.* Structure of alginate gels: Interaction of diuronate units with divalent cations from density functional calculations. Biomacromolecules 2012; 13: 1899-07.
[153] Lee KY, Mooney DJ. Alginate: properties and biomedical applications. Progr Polym Sci 2012; 37: 106-26.
[154] Kster KH, Sele V, Schwartz M, *et al.* Blood-volume changes after infusion of dextran solutions. Lancet 1957; 270: 262-65.
[155] Whistler RL, DeMiller JN, Ed. Pullulan. San Diego: Academic Press 1993; pp.447-460.
[156] Callant D, Schacht E. Macromolecular prodrugs of 5-aminosalicylic acid. 1 Azo-conjugates. Makromol Chem 1990; 191: 529-36.

[157] Kanke M, Tanabe E, Katayama H, *et al.* Application of curdlan to controlled drug delivery. III. Drug-release from sustained-release suppositories *in vitro*. Biol Pharm Bull 1995; 18: 1154-58.

[158] Scomparin A, Salmaso S, Bersani S, *et al.* Novel folated and non-folated pullulan bioconjugates for anticancer drug delivery. Eur J Pharm Sci 2011; 42: 547-58.

[159] Vasi AM, Popa MI, Butnaru M, *et al.* Chemical functionalization of hyaluronic acid for drug delivery applications. Mater Sci Eng C 2014; 38: 177-85.

[160] Santos H, Veiga F, Pina ME, *et al.* Compaction, compression and drug release properties of diclofenac sodium and ibuprofen pellets comprising xanthan gum as a sustained release agent. Int J Pharm 2005; 295: 15-27.

[161] Agnihotri SA, Aminabhavi TM. Development of novel interpenetrating network gellan gum-poly(vinyl alcohol) hydrogel microspheres for the controlled release of carvedilol. Drug Develop Ind Pharm 2005; 31: 491-03.

[162] Balazs EA. Hyaluronan as an ophthalmic viscoelastic device. Curr Pharm Biotechnol 2008; 9: 236-38

[163] Ceulemans J, Inge Vinckier I, Ludwig A. The use of xanthan gum in an ophthalmic liquid dosage form: Rheological characterization of the interaction with mucin. J Pharm Sci 2002; 91: 1117-27.

[164] Kesavan K, Nath G, Pandit J. Preparation and *in vitro* antibacterial evaluation of gatifloxacin mucoadhesive gellan system. Daru 2010; 18: 237-46.

[165] Dausse Y, Grossin L, Miralles G, *et al.* Cartilage repair using new polysaccharidic biomaterials: macroscopic, histological and biochemical approaches in a rat model of cartilage defect. Osteoarthritis Cartilage 2003; 11: 16-28.

[166] Lee MW, Chen HJ, Tsao SW. Preparation, characterization and biological properties of gellan gum films with 1-ethyl-3-(3-dimethylaminopropyl) carbodiimide cross-linker. Carbohydr Polym 2010; 82: 920-26.

[167] Chen G, Ito Y, Imanishi Y, *et al.* Photoimmobilization of sulfated hyaluronic acid for antithrombogenicity. Bioconjug Chem 1997; 8: 730-34.

[168] Langer R, Vacanti JP. Tissue engineering. Science 1993; 260: 920-26.

[169] Edouard P, Rannou F, Coudeyre E. Animal evidence for hyaluronic acid efficacy in knee trauma injuries. Review of animal-model studies. Phys Ther Sport 2013; 14: 116-23.

[170] Jeong CG, Francisco AT, Niu Z, *et al.* Screening of hyaluronic acid–poly(ethylene glycol) composite hydrogels to support intervertebral disc cell biosynthesis using artificial neural network analysis. Acta Biomater 2014; 10: 3421-30.

[171] Oliveira JT, Santos TC, Martins L, *et al.* Performance of new gellan gum hydrogels combined with human articular chondrocytes for cartilage regeneration when subcutaneously implanted in nude mice. J Tissue Eng Regen Med 2009; 3: 493-500.

[172] Hao XJ, Silva EA, Mansson-Broberg A, *et al.* Angiogenic effects od sequential releas4e of VEGFA(165) and PDGF-BB with alginate hydrogels after myocardial infarction. Cardiovasc Res 2007; 75: 178-85.

[173] Huang X, Lowe TL. Biodegradable thermoresponsive hydrogels for aqueous encapsulation and controlled release of hydrophilic model drugs. Biomacromolecules 2005; 6: 2131-39.

[174] Lee JS, Jung YJ, Doh MJ, *et al.* Synthesis and properties of dextran nalidixic acid ester as a colon specific prodrug of nalidixic acid. Drug Dev Ind Pharm 2001; 27: 331-36.

[175] Kanke Kanke M, Tanabe E, Katayama H, *et al.* Application of curdlan to controlled drug delivery. III. Drug-release from sustained-release suppositories *in vitro*. Biol Pharm Bull 1995; 18: 1154-58.

[176] Fundueanu G, Constantin M, Ascenzi P. Preparation and characterization of pH- and temperature-sensitive pullulan microspheres for controlled release of drugs. Biomaterials 2008; 29: 2767-75.

[177] Suginoshita Y, Tabata Y, Matsumura T, *et al.* Liver targeting of human interferon-β with pullulan based on metal coordination. J Control Rel 2002; 83: 75-88.

[178] Yang WZ, Chen HL, Gao FP, *et al.* Self-aggregated nanoparticles of cholesterol-modified pullulan conjugate as a novel carrier of mitoxantronep. Curr Nanosci 2010; 6: 298-06.

[179] Mocanu G, Constantin M, Fundueanu G, *et al.* Macromolecular prodrugs synthesis and hydrolytic behavior of three different drugs bounded on chloroacyl or chlorodesoxy pullulan. STP Pharma Sci 2000; 10: 439-44.

[180] Vasiliu S, Popa M, Luca C. Evaluation of retention and release processes of two antibiotics from the biocompatible core-shell microparticles. Eur Polym J 2008; 44: 3894-98.

[181] Vasiliu S, Popa M, Rinaudo M. Polyelectrolyte capsules made of two biocompatible natural polymers. Eur Polym J 2005; 41: 923-32.
[182] Ruissen ALA, van der Reijden WA, Van't Hof W, *et al.* Evaluation of the use of xanthan as vehicle for cationic antifungal peptides. J Control Rel 1999; 60: 49-56.
[183] Billa N, Yuen KH, Khader MAA, *et al.* Gamma-scintigraphic study of the gastrointestinal transit and *in vivo* dissolution of a controlled release diclofenac sodium formulation in xanthan gum matrices. Int J Pharm 2000; 201: 109-20.
[184] Talukdar MM, Vinckier I, Moldenaers P, *et al.* Rheological characterization of xanthan gum and hydroxypropylmethyl cellulose with respect to controlled-release drug delivery. J Pharm Sci 1996; 85: 537-40.
[185] Pongjanyakul T, Puttipipatkhachorn S. Xanthan–alginate composite gel beads: Molecular interaction and *in vitro* characterization. Int J Pharm 2007; 331: 61-71.
[186] Dumitriu S, Dumitriu M, Teaca G. Bioactive polymers 65 – studies of cross-linked xanthan hydrogels as supports in drug retardation. Clin Mater 1990; 6: 265-76.
[187] Chellat F, Tabrizian M, Dumitriu S, *et al. In vitro* and *in vivo* biocompatibility of chitosan-xanthan polyionic complex. J Biomed Mater Res 2000; 51: 107-16.
[188] Chellat F, Tabrizian M, Dumitriu S, *et al.* Study of biodegradation behavior of chitosan–xanthan microspheres in simulated physiological media. J Biomed Mater Res 2000; 53: 592-99.
[189] Ughini F, Andreazza IF, Ganter JLMS, *et al.* Evaluation of xanthan and highly substituted galactomannan from *M. scabrella* as a sustaincd rclcasc matrix. Int J Pharm 2004; 271: 197-05.
[190] Vasiliu S, Bunia I, Racovita S, *et al.* Adsorption of cefotaxime sodium salt on polymer coated ion exchange resin microparticles: Kinetic, equilibrium and thermodynamic studies. Carbohydr Polym 2011; 85: 376-87.
[191] Lungan MA, Popa M, Desbrieres J, Racovita S, Vasiliu S. Complex microparticulate systems based on glycidyl methacrylate and xanthan. Carbohydr Polym 2014; 104: 213-22.
[192] Miyazaki S, Aoyama H, Kawasaki N, *et al. In situ*-gelling gellan formulations as vehicles for oral drug delivery. J Control Rel 1999; 60: 287-95.
[193] Ohkawa K, Kitagawa T, Yamamoto H. Preparation and characterization of chitosan-gellan hybrid capsules formed by self-assembly at an aqueous solution interface. Macromol Mater Eng 2004; 289: 33-40.
[194] Maiti S, Singha K, Ray S, *et al.* Adipic acid dihydrazide treated partially oxidized alginate beads for sustained oral delivery of flurbiprofen. Pharm Develop Technol 2009; 14: 461-70.
[195] Bouhadir KH, Alsberg E, Mooney DJ. Hydrogels for combination delivery of antineoplasic agents. Biomaterials 2001; 22: 2625-33.
[196] Zhang X, Hui Z, Wan D, *et al.* Alginate microsphere filled with carbon nanotube as drug carrier. Int J Biol Macromol 2010; 47: 389-95.
[197] Wang FQ, Li P, Zhang JP, *et al.* A novel pH-sensitive magnetic alginate-chitosan beads for albendazole delivery. Drug Dev Ind Pharm 2010; 36: 867-77.
[198] Ishak RAH, Awad GAS, Mortada ND, *et al.* Preparation, *in vitro* and *in vivo* evaluation of stomach-specific metronidazole loaded alginate beads as local anti-*Helicobacter pylori* therapy. J Control Rel 2007; 119: 207-14.
[199] Chang CH, Lin YH, Yeh CL, *et al.* Nanoparticles incorporated in pH-sensitive hydrogels as amoxicillin delivery for eradication of *Helicobacter pylori*. Biomacromolecules 2010; 11: 133-42.
[200] Pasparakis G, Bouropoulos N. Swelling studies and *in vitro* release of verapamil from calcium alginate and calcium alginate-chitosan beads. Int J Pharm 2006; 323: 34-42.
[201] Wittaya-areekul S, Kruenate J, Prahsarn C. Preparation and *in vitro* evaluation of mucoadhesive properties of alginate/chitosan microparticles containing prednisolone. Int J Pharm 2006; 312: 113-18.
[202] Almeida PF, Almeida AJ. Cross-linked alginate-gelatine beads: a new matrix for controlled release of pindolol. J Control Rel 2004; 97: 431-39.
[203] Stops F, Fell JT, Collett JH, *et al.* Floating dosage forms to prolong gastro-retention. The characterization of calcium alginate beads. Int J Pharm 2008; 350: 301-11.
[204] Taha MO, Nasser W, Ardakani A, *et al.* Sodium lauryl sulfate impedes drug release from zinc-crosslinked alginate beads. Switching from enteric coating release into biphasic profiles. Int J Pharm 2008; 350: 291-00.

[205] Mladenovska K, Raicki RS, Javenik EI, *et al.* Colon-specific delivery of 5-aminosalicylic acid from chitosan-Ca-alginate microparticles. Int J Pharm 2007; 342: 124-36.
[206] Ciofani G, Raffa V, Pizzorusso T, *et al.* Characterization of an alginate-based drug delivery system for neurological applications. Med Eng Phys 2008; 30: 848-55.
[207] Moebus K, Siepmann J, Bodmeier R. Alginate-poloxamer microparticles for controlled drug delivery to mucosal tissue. Eur J Pharm Biopharm 2009; 72: 42-53.
[208] Jana S, Gandhi A, Sen KK, *et al.* Natural polymers and their application in drug delivery and biomedical field. J Pharm Sci Tech 2011; 1: 16-27.
[209] Pang YN, Zhang ZR, Pang QJ, *et al.* Study on the enzyme dependant colon targeting prodrug dexamethasone succinate dextran and its tablets. Acta Pharm Sinica 2001; 36: 625-30.
[210] McLeod AD, Friend DR, Tozer TN. Glucocorticoid-dextran conjugates as potential prodrugs for colon-specific delivery: Hydrolysis in rat gastrointestinal tract contents. J Pharm Sci 1994; 83: 1284-88.
[211] Gervalase C, Avramoglou T, Crepin M, *et al.* Growth inhibition of human melanoma tumor cells by the combination of sodium phenylacetate (NaPA) and substituted dextrans and one NaPA-dextran conjugate. Anti-Cancer Drugs 2002; 13: 37-45.
[212] Nomura T, Saikawa A, Morita S, *et al.* Pharmacokinetic characteristics and therapeutic effects of mitomycin C-dextran conjugates after intratumoural injection. J Control Rel 1998; 52: 239-52.
[213] Bastow ER, Byers S, Golub SB, *et al.* Hyaluronan synthesis and degradation in cartilage and bone. Cell Mol Life Sci 2008; 65: 395-13.
[214] Qiu L, Li Z, Qiao M, *et al.* Self-assembled pH-responsive hyaluronic acid–g-poly(L-histidine) copolymer micelles for targeted intracellular delivery of doxorubicin. Acta Biomater 2014; 10: 2024-35.
[215] Park JH, Cho HJ, Yoon HY, *et al.* Hyaluronic acid derivative-coated nanohybrid liposomes for cancer imaging and drug delivery. J Control Rel 2014; 174: 98-08.
[216] Karbownik MS, Nowak JZ. Hyaluronan: Towards novel anti-cancer therapeutics. Pharmacol Rep 2013; 65: 1056-74.
[217] Li J, He Y, Sun W, *et al.* Hyaluronic acid-modified hydrothermally synthesized iron oxide nanoparticles for targeted tumor MR imaging. Biomaterials 2014; 35: 3666-77.
[218] Yang X, Li Y, Li M, *et al.* Hyaluronic acid-coated nanostructured lipid carriers for targeting paclitaxel to cancer. Cancer Lett 2013; 334: 338–45.
[219] Son YJ, Yoon IS, Sung JH, *et al.* Porous hyaluronic acid/sodium alginate composite scaffolds for human adipose-derived stem cells delivery. Int J Biol Macromol 2013; 61: 175-81.
[220] Kim H, Park HT, Tae YM, *et al.* Bioimaging and pulmonary applications of self-assembled Flt1 peptide-hyaluronic acid conjugate nanoparticles. Biomaterials 2013; 34: 8478-90.
[221] Maltese A, Borzacchiello A, Mayol L, *et al.* Novel polysaccharides-based viscoelastic formulations for ophthalmic surgery: Rheological characterization. Biomaterials 2006; 27: 5134-42.
[222] Arshinoff SA, Jafari M. New classification of ophthalmic viscosurgical devices-2005. J Cataract Refract Surg 2005; 31: 2167-71.
[223] Chung JH. Therapeutic effect of 1% Na-hyaluronan on corneal wound healing. In: Kennedy JF, Phillips GO, Williams PA, Hascall VC, Eds. Hyaluronan, Volume 2, Biomedical, medical and clinical aspects. Cambridge, Woodhead 2002; pp. 129-36.
[224] Wright B, Mi S, Connon CJ. Towards the use of hydrogels in the treatment of limbal stem cell defficiency. Drug Discov Today 2013; 18: 79-86.
[225] Acosta MC, Gallar J, Belmonte C. The influence of eye solutions on blinking and ocular comfort at rest and during work at video display terminals. Exp Eye Res 1999; 68: 663-69.
[226] Langer K, Mutschler E, Lambrecht G, *et al.* Methylmethacrylate sulfopropylmethacrylate copolymer nanoparticles for drug delivery: Part III: Evaluation as drug delivery system for ophthalmic applications. Int J Pharm 1997; 158: 219-31.
[227] Saettone MF, Monti D, Torracca MT, *et al.* Mucoadhesive ophthalmic vehicles: Evaluation of polymeric low-viscosity formulations. J Ocul Pharmacol 1994; 10: 83-92.
[228] Bucolo C, Mangiafico P. Pharmacological profile of a new topical pilocarpine formulation. J Ocul Pharmacol Ther 1999; 15: 567-73.

[229] Bernatchez SF, Tabatabay C, Gurny R, *et al.* Sodium hyaluronate 0.25% used as a vehicle increases the bioavailability of topically administered gentamicin. Graefes Arch Clin Exp Ophtalmol 1993; 231: 157-61.
[230] Herrero-Vanrell R, Fernandez-Carballido A, Frutos G, *et al.* Enhancement of the mydriatic response to tropicamide by bioadhesive polymers. J Ocul Pharmacol Ther 2000; 16: 419-28.
[231] Doi K, Aki H. Ophthalmic composition containing xanthan gum and amino acid. WO 2006035969 (**2006**).
[232] Llamas-Moreno JF, Baiza-Durán LM, Saucedo-Rodríguez LR, *et al.* Efficacy and safety of chondroitin sulfate/xanthan gum *versus* polyethylene glycol/propylene glycol/hydroxypropyl guar in patients with dry eye. Clin Ophtalmol 2013; 7: 995-99.
[233] Mazzone MG, Paladino G, Marino C, Peri O. Re-epithelializing pharmaceutical compositions containing xanthan gum. EP 1358883 (**2003**).
[234] Faraldi F, Papa V, Santoro D, *et al.* A new eye gel containing sodium hyaluronate and xanthan gum for the management of post-traumatic corneal abrasions. Clin Ophtalmol 2012; 6: 727-31.
[235] Bhowmik M, Kumari P, Sarkar G, *et al.* Effect of xanthan gum and guar gum on *in situ* gelling ophthalmic drug delivery system based on poloxamer-407. Int J Biol Macromol 2013; 62: 117-23.
[236] Osmałek T, Froelich A, Tasarek S. Application of gellan gum in pharmacy and medicine. Int J Pharm 2014; 466: 328-40.
[237] Balasubramaniam J, Kant S, Pandit JK. *In vitro* and *in vivo* evaluation of the Gelrite gellan gum-based ocular delivery system for indomethacin. Acta Pharm 2003; 53: 251-61.
[238] Rozier A, Mazuel C, Grove J, *et al.* Functionality testing of gellan gum, a polymeric excipient material for ophthalmic dosage forms. Int J Pharm 1997; 153: 191-98.
[239] Sanzgiri YD, Maschi S, Crescenzi V, *et al.* Gellan-based systems for ophthalmic sustained delivery of methylprednisolone. J Control Rel 1993; 26: 195-01.
[240] Dickstein K, Hapnes R, Aarsland T. Comparison of aqueous and gellan ophthalmic timolol with placebo on the 24-hour heart rate response in patients on treatment for glaucoma. Am J Ophthalmol 2001; 132: 626-32.
[241] Tayel SA, El-Nabarawi MA, Tadros MI, *et al.* Promising ion-sensitive *in situ* ocular nanoemulsion gels of terbinafine hydrochloride: Design, *in vitro* characterization and *in vivo* estimation of the ocular irritation and drug pharmacokinetics in the aqueous humor of rabbits. Int J Pharm 2013; 443: 293-05.
[242] Sajadi Tabassi SA, Tafaghodi M, Jafari MR. Dextran microspheres containing tetanus toxoid as a potential delivery system for nasal immunization: *In vitro* and *in vivo* characterization. Iranian J Basic Med Sci 2004; 7: 88-98.
[243] Bertram U, Bodmeier R. *In situ* gelling, bioadhesive nasal inserts for extended drug delivery: *In vitro* characterization of a new nasal dosage form, Eur J Pharm Sci 2006; 27: 62-71.
[244] Fatnassi M, Jacquart S, Brouillet F, *et al.* Optimization of spray-dried hyaluronic acid microspheres to formulate drug-loaded bone substitute materials. Powder Technol 2014; 255: 44-51.
[245] Tiantian Y, Wenji Z, Mingshuang S, *et al.* Study on intralymphatic-targeted hyaluronic acid-modified nanoliposome: Influence of formulation factors on the lymphatic targeting. Int J Pharm 2014; 471: 245-57.
[246] Liu S, Jin M, Quan Y, *et al.* Transdermal delivery of relatively high molecular weight drugs using novel self-dissolving microneedle arrays fabricated from hyaluronic acid and their characteristics and safety after application to the skin. Eur J Pharm Biopharm 2014; 86: 267-76.
[247] Li J, Kamath K, Dwivedi C. Gellan film as an implant for insulin delivery. J Biomater Appl 2001; 15: 321-43.
[248] Firestein GS. Envolving concepts of rheumatoid arthritis. Nature 2003; 423: 356-61.
[249] Marshall SA, Lazar GA, Chirino AJ, *et al.* Rational design and engineering of therapeutic proteins. Drug Discov Today 2003; 8: 212-21.
[250] Putney SD, Burke PA. Improving protein therapeutics with sustained-release formulations. Nat Biotechnol 1998; 16: 153-57.
[251] Jung YS, Park W, Na K. Temperature-modulated noncovalent interaction controllable complex for the long-term delivery of etanercept to treat rheumatoid arthritis. J Control Rel 2013; 171: 143-51.
[252] Miltner O, Schneider U, Siebert CH, *et al.* Efficacy of intraarticular hyaluronic acid in patients with osteoarthritis-a prospective clinical trial. Osteoarthritis Cartilage 2002; 10: 680-86.

[253] Yu CJ, Ko CJ, Hsieh CH, *et al.* Proteomic analysis of osteoarthritic chondrocyte reveals the hyaluronic acid-regulated proteins involved in chondroprotective effect under oxidative stress. J Proteomics 2014; 99: 40-53.

[254] Oliveira MZ, Albano MB, Namba MM, *et al.* Effect of hyaluronic acids as chondroprotective in experimental model of osteoarthrosis. Rev Bras Ortop 2014; 49: 62-68.

[255] Price RD, Berry MG, Navsaria HA. Hyaluronic acid: the scientific and clinical evidence. J Plast Reconstr Aesthet Surg 2007; 60: 1110-19.

[256] Goldberg VM, Buckwalter JA. Hyaluronans in the treatment of osteoarthritis of the knee: evidence for disease-modifying activity. Osteoarthritis Cartilage 2005; 13: 216-24.

[257] Ghosh P, Guidolin D. Potential mechanism of action of intra-articular hyaluronan therapy in osteoarthritis: Are the effects molecular weight dependent? Semin Arthritis Rheum 2002; 32: 10-37.

[258] Maxson S, Lopez EA, Yoo D, *et al.* Concise review: role of mesenchymal stem cells in wound repair. Stem Cells Transl Med 2012; 1: 142-49.

[259] Collins MN, Birkinshaw C. Hyaluronic acid based scaffolds for tissue engineering-A review. Carbohydr Polym 2013; 92: 1262-79.

[260] Park H, Choi B, Hu J, *et al.* Injectable chitosan hyaluronic acid hydrogels for cartilage tissue engineering. Acta Biomater 2013; 9: 4779–86.

[261] Rinaudo M. Main properties and current applications of some polysaccharides as biomaterials. Polym Int 2008; 57: 397-30.

[262] Cencetti C, Bellini D, Pavesio A, *et al.* Preparation and characterization of antimicrobial wound dressings based on silver, gellan, PVA and borax. Carbohydr Polym 2012; 90: 1362-70.

[263] Queen D, Orsted H, Sanada H, *et al.* A dressing history. Int Wound J 2004; 1: 59-77.

[264] Wiegand C, Heinze T, Hipler UC. Comparative *in vitro* study on cytotoxicity, antimicrobial activity and binding capacity for pathophysiological factors in chronic wounds of alginate and silver-containing alginate. Wound Repair Reqen 2009; 17: 511-21.

[265] Yeo Y, Kohane DS. Polymers in the prevention of peritoneal adhesions. Eur J Pharm Biopharm 2008; 68: 57-66.

[266] Li L, Wang N, Jin X, *et al.* Biodegradable and injectable *in situ* cross-linking chitosan-hyaluronic acid based hydrogels for postoperative adhesion prevention. Biomaterials 2014; 35: 3903-17.

[267] Lee KY, Mooney DJ. Hydrogels for tissue engineering. Chem Rev 2001; 101: 1869-80.

[268] Ma PX. Scaffolds for tissue fabrication. Materials Today 2004; 7: 30-40.

[269] Ferreira LS, Gerecht S, Fuller J, *et al.* Bioactive hydrogel scaffolds for controllable vascular differentiation of human embryonic stem cells. Biomaterials 2007; 28: 2706-17.

[270] Lobmann R, Pittasch D, Muhlen I, *et al.* Autologous human keratinocytes cultured on membranes composed of benzyl ester of hyaluronic acid for grafting in nonhealing diabetic foot lesions: a pilot study. J Diabetes Complications 2003; 17: 199-04.

[271] Bourguignon LY. Matrix hyaluronan-activated CD44 signaling promotes keratinocyte activities and improves abnormal epidermal functions. Am J Pathol 2014; 184: 1012-19.

[272] Skaalure SC, Dimson SO, Pennington AM, *et al.* Semi-interpenetrating networks of hyaluronic acid in degradable PEG hydrogels for cartilage tissue engineering. Acta Biomater 2014; 10: 3409-20.

[273] Yamane S, Iwasaki N, Majima T, *et al.* Feasibility of chitosan-based hyaluronic acid hybrid biomaterial for a novel scaffold in cartilage tissue engineering. Biomaterials 2005; 26: 611-19

[274] Segura T, Chung P, Shea LD. DNA delivery from hyaluronic acid-collagen hydrogels *via* a substrate-mediated approach. Biomaterials 2005; 26: 1575-84.

[275] Lam J, Truong NF, Segura T. Design of cell–matrix interactions in hyaluronic acid hydrogel scaffolds. Acta Biomater 2014; 10: 1571-80.

[276] Peroglio M, Eglin D, Benneker LM, *et al.* Thermoreversible hyaluronan-based hydrogel supports *in vitro* and *ex vivo* disc-like differentiation of human mesenchymal stem cells. Spine J 2013; 13: 1627-39.

[277] Freymann U, Endres M, Goldmann U, *et al.* Toward scaffold-based meniscus repair: effect of human serum, hyaluronic acid and TGF-ß3 on cell recruitment and re-differentiation. Osteoarthritis Cartilge 2013; 21: 773-81.

[278] Smith AM, Shelton RM, Perrie Y *et al.* An initial evaluation of gellan gum as a material for tissue engineering applications. J Biomater Appl 2007; 22: 241-54.

[279] Tang Y, Sun J, Fan H, *et al.* An improved complex gel of modified gellan gum and carboxymethyl chitosan for chondrocytes encapsulation. Carbohydr Polym 2012; 88: 46-53.

[280] Coutinho DF, Sant SV, Shin H, *et al.* Modified gellan gum hydrogels with tunable physical and mechanical properties. Biomaterials 2010; 31: 7494-02.

[281] Silva-Correia J, Oliveira JM, Caridade SG, *et al.* Gellan gum-based hydrogels for intervertebral disc tissue-engineering applications. J Tissue Eng Regen Med 2011; 5: 97-07.

[282] Park DJ, Choi BH, Zu SJ, *et al.* Injectable bone using chitosan-alginate/mesenchymal stem cells/BMP-2 composites. J Cranio Maxill Surg 2005; 33: 50-54.

[283] Suzuki K, Suzuki Y, Tanihara M, *et al.* Reconstruction of rat peripheral nerve gap without sutures using freeze-dried alginate gel. J Biomed Mater Res 2000; 49: 528-533.

[284] Dvir-Ginzberg M, Gamlichi-Bonshtein I, Agbaria R, *et al.* Liver tissue engineering whitin alginate scaffolds: effects of cell-seeding density on hepatocyte viability, morphology and function. Tissue Eng 2003; 9: 757-66.

[285] Gardner DK, Lane M, Stevens J, *et al.* Changing the start temperature and cooling rate in a slow-freezing protocol increases human blastocyst viability. Fertil Steril 2003; 79: 407-10.

[286] Palasz AT, Rodriguez-Martinez H, Beltran-Breña P *et al.* Effects of hyaluronan, BSA, and serum on *bovine embryo in vitro* development, ultrastructure, and gene expression patterns. Mol Reprod Dev 2006; 73: 1503-11.

[287] Butler JE, Hammond TH, Gray SD. Gender-related differences of hyaluronic acid distribution in the human vocal fold. Laryngoscope 2001; 111: 907-11.

[288] Kanchana S, Arumugam M, Giji S, *et al.* Isolation, characterization and antioxidant activity of hyaluronic acid from marine bivalve mollusc Amussium pleuronectus (Linnaeus, 1758). Bioact Carbohydr Dietary Fibre 2013; 2: 1-7.

[289] Ueno T, Hashimoto O, Tamaki S, *et al.* Clinical applications of serum hyaluronan for liver diseases and its significance. In: Kennedy JF, Phillips GO, Williams PA, Hascall VC, Eds. Hyaluronan, Volume 2, Biomedical, medical and clinical aspects. Cambridge, Woodhead 2002; pp. 445-52.

CHAPTER 9

Bio-Interfacial Properties of Environmentally Responsive Biomaterials

Charu Tyagi[1,2,*], Lomas K. Tomar[1] and Divya Bijukumar[1]

[1]University of the Witwatersrand, Faculty of Health Sciences, Department of Pharmacy and Pharmacology, 7 York Road, Parktown 2193, Johannesburg, South Africa and [2]V.S.P.G.(P.G) College, C.C.S University, Meerut 250004, U.P., India

Abstract: Environmentally responsive or smart biomaterials have emerged as an answer to the pursuit and growing need for efficient bio-mimicking stimuli responsive systems. Their efficiency resides in the ability to sense a given perturbation in the biological environment through interfacial interaction and respond intelligently, as per biological need, through switching of their physicochemical properties. The distinct environmental changes that could trigger such responses include a change in temperature, pH, ionic strength, light, electric and magnetic field. Polymers with the plethora of possible chemistry provide a drive for exhaustive exploration and are therefore the focal of environment responsive biomaterials. Advancement in chemistry combined with biology provides a better approach to synthesize novel polymers, incorporating functionalities and developing supramolecular architectures which can impart new bio-interfacial properties. The leading avenues of environmentally responsive polymers application are tissue engineering, regenerative medicines, integrated bio-sensors, drug delivery and multi-responsive materials but are not limited to them. Development of single, dual or even multifunctional materials can be possible through appropriate design, synthesis method and the combination of the building blocks. The chapter reviews an array of polymeric materials responsive to internal stimuli with key insight into the recent strategies employed for their synthesis and modification for specific biomedical application. The chapter further presents a close discussion on the interfacial characteristics of these polymers that enable reversible switching of physicochemical properties in response to different environmental cues prevailing in a biological system.

Keywords: Bio-interface, stimuli responsive, switching physiochemical properties, conjugation, biomaterials.

INTRODUCTION

The fundamental macromolecules of a biological system like nucleic acids and proteins exhibit smart adaptability to the variations in physiological environment

***Correspondence author Charu Tyagi:** University of the Witwatersrand, Faculty of Health Sciences, Department of Pharmacy and Pharmacology, 7 York Road, Parktown 2193, Johannesburg, South Africa; Tel: +27-767978187; E-mail: charu.tyagi@wits.ac.za

Viness Pillay, Yahya E. Choonara and Pradeep Kumar (Eds.)

through conformational changes in their structure thus synchronizing the regulatory needs of the system. Such naturally existing stimuli responsive polymeric system although has complicated macromolecular and component integrity but form the basis for fabricating, designing and decorating biomaterials that could mimic this adaptive property. These biomaterials instilled with adaptability to perturbations in the biological environment forms the next generation of intelligent system with repertoire of smart biomedical applications. Polymers with plethora of possible inter-/intra- structural and functional chemistries are the most promising building blocks of these exciting class of environmentally responsive biomaterials also referred to as 'smart', 'intelligent' or 'stimuli responsive'. These smart systems are capable of sensing a small variation in environment (stimuli) that stimulates macroscopic changes in their physical (structure) and/or chemical (functionality) properties directing the desired response in accordance to their design for specific application.

Environmental changes or stimuli that could trigger a smart polymeric biomaterial (SPB) are broadly categorized as physical (temperature, light, magnetic/electrical field, ultrasound), chemical (pH, ionic, redox) and biological (enzymes, glucose, cells or other biomolecules) [1]. These stimuli can alternatively be classified as external - when applied from outside (light, temperature, magnetic/electrical field and ultrasound) or internal - arising/prevailing within the biological compartments or regions (temperature, pH, ionic, redox and biological molecules). SPBs designed for responding to internal biological stimuli or through interaction with the biomolecule has obvious advantages of self-control. External stimuli on the other hand impart greater freedom and more flexible tuning of response site, degree and time through the use of external remote.

There has been an enormous interest in the field of environmentally responsive biomaterials but it has recently grown with new technical advancements in polymer science coupled to the molecular sciences, generating an array of novel SPBs in the last two decades. Responsive polymer constructs have been crafted *via* strategic designing, synthesis and combination of building blocks. Synthesis of polymers with defined architecture, predetermined and reproducible molecular weight, narrow polydispersity and end-group fidelity has been significantly achieved *via* advancement in controlled radical polymerization (CRP) techniques - reversible addition-fragmentation chain transfer (RAFT) polymerization, atom transfer radical polymerization (ATRP), nitroxide mediated radical polymerization (NMP), stable free radical polymerization (SFRP), ring opening metathesis polymerization (ROMP) [2, 3]. The functional groups seeded on or within the polymeric chain are the deciding factor for polymer response to a given stimuli. Additionally, biological molecules like receptors, enzymes, antigen, glucose, *etc* are attached or incorporated to generate biologically responsive SPB. Incorporated functionalities during the polymerization process or post-polymerization, yields a variety of SPBs with enormous biomedical applications like drug delivery, tissue engineering, regenerative medicines, integrated bio-sensors and multi-responsive materials.

Based on the physical attributes, SPBs could be: i) linear and solubilized, such polymers undergo phase-transition of sol-gel under the influence of stimuli. A tunable balance of hydrophobic and hydrophilic substitutes governs this phase-transition; ii) cross-linked networks, polymeric chains reorganize on application of stimuli switching between collapsed and expanded state; or iii) surface engineered, involves grafting or adsorption of polymeric chains/molecules onto the bulk polymer. There is a hydrophilic-hydrophobic interface switching due to extended-collapsed state of surface chains triggered by the stimuli. The physical sate and architecture of SPB decide the spatial mobility of polymeric chains and consequently the required energy for responding to a given stimuli. Fig. (**1**) illustrates four major physical states of such systems: polymeric solutions, surfaces and interfaces, gels and solids [4]. The mobility of polymeric chains decreases from solutions to surfaces and interfaces or gel to solid states on account of spatial restrictions imposed due to smaller displacement vectors in x, y and z directions [5]. Subsequently, a system with higher solvent content requires minimal energy input to attain the responsive state whereas the restricted chain mobility manifested by anchoring one end of the surface polymers demands increased energy inputs to achieve a transition state. Attaining a responsive solid is greater challenge due to limited free volume [5]. As dictated by the composition and architecture of polymer-containing responsive systems, the type of stimuli can induce alteration in polymeric chains which include changes in conformation, solubility, polarity, charge, degradation, bond cleavage and self assembly into micelles/vesicles [6]. This interfacial molecular change translates into actual response which could be manifested in the form of change in size/dimension of individual chain, attaining secondary structure, inter-/intra- molecular interaction, all of which is linked to association/destruction of weak interactive forces, osmotic interactions or other reactions like acid-base or redox.

SPB related research for biomedical applications has recently picked-up momentum towards developing dual or multi-stimuli responsive systems whereby a constellation of functionalities that respond to different stimuli form a part of single SPB. The impact of combining different functionalities could be; i) parallel, orthogonal system where individual responses are not affected by the presence of one another, ii) serial, response of one group amplifies the stimuli responsiveness of the fellow group, or iii) casual, interestingly the stimuli generated response of one functionality acts as stimuli for second group of functionality [7]. It could be interplay of combination of chemistries and biological entities in a single polymer has been implemented in designing future SPBs exhibiting multifacet behaviors for a series of presented stimuli.

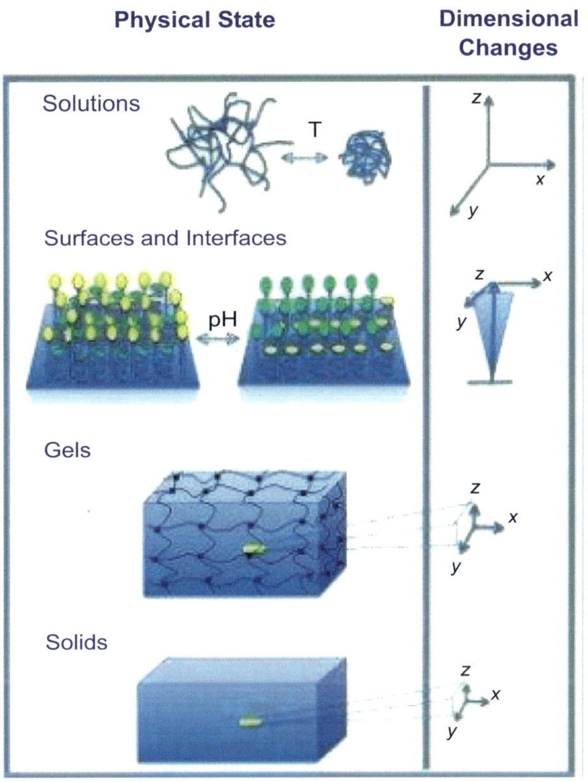

Figure 1: Physical or chemical stimuli induced dimensional changes in different physical states of polymer solutions.

The chapter provides a brief introduction on the environmentally responsive polymer biomaterials, their design and preparation strategies for dynamic response to internal stimuli of biological compartments and region. The bio-interfacial properties and interplay in biological milieu. The chapter is divided into sub-heads of internal stimuli under which a comprehensive description is provided on fabrication and modification of respective smart polymer biomaterials, their interaction and adaptation at biological interface as well as biomedical applications are discussed in detail.

INTERNAL STIMULI RESPONSIVE

Internal stimuli responsive polymer systems are of specific interest for biomedical applications where they are to be intelligently responsive and functional within the dynamic biological system. These are the next-generation stimuli-responsive systems because of self-regulation characteristic and therefore better applicability in therapeutics, tissue-engineering and regenerative medicines. For these reasons the chapter only focuses on the internal-stimuli that are actually and naturally presented to the biomaterial.

pH Responsive

Biological systems present a range of pH variations along the GI tract from stomach (pH 1-3) to the intestine (pH 5-8), in various body tissues down to the cellular compartments like endosomes (pH 5.0-6.5) and lysosomes (pH 4.5-5.0) [8]. The pH in tumor tissue is slightly acidic (pH 6.5-7.2) and that of chronic wound also range between 5.4 and 7.4 [9]. This variation of pH in different compartments or region of the body can be utilized with concomitant exploitation of the pH-vulnerability of polymer functionalities for the designing of SPB. The key challenge in fabricating such materials is infusing in and balancing of contradictory properties that render them stable at physiological pH (pH 7.4) during circulation and extremely responsive to the slight variation of this pH at the target site of the action.

The pH sensitive SPBs are basically polyelectrolytes bearing weak acid or basic moieties in the polymeric backbone or as pendant chains. As a function of change in environmental pH (in aqueous environment of H_2O or solvent) these moieties either protonate (accept proton) or deprotonate (donate proton) thereby ionizing as in monoacids and monobases. But a thorough ionization of such materials is not achieved due to the electrostatic effect imposed by other existing ionized groups in the polymeric structures. The changes on polymer chains generated due to the ionized sate in aqueous environment builds-up an electrostatic repulsion between these chains with consequential extension of the chains thus increasing the hydrodynamic volume of the polymer. Neutralization of the charges reversibly impact in folding and coiling of polymer chains. This reversible phase switching between extended and collapsed state is modulated by the osmotic pressure exerted by the mobile counter ions that neutralizes the charge of the polymeric network [10]. Such conformational changes in polymer structure leads to swelling or dissolution. The two categories of pH responsive polymers are: i) polyacid polymers with anionic groups that remain protonated (unionized) at low pH exhibiting unswollen state. Whereas at higher pH value the polymer shrinks as negative charges develop on the surface due to proton release. The polymeric chain configuration of a weak polyacid is function of polymer pKa that varies from 3 to 11[11]. To exemplify the process, poly(acrylic acid) (PAA) with pKa around 5 display collapsed compact structure in a biological environment with pKa below 5 and a stretched out expanded conformation as it reaches the biological environment with pKa above 5 [12]. Other polyacid polymers with significant pH sensitivity are poly(methacrylic acid) (PMAA) and polysulfonamides [13]; ii) polybasic polymers acquire a swollen state in low pH environment because the ionization of cationic groups increases with a decrease in pH. At higher pH values polymer chain interaction increases as the charge repulsion is decreased resulting in reduced hydrodynamic diameter of the polymer. Cationic pH responsive polymers include poly(N,N-diakyl aminoethyl methacrylate) (DAAEMA), poly(L-lysine) (PLL), poly(ethylenimine) (PEI) and chitosan [14-16].

Modulating the pH window of effective phase transition for a responsive polymer requires careful selection of polyacid or polybase with overall pKa tallying the desired pH range and regulated incorporation of hydrophobic moieties within the polymeric backbone. As the electrostatic repulsive forces disappear upon neutralization of the ionizable groups, the hydrophobic interactions dominate. A more compact globular conformation can be realized through introduction of hydrophobic structures in the design of polymer. A range of pH-responsive polymer structures including homopolymers, block copolymers (amphiphilic block copolymers, di- and tri-block copolymers), micro- and nano- particles/gels, hydrogels and polymer brushes have been prepared through conventional or controlled radical polymerization techniques. Emulsion polymerization is the widely adopted methodology for vinyl based microgels preparations [17]. Microgels are crosslinked particles with tridimensional polymer network like hydrogels. pH-responsive smart gels manifest porosity changes as a consequence of developed electrostatic repulsion between ionized chains due to alteration in environmental pH. The amount or number of pendant acidic/basic groups within the network controls the degree of ionization of such gels. As a consequence the hydrophilicity of the network elevates resulting in greater swelling ratio. Microgels with acidic or basic pendant functionalities display ionization at high and low environmental pH, respectively.

Block copolymer with self-assembling attribute forms an interesting candidate for the preparation of soft materials responsive to pH variations [18]. Amphiphilic block copolymers containing ionizable moieties in the backbone and as pendants can respond to the change in pH. Such structures can self-assemble as micelles and therefore form interesting candidate for therapeutic and gene delivery. Drug micro-encapsulation into amphiphilic copolymer micelles with core-shell structure is a well researched domain. Testifying the domain, a crosslinked nanogel of pH-sensitive polyamine core and tethered poly(ethylene glycol) (PEG) chains showed collapsed structure at high/neutral pH and conversely demonstrated an increased volume transition with decrease in pH. These nanogels have been used for the delivery of cancer drug doxorubicin (DOX) [19].

Polymer brushes with controlled and reversible changes in chain conformation and surface energy are an exciting polymeric architecture for designing surface assemblies. A weak polyelectrolyte is incorporated in the surface grafted polymer chains of pH-responsive polymer brushes. Existing ionizable pendant groups undergo a rapid alteration in the net charge sourcing changes in the hydrodynamic volume of the polymer chains which ensues from changes in the osmotic pressure exerted by mobile counterions neutralizing the polymer charges [20].

These surface assemblies exhibit reversible switching between adsorption and release of macromolecules or nanoparticles for application in substrate-mediated drug delivery [20]. Polymers like poly(styrene-b-2-vinylpyridine) (P2VP), PAA, poly(N,N-dimethyl acrylamide (PDMA), ethylene methacrylate (EMA), poly(dimethylaminoethyl methacrylate) (PDMAEMA) can be surface grafted adopting 'grafting from' or 'grafting to' approach for preparing pH-responsive polymer brushes [20-22].

As an example of employing pH-responsive linkers for conjugating drug to nanocarriers, Lee *et al.* fabricated micelles from a double-hydrophilic hyperbranched copolymer, poly(ethylene glycol-hb-glycerol) (PEG-hb-PG) [23]. Acid cleavable hydrazone linkages were utilized for conjugating hydrophobic drug DOX to the hyperbranched PG units which in turn rendered hydrophobicity to these units. The amphiphilic copolymer aggregated into micelles in the aqueous environment thus protecting the drug in physiological milieu. In the endosomes of human cervical cancer HeLa cells, the low pH (pH 5) of compartment triggered the cleavage of hydrazone linkages, releasing the drug and disassembling micelles into PEG-hb-PG unimers [23].

Anionic polyelectrolytes can be tailored to activate and destabilize the endosomal phospholipid membrane under acidic pH of the organelles. This strategy aims at enhancing transport of biomolecules from endosomes (result of endocytotic uptake) to the cytosol. Cationic polymers like PEI, polyamidoamine dendrimer, PLL and poly(β-amino esters) have materialized gene transfection. Positively charged polymer-DNA complexes bind into vascular or cardiac cells *via* pH-responsiveness to the negatively charged cell membrane and is endocytosed [24, 25].

Exemplifying the strategy of responsive assembly design under which the pH-triggers a conformational or hydrophilicity alteration in the copolymer or a segment of it. Such alterations leads to the disassembly of polymer structure into its corresponding unimers, Manganiellow *et al.* demonstrated endosomal pH-responsiveness of dicthylaminocthyl methacrylate (DEΛEMA) and butyl methacrylate (BMA) copolymers for nucleic acid delivery. The research group further established tweaking of copolymeric architecture in a way that the released unimers function to destabilize the endosomal membrane enabling enhanced cytoplasmic delivery of the cargo [26].

Lee *et al.* demonstrated the relevance of polymer brushes grafting density in pH-responsive applications [27]. They prepared loosely grafted PAA polymer brushes which exhibited a conformation transition from globule to extended state upon change in pH from acidic to basic. Unlike these, the 100% grafted PAA brushes maintained fully extended conformation in a broad pH range (pH 2-9) because of the steric repulsion between densely grafted side chains.

Biofunctionalized synthetic copolymers with pH-responsive behaviour are prepared for improved cell toxicity. Sevimli *et al.* reported amphiphilic copolymers of poly(DMAEMA-co-cholesterylMA) with varying cholesterol units [28]. The copolymer underwent pH-induced phase transition due to protonation of the amine group in an aqueous media and displayed decreased cell toxicity with increase in cholesterol moieties.

Temperature Responsive

Temperature-responsive SPBs are the widely studied group of stimuli-responsive polymers. Temperature-responsive SPBs have a sensitive balance between hydrophilic and hydrophobic moieties and small changes in this balance can create new architecture. One of the inimitable features of temperature-responsive polymers is their critical solution temperature. At this temperature, there is a discontinuous phase transition of the polymers or polymer and other solution according to their composition [10]. If the polymer has one phase (isotropic state), below a specific temperature and become insoluble above that temperature (anisotropic state), then these polymer have a lower critical solution temperature (LCST). Else it is called a higher critical solution temperature (HCST). Thermodynamically the polymer aggregation at LCST is related to the positive change in entropy (ΔS) of the polymer chains and water. Even though, negative excess entropy of mixing is prerequisite for an LCST, the interaction between the polymer and water may also be accountable. Additionally, intermolecular interactions such as hydrogen bonding and hydrophobic forces can also engender random coil to helix conformation or association/dissociation of different pendant groups to generate gel networks.

Temperature-responsive polymers which are the cardinal part of temperature-responsive -SPBs can be broadly grouped into three categories based on their mechanism of thermo responsiveness and the mode of synthesis/modification: i) based on LCST, polymers with LCST show phase transition from soluble to insoluble below and above the LCST, respectively. This group include poly(N-substituted acrylamide), poly(N-vinylisobutylamide), poly(vinyl methyl ether), poly(N-vinylcaprolactam), poly(dimethylaminoethyl methacrylate), poly (N-isopropylacrylamide) *etc*. One of the most extensively studied polymer is poly (N-isopropylacrylamide) (PNIPAAm); ii) on the basis of amphiphilic balance, a number of amphiphilic polymers have thermo-sensitive micellation behaviour and form hydrogel above critical micelle temperature. These copolymers sol-gel transition oscillates below and above this temperature. Interesting temperature-responsive hydrogels biocompatible and biodegradable with attributes include PEG based copolymers such as, [PEG-PLGA-PEG], [PEG-PCL-PEG], PLGA-PEG-PLGA, PCL-PEG-PCL *etc.*; iii) biopolymers and their derivatives, some natural polymers or their derivatives show thermogelling properties. These include modified dextran, gelatin, agarose, gellan benzyl ester and xyloglucan.

Cellulose derivative such as methyl cellulose has thermoreversible gelation property in aqueous solution, in the range of 60-80°C. The incorporation of PEG with chitosan, a natural polymer, offers thermosensitive hydrogel property to chitosan without any cross linker and has got enormous application in tissue engineering and drug delivery. Thermosensitive hydroxybutyl modified chitosan was established for hMSC proliferation for the treatment of degenerative disc disease. PNIPAAm has remarkable interfacial property with living cells. Above LCST (32°C) PNIPAAm shows hydrophobic behaviour and the cells preferentially adhere to those surfaces. Whereas, below LCST PNIPAAm is hydrophilic and results in cell detachment from the polymer surface [29]. This property of the polymer can be used for cell recovery without mechanical or enzymatic methods. Recently Idota *et al.* reported galactose incorporated copolymer brushes comprising of PNIPAAm and 2-lactobionamidoethyl methacrylate (LAMA) [30]. This thermosensitive sugar based SPB successfully demonstrated hepatocyte cell adhesion above LCST and cell detachment below LCST (Fig. **2**).

Figure 1 from Naokazu Idota et al 2012 Sci. Technol. Adv. Mater. 13 064206

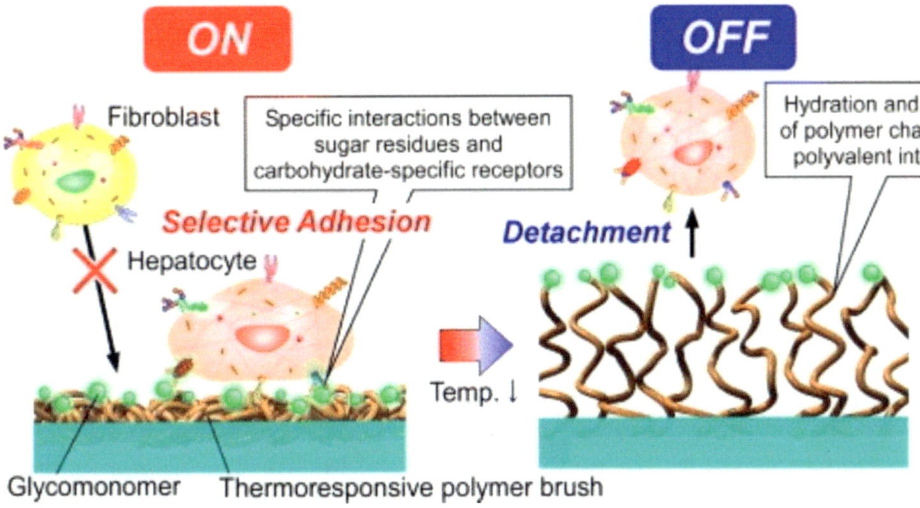

Figure 2: Schematic illustration of hepatocyte cell adhesion above LCST and cell detachment below LCST.

The bio-interfacial property of certain SPBs has been exploited for their use as molecular gate and switches. Hoffman *et al.* reported the ability of PNIPAAm to act as a molecular gate by its temperature sensitive swelling property [31]. PNIPAAm was linked to streptavidin near to the active site for biotin binding and when the temperature rose above LCST, the polymer collapsed and covered the biotin binding site preventing the biotin molecule from binding to streptavidin. This same mechanism can be utilized for the release of biotin from streptavidin during temperature change. Similarly, protein folding can be controlled with the help of thermosensitive polymer. For example, Kuboi *et al.* reported that PPO-Ph-PEG can increase the refolding yield of carbonic anhydrase enzyme [32]. The local hydrophobicity of PPO-Ph-PEG increased to 45°C and after that had a sharp increase. Thus after the enzyme refolded to native state, it will release from polymer because of the less local hydrophobicity. PNIPAAm was also reported as autonomous flow controlling coating in microfluidics primarily because of the thermosensitivity. Correspondingly these thermosensitive polymers have an important role in tissue engineering and drug delivery. In tissue engineering, delivering cells for tissue repairing or regeneration can be achieved by using appropriate stimuli sensitive polymer since it has an aqueous interior environment. Kim *et al.* reported that cells adhered on PNIPAAm at above LCST can release chondrocyte cells below LCST showing desirable characteristics in tissue engineering [33]. *In situ* gel forming thermosensitive polymers have been exploited for drug delivery applications based on the LCST.

Ion Responsive

The polymers with ionizable moieties have inherent property of responding to ionic strength. Both anionic and cationic groups are incorporated together to obtain amphiphilic polymers. The ratio of existing ionic charges can be regulated through polymerizing strategies or/and ionic charges in the aqueous environment [34, 35]. Such SPBs display unusual rheological behaviors due to coulombic interactions between oppositely charged species. Accesses of either of the groups confer on the copolymer the proteolytic properties. With the ratio of cationic to anionic charges approaching unity, attractive coulombic interaction dominate that renders the polymer insoluble in deionized water. Shielding these attractive interactions by the addition of a critical concentration of electrolyte (as salts) incites phase-transition solubility of these material [34-36]. Variations in ionic strengths affect the chain lengths of the polymer, its solubility and fluorescence quenching kinetics of chromophores bound to electrolytes [34, 37, 38]. A polyphophazene-functionalized diaminobutane poly(propyleneimine) (DAB-PN) ion-responsive dendrimer structures have been prepared for hydrophobic drug delivery. The ethyleneoxy groups in polyphosphazene chains get complexed with cations (like Na^+ and K^+) presented in the solutions which trigger swelling behavior of the SPB [39].

Redox Responsive

Redox responsive biomaterials contain a labile group that can be oxidized or reduced by the action of redox species present in biological compartments (intracellular/extracellular). Such materials are usually biodegradable and are gaining special importance for controlled delivery of therapeutics. They utilize the reducing environment, such as those of endosomes, and higher concentrations of reducing agent glutathione in cytosol and cell nucleus as compared to the extracellular environment. Other SPBs are designed to be responsive to oxidizing agents like H_2O_2 and super oxide anions that are present in high concentrations in the inflammatory sites and cancer tissues due to the bubbling enzymatic activities in these regions. In general, the redox-responsive SPB designing involves incorporation of disulphide linkages or chemical species with multiple oxidation states. The disulphide chemistry which shows instability in reducing environment has prominently been explored for therapeutical delivery applications. When exposed to reductive amino acids base molecules like cysteine or glutathione (as found in biological environment) the linkage cleave to thiol groups resulting in polymer degradation or release of linked molecules [40]. Glutathione-responsive assemblies have huge potential in delivering therapeutics (drugs, peptides, proteins, oligonucelotides and DNA) to cytosol and nucleus - the sites where most therapeutics are active. The disulphide linkages can be incorporated directly into the polymeric side chains or the polymeric backbone through use of a suitable monomer and an initiator or chain transfer agent [41]. Glutathione responsive block copolymers usually have hydrophobic and hydrophilic blocks linked through disulphide groups creating shell-shedable micelles. Alternatively thiol-reducible linker can be used to crosslink micelle core or shell or to simply conjugate drug molecule. Such structures get destabilized in a glutathione environment releasing the loaded cargo.

Redox responsive ter polymers have been prepared by copolymerization of a pyridyl disulphide containing acryloyl monomer with MAA and BA. The system presented with membrane disruptive properties essential for endosomal escape in gene delivery [41, 42]. Dong and co-workers used the disulphide redox responsiveness for inducing morphological transitions of block copolymer aggregates. The micellar core comprised of polyelectrolyte complex of positively and negatively charged chains and the corona build-up with disulphide linked PEG. Removal of PEG corona shield in response to dithiothreitol presence leaves behind homopolymer complexes which transform into a vesicular structure - the driving force being the decreased free energy of the corona (Fig. **3**) [43].

The oxidative environment in many diseased states has prompted the designing of amphiphilic block copolymers with hydrophobic block of propylene sulfide (PPS). The sulfide groups get oxidized to hydrophilic sulfoxide and sulfone groups destabilizing the hydrophobic-hydrophilic balance and therefore resulting in a structural disruption [44]. Poly(NIPAAm-co-RU$(bpy)_3$) has been used by Yoshida and co-workers as redox-responsive copolymeric materials that display periodic redox change of RU(bpy)3 into an oxidized state thus capable of generating a chemical wave [45].

Ren *et al.* prepared selenium-containing block copolymers with fully recoverable redox-responsive properties. The copolymer forms spherical micelles in an aqueous environment with hydrophobic cores. In an oxidative environment of water, the selenide groups of core oxidized to selenoxide increasing the hydrophilicity of otherwise hydrophobic block and resulting in micellar disassembly. Addition of reducing agent has a reversal effect of regenerating the micelles [46].

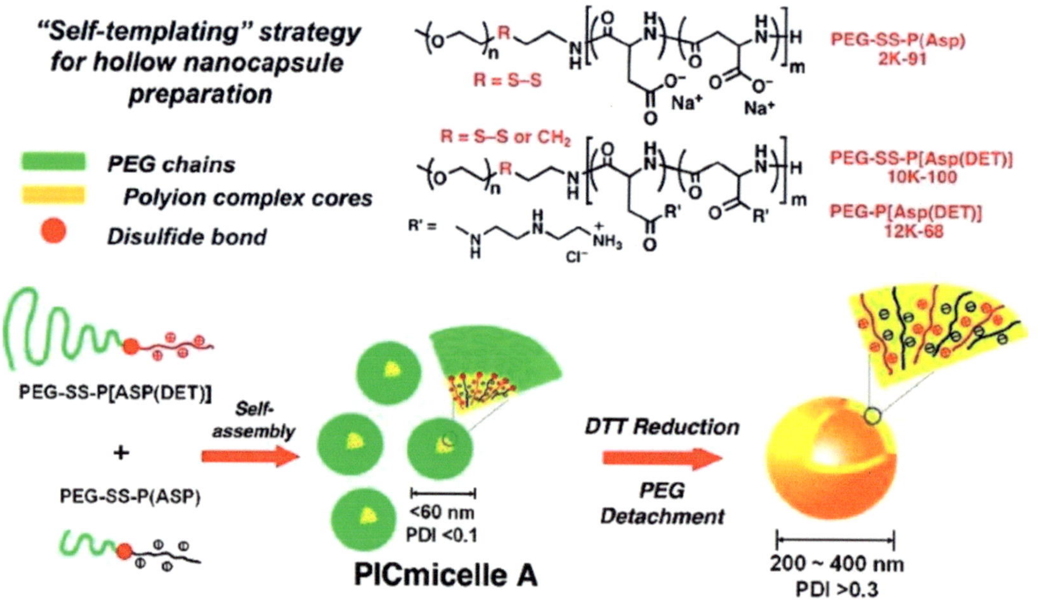

Figure 3: Preparation of hollow nanocapsule vesicles from redox-responsive block copolymers. The addition of reducing agent Dithiothreitol (DTT) into a hetero-PEG-detachable polyelectrolyte complex micelle solution resulted in a morphology transition to polymeric vesicles. (Reprinted with permission from [43]).

Biologically Responsive

This class of SPB is bestowed with unique feature of interacting with biological molecules to elicit a dynamic response that regulates the biological functions. They form the most versatile class of stimuli responsive SPB because of the specificity of response towards a given biomolecule thus providing a tailored response to biological demands. Design and fabrication of biologically responsive polymers generally adopts either of the two approaches. One is the integration of functional groups that are recognized to react with a given specific biomolecule and the other implies conjugating the polymeric structure to a complete or partial (fragmented, sequenced) biological entity. Advancements are underway to explore potential stimuli that would inspire the development of novel biologically responsive SPBs. The so far researched stimuli include enzymes, receptors, antigen/antibodies, glucose, glutathione and inflammatory metabolites.

Enzymes Responsive

Enzymes are the exclusive biomolecules that regulate the interplay of biological mechanisms within the living system. Remodeling of the bimolecules like ECM is governed by enzymatic action and researchers have focused on harnessing this intrinsic catalytic property of enzymes as stimuli to remodel or functionalize the SPBs for desired *in vivo* action. Enzyme responsive (ER) SPBs are remarkable since the enzymes exhibit highly selective reactivity, can be operated under mild *in vivo* environment and are indispensable component of the biological pathways. Moreover, certain enzymes are localized in specific body locations or under specified condition such as azoreductases mainly confined in the digestive system and enzymes like phosphatases overexpressing in diseased cell [47].

Preparation of ER-SPBs requires integration of enzymatic substrate or ER-functionality into polymeric structures which could be a homopolymer or copolymer construct of purely synthetic or hybrid (synthetic-natural) origin. Such enzyme-labile features can be interspersed within the main structure or could act as linkers for conjugating specific active/biomolecules or otherwise could simply elicit enzyme activity for polymerization or self assembly. Realization of these biohybrid copolymers has been the result of integrated advanced research in controlled polymerization techniques, solid phase synthesis of biomolecules and bioconjugation chemistry.

Introduction of ER cues to a hydrogel imbibes the latter with functionalities that further complements their structural similarity to the ECM. Further, these functionalities can endow them with drug release characteristics triggered in presence of specific enzymes. ER-hydrogels could either be assembled or degraded *in situ* in response to enzymatic action. For hydrogel formation polymer strand could be cross-linked through enzymatic action (transglutaminase) or reactive groups could be generated enzymatically on the polymeric strand (horseradish peroxidase and tyrosinase) that facilitate their crosslinking [48, 49]. Degradation of ER-SPBs could be direct for naturally derived hydrogels (dextranase degrading CL dextran) or would require presence of enzymatically cleavable moieties like short peptide sequences in case of artificial polymers. Peptide sequences can be handpicked and tailored to match the target enzyme. Hydrogel with degradable susceptibility to matrix metalloproteinases, elastase and other enzymes have been reported [50, 51]. Enzymes, apart from aggregating-degrading the polymer chain of hydrogel, can also alter the structural conformation of hydrogel particles by changing the overall charge and consequently the degree of swelling (trypsin, elastase and matrix metallo protienases). Enzymatic cleavage of hydrogel crosslinks also has similar effect [52-54].

Polymers or small molecules can self-assemble into supramolecular constructs with architecture far differing from their building units. The assembly, disassembly or change in morphology of these supramolecular architectures can be enzymatically regulated providing application of ER supramolecular architecture in drug delivery and regenerative medicine. Polymeric supramolecular architecture can be assembled as enzymatic response through the use of double-hydrophilic block polymers. Masking or dampening the hydrophilicity of one polymer block will translate it to polymer amphiphile that can self assemble. Incorporation of phosphate groups to the side chains of one polymer block incites phosphatase enzyme mediated dephosphorylation in biological environment rendering the block less hydrophilic leading to self assembling of the polymer [55]. A similar micelle construct is of PEG-poly(lysine) block polymer formed due to association with ATP (in ATP rich environment) which renders the poly(lysine) block hydrophobic. The enzymatic hydrolysis of ATP by phosphatase destroys the micelle due to removal of poly(lysine) hydrophilic characteristic [56]. In another degradation approach the hydrophilic and hydrophobic blocks of an amphiphilic block copolymer can be linked through enzyme-sensitive linker like ester links or short peptide sequences that can be cleaved by esterases and proteases [57, 58]. In yet another approach, employing enzymatic action can extend or reduce the polymer chain lengths, altering the superamolecular size and shape. Even a minute alteration, as of a single amino acid (one phosphate group removal by a phosphatase), in the structure can transmit change of spherical aggregation to micelles [59, 60]. The most attractive feature of these structures is the multiple-time reversibility of their conformation.

Solid randomly aggregated polymeric particles or hollow polymer capsules are extensively utilized drug delivery vehicles. The degradation or disassembly of these particles can be mediated by enzymes through protease activity on polymer/polypeptide conjugates or enzymatic cleavage of enzyme-sensitive crosslinks in the structure. ER-SPB are majorly designed for protease responsiveness although phosphatase has also been employed in some case to alter the electrostatic interaction of polymer aggregates in the particle [61]. Chitosanase and α-chymotrypsin mediated polymer degradation is applied in case of polymeric capsules [62].

Self-immolative polymers are other metastable polymers that disassemble into its constituent units upon removal of stable-end group [63]. The possible designing of these end-groups to be enzymatically cleavable give rise to this young class of ER self-immolative polymers. Penicillin G amidase and catalytic antibody 28C2 cleavable end-groups have been tagged to design linear polymers, dendrimer-like oligomers and comb polymers that disassemble to release therapeutics/ molecules when the latter are attached to the polymer *via* similar metastable chemistry [64-66].

Antigen Responsive

Similar to ER-SPBs, the antigen responsive SPBs display extreme specificity and remarkable responsiveness due to high affinity and specificity of antigen-antibody interaction that involves non-covalent forces like van der Waals, electrostatic, hydrophobic and hydrogen bonding. Antigen responsive hydrogels can be designed in three possible ways: i) physical entrapment of antibodies in hydrogel network, ii) conjugating antibody to the hydrogel network through chemical bond and iii) cross-linking the hydrogel network with antigen-antibody pairs [67]. An antigen responsive semi-IPN hydrogel with reversible responsiveness has been prepared by Miyata *et al.* They introduced vinyl functionality to the antigen (rabbit IgG) and antibody (goat anti-rabbit IgG), the latter was then polymerized to acrylamide and finally in its presence the functionalized antigen was polymerized to acrylamide and MBA. Free antigen (rabbit IgG) added to the resulting semi-IPN participated in competitive binding to the hydrogel antibody thereby destabilizing the antigen-antibody crosslinks in semi-IPN causing hydrogel swelling. Protein permeation into the semi-IPN was realized in pulses by changing the antigen concentration stepwise [68].

Inflammation Responsive

The process of inflammation in a biological system is initiated by T-and B-lymphocytes and further amplified and perpetuated by polymorphonuclear leukocytes (PMNs) and macrophages. The chemical mediators like proteolytic enzymes, arachidonic acid metabolites and oxygen metabolites causing the tissue damage has been recognized as putative stimuli for inflammation responsive SPBs. Reactive oxygen metabolites released by the PMNs and macrophages during the initial phase of inflammation has actually been targeted for designing inflammation responsive SPBs [69]. Hyaluronic acid hydrogels with glycidyl ether cross linking degrades in response to hydroxyl radical *in vitro* and in response to inflammation upon implantation *in vivo* [70]. Responsive SPBs infused with linkers specifically degradable by elastase of neutrophils have been reported for smart actions at inflammation sites where neutrophils activity is observed [71].

Glucose Responsive

Glucose responsive polymers are interesting candidates for developing self-regulatory modules for insulin delivery. A continuous monitoring of blood sugar level in diabetes mellitus patient is taken over by these smart systems which automatically respond to the rise in glucose level by delivering insulin at the site. The design of glucose responsive SPBs is based on either of the three basic approach; enzymatic oxidation of glucose by glucose oxidase (GOx), glucose binding with concanavalin A (Con A) and reversible covalent linking of glucose to boronic acids. Based on first approach the GOx is conjugated to smart pH sensitive polymers. GOx oxidises the glucose into gluconic acid and hydrogen peroxide (H_2O_2). The response of the attached stimuli sensitive polymer is not dictated by direct interaction with glucose but rather with its degradative products. The released gluconic acid alters the environmental pH which dictates a volume transition in responsive polymer of the SPB resulting in drastic conformational changes in polymer structure which consequently leads to insulin release as per the design. In a typical example of diabetes therapy cellulose membrane was crosslinked to GOx conjugated PAA. The negatively charged carboxylate groups on polymeric chains at a neutral and alkaline pH exert electrostatic repulsion between the chains. This results in an extended state of the chain and thereby occlusion of membrane pores. With change of environment to acidic pH, because of the enzymatic action in presence of glucose, the charge over the chain surface neutralized due to protonation of carboxylate moieties. Subsequently the chains achieved a collapsed state which facilitates insulin release from the membrane (Fig. **4**) [72]. On similar lines GOx containing hydrogel have been reported with (PMAA)-graft ethylene glycol and NIPAAm copolymerized with MAA [73, 74]. Conversely polymers with polybasic groups that develop positive charge under acidic environment have been employed as insulin depot which release insulin due to swelling which being a consequence of electrostatic repulsion in chain at lower pH. Additionally researchers express possibility of devising polymers with the adaptability to H_2O_2-another catalytic by-product of glucose [75]. The second approach makes use of glycopolymer-lectin complexes for glucose responsiveness. The system deciphers specific binding of lectins with carbohydrates [76]. Glycopolymers aggregate in presence of multivalent lectins which could be disrupted in the environment of competitively binding saccharides like glucose. Con A is the abundantly used lectin for developing such glucose responsive SPBs. A Con A-glycogen gel system has been prepared that exhibit gel-sol transition in glucose environment because of the binding preference of Con A for free glucose compared to glycogen gel. Authors express potentiality of devising hybrid pancreas substitute by using glucose responsive SPBs and constitutively secreting cells to better mimic insulin release physiologically [77].

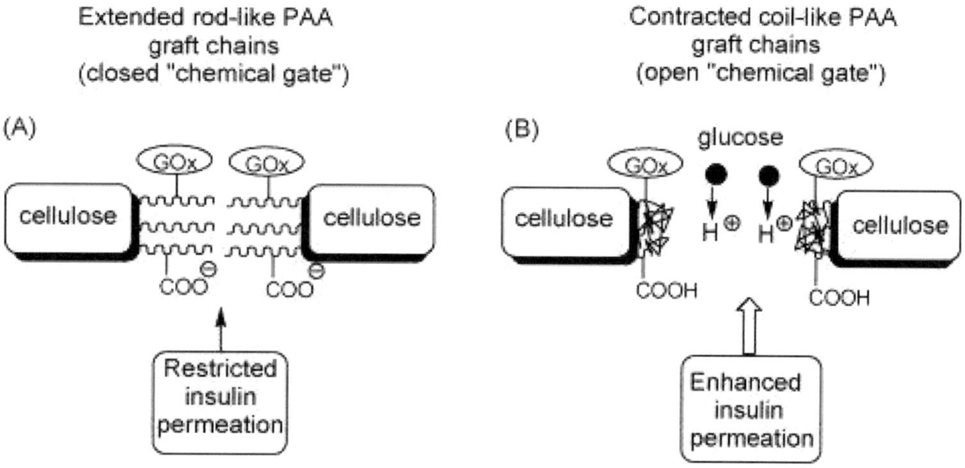

Figure 4: (**a**) Extended PAA graft chains because of the prevailing electrostatic repulsion in the glucose molecule absence results in closed 'chemical gate' restricting the membrane permeability. (**b**) Contracted or collapsed PAA graft chain confirmation because of the reduced repulsive forces between the chains on addition of glucose molecule which lowers the pH that enhance permeation.

The last approach taps the property of boronic acid to reversibly complex with sugar in addition to tunable water solubility with changes in pH or diol concentration [78]. Since this category of glucose responsive materials with boronic acid moiety is not used for developing biomaterials but limited to industrial application, they are not discussed in detail.

Dual and Multiple Responsive

The complexity of a biological system and its dynamic environment pose a pressing demand for intricate designing of intelligently responding systems. The SPBs integrated with consortium of functionalities, to confer sensitivity for dual- or multiple-stimuli encountered within the biological compartments, are fascinating as well as challenging. Such materials are gaining an increased impetus as they can exert a better spatial and temporal self-regulated control over desired/required response to the biological demands. These SPBs not only offer superior efficacy and performance in biological environments but also facilitates the preparation of complex structures and/or biomolecule/therapeutic loading and conjugation. The designing strategy of these biomaterials would allow simultaneous or sequential response to all stimuli at a single location or in different compartment/settings. Literature reports on the performance of SPBs prepared for different combinations of internal stimuli like pH-temperature, pH-redox, temperature-redox, pH-temperature-redox, double pH, pH-diols, temperature-enzymes and pH-biomolecules. An interesting and most researched combination of dual internal stimuli is the pH-temperature. The design template for such SPBs requires integration or implantation of pH-responsive units like weak acids in thermo-responsive polymers like PNIPAm, imparting pH dependent LCST of the copolymer. Yamamoto *et al.* prepared temperature-pH dual responsive molecular brushes by 'grafting from' PBIEM macroinitiators the MEO2MA unit and MAA/DMAEMA unit respectively in the side chains using ATRP technique. They imparted thermo and pH dual sensitivity to the brushes. Due to stronger ionization of MAA, molecular brushes showed LCST at pH 4 and 7 but not at pH 9 [79].

Many tumor drug release copolymeric constructs are based on this subtle pH change dictated tuning of phase transition. As an example, a self-assembled core-shell nanoparticles from poly(-NIPAAm-co-N,N-dimethylarylamide-co-10-undecenoic acid) ((NIPAAm-co-DMAAm-co-UA) terpolymer resonated good stability at physiological conditions (pH 7.4) but destabilized and precipitated in low pH conditions as observed in tumor tissues releasing the DOX [80]. In a further progress by the same group, cholesterol was conjugated to the hydrophobic segment of (P(NIPAAm-co-DMAAm-co-UA)) and folic acid to free amine group [81]. The resultant nanoparticles were stable and released DOX into the cancer cells which overexpressed the receptors for folic acid. A micellar construct of a mixture of methoxy-PEG-b-P(N-(2-hydroxypropyl) methacrylamide dilactate)-co-(N-(2-hydroxypropyl) methacrylamide-co-histidine) (mPEG-b-P(HPMA-Lac-co-His)), mPEG-b-PLA and cy5.5-PEG-PLA copolymers responsive to the same dual-stimuli presented specific DOX targeting efficiency and improved anti-tumor activity in balb-c/nude mice bearing human cervical tumors [82]. Researchers attributed this efficiency of dual-responsive nanoparticles to their tumor specific accumulation, enhanced permeation through tumor site and tumor pH-triggered release. Nanoparticles and micelles with endosomal and lysosomal pH modulated drug release are also reported in the literature [83, 84].

Another exciting intrinsic stimuli duo could be pH and redox. As stated earlier they co-exist in all cancer cells and many pathological environments. Designing and collating such biomaterials facilitate their preparation in aqueous media *via* change of pH, increased physiological stability through the use of disulfide linkers, enhance cellular uptake through surface charge reversal in effective pH range, therapeutic delivery in cell compartments like endosomes/lysosomes and/or functionally important fast release in cystosol and cell nucleus. A therapeutic delivery of drug molecule has been achieved by deploying dual-stimuli responsive delivery system. A copolymeric system of pH-sensitive hydrophobic BA and glutathione-responsive pyridyl disulfide acrylate was prepared and the drug release kinetics was fine tuned by symmetrically varying the composition of two moieties or by modifying both [85].

Thermo and redox-responsive biomaterials have also been developed. Phillips and Gibson reported PNIPAAm precursors or macromers incorporated with disulfide bonds that degrade selectively at intracellular concentration of glutathione and while being stable at lower extracellular glutathione concentrations. The PNIPAAm fragments thus generated exhibit molecular weight-dependent cloud point that could switch off the temperature dependent behavior of the SPBs by glutathione gradient [86]. Morimoto *et al.* developed nanogel systems based on pullulan. By increasing its solution temperature from 25 to 50°C a light grafting thiol-terminated PNIPAm was obtained and the thiol-ends were further crosslinked oxidatively [87]. On similar principle a nanogel, dextran grafted with thiol terminated PNIPAAm, has been prepared [88]. These nanogel structures could be disrupted by change in temperature or *via* exposure to reducing agents.

Dual pH-responsive polymers have also been designed and prepared for improved site specific action of the SPBs. Almutairi *et al.* simultaneously incorporated β-amino ester and ketal bonds - both of which are pH-labile - within the polymer backbone. The SPBs demonstrated physiological stability but a dramatic and fast degradation at pH 5 due to sequential play of pH-labile functionalities. The pH-change triggered response or solubility switch of β-amino esters that further presented a controlled degradation of ketal moieties promoting rapid degradation of polymer naoparticles which according to the investigators may elevate the osmotic pressure of endosomal compartments facilitating therapeutic release in cytosol [89]. Wang and group designed and developed poly(PPC-Hyd-DOX-DA) polymer-drug conjugate that had PEG shielding with negative surface charge at physiological pH 7.4. With change in pH to 6.8 the surface charge reversed on account of amide bond cleavage and at another responsive pH 5.0 - as in endosome and lysosomes intracellular compartments - the cleavage of hydrazone bonds was initiated. Fig. (**5**) schematically illustrates the functioning of these (PPC-Hyd-DOX-DA) polymer-drug conjugates [90].

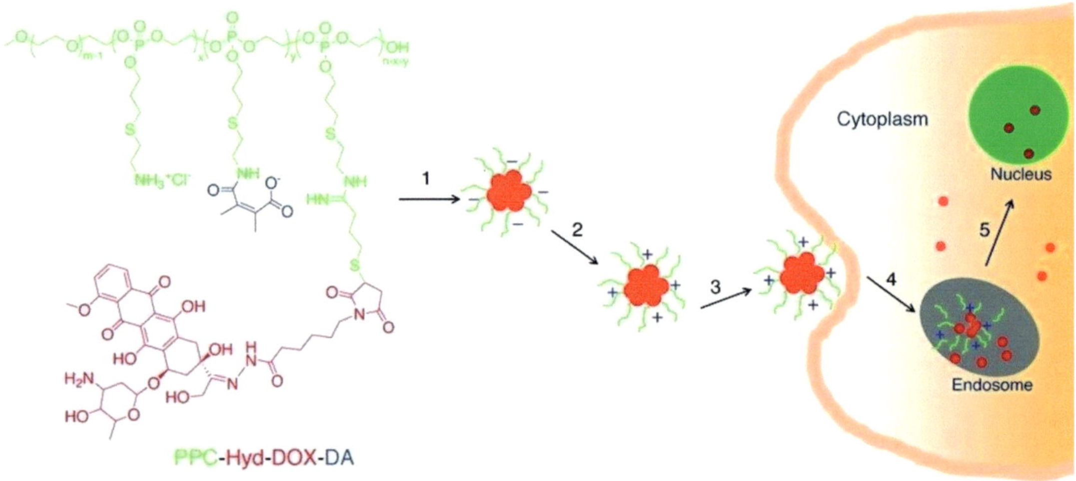

Figure 5: Chemical structure of the dual pH-responsive polymer-doxorubicin (dox) conjugate (ppc-hyd-dox-da) and schematic illustration of its pH triggered cellular internalization and intracellular drug release.(The conjugate self-assembles into negatively charged NPs in water (**1**), and the NPs become positively charged at the extracellular pH (**2**). The positive NPs enter the cells by endocytosis (**3**,**4**), and DOX is released after cleaving the hydrazone bond at endo/lysosomal pH and diffuses into nuclei (**5**). (Reprinted with permission from [90]).

Multi-responsive SPBs with biomedical applications are enthusiastically being designed and developed. A triple sensitive micelle system with acid responsive hydrophobic copolymer of tetrahydropyran-protected 2-hydroxyethyl methacrylate (THP-protected HEMA), thermo-responsive hydrophilic polymer PNIPAAm and redox-responsive disulfide-linkers has been reported [91]. The micelle particles were designed to function *via* three stimuli-triggers as: i) temperature rise above LCST, solubilizing the hydrophilic shell leading to micelle nanoparticle nanoprecipitation. ii) lower pH, transforming the hydrophobic core into hydrophilic leading to micellar nanoparticle dissolution and iii) reducing agent attacking the linkages releasing individual copolymeric blocks leading to disruption of the micellar assembly. The SPB allowed fine tuning of the release kinetics of hydrophobic therapeutic load. The research group confirmed the significantly accelerated complete release of Nile Red at 3.2mM glutathione concentration coexisting with pH 5 compared to otherwise physiological glutathione concentration of 3.2mM and pH 7.4. Li *et al.* reported similar tri-stimuli sensitive nanogels prepared by miniemulsion copolymerization of monomethyl oligo(ethylene glycol) acrylate (OEGA), 2-(5,5-dimethyl-1,3-dioxan-2-yloxy) ethyl acrylate (DMDEA) [92]. Supramolecular vesicles based on host-guest complex formation between p-sulfonatocalix(4)arene (C4AS) and asymmetric viologen (MVC12) has been designed as multi-stimuli responsive. The vesicles' structural integrity was triple controlled and was sensitive to - temperature rise to 70°C, competitive CD addition or reduction potential application. The DOX.HCl-loaded vesicles proved to be extremely stable towards load leakage at room temperature and were reported to rapidly release complete pay load upon temperature increase or CD addition. Researchers established high anti-tumor activity of these triple-responsive vesicles by cellular studies with HepG2 cells [93].

CONCLUSION

Introduction of novel polymers and advances in polymeric techniques has given rise to the fascinating class of environmentally responsive biomaterials. The polymer chemistry has a trail of functionalities that could be engineered into complex biomaterials. Production of these materials with precisely controlled and reproducible properties - which is requisite when preparing a biomaterial - has been actualized recently. These techniques have been outlined in the chapter. Further, advancement in molecular biology and conjugation chemistry has produced novel polymer-bioconjugates that complements the functionality of stimuli responsive polymer biomaterials, providing better adaptability at dynamic biological interface. It has been detailed how stimuli modulates the polymer chain dynamics, molecular interactions at the biological interface and within the chain and also the molecular recognition. The change in energy levels for conformational changes has been briefed. Response to the stimulus can be manifested as changes in shape, surface characteristics, solubility, formation of an intricate molecular assembly, a sol-to-gel transition and others. Internal stimuli responsive biomaterials have been addressed in the chapter as these are self-regulatory and more specific in action compared to the external stimuli. Dual and multi-stimuli responsive polymer biomaterials have been discussed but they are still in infancy. Recognition of novel stimuli and detailed understanding of the molecular mechanism of biological interactions needs to be weaved into intricate biomaterial designing for actualizing an implementable concept of multi-stimuli responsive biomimicking biomaterials.

ACKNOWLEDGEMENTS

The study was supported by the National Research Foundation (NRF) of South Africa.

CONFLICT OF INTEREST

The authors confirm that this chapter content has no conflict of interest.

REFERENCES

[1] Delcea M, Mo¨hwald H, Skirtac AG. Stimuli-responsive LbL capsules and nanoshells for drug delivery. Adv Drug Deliv Rev 2011; 63: 730-47.

[2] Iha RK, Woole K.L, Nystrom AM, Burke DJ, Kade M J, Hawker CJ. Applications of orthogonal "click" chemistries in the synthesis of functional soft materials Chem Rev 2009; 109: 5620.

[3] Hawker CJ, Bosman AW, Harth E. New polymer synthesis by nitroxide mediated living radical polymerizations. Chem Rev 2001; 101: 3661-88.

[4] Fang L, Marek WU. Recent advances and challenges in designing stimuli-responsive. Prog Polym Sci 2010; 35: 3-23.

[5] Marek WU. Stratification, stimuli-responsiveness, self-healing, and signaling in polymer networks. Prog Polym Sci 2009; 34: 679-87.

[6] Schmaljohann D. Thermo- and pH-responsive polymers in drug delivery. Adv Drug Deliv Rev 2006; 58: 1655-70.

[7] Schattling P, Florian D and Theato P. Multi-stimuli responsive polymers - the all-in-one Talents. Polym Chem 2014; 5: 25-36.

[8] Dissemond J, Witthoff M, Brauns TC, Haberer D, Goos M. pH values in chronic wounds. Evaluation during modern wound therapy. Hautarzt 2003; 54: 959-65.

[9] Rofstad EK, Mathiesen B, Kindem K and Galappathi K. Acidic extracellular pH promotes experimental metastasis of human melanoma cells in athymic nude mice Cancer Res 2006; 66: 6699-707.

[10] Gil ES, Hudson SM. Stimuli-reponsive polymers and their bioconjugates. Prog Polym Sci 2004; 29: 1173-222.

[11] Park SY, Bae YH. Novel pH sensitive polymers containing sulfonamide groups. Macromol Rapid Comm 1999; 20: 269-73.

[12] Li G, Song S, Guo L and Ma S. Self-assembly of thermo- and pH-responsive poly(acrylic acid)-*b*-poly(*N*-isopropylacrylamide) micelles for drug delivery. Polym Sci Part A: Polym Chem 2008; 46: 5028-35.

[13] Kang S I, Bae Y H. pH-Induced solubility transition of sulfonamidebased polymers. J Controlled Release 2002; 80: 145-55.

[14] Sideratou Z, Tsiourvas D, Paleos CM. Quaternized poly(propylene imine) dendrimers as novel pH-sensitive controlled-release systems. Langmuir 2000; 16: 1766-69.

[15] Burke SE, Barrett CJ. pH-responsive properties of multilayered poly(L-lysine)/hyaluronic acid surfaces. Biomacromolecules 2003; 4: 1773-83.

[16] Toncheva V, Wolfert MA, Dash PR, Oupicky D, Ulbrich K,Seymour LW. Novel vectors for gene delivery formed by self-assembly of DNA with poly(L-lysine) grafted with hydrophilic polymers. Biochem Biophys Acta 1998; 1380: 354-68.

[17] Chuang C-Y, Don T-M, Chiu W-Y. Synthesis of chitosan-based thermo- and pH-responsive porous nanoparticles by temperature-dependent self-assembly method and their application in drug release. J Polym Sci Part A: Polym Chem 2009; 47: 5126-36.

[18] Xu C, Wayland BB, Fryd M, Winey KI, Composto RJ. pH Responsive Nanostructures Assembled from Amphiphilic Block Copolymers. Macromolecules 2006; 39: 6063-70.

[19] Oishi M, Nagasaki Y. Synthesis, characterization, and biomedical applications of core-shell-type stimuli-responsive nanogels - Nanogel composed of poly[2-(N,N-diethylamino)ethyl methacrylate] core and PEG tethered chains. React Funct Polym 2007; 67: 1311-29.

[20] Chen T, Ferris R, Zhang J, Ducker R, Zauscher S. Stimulus-responsive polymer brushes on surfaces: Transduction mechanisms and applications. Prog Polym Sci 2010; 35: 94-112.

[21] Minko J. Responsive Polymer Brushes. Macromol Sci Part C: Polym Rev 2006; 46: 397.

[22] Kim B, Lam CN, Olsen BD. Nanopatterned protein films directed by ionic complexation with water-soluble diblock copolymers. Macromolecules 2012; 45: 4572.

[23] Lee S, Saito K, Lee H-R, *et al.* Hyperbranched double hydrophilic block copolymer micelles of poly(ethylene oxide) and polyglycerol for pH-responsive drug delivery. Biomacromolecules 2012; 13: 1190.

[24] Jeon O, Yang HS, Lee T-J, Kim B-S. Heparin-conjugated polyethylenimine for gene delivery. J Controlled Release 2008; 132: 236-42.

[25] Katz MG, Swain JD, Tomasulo CE, Sumaroka M, Fargnoli A, Bridges CR. Current strategies for myocardial gene delivery. J Mol Cell Cardiol 2011; 50: 766-76.

[26] Manganiello MJ, Cheng C, Convertine AJ, Bryers JD and Stayton PS. Diblock copolymers with tunable pH transitions for gene delivery. Biomaterials 2012; 33: 2301.

[27] Lee H-i, Boyce JR, Nese A, Sheiko SS, Matyjaszewski K. pH-induced conformational changes of loosely grafted molecular brushes containing poly(acrylic acid) side chains. Polymer 2008; 49: 5490-6.

[28] Sevimli S, Sagnella S, Kavallaris M, Bulmus V, Davis TP. Synthesis, self-assembly and stimuli responsive properties of cholesterol conjugated polymers. Polym Chem 2012; 3: 2057.

[29] Okano T, Tamada N, Sakai H, Sakurai Y. A novel recovery system for cultured cells using plasma-treated polystyrene dishes grafted with poly(N-isopropylacrylamide). J Biomed Mater Res 1993; 27: 1243-51.

[30] Idota N, Ebara M, Kotsuchibashi T, Narain R and Aoyagi T. Novel temperature-responsive polymer brushes with carbohydrate residues facilitate selective adhesion and collection of hepatocyte. Sci Technol Adv Mater 2012; 13: 4206.

[31] Hoffman AS. Bioconjugation of intelligent polymers and recognition proteins for use in diagnostics and affinity separations. Clin Chem 2000; 46: 1478-86.

[32] Kuboi R, Morita S, Ota H, Umakoshi H. Protein refolding using stimuli-responsive polymer-modified aqueous two-phase systems. J Chromatogr B 2000; 743: 215-23.

[33] Kim M.R, Jeong JH, Park TG. Swelling induced detachment of chondrocytes using RGD-modified poly(N- isopropylacrylamide) hydrogel beads. Biotechnol Prog 2002; 18: 495-500.

[34] Kathmann EEL, White LA, Cormick CL. Electrolyte- and pH-responsive zwitterionic copolymers of 4-[(2-acrylamido-2-methylpropyl)- dimethylammonio]butanoate with 3-[(2-acrylamido-2- methyl-propyl)dimethylammonio] propanesulfonate. Macromolecules 1997; 30: 5297-304.

[35] Corpart J, Candau F. Aqueous solution properties of ampholytic copolymers prepared in microemulsions. Macromolecules 1993; 26: 1333-43.

[36] Salamone JC, Rodriguez EL, Lin KC, Quach L, Watterson AC, Ahmed I. Aqueous salt absorption by ampholytic polysaccharides. Polymer 1985; 26: 1234-8.

[37] Szczubialka K, Jankowska M, Nowakowska M. 'Smart' polymeric nanospheres as new materials for possible biomedical applications. J Mater Sci Mater Med 2003; 14: 699-703.

[38] Morrison ME, Dorfman RC, Clendening WD, Kiserow DJ, Rossky PJ, Webber SE. Quenching kinetics of anthracene covalently bound to a polyelectrolyte: effects of ionic strength. J Phy Chem 1994; 98: 5534-40.

[39] Cho SY, Allcock HR. Macromolecules dendrimers derived from polyphosphazene–Poly(propyleneimine) systems: encapsulation and triggered release of hydrophobic guest. Macromolecules 2007; 40: 3115-21.

[40] Matsumoto S, Christie RJ, Nishiyama N, *et al.* Environment-responsive block copolymer micelles with a disulfide cross-linked core for enhanced siRNA delivery. Biomacromolecules 2008; 10: 119-27.

[41] Bulmus V, Woodward M, Lin L, Murthy N, Stayton P, Hoffman A. A new ph-responsive and glutathione-reactive, endosomal membrane-disruptive polymeric carrier for intracellular delivery of biomolecular drugs. J Controlled Release 2003; 93: 105-20.

[42] El-Sayed MEH, Hoffman AS, Stayton PS. Rational design of composition and activity correlations for ph-sensitive and glutathionereactive polymer therapeutics. J Controlled Release 2005; 101: 47-58.

[43] Dong W-F, Kishimura A, Anraku Y, Chuanoi S, Kataoka K. Monodispersed polymeric nanocapsules: spontaneous evolution and morphology transition from reducible hetero-peg picmicelles by controlled degradation. J Am Chem Soc 2009; 13: 3804-5.

[44] Napoli A, Valentini M, Tirelli N, Muller M, Hubbell JA. Oxidation-responsive polymeric vesicles. Nat Mater 2004; 3: 183.

[45] Yoshida R, Yamaguchi T, Kokufuta E. Self-oscillation of polymer gels coupled with the Belousov-Zhabotinsky reaction. J Artif Organs 1999; 2: 135-40.

[46] Ren HY, Wu T, Ma N, Xu HP, Zhang X. Side-chain selenium-containing amphiphilic block copolymers: redox-controlled self-assembly and disassembly Soft Matter 2012; 8: 1460.

[47] Miyata T, Uragami T and Nakamae K. Biomolecule-sensitive hydrogels. Adv Drug Deliver Rev 2002; 54: 79-98.

[48] Fuchsbauer HL, Gerber U, Engelmann J, Seeger T, Sinks C, Hecht T. Influence of gelatin matrices cross-linked with transglutaminase on the properties of an enclosed bioactive material using beta-galactosidase as model system. Biomaterials 1996; 17: 1481-8.

[49] Chen TH, Embree HD, Wu LQ, Payne GF. *In vitro* protein-polysaccharide conjugation: Tyrosinase-catalyzed conjugation of gelatin and chitosan. Biopolymers 2002; 64: 292-302.

[50] Kim S, Healy KE. Synthesis and characterization of injectable poly(Nisopropylacrylamide- co-acrylic acid) hydrogels with proteolytically degradable cross-links. Biomacromolecules 2003; 4: 1214-23.

[51] Yang JY, Jacobsen MT, Pan HZ, Kopecek J. Synthesis and characterization of enzymatically degradable PEG-based peptide-containing hydrogels. Macromolecular Bioscience 2010; 10: 445-54.

[52] Thornton PD, Mcconnell G, Ulijn R V. Enzyme responsive polymer hydrogel beads. Chem Commun 2005; 47: 5913-5.

[53] Mcdonald TO, Qu HL, Saunders B, Ulijn RV. Branched peptide actuators for enzyme responsive hydrogel particles. Soft Matter 2009; 5: 1728-34.

[54] Patrick AG, Ulij RV. Hydrogels for the detection and management of protease levels. Macromolecular Bioscience 2010; 10: 1184-1193.

[55] Woodcock JW, Jiang XG, Wright RAE, Zhao B. Enzyme induced formation of thermoreversible micellar gels from aqueous solutions of multiresponsive hydrophilic ABA triblock copolymers. Macromolecules 2011; 44: 5764-75.

[56] Wang C, Chen QS, Wang ZQ, Zhang X. An enzyme-responsive polymeric superamphiphile. Angewandte Chemie-International Edition 2010; 49: 8612-5.

[57] Ge J, Lu DN, Yang C, Liu Z. A lipase-responsive vehicle using amphipathic polymer synthesized with the lipase as catalyst. Macromol Rapid Commun 2011; 32: 546-50.

[58] Castelletto V, Mckendrick JE, Hamley IW, Olsson U, Cenker C. PEGylated amyloid peptide nanocontainer delivery and release system. Langmuir 2010; 26: 11624-7.

[59] Caponi P-F, Qiu X-P, Vilela F, Winnik FM, Ulijn RV. Phosphatase/ temperature responsive poly(2-isopropyl-2-oxazoline). Polym Chem 2011; 2: 306-8.

[60] Alemdaroglu FE, Wang J, Borsch M, Berger R, Herrmann A. Enzymatic control of the size of DNA block copolymer nanoparticles. Angewandte Chemie-International Edition 2008; 47: 974-6.

[61] Koga H, Toita R, Mori T, Tomiyama T, Kang JH, Niidome T, Katayama Y. Fluorescent nanoparticles consisting of lipopeptides and fluorescein modified polyanions for monitoring of protein kinase activity. Bioconjugate Chem 2011; 22: 1526-34.

[62] Itoh Y, Matsusaki M, Kida T, Akashi M. Time-modulated release of multiple proteins from enzyme-responsive multilayered capsules. Chem Lett 2008; 37: 238-9.

[63] Blencowe CA, Russell AT, Greco F, Hayes W, Thornthwaite DW. Self-immolative linkers in polymeric delivery systems. Polym Chem 2011; 2: 773-90.

[64] Weinstain R, Baran PS, Shabat D. Activity-linked labeling of enzymes by self-immolative polymers. Bioconjugate Chem 2009; 20: 1783-91.

[65] Haba K, Popkov M, Shamis M, Lerner RA, Barbas CF, Shabat D. Single-triggered trimeric prodrugs. Angewandte Chemie-International Edition 2005; 44: 716-20.

[66] Weinstain R, Sagi A, Karton N, Shabat D. Self-immolative comb-polymers: Multiple-release of side-reporters by a single stimulus event. *Chemistry*-a European Journal 2008; 14: 6857-61.

[67] Lu Z, Kopeckova P, Kopecek J. Antigen responsive hydrogels based on polymerizable antibody fab fragment. Macromol Biosci 2003; 3: 296-300.
[68] Miyata T, Asami N, Uragami T. A reversibly antigen-responsive hydrogel. Nature 1999; 399: 766-9.
[69] Rao N A. Role of oxygen radicals in retinal damage associated with experimental uveitis. Trans Am Opth Soc 1990; 87: 797-850.
[70] Yui N, Nihira J, Okano T, Sakurai Y. Regulated release of drug microspheres from inflammation responsive degradable matrices of crosslinked hyaluronic acid. J Controlled Release 1993; 25: 133-43.
[71] Fleige E, Quadir MA, Haag R. Stimuli-responsive polymeric nanocarriers for the controlled transport of active compounds: Concepts and applications. Adv Drug Deliver Rev 2012; 64: 866-84.
[72] Ito Y, Casolaro M, Kono K, Imanishi Y. An insulin-releasing system that is responsive to glucose. J Controlled Release 1989; 10: 195-203.
[73] Hassan CM, Doyle FJ, Peppas NA. Dynamic behavior of glucoseresponsive poly(methacrylic acid-g-ethylene glycol) hydrogels. Macromolecules 1997; 30: 6166-73.
[74] Huang HY, Shaw J, Yip C, Wu XY. Microdomain pH gradient and kinetics inside composite polymeric membranes of pH and glucose sensitivity. Pharm Res 2008; 25: 1150-7.
[75] Uchiyama T, Kiritoshi Y, Watanabe J, Ishihara K. Degradation of phospholipid polymer hydrogel by hydrogen peroxide aiming at insulin release device. Biomaterials 2003; 24: 5183-90.
[76] Brownlee M, Cerami A. A glucose-controlled insulin-delivery system: semisynthetic insulin bound to lectin. Science 1979; 206: 1190-1.
[77] Cheng SY, Gross J, Sambanis A. Hybrid pancreatic tissue substitute consisting of recombinant insulin-secreting cells and glucoseresponsive material. Biotechnol Bioeng 2004; 87: 863-73.
[78] James TD, Shinkai S. Artificial receptors as chemosensors for carbohydrates. Top Curr Chem 2002; 218: 159-200.
[79] Yamamoto S-I, Pietrasik J, Matyjaszewski K. Temperature- and pH-responsive dense copolymer brushes prepared by ATRP. Macromolecules 2008; 41: 7013-20.
[80] Soppimath KS, Tan DCW, Yang YY. pH-triggered thermally responsive polymer core-shell nanoparticles for drug delivery. Adv Mater 2005; 17: 318-23.
[81] Soppimath KS, Liu LH, Seow WY, *et al.* Multifunctional core/shell nanoparticles self-assembled from pH-induced thermosensitive polymers for targeted intracellular anticancer drug delivery. Adv Mater 2007; 17: 355-62.
[82] Chen YC, Liao LC, Lu PL, *et al.* The accumulation of dual pH and temperature responsive micelles in tumors. Biomaterials 2012; 33: 4576-88.
[83] Lo CL, Lin KM, Hsiue GH. Preparation and characterization of intelligent coreshell nanoparticles based on poly(D, L-lactide)-g-poly(N-isopropyl-acrylamide- co-methacrylic acid). J Controlled Release 2005; 104: 477-88.
[84] Li W, Li JF, Gao J, *et al.* The fine-tuning of thermosensitive and degradable polymer micelles for enhancing intracellular uptake and drug release in tumors. Biomaterials 2011; 32: 3832-44.
[85] Miyata K, Kakizawa Y, Nishiyama N, *et al.* Block Catiomer Polyplexes with Regulated Densities of Charge and Disulfide Cross-Linking Directed To Enhance Gene Expression. J Am Chem Soc 2004; 126: 2355-61.
[86] Phillips DJ, Gibson MI. Biodegradable poly(disulfide)s derived from RAFT polymerization: monomer scope, glutathione degradation, and tunable thermal responses. Biomacromolecules 2012; 13: 3200-8.
[87] Morinloto N, Qiu XP, Winnik FM, Akiyoshi K. Dual stimuli-responsive nanogels by self-assembly of polysaccharides lightly grafted with thiol-terminated poly(N-isopropylacrylamide) chains. Macromolecules 2008; 41: 5985-7.
[88] Lv WP, Liu SQ, Feng WQ, *et al.* Temperature- and redox-directed multiple self assembly of poly(N-isopropylacrylamide) grafted dextran nanogels. Macromol Rapid Commun 2011; 32: 101-7.
[89] Sankaranarayanan J, Mahmoud EA, Kim G, Morachis JM, Almutairi A. Multiresponse strategies to modulate burst degradation and release from nanoparticles. ACS Nano 2010; 4: 5930-6.
[90] Du JZ, Du XJ, Mao CQ, Wang J. Tailor-made dual pH-sensitive polymer-doxorubicin nanoparticles for efficient anticancer drug delivery. J Am Chem Soc 2011; 133: 17560-3.
[91] Klaikherd A, Nagamani C, Thayumanavan S. Multi-stimuli sensitive amphiphilic block copolymer assemblies. J Am Chem Soc 2009; 131: 4830-8.

[92] Qiao ZY, Zhang R, Du FS, Liang DH, Li ZC. Multi-responsive nanogels containing motifs of ortho ester, oligo(ethylene glycol) and disulfide linkage as carriers of hydrophobic anti-cancer drugs. J Controlled Release 2011; 52: 57-66.

[93] Wang K, Guo DS, Wang X, Liu Y. Multistimuli responsive supramolecular vesicles based on the recognition of p-sulfonatocalixarene and its controllable release of doxorubicin. ACS Nano 2011; 5: 2880-94.

CHAPTER 10

Bio & Polymeric Materials Mimicking the Extra-Cellular Microenvironment

Riaz A. Khan[1,2,*]

[1]*Department of Medicinal Chemistry, College of Pharmacy, Qassim University, Qassim 51452, KSA and* [2]*Department of Chemistry, MRIU, Aravali Hills, National Capital Region, Faridabad, HR 121 001, India*

Abstract: Providing cyto and, biocompatible, biodegradable, permissive, elastic and, three-dimensional extra-cellular microenvironment with bio-mimicking capabilities is a daunting challenge in contemporary tissue bioengineering. The ability to guide cell differentiation and proliferation with adequate adaption to the desired functional tissue or, organ forming with positive affinity for extracellular support in regeneration and, enough spatio-temporal feasibility for newly constructs and advancing cell mass are amongst the most desirables towards producing functional bioconstructs in the modern tissue bio-engineering realm. The scaffold providing extracellular microenvironment with its desired porosity and surface properties is the main driving force that primarily influences the cell bio-engineering outcomes. The physical factors at the localized cellular microenvironment, individual cell's shape with the overall cellular-mass geometry at its generation and, propagation stages, biomatrix's response(s) and interfacial interactions feasibilities between the biomass and the extracellular matrix-support platform along with the molecular recognition capabilities on both sides with the involved and, mutually related bio-mechanics as well as intra-entities environmental forces of physico-mechanical nature and, the surface specifications of the matrix-support, be it the nano or micro-topographical features of the extracellular support-matrix or, the force(s) responsible for interplay among various factors causing cellular and sub-cellular level contractions are very important and crucial deciding factors in preparative-designs and outcomes of the tissue bioengineering matrix support platform. The external resisting factors originating from the support matrix constituting the extracellular microenvironment with potent growth-factors mediated signaling as well as presence of nutrients, biochemicals, gases and their diffusions have, at all times, impacted very strong influences on regulating the overall cell-fate in the extracellular microenvironment towards generating the desired biomass. Various matrix-support or the scaffolds for encompassing the extracellular microenvironment have been developed. These materials are far from perfect or practical enough for various characteristically defined outcomes, hence fewer types of scaffold materials fit for majority of functional characteristics. Several bioengineering tools and, nano and micro-scale specifications laden entities, materials and, devices of biomaterial and

***Corresponding author Riaz A. Khan:** Medicinal Chemistry Department, College of Pharmacy, Qassim University, Qassim 51452, KSA; MRIU, National Capital Region, Aravali Hills, Sector 43, Faridabad HR 121 001, India; E-mail: Kahnriaz@gmail.com

Viness Pillay, Yahya E. Choonara and Pradeep Kumar (Eds.)

synthetic origins are being employed successfully to develop scaffold and its materials to fulfill the essentials in tissue bioengineering domain towards producing various bioconstructs and tissue types by providing desired scaffold. The employed, natural or synthetic, molecular devices, materials and chemical entities, mostly polymeric in nature, with proven intrinsic properties and extrinsic physical factors significantly playout in practical tissue-engineering situations recapitulate the critical biochemical and biophysical aspects of the developing cell types in the generated extracellular microenvironment for desired tissue bioengineering constructs outcomes through the environment of the scaffold. Some of these materials, entities and devices in-use now include bio-compatible polymeric hydrogels, single and, composite synthetic and natural polymeric materials in random or, designated/patterned formations or, in layered/arrayed scaffoldings, the nanofibers, semi and fully-lithographed polymeric material devices, proteins, polypeptides, antibiotic-based hydrogels as well as scaffolds produced through a variety of techniques including photo-reactivity with their inherent specifications for producing the intended scaffold are worth mentioning.

Keywords: Extracellular Microenvironment, Biomaterials, Biopolymers, Synthetic Polymers, Homo & Hetero-Polymers, 3D-Scaffold, Functional Tissue Bioengineering, Collagen, Chondroitin Sulfate, PEG, Hydrogels, Nanofiber, Electrospinning.

INTRODUCTION: SIGNIFICANCE, SCOPE & OUTREACH

Biological tissues are result of defined arrangement with specific spatial organization of group of cells with extracellular microenvironment holding therein various soluble factors of biochemical origins. These micro and, nano-scale biochemical as well as biological entities organized into regional tissue architecture create highly complex and heterogeneous cell mass as tissues alongside the cellular microenvironment with its functional components (Table **1**) [1]. To artificially generate functional tissue equivalents, engineered materials need to mimic the functional competency through structural, chemical and cellular equivalence with high order of competitiveness at par with the natural conditions and materials. For recapitulating the unique microenvironment outside, the emerging cells need to be at its best of interaction efficiency and at higher orders of cellular engagements to generate the artificial cellular mass in an artificially set-up environment. The extracellular microenvironment materials can be derived either from nature or synthesized in the laboratory using a variety of chemical approaches. The advances in cell culture, stem cell technology and nano-biotechnology have provided much impetus to improve the existing state of the methodology and design materials for development to meet interdisciplinary bio-medical requirements and, other applications thereof in dealing with the extracellular micro-environmental platform generation. The corresponding materials fabrication for use in achieving feasible outcomes in the progression and development of intended cellular mass and desired organelles with anatomical and functional compatibility at the experimental and clinical levels, both inside and outside of the body [2-6] is very important. This feat requires design and development of new, biocompatible three dimensional (3D) platforms or, the desired 3D-scaffold which holds the key to shaping of the cellular progression's outcome. The operational feasibility of the scaffold demands that levels of outreach of nutrients, diffusion of gases, growth and other soluble factors are made available at the optimum with continuous flow status as well as in standby conditions in the scaffold for the generated and developing cellular mass so as to share raw materials and biochemicals as and when required. The scaffold needs to be receptive, appropriately elastic, biocompatible and, biodegradable at the appropriate and synchronized rate to make space for the developing biomass [7-12] in the medium so as to keep the pace and, space-balance of cell mass generation. The scaffold needs to be biodegradable or, the harvesting of the biomass from scaffold feasible enough, although former condition preferred except for certain implants.

Table 1. Major extra-cellular microenvironment components and their principal functions at the site*.

Major Extracellular Microenvironment Components	Principal Functions
Adhesive Molecules	Mediation of trans-membrane signals; cell adhesion to extracellular microenvironment-based matrix
Collagen	Cell-extracellular microenvironment-based matrix interactions; provides tensile strength
Elastin	Provides elasticity to extracellular microenvironment-based matrix
Fibronectin	Cell and matrix interactions; cell proliferation and cell migration
Fibrinogen	Cell proliferation & migration; hemostasis
Hyaluronan	Cell-matrix interactions; cell proliferation & migration
Laminin	Membrane component; cell migration
Proteoglycan	Cell-extracellular microenvironment-based matrix interactions; cell proliferation & migration

*adopted from reference [1].

Therefore, ideally, at all levels, cell signaling and angiogenesis in environments of cell growth at its early and at later stages of development are highly desirable. The cells adhesions are in play from the outset and are in need to be firm in nature to guide further proliferation and advancing of cell differentiation for desired tissue and/or organ/organelle formation. Primarily, the approach involves the use of polymeric scaffolds with superior 3D-architecture suitable for cell differentiation providing desired levels of physical connectivity and space for the purpose of cell differentiation and interlinked proliferation to produce the desired 3D-biomass. The live-in cellular interactions at the interface of extracellular material prompted through intelligent molecular recognition at the material's surface or vague interaction and progression of the desired biological mass within the material-supported microenvironment finds its levels of tolerability through biocompatible interactives in the medium *via* feed-back loops formed of soluble factors, signaling pathways put in place by the advancing biomass in the extracellular microenvironment matrix and, the biochemical entities of growth factors which is well-supported by the cell signaling and, advances in complexity and constitution with time. The mechanical properties of the extracellular support material in terms of strength, stiffness and elasticity pave the way for a suitable surface that supports the initiative nucleation, cell divisions, cell adhesions and eventual proliferation of the advancing cell-mass through the interactive, operational contractibility, compression, mechano-transduction and fluid shear stress of the medium. Necessarily, the scaffold is required to biodegrade in a compatible manner in a short period of time with-out producing toxic waste(s). The 3D-network of interlinked pores in the purported scaffold provides the required space and creates a permissive microenvironment for the materials and gases' two-way transport in supply and waste disposal, signaling phenomenon, cell division and advancements towards the desired biomass production with functional and topographical features included as intended. The scaffold's pore size, 3D-architecture, pores' interconnection abundance, surface's minute topography at nano and micro-meters scales, physico-mechanical properties as well as bio-chemical environment play substantial role in providing required conditions at the cellular and subcellular environmental levels of initiation, propagation and proliferation and, towards functioning feasibility of the generated extracellular microenvironment. The scaffold supported extracellular microenvironment also contributes in maintaining the cytoskeletal tension and homeostasis at various stages of the biomass production and development. The continuous and dynamic change in the chemo-biological and, physico-mechanical arena which surrounds the advancing biomass in extracellular microenvironment with the surrounding/existent fluidity is crucial in the overall development of bioconstructs. The choice of scaffold material which itself can be manipulated through micro and nano-techniques and tools thereof, in addition to its physico-chemical and mechanical controls in design, production and maneuverability during use in experimental and clinical settings for best of the operational feasibility, is another factor [13-21] affecting the tissue-engineering outcomes.

STRUCTURE-FUNCTION REQUIREMENTS: PHYSICO-MECHANICAL & BIOCHEMICAL CHARACTERISTICS OF AN IDEAL SCAFFOLD HOLDING THE DESIRED MICROENVIRONMENT

The physico-chemical prerequisites and desirables have been understood to an extent with use of different materials of synthetic and natural origins in various settings. A number of single and, composite materials have been studied for the purpose. The biodegradable polymers, both natural and, synthetic, as a single scaffold material can be recommended, although, mostly as a trial and hit option to test initial feasibility. The structural properties dictate the polymeric preparations to be used for scaffold's format, *e.g.*, high water content hydrogels, fibrous, arrayed or amorphously generated scaffolds designs, as the need/specification might be assessed or understood. The choice also dictates the use of synthetic or natural materials with a view on the outcome and, working feasibility. The physico-chemical properties are one of the main deciding factors in choosing of the material types. The physico-chemical properties of polymers have been estimated through various techniques including Differential Scanning Calorimetry (DSC) to predict decomposition and on-site proteins' denaturation possibilities. For example silk-fibrin and type I collagen's blend for desired characteristics have been utilized based on the requirements and, physico-chemical properties determination outcome from the initial results in producing workable scaffold [22-30].

The irreversible endothermic transitions, enthalpy, Glass Transition temperature (Tg), recommended storage conditions and, temperature are also some of the important parameters in turning out a workable scaffold [22-24, 31] which is based upon the polymeric characteristics as well as demands of the polymer-outcomes. One of the important aspects in raw materials' supply and its outreach for soluble liquid, especially in first few seconds for glucose-feed through absorption in the scaffold, is recommended to be at 6.3 x 10^{-10} meters^2/second of concentration gradient with a gradual increase to meet the further developing demands of the advancing cellular biomass in the polymeric scaffold supported extracellular microenvironment over the time and this factor has defined several polymer-based scaffold's workings. It is among important considerations for feed supply through (properly located) channels' diffusion to the differentiating cellular mass and linking the supply to the gradient between the start and the end with the near, far and of-site locations along with materials transport in the channels for meeting the requirements of the growing cells [32]. This plays crucial role in survival of the scaffold and ability of the bio or synthetic material to provide functional scaffold. Nonetheless, the morphological characteristics of the designed/produced scaffolds are heavily dependent upon factors relating to the polymer's chemical structure, physico-chemical properties and production technique/methodology used to prepare the scaffold. The chemistry immensely contributes towards providing the desirable physico-mechanical characteristics to the scaffold material. The prepared scaffold is required to have a 3D-structure format with inter-connected, near evenly distributed with uniform pores making internal channels [33, 34] encompassing large surface areas to support the intended biomass. The morphology is an outcome of the chemical structures of the residual entities, polymeric back-bone and, functional groups availability in the molecular structure of the starting material including other factors exhibited out during preparation and the technique adopted for preparation of the scaffold. Moreover, the scaffold is also required to be non-antigenic, non-immunogenic, bio-compatible and, bio-degradable wherein the dissolution/ degradation rate is superior to the pace of new tissue emergence to make space for it [35-38]. The degradation rate preferably needs to be slightly higher so as not to deform the bio-construct's outcomes. Other considerations include a mild preparation protocol to produce the scaffold, from the polymeric material to avoid denaturation which may lead to short-comings in the final shape and characteristics outcomes of the scaffold whose properties which may render them unsuitable for scaffold-platform work. Additionally, the cyto-compatibility for cell-culture introduced cell adhesion ligands with desired levels of visco-elasticity, compression characteristics and, mechanical strength to properly withstand the nucleation, initiation, cell differentiation, cell proliferation and, at last the bio-degradability is of immense value [39] for the prepared scaffold. The cells and scaffold-matrix interaction capability phenomenon [34, 40-44], bio-absorptivity, affinity for surroundings, immunogenicity [40, 45], gel-forming ability, high adsorption capacity for soluble & liquid state materials in the extracellular microenvironment matrix, bacterio-fungistatic, nontoxicity in nature [46, 47], chemical stability with adequate chemical crosslinking (with hexamethlyne diidocyanate or glutaraldehyde as cross-linker), avoidance to proteins' denaturation (crosslinking dependence of the polymer material), wettability, water absorbance capacity/swelling capacity of polymer (cross-linking & structural groups dependence) [48, 49], as well as the physical, mechanical and thermal properties for physical stability, shrinkage, and expansion properties [50-54] are among a list of high desirables for scaffold materials intended for use in tissue bioengineering. The disinfection [55, 56] prior to preparation is also recommended. There are specific considerations and a list of different structural and functional requirements are summarized in Table **2**. Moreover, the porosity and pore architecture also plays crucial role in working of the

Table 2. Structural and functional requirements for scaffold.

Structural & Functional Requirements	References
Inter-connected & uniform pore sized; large surface area; 3D-structure format; non-antigenic; bio-compatible in all aspects, biocompatibility for avoiding the unwanted host-tissue response to the intended implant; biodegradability characteristics along with the degradation-rate of the scaffold matching the rate of new tissue formations	[37, 38, 156, 176, 177, 191, 214, 238]
Glucose diffusivity (for hydrated polymer), comparatively near to the diffusion value of nutrient (glucose) in pure-water medium (6.9 x 10^{-10} meters^2/second)	[32]
Mild preparation steps; cyto-compatible for cell adhesion ligands; optimal viscoelasticity; compression characteristics; mechanical strength to properly withstand nucleation, initiation, cell differentiation, cell proliferation and, bio-degradability	[40, 61, 77, 95, 126, 127, 133, 239, 247, 260, 265, 344, 383]
Cell-matrix interaction capability; highly compatible surface chemistry to allow cell attachment, cell migration, cell proliferation and, cell differentiation	[39, 41-44, 46, 52]
Bioabsorbable in nature; affinity for surroundings; non-immunogenic or, low immunogenicity	[40, 45, 47, 118, 148]
Excellent gel-forming characteristics; high adsorption capacity; bacterio-fungistatic, bactericidal, fungicidal and non-toxic in nature	[46, 47]
Chemically stable, appropriate chemical-crosslinking (with hexamethlyne-di-iodo-cyanate or, glutaraldehyde); proteins' denaturation-avoidance properties (crosslinking dependence); Water retaining/absorbance capacity, swelling & wettability properties of polymer (cross-linking & structural groups dependence)	[142, 265, 273]
Physical, mechanical (toughening and fracture-avoidance), thermal properties for physical stability; shrinkage and, scaffold's expansion; appropriate levels of mechanical strength to maintain the chemical and physical structure as well as functional viability immediately after cell implantation and, during the re-modelling of the intended implants; abilities to withstand shear stress	[19, 31, 51-54, 59-60, 149, 162, 238]
Disinfection, preferably chemical disinfection (e. g., Ethanol)	[50, 55, 56]
Adequately interconnected pores of appropriate size to support cellular infiltration and vascularization; porous sizes >20 μm; pores with mean sizes > 125 μm; high degree of porosity: appropriately at 80-95%	[62-66, 162]
Support the formation of extracellular microenvironment through promotion of cellular functions	[51, 77]
Biomimetism	[15, 79, 156, 176, 177]
Enthaply 200 J/g, may increase slightly with water-wetting	[22, 23
Glass Transition temperature (Tg) ~70 °C for Carbohydrate-based dry polymers; Maximum Storage Temperature ~30 $^{o}C_{max}$	31]

scaffold which is responsible for holding the extracellular microenvironment [57, 58]. A very small pore inhibits mass and material transport and, thus hinders nourishment and growth of the intended biomass in the culture media. A small enough pore may also give rise to a biofilm and consequentially stop all movements of materials including the gases through these pore channels [59-61]. On the other hand, a large pore will be providing substantially reduced interfacial area to cells which will be ineffective for cell adhesion and thus, effectively curtailing the functions intended from the scaffold. Therefore, an assortment of small and large pore seems best for the purpose [62-64]. Furthermore, the creation of the microenvironment for the developing biomass is provided by the scaffold-based material through its strength, elasticity, compression and contractibility. The other factor responsible for better working of the extracellular microenvironment is the generation of vascular network within the structure of the interconnected pores of the scaffold material in the extracellular microenvironment which facilitates the two-way transport of both the gaseous substances and nutrients and, at the same time works as waste disposal channel for the cellular metabolic outcomes for the advancing cellular biomass. Again, the angiogenesis is one of the main factors playing part in keeping alive the cell growth through nutritional supply chain and wastes disposal. Another extremely important feature to improve the cell-scaffold interaction is scaffold's surface property that can be synthetically or, naturally incorporated to improve the bio and cyto-compatibility [65] of the scaffold. The desirable supporting scaffold is a biodegradable, artificially synthesized 3D-support material emulating the natural extracellular micro-environment [66], whether it is chemically made up of synthetic materials or natural polymers of animal or plant origins, it has much to offer.

Moreover, the inclusion of macromolecular entities of natural origins including growth factors are put into play along with the seeded cellular mass which can work as regulators during the organogenesis/cellular mass formation. The growth factors and several other micro and macromolecular entities of pertinent interest can also be introduced into the scaffold along with the seeded cells to support the process and regulate the cellular functions during tissue or organ formations [67]. The cell's response to topographical features on the scaffold material's surface and behavior in specific chemical conditions characterizes the course to be undertaken and sets the agenda for extracellular microenvironment generation. The inter-facial responses within the scaffold surface represent the main driving force that influences the orientation and composition of biological materials. These responses control the release of bio-factors influencing cell growth, differentiation and cell motility in the extracellular microenvironment. There are known instances where scaffolds have been modified with type I collagen to mimic the extracellular microenvironment or, the plasma aided processing of the scaffold preparation have been employed [65] to modify the surface chemistry of the material for better results. These modifications by introducing required extracellular microenvironment components and introduction of viable chemical groups as binding sites for freshly generating new cells may sufficiently improve and accelerate the cells adhesion and colonization inside the scaffold-platform. The structural parameters of the scaffolds defining its pore size, interconnectivity, porosity and, the chemical surface viability dictates the very first steps of cell recognition and adhesion in no uncertain terms for future acceptability of the scaffold for tissue bio-engineering work and these aspects need to be given proper consideration while designing or, choosing an already existing material for the tissue bio-engineering needs.

TRADITIONAL SCAFFOLDS *VERSUS* 3D-SCAFFOLDS: FUNCTIONAL FEASIBILITIES AND IMPACTS

The traditional approach of single cell type population culturing on two-dimensional scaffolds as nearly inert platforms, usually made-up of substances like polystyrene and other similar material types, have been replaced with the 3D-scaffolds to overcome the inter-responsive and functional shortcoming of 2D-substrate based techniques for better processing and feasible outcomes. Current culture methods involve encapsulation of cells in various conditions and characteristics levels of scaffold including in micro-porous [68-70] arena, as nano-fibrous structures [71, 72] and, in hydrogel like scaffold's [73] materials. The micro-porous scaffold allows convenient entrapment of cells because of its greater-sized porosities of ~100 μm in comparison to the average cells size with diameter of about 10 μm. The 2D-scaffold is not much effective for driving the contemporary cell culture demands. The 2D-structures lack mechanical strength required for the desired levels of mechano-transduction during cell growth and, other inter-linked and, consequential phenomenon in tissue bioengineering conditions. More insight has been provided into the culturing and role of 2D-scaffolds [74] which differentiates and sets apart the characteristics and requirements for a 3D-scaffold, the contemporary tissue culturing scaffold. The 2D-scaffolds based cell cultures have highlighted the performance levels and, dependency of functional variations and feasibility of support materials for culture among cells where the functional viability and interactive adjustments with the extracellular microenvironment is of utmost demarcation to distinguish the scaffolds dimensions. It has been recently demonstrated that cells often exhibit unnatural behavior when placed outside of their native environment where these cells became confined to a monolayer formation as in a 2D-scaffold which is a limiting factor. It has been shown that epithelial cell culture of human breast origins develop tumor like characteristics when cultured outside in traditional media. However, the same type of cells develop like their natural counterparts in a 3D-scaffold setting [75]. Interestingly, an increased embryonic stem cells chondrogenesis was also noticed in a 3D-scaffold which is a supportive proof for the extent of the functional competitiveness of 3D-scaffolds [76]. Therefore, tissue culturists are more interested in 3D-matrices for proper handling of the culturing process and their outcomes. The microenvironment profoundly contributes in the spatial and temporal cell signaling portals, in effect, directing the cell phenotype during culture's progression. It has been suggested that the phenotypes dominate genotypes probably because of the extracellular matrix's structure-related characteristics, especially the morphology [75]. Therefore, for culturing purposes, a cell is no longer treated as an isolated, single-character or, a solitary entity ready to display and, function out its genetic materials manifestation in a given environment, rather, it is seen in the context of the defining elements and, characteristic factors of the extracellular materials properties and its competitiveness to achieve desired culture outcomes which needs to be with assured presence and supply of growth factors, cellular signaling substances and parametric levels of signaling efficiency in the environment as evident through the outcome/progression of the culture and, availability of other biomolecules that take part in regulating tissue-mass/organelle/organ in the extra-cellular matrix. The ultimate functional behavior of the culture product is an indication of the competitiveness of the scaffold (Fig. **1**). This dynamism, as a consequence of extracellular microenvironment feedback, manipulates the intra-cellular information feedback and, in turn the molecular signaling processes influence the phenotypic fate of the culture and, the involved protein expression in the processes, probably through modification(s) in genotypic considerations [77-80] which play out their part. Therefore, the difference in the cells behavior in cultures of 2D and 3D-matrices is an ultimate outcome of the differences in the genetic expressions which is, actually, a result of the cell experiences in the extra-cellular microenvironment in 2D or, 3D-scaffolds. There is considerable knowledge suggesting that cells can sense multiple features of the extracellular matrix, integrate that information and, respond accordingly. The cell in the body resides in and interacts within a nanoscale structured 3D-extracellular microenvironment network. Increasing evidences have shown that the cellular responses to the 3D-scaffold based microenvironment are significantly different and, superior to the 2D-scaffold's microenvironment.

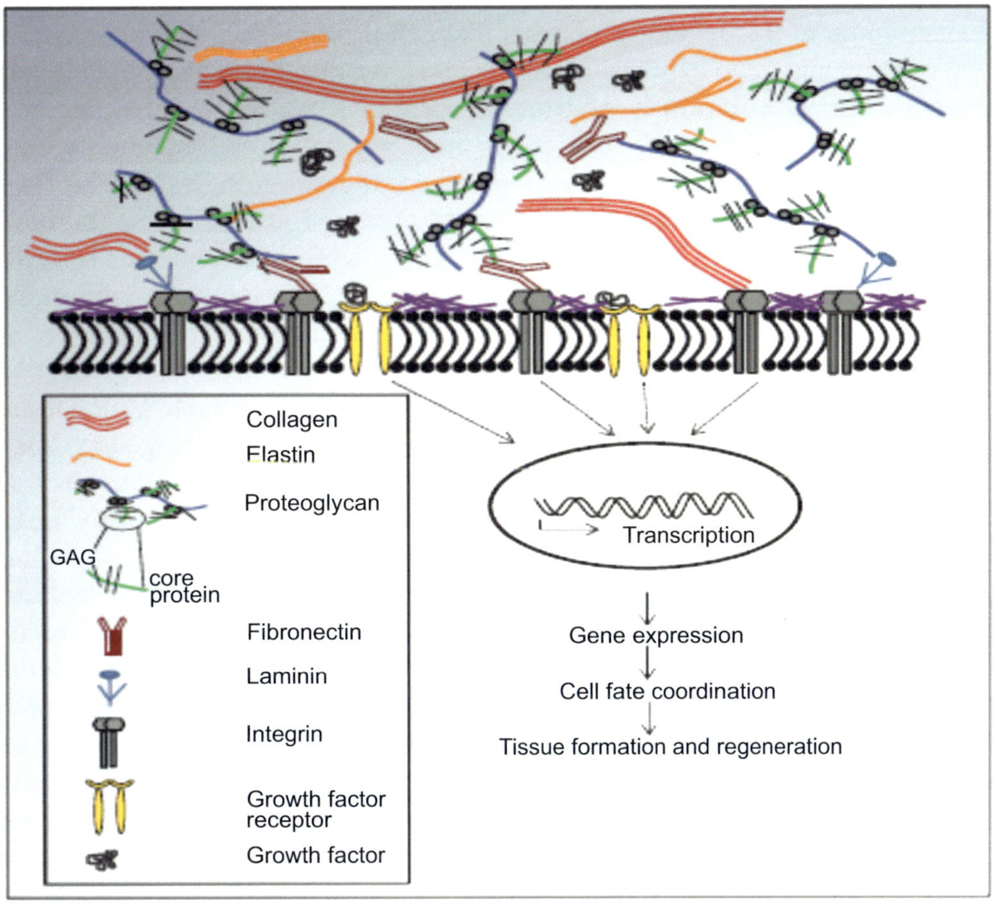

Figure 1: Schematic illustration of key extracellular microenvironment components and their cellular interactions; adopted from reference [279]: G Papavasiliou, S Sokic, M Turturro (2012). Chapter: Synthetic PEG Hydrogels as Extracellular Matrix Mimics for Tissue Engineering Applications, Biotechnology - Molecular Studies, and Novel Applications for Improved Quality of Human Life, Reda Sammour (Ed.), ISBN: 978-953-51-0151-2, ©InTech, http://www.intechopen.com/books/biotechnology.

The 2D surroundings lead to unnatural, polarized integrin binding and mechano-transduction, affecting the intracellular signaling and phenotypic outcomes [80-82]. The scaffold's hydrophilicity also decides the interaction outreach with extracellular materials' soluble factors, whereby the concentration of soluble factor(s) influences the cell migration, cell interaction, cellular communication and, cell differentiation. The 2D-scaffold based extracellular microenvironment possesses a spatial gradient which affects the culture outcome adversely [83].

However, topological/morphological factors *alone* have not been shown to influence some of the cellular processes like cell proliferation, cell apoptosis, cell differentiation and gene expression [80, 84, 85]. Again, the traditional cell culture techniques almost completely restrict the cell to a nearly scalar/single phasic direction of environment and, restrict the multi-phasic, more elaborate, spacious, inter-connected, complex and, spatial varieties of morphologies as observed in 3D-scaffold. Thus, there are clear differences observed in cellular migration, cell proliferation between a 2D and, 3D-extracellular microenvironments. The cell in a planar space feels nearly no resistance in migration from the surrounding matrix and materials and, moves wayward as though in a tunneling way. The 3D-extracellular microenvironment recapitulates the critical mechanical and biochemical characteristics as present in any native extra-cellular material and it facilitates *nearly* all the processes and, movements in cell-culturing stage such as cell migration, tissue build-up and, tissue organization as nearly equivalent as in the natural settings. In addition, the cell signaling which is primarily integrin based, regulates the surface responsiveness of certain cell-types and, can be linked to extra-cellular materials presence/control including attenuation of the features of skeletal specifications of the cellular bodies at the start-out of the extracellular microenvironment while formation/evolution of the ion channels at the initial stages. The relative perturbations and displacement of channels with respect to extracellular microenvironment and, the exo-cytoskeleton is playing a part in gating of the channels [86, 87]. These channels are activated through external forces as well as intra-cellular cytoskeletal contractibility in the surrounding environment [88-90]. Thus, the mechanical forces at play have the potential to regulate the fate of the cells through modulating the channels, materials and, ions' presence. The downstream effect of the activation of these mechano-sensitive ion channels lies in the change of the cytoplasmic Ca^{2+} concentrations as well as their oscillators/oscillations [89]. The Ca^{2+} oscillations have been observed in certain cells which are influenced by substrate stiffness [91] which, in-turn, controls the tissue-bioengineering outcomes inside the scaffold.

SCAFFOLD DESIGNING: POLYMER TYPES & BIO-APPLICATIONS

In natural settings, the requisite physico-mechanical properties for better tissue bioengineering outcomes are drawn from the structural architecture of the scaffold which is made up of mostly proteins such as fibronectin, laminin and collagen. The structural backbones of these proteins are chemically complex skeleton that provide the required mechanical strength to the extracellular microenvironment in combination with other factors, and cells sense these bio-mechanistic phenomenon, interactions and associated changes through receptor-binding phenomenon between the cell-surface integrins and the binding entities of the extracellular microenvironment. The well-hydrated proteoglycan molecules fill the interstitial spaces of the proteins structural backbone along with sequestering of the soluble bio- and other bio-macromolecular entities comprising of growth factors, small integrin-binding glycoproteins and various other involved motifs to produce the native microenvironment. The release of the signaling molecules to advance the desired cell proliferation and availability of growth factors allow cell migration and other cellular functions in the extracellular matrix area. The microenvironment also accommodates some of the cellular functions *via* extracellular matrix cleaving proteins like metallo-proteinases and the deposits of natural extracellular microenvironment ingredients which are also controlled by the integrin-mediated signaling pathways in the natural settings [92]. However, in artificial settings, the scaffold's chemo-mechanical properties are pre-set with nearly no user/cell-defined controls. In order to mimic the natural extracellular microenvironment, it is pertinent to fabricate materials of which chemo-mechanical properties can be synchronized with cell progression during all times and at all levels and, scales of cell culture by the handler and /or by the culturing cells. The need is to prepare the scaffold that bio-mimics the natural settings. Recent works in scaffold engineering have demonstrated that 3D-synthetic microenvironments can be designed to promote cell viability and to direct the cell adhesion, differentiation, proliferation and migration *via* mechanical and biochemical controls, only up to an extent [93-97]. However, properly designed materials are bridging the functional gaps between the natural and artificial scaffold-platforms desired characteristics and these newly designed and fabricated 3D-scaffolds are being tailored to specific cell culture needs which are providing well-defined extracellular microenvironment mimics. *In vivo*, the extracellular microenvironment facilitates and provides sites for binding ligands for cell adhesion that connect cells' cytoskeletons to the extra cellular microenvironment [98]. These filled-in ligands in the extracellular microenvironment encourage integrin-binding events that signal the mechanics of the extracellular microenvironment to the cell and vice-versa and, direct the cell fate through intra-cellular signaling pathways [99]. The integrin-binding events play a critical role not only in cell adhesion but nearly in all cellular processes [100]. In the simplest terms, these binding ligands are also recapitulated in synthetic scaffolds by physically entrapping the extracellular microenvironment macromolecules. These macromolecules, especially proteins have improved the cell viability and cellular functions to a large extent [101]. However, the entrapped proteins, over time, denature, aggregate and start to possess new binding motifs. These may also spread-out randomly in the extracellular microenvironment which can confound the overall functioning of the extracellular matrix. Therefore, a more pragmatic approach is required for facilitating conditions in the artificial scaffolds for better cell signaling through macromolecule mimics. Hence, more recent approaches are in trials to overcome these hurdles in scaffold designing and, these are based on rapid prototyping [102], computer-aided tissue engineering with a view on biomimetic modeling [103, 104] and, characterizations of the effective mechanical properties through computations [105] as well as use of the algorithm for generating complex polyhedral scaffolds capable of replicating the natural settings of extracellular microenvironment [106].

DESIGNING OF SCAFFOLD MATERIAL: CREATING INTERFACE WITH THE NATURAL TISSUE

Biopolymeric and, other synthetic and natural-origin materials are being used to recreate the natural extracellular microenvironment in a controlled/ limited sense of functional specifications through fabrication of various artificial scaffolds. In this connection, an understanding of the interactions among the molecular entities and, surrounding cells including intercellular interactions within the extracellular microenvironment may provide the elements for designing and, bio-engineering the artificial extracellular microenvironment to guide cells behavior outside of the body and, also at the interface of the natural and, artificial settings in tissue-mass generation in artificial settings. Again, over increasing evidences suggest that the extracellular microenvironment provides the basic necessities to establish, almost precisely, the correct tissue development and, the cells and, scaffold's interactive functions in and, outside of the biosystem governed through the extracellular microenvironment. However, at present, synthetic extracellular microenvironment assembly is poorly understood in this context and, an all-round efforts are underway to comprehend these aspects in non-natural settings to bioengineer synthetic extracellular-microenvironment to work towards replacement/modifications of damaged/lost tissue(s) [107-109]. The biomaterials' interface with the natural tissue, formation of new tissue in the biomaterial-based scaffold, capability of scaffold in responding to diverse signals from the cells in its vicinity within the scaffold and/or improving of the regenerative capabilities of the damaged tissue, certainly, are among the critical areas of recent tissue bioengineering advancements where a number of polymers have been evaluated for the purpose of scaffold fabrications (Table **3**). Thus, the artificial extracellular microenvironment is essentially a structure that is perpetually receiving responses during cell development, cell

Table 3. Scaffold materials of natural & synthetic origins and their bio-applications.

Materials	Bio-applications	References
PEG Tail-functionalized-PEG PEG-pendant-RGD PEG-insert-Chondroitin Sulfate PEG-blend-PLGA PEG-blend-PLA	Guided Cell Growth, Mesenchymal Stem Cells, Tissue Bioengineering Neural Bioengineering Chondrogenesis Bone Regeneration Tissue Bioengineering	[120, 132, 279, 289, 290, 297, 379, 384, 396]
PVA PVA-Chitosan PVA-Collagen-HyAp (Hydroxyapatite) PVA-HyAp PVA-Alginate	Tissue Bioengineering Cartilage Regeneration Bone Tissue Bioengineering Bone Tissue Bioengineering Tissue Bioengineering	[346, 368, 373, 374, 372]
Alginate Alginate-RGD-Immobilized	Bone Marrow Cell Growth Bone Substitute MSCs Chondrogenesis	[140-142, 263, 294]
Collagen Collagen–mimic-PEG–structure-inserts Collagen-mimic-peptide (CMP) Collagen-Hyaluronan Collagen-P-L-lactide-co-Caprolactone	Tissue Bioengineering 313 Endothelial Cell Growth Bone Regeneration Enhanced Chondrogenesis Mesenchymal Stem Cells Invertebral Disc Tissue Bioengineering Tissue Bioengineering Smooth Muscle Cells	[23, 33, 48, 64, 120, 121, 172, 182, 296, 313, 316, 321, 353, 378, 380, 381]
Chitosan-Silk Chitosan-PVA-PLGA; Chitosan-Hydroxyapatite–PEO; Chitosan-Gelatin-AgNPs; Chitosan-HyAp (Hydroxyapatite); Chitosan-Poly-L-lactide-ε-Caprolactone Chitosan-Gelatin Chitosan-Gelatin-Hyaluronic Acid Chitosan-Gelatin-Hyaluronan	Neural Tissue Bioengineering Tissue Bioengineering Tissue Bioengineering Artificial Skin, Cartilage Regeneration	[22, 24-29, 40, 45, 46, 175, 261, 324, 345]
Hydroxyapatite	Bone Tissue Bioengineering	[349, 353, 391]
Silk-Fibrin	Tissue Bioengineering, Nerve Regeneration	[41-44]
Poly-L-Caprolactone-HyAp Poly-2-Hydroxy-Ethyl Methacrylate Poly-Ethylene-Terephthalate (PET) & Poly-Urethane (PU)	Bone Tissue Bioengineering Orthotopic Transplant Tracheal Scaffold Vascular Tissue	[201, 271, 278, 339, 359, 384]

Table 3: contd…

Poly-Urethane & Poly-(Ethylene-Glycol)-Mono-Methyl-Ether-Methacrylate (PEGMA)	Bioengineering	
Poly-lactides & Poly-glycolides	Tissue Bioengineering	[60, 152, 180, 207, 237, 274, 276-279, 323, 357]
Fibrin	Cartilage Bioengineering Stem Cell Differentiation Neural Progenitor Spinal Cord Injury Peripheral Nerve Injury	[259, 284, 333, 334, 336, 337]
Gelatin Gelatin-Fibrin Gelatin-PHEMA Gelatin-Poly-caprolactone	Neural Tissue Bioengineering Wound Repair	[77, 371, 305, 183, 304]
Antibiotic-based Hydrogels	3D-Tumor Modeling	[293]
Nano-fibrous Materials	Neural Tissue Bioengineering	[326, 330, 332]

differentiation, cell proliferation and tissue bio-mass generation processes [107]. Hence, the importance of biomaterials and their interactions with biological system is noteworthy. In the living biological environment, the natural extracellular matrix's degradation is controlled at multiple levels and its dysregulation results in extracellular microenvironment damage. The matrix metallo-proteinases are proteolytic enzymes that play major roles in the degradation and, functioning of natural extracellular microenvironment. This insight into natural setting offers exciting possibilities in the tissue bio-engineering realm where it is becoming highly possible to produce scaffolds, through various preparative techniques, that can emulate the natural extracellular microenvironment and, respond to the dynamical changes of the extracellular microenvironment's in competitive manner at all levels of tissue biomass development in the desired artificial settings.

Moreover, the structurally and, functionally perfect as well as time-resolved signaling for extracellular microenvironment modeling is highly desirable characteristic for designing artificial scaffold. Biomaterials that focus on mimicking individual aspects of extracellular microenvironment's functions may well serve the field of tissue bio-engineering for some time to come to help in the building of more sophisticated materials that hopefully will elevate the overall tissue bioengineering performances with advent of new and, competitive materials or, natural, synthetic and mixed characteristics and, of various origins. Indeed, there have been need to incorporate these subtleties in artificial scaffold and, in turn in the extracellular microenvironment which can be achieved, though in part or full, through more bio-friendly scaffolds which encourages, promotes and, involves itself in directing/providing receptive microenvironment. The advancing cell which creates its own niche in the scaffold based microenvironment, lays down the intricacies and, elements of cell signaling with proper mechanical support at the very initial stages of the cell involvement with outside environment in its proximity, though, up to some extent, in comfort of its own extra-cellular excreted material which is made available by the cell for its advancement and it includes the protein fibrils, growth factors, soluble biochemical materials and, the macromolecules. Since recently, there has been much direct approach in designing more naturally near, controllable and complex scaffolds, the image-based design approach and manufacturing of craniofacial scaffolds [110] along with molding of the internal architecture design with desired elasticity and porosity of the scaffold (to match natural bone) [111], cartilage-mimics through gels [112], osteogenesis and, chondrogenesis phenotypic expression controls through carrier geometry [113] and, understanding of the molecular interactions of polymer [114] have made, up to an extent, biomaterials designing and preparations possible to interface with the natural tissue. Thus, the development of biomaterials is also providing powerful means and tools to understand the complexity of cell signaling in isolated instances of cell mass development which may emulate the natural extracellular microenvironment. The outcome and, their influence on the cell renewal and, cell differentiations as part of the tissue bio-engineering process. This ability to mimic the natural extracellular microenvironment using 3D-scaffold's based artificial extracellular microenvironment through cyto, biocompatible and biodegradable (polymeric) material with understanding of the cellular interactions from computographics and computational designing inputs may provide a blueprint for further building the instructive, dynamic and, interactive (artificial) scaffold that can more precisely and, accurately direct cell differentiation in a desired/dictated way [115].

FUNCTIONAL TISSUE BIOENGINEERING: TOWARDS BIO-RESPONSIVE SCAFFOLDS

Several polymeric materials have been investigated for their feasibility towards creating desired extracellular microenvironment-mimics through scaffold designing. An approach is in place which involves use of different polymeric materials, both natural and synthetic with homogeneous and heterogeneous structures of single and blended composite/singular polymeric entities. The polymeric material's characteristics dictate the scaffold's outcomes and, efforts are still underway (*ca* 2015) to achieve complete control in all aspects of the scaffold designs and, their eventual outcome. An artificial scaffold that possesses various natural extra-cellular microenvironments' characteristics to better mimic the natural extracellular microenvironment was bioengineered with encouraging results. The chondroitin sulfate (CS) is a sulfated glycosaminoglycan (GAG) made-up of multiple chains of alternating amino-acetyl sugars along with carboxylated sugar molecules namely, n-acetyl- galactosamine and, glucuronic acid. A chondroitin glycan chain may contain over 100 sugar units each of which is sulfated at various positions and, in different molecular ratios/stoichiometric quantities. This natural biopolymer (CS) is an important structural component (Fig. **2**) in the extracellular microenvironment preparation and provides desired physico-mechanical support to cells. it also played part in regulating the intracellular communications [116]. The CS is also one of the major molecules present during natural pre-cartilage condensation process and, thereby suggested that CS plays significant roles during chondrogenesis [117]. A 3D-scaffold was

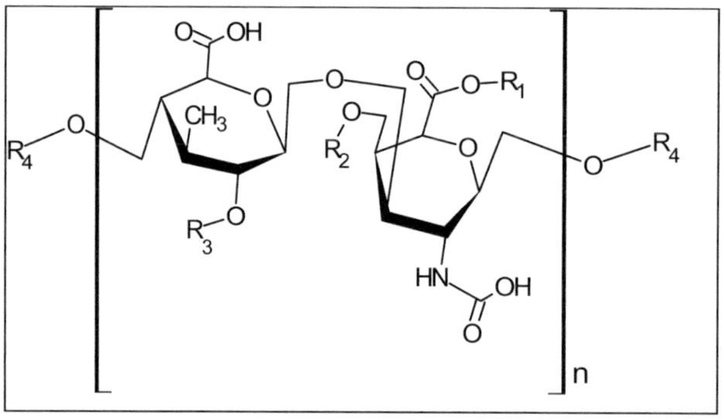

Figure 2: Structure of the monomeric unit in a chondroitin sulfate chain. Chondroitin-4-sulfate: $R_1 = H$; $R_2 = SO_3H$; $R_3 = H$. Chondroitin-6-sulfate: $R_1 = SO_3H$; R_2, $R_3 = H$; R_4 is the chain extender for the polymeric unit.

prepared that composed of CS and PEG and, which closely mimicked the native cartilage extracellular microenvironment [118]. These scaffolds provide an alternative extracellular microenvironment for cells growth and enhance the clustering of cells (mesenchymal), probably through molecular sequestrations also. These sequestrations improve the chondrogenic cell differentiation and increase the cartilage extracellular matrix deposition in the scaffold infrastructure [119]. Furthermore, the collagen-mimetic synthetic peptide (CMP) is the other material that forms triple helices and can function in assisting roles by binding to the natural extracellular microenvironment [120]. On design levels, the CMPs cross-linked to PEG provide comparable extracellular microenvironment sufficient for chondrocyte maintenance. It also enhanced the chondrogenesis in artificial settings [121, 122].

Furthermore, with the aid of protein bio-engineering, active peptide sequences from desired proteins can be incorporated into synthetic scaffold under preparation stages of the polymeric material [123]. This facilitates the introduction of specific binding domains into the otherwise bioinert scaffold-matrix. The binding interactions of adhesive peptide sequences, *e.g.*, RGD and IKVAV, to name a few, have helped in understanding the cell functions [124] in greater details. In PEG based scaffolds, the incorporation of RGD as pendant, a well-known binding ligand tripeptide of fibronectin protein, has clearly shown to increase the feasibility of adhesion of entrapped cells [125-127]. Further studies on RGD incorporated-polymeric scaffold of synthetic origin through designer synthetic methodology have indicated their versatility for cell clustering as well as tethering with proper and, at the desirable levels of ligand density while cell proliferation and, cell migration is going on [128, 129]. Novel chemical reactions leading to polymerization, such as, photo-initiated thiol-acrylate and, thiol-ene chemistries are now allowing facile incorporation of peptide segments within the synthetic scaffold which are being prepared for use as scaffold matrix for generating desired extracellular microenvironment [130-132]. These modified scaffolds have improved the cell adhesion and, driven the desired responses into the intended culture [133]. Another interesting phenomenon/factor contributing in artificial functional tissue bioengineering is the nanoscale cues which seems to be regulating tissue constructs more precisely than before. Hence, a more cautious approach has to be pursued with the nano-scale designing of the scaffold or, application/loading and, interspersion of nanoparticles in the scaffold matrix [134]. The extracellular matrix molecules interact with the cellular receptors in finding further advances in roles at the scaffold levels and, at cell interaction/conjunction sites [135, 136]. The topography at the micro and nano-scale, up to an extent, controls the functions of the cells [137-139] and, these aspects can be taken care by the chemical structure's specifications and, chemical affinity of the polymer in extracellular microenvironment generation [138].

Importantly, any immobilization at the structural levels of the polymer scaffold with entities of affinity, within the culture media and their internalized components, favor the cell growth. Examples of alginate sulfation, RGD immobilization and, cross-linking levels of the polymer substrate have been found to add to cell adhesion and culture growth very well [140-143].

CELL SIGNALING IN SCAFFOLD'S MICROENVIRONMENT: EVOLUTION OF BIORESPONSE

The scaffold proteins play critical role in providing relevant micro- and, macro-molecular components and, most scaffolds use tethering mechanism to interact between individual molecules. These proteins exert allosteric control on interacting molecular entities and other substances and, are under regulatory mechanism within the scaffold. They offer a simple, flexible strategy for regulating selectivity in cell signaling pathways, shaping out-put behavior along with capabilities for receiving new responses including from the pre-existing signaling components in the extra-cellular microenvironment. Therefore, these proteins can be regulated to remodel the cellular behavior in an effective manner. As the cellular differentiation requires capabilities to sense and respond to a variety of extra-cellular signals originating in the extracellular microenvironment, the extra-cellular proteins play important role in majority of signaling processes. The seeded cells exposed to high doses of chemotropic growth-factors may undergo cell division arrest and, at intermediate dose levels these cells may become elongated in shape and, can divide in the direction of the concentration gradient of the growth-factors. The MAP (mitogen activated protein) kinases, Fus3 and Kss1, promote cell elongation and only the Fus3 promotes chemotropic growth. The Kss1 is activated rapidly and, with a graded dose-response profile, Fus3 is activated slowly but exhibits steeper dose-response relationship. Meanwhile, the Fus3 requires the scaffold protein Ste5 and, when binding to Ste5 is abrogated, Fus3 behaves like Kss1 and, cells no longer respond to a gradient in the natural settings. The scaffold proteins serve to modulate the temporal and, dose-response behavior of the MAP kinases [144, 145]. Moreover, contrary to the given understanding of multicomponent signal transduction of biological systems, emerging data suggests otherwise which also applies about the scaffolding proteins' signaling phenomenon. The signal transduction in an scaffold [146] and, some chemo-modeling studies dealing with multi-phosphorylation sites in scaffold binding and, cell signaling have been attempted with a view to understand the intricacies of the signaling phenomenon in scaffold [147] between the embedded proteins and developing cellular mass. The recent observation suggests that scaffolding limits the possibilities for signal amplification. However, methods currently available to assess signal transduction are too crude to know the precise role(s) of the scaffold in terms of its physical, morphological and chemo-biological impact on signaling. The studies on the MAP kinase pathway, Ste5, indicated functions control beyond passive structural elements. The identification of new MAP kinase pathway suggests existence of multiple 'signalosomes' or 'transducisomes' as authors have named them, in controlling the signaling in the scaffold [148].

FABRICATION OF SCAFFOLDS: TECHNIQUES & METHODOLOGIES

A well-designed, porous and elastic 3D-scaffold is one of the fundamental requirement and desirable tool to guide biomass formation both *in vitro* and *in vivo* conditions. Use of the biodegradable material, preferably polymers of synthetic origins, wherein scaffold is integrated into cells regenerate tissues [149], have shown desired results and gained much ground. The selection of scaffold material and, scaffold structural type is crucial for enabling the culturing cells to behave in an intended manner in the fabricated scaffold-matrix to generate the expected tissue bioengineering outcomes in terms of desired shape, size and, functional behavior [150-155]. To this effect, multiple preparation methods have been reported which fabricate scaffolds of desired architecture encompassing requisite and, intended characteristics. Among various techniques employed to fabricate a scaffold, leaching [156, 157], phase separation [158, 159] and textile-based weaving technologies are conspicuous. However, scaffolds fabricated through these initial methods do not sufficiently and appropriately mimic the natural environment and the extracellular matrix in terms of their structural outcomes with desired physico-mechanical properties and, their chemical characteristics which is precisely among the reasons for sub-optimal outcome in generating functional tissues in a lab and/or at commercial scales today [160-162]. Therefore, new design concepts, improved manufacturing technologies for materials' superior chemistry and, better scaffold fabrication methods are continuously sought to improve the functional characteristics and, 3D-architectural differentiation for enhanced and, improved culture spaces as well as remarkable functional feasibility of the scaffold. Based on polymeric material's physico-chemical characteristics and treatment methods to produce the scaffold's basic platform, several new techniques have been introduced. These techniques include gas based foaming, phase separation, solvent-casting of the scaffold material, electrospinning, fiber-mesh production, fiber bonding, particulate leaching, porogen leaching, rapid prototyping, melt molding, self assembly, soft lithography, membrane lamination, hot embossing and freeze drying methods. During the initial developmental stages in tissue bioengineering, collagen and, polyesters were considered as the material-of-choice. Further advancement was dictated by the need to pre-arrange generating cells/tissue in a desired 3D-configuration to meet the shape, size as well as functional viability levels of the grown bio-material. The need for more established scaffold material and its preparative techniques, to produce tissue bio-engineering platforms capable of providing naturally *at par* feasibilities for tissue bioengineering, arose to control the fabrication reproducibility, economic feasibility and ease of bulk preparation of the scaffold platform for (any) commercial purposes. The product for scaffold-platforms generated through use of different fabrication methodologies that support/tone up or, do not deactivate the biological activity of the bioactive molecule within the scaffolds, have been designed, produced and used judiciously in demonstrable manner. A scaffold can be made by different fabrication techniques and, a specific technique may, or is expected to, provide pre-set 3D-architecture with suitable physico-mechanical and chemical characteristics suitable for holding micro bio-environments for cells and tissues bio-engineering functions nearly or, *at par* with the natural material. The choice of fabrication technique for scaffold preparation is dependent upon the surface characteristics and other physico-chemical properties of the polymeric materials employed. The fabrication technique's outcome of the scaffold-platform is dependent on the technique itself whereby the proposed structure, properties and functional mimics of the natural extracellular microenvironment is dependent on the processing of the polymer. Each technique has its own merits and, use of suitable methodology is a must to prepare specific tissue or its repairs using the prepared scaffold. The understanding of the polymeric material's structure and behavior during fabrication process and, its (fabrication technique's) impact on the polymer's structural changes of chemical and non-chemical sorts to affect and, produce the desired scaffold with the pre-conceived properties in architecture and, functional feasibility with any variability of fixed quotient of the scaffold properties during its preparation and tissue bio-engineering use is a must [163-166]. A number of common polymeric materials, their scaffold preparation methodology and, scaffold's characteristics are summarized in Table **4** [167-190].

Table 4. Common polymeric materials, their scaffold preparation methodology & prepared scaffold's characteristics*.

Polymer	Methodology	Maximum Porosity & Pore Diameter	Applications	References
PLLA PLGA PGA-PLLA PCL-Collagen PCL-Chitosan Silk-Hydroxyapatite (HAP)	Electrospinning	95%, 1100 nm	Skin Cartilage Vascular Nerve Bone	[167-169] [170] [171] [172] [173, 174] [175]
PLLA-HAP PLGA Chitosan	Phase Separation	98%, 500 nm	Bone	[176, 177] [175] [178, 179]
PLLA-Chitosan Collagen-GAG PHEMA-Gelatin	Freeze-Drying	80%, 450 nm	Tendon Skin Bone, Tendon	[180, 181] [182] [183]
Peptides	Self-Assembly	90%, 300 nm	Cartilage	[184-190]

*adopted from reference [166].

Another aspect needing attention is a plethora of plausible interactions that will play-out between different cell types and the scaffold. Seeded cells, various soluble factors and typical behavior of cells under physical duress, *e.g.*, mechanical shear, compression and vibrations which play crucial roles in the interaction feasibility are important aspects for consideration. The need to control these aspects of design and functions rests on the polymer's structure and its post-fabrication behavior. Various fabrication techniques have been developed to improve the scaffold materials pores distribution phenomenon, porosity and pores' interconnectivities. The pores specifications are the desirable and defining criteria for all cell culture media. Recent advances in design by computational topology estimation as well as advances in fabrication techniques have made it possible to design and create scaffolds with controlled pore architecture [149]. Utility of each fabrication technique is nearly dependent upon the design components as per the requirements defined by intended outcome which may/may not be specific to every single tissue bioengineering application in different ways, including for the chemical composition, mechanical strength, degradation profile, nutrient transport capabilities and capacity and cellular organization capabilities of the intended scaffold matrix. The following Table **5** enlists the commonly used scaffolds fabrication techniques [191-229] which have potential tissue bio-

Table 5. Commonly used scaffolds fabrication technologies*.

Serial	Methodology	Merits	Demerits	References
1	Solvent Casting	Control over porosity, pore size and, crystallinity	Limited mechanical property, residual solvents and porogen material	[191-193]
2	Porogen (Salt, Sugar & Wax) Leaching	Control over porosity and pore geometry	Inadequate pore size and pore interconnectivity	[194, 195]
3	Gas Foaming	Free of harsh organic solvents, Control over porosity and pore size	Limited mechanical property, inadequate pore interconnectivity	[196-198]
4	Self Assembly	Control over porosity, pore size and fiber diameter	Expensive material, complex design parameters	[199-202]
5	Electrospinning	Control over porosity, pore size and fiber diameter	Limited mechanical property, pore size decrease with fiber thickness	[203-205]
6	Phase Separation	No decrease in the activity of the molecule	Difficult to precisely control scaffold morphology	[206, 207]
7	Rapid Prototyping	Excellent control over geometry, porosity, no supporting material required	Limited polymer type, highly expensive equipment	[208, 209]
8	Fiber Mesh	Large surface area for cell attachment, Rapid nutrient diffusion	Lack the structural stability	[197, 210-211]
9	Fiber Bonding	High surface to volume ratio, high porosity	Poor mechanical property, limited applications to other polymers	[210, 212-214]
10	Melt Molding	Independent control over porosity and pore size	Required high temperature for non-amorphous polymer	[215, 216]
11	Membrane Lamination	Provide 3D-matrix	Lack required mechanical strength, inadequate pore interconnectivity	[191, 196, 217]
12	Freeze Drying	High temperature and separate leaching step not required	Small pore size and long processing time	[218, 219]
13	Soft Lithography	Micro-contact printing (μCP), Patterning substrates for cell culture, 30 nm resolution	Rigid surface, capillary force required, Micro-molding in capillaries (MIMIC), 1 μm based on capillary force	[220-225]
14	Embossing	High-aspect ratio product	Heat applied, Temperature above Tg	[226-229]

*adopted from reference [163].

engineering applications with some of the scaffold's outcome characteristics as observed after the fabrication. Various other non-cellular, cellular and, hybrid cell/scaffold's fabrication techniques [230-244], some of which are under further advancements, are also summarized in Table **6** to provide glimpse of the technique, methodology and their merits.

BIO AND CHEMO-SENSING IN THE SCAFFOLD: TOWARDS INTERACTIVE BIOENGINEERED PLATFORMS

Assessing the 3D-scaffold's microstructure, nano-scale topography and, geometric availability within the platform through use of multi-photon fluorescence microscopy [245] and applications of enhanced darkfield optical microscopy coupled with high resolution hyper-spectral imaging and, dual-mode fluorescence imaging (CytoViva®) have been employed in imaging of nanoparticle laden hydrogels with enhanced penetration into (3D) cell culture platforms using layered/embed nanoparticles which have provided some details into sensing aspects [246-249]. Oxygen-sensing fluorescent microparticles which are capable of measuring oxygen concentration in microenvironments in standardized cell culture and transparent 3D-cell scaffolds are emerging fast. The synthesis of these microparticle involve poly(dimethyl siloxane) to encapsulate silica nanoparticles which are bound with oxygen-sensitive entities termed as luminophore along with a reference standard to facilitate the measurement of oxygen-levels inside the cell culture platform [250].

Table 6. Comparisons of acellular, cellular and hybrid cell-scaffold fabrication technologies*.

Type	Status Highlights	General Technique	Methodology	Advantage	Disadvantage	References
Acellular 3D-Scaffolds	Automated	Heat-Mediated Fabrication	Micro-molding	Simple, Reusable models, temperature above Tg, Use of vacuum to prevent air trapping	Lesser control in cell distribution, Cell seeding required, applicable to thin membranes, manual alignment & contiguous cell layer needed	[230, 231]
			Selective LASER sintering	High porosity	High temperature & powder	[232-233]

Table 6: contd…

					entrapment possible	
			Fused deposition modeling	No trapped solvent/particles	High temperature during processing	[234]
			3D-Plotting	Agar, Gelatin use possible	Low resolution	[235]
		Light-Mediated Fabrication	Stereo-lithography	Easy to achieve desired features for scaffold	Photo-sensitive material required	[236]
		Adhesive Mediated Fabrication	3D-Printing	High porosity	Limited choice of material, organic solvent present as binder	[237]
			Pressure Assisted Micro-Syringing	Not subjected to heat, resolution high	Viscosity dependent	[238]
		Indirect Molding Fabrication	Matrix Molding	Ease of biological material (Collagen) use	Weak mechanical properties in output scaffold	[239, 240]
Cellular 3D-Scaffolds	Automated	Cellular Assembly	Organ Printing	Cell aggregates, multiple cell type incorporation possible	Dependent on self assembly; sterility, pH & temperature controls required	[241, 242]
			Laser- Guided Deposition	Precise cell placement possible	Structural support lacking, LASER assisted deposition not yet applicable for 3D-format	[243]
Hybrid Scaffolds	Non-Automated	Hybrid Cell-Scaffold Assembly	Hydrogel Photo-patterning	Versatile, incorporation of peptides, living cells in scaffolds	Cells exposed to UV light, Limited diffusion owing to hydrogel's pore size	[244]

*adopted from reference [164].

Another assessing technique which utilizes the photo-polymerization of poly(2-hydroxy-ethyl-methacrylate (PHEMA) and poly-(meth-acryloyloxy)-ethyl-trimethyl ammonium chloride (PMEATAC) monomers in preparing the hydrogel scaffold with efficient oxygen-sensors through luminescence-based oxygen sensing capabilities have opened the possibility of 3D-mapping of oxygen levels during cell growth. Tris (4,7-diphenyl-1,10-phenanthroline)-ruthenium chloride [Ru $(dpp)_3$] was coated on the PHEMA–PMEATAC 3D-scaffold in a layer-by-layer assembly wherein PHEMA–PMEATAC copolymer was coated on top of the $Ru(dpp)_3$ layers as a protective layer. The fluorescence emission from $Ru(dpp)_3$ is dynamically quenched by oxygen's presence. By measuring the emission intensity the local oxygen levels of the scaffold were estimated. Since the hydrogel scaffolds are nearly transparent, the 3D-fluorescence intensity of the scaffold platform can also be mapped by confocal microscopy. Human bone-marrow stromal cells (HS-5) were successfully cultured on this type of oxygen sensing scaffold and the observed $Ru(dpp)_3$ emission intensity from the scaffold was found to be stronger in cells-rich area which indicated a lower oxygen level due to consumption of oxygen by the cells [251].

THE HYDROGELS: SIMPLE BUT COMPETENT 3D-CELL CULTURE PLATFORMS

The artificial scaffold derived from biomimetic, synthetic and natural polymeric materials, at present, offer the best solution for *ex-vivo* cell culturing and, have been used for clinical as well as non-clinical applications and, in studying the cell functions. In order to properly mimic the native/ natural extracellular microenvironment, some of these complexities need to be understood and resolved for full structural and functional viability of the employed polymeric materials of synthetic and natural origins [252-256]. The hydrogels, one of the feasible materials for biomimec-extracellular microenvironment, are 3D-materials with inherent ability to absorb large amounts of water while maintaining their stability and 3D-formations. They nearly emulate the molecular characteristics, morphological architecture and, physico-mechanics of the natural extracellular microenvironment [257]. Due to their inherent abilities to simulate the nature of most of the soft tissues, the hydrogels have proven to be the highly attractive material for the purpose. The reticulation, proper cross-linking and chain structures add to the holding of high water contents in the hydrogel which is good for facile transport of gases, especially oxygen, nutrients and metabolic waste disposal with two-way transport feasibility [258]. The feasibility to transport the soluble materials and the outreach of biochemical factors has made the hydrogel as the ideal design format for scaffolds towards producing artificial extracellular microenvironment. The hydrogels can be formed under mild, cyto-compatible conditions and can be easily modified to possess cell adhesion ligands (a two way interaction activity output between cells and, scaffold), desired viscoelasticity, compression characteristics, and mechanical strength to withstand properly the nucleation, initiation, cell differentiation, cell proliferation and most importantly, bio-degradability processes [259]. Developing the cyto and, bio-active hydrogels for 3D-cell culture is a bioengineering challenge which requires the controls of the physical and, chemical properties on scales from sub-microns to centimeters and on time scales from seconds to weeks.

In order to develop fully functional artificial extracellular microenvironment, the gel's physico-mechanical properties, ligand's adhesion properties, adhesion feasibility and, inoculation of growth factor(s) are first considered. The transport capabilities, overall self-maintenance and degradation parameters are finely adjusted to the given culture's needs in a cyto, biocompatible as well as cost-effective way [260]. Further, the gels' chemistries and, their degradation products should not exert deleterious effect on entrapped cells and growths of the desired biomass at its site. However, in an artificial hydrogelled condition, the matrix's available subcellular porosities will work as filter and barrier for cells migration, cells proliferation, cells differentiation as well as towards the outreach and levels of the distribution of soluble factors in the extracellular microenvironment. These artificial, interconnected scaffold networks often do not possess the fibrillar structure of the extracellular microenvironment protein backbone as is the case with the natural material. To address these shortcomings, nanofabrication techniques have been employed. Therefore, artificial polymeric or, non-polymeric synthetic scaffolds need to be developed which can self-assemble to provide an alternative biomimec extracellular microenvironment. These kinds of hydrogels can be formulated from a variety of materials having both synthetic and natural origins to offer broad range of physico-mechanical, biochemical and chemical properties (for scaffold) which can translate into proper and, competitive extracellular microenvironment development inside the scaffold. These gel scaffold are typically constituted of extracellular microenvironment components and bio-based proteins such as collagen, fibrin, hyaluronic acid [261-264] and matrigel. Materials originated from other natural sources, *e.g.*, chitosan, alginate and silk fibrils have also been used for this purpose [265, 266]. Since these materials are derived from natural sources, they are considered biocompatible by default [267, 268]. These scaffolds promote cells growth and are considered advantageous for cell proliferation. However, such scaffolds are complex and are often ill-defined, thereby restricting their usefulness. Moreover, the tuning of the physico-mechanical properties and biochemical desirability in their preparation is still unclear or, cumbersome and, there are risks associated with contamination and/or premature degradation due to poorly defined characteristics of the materials. These hydrogels may also show batch-to-batch variability in preparation and performance which further limits their functioning as a desired extracellular microenvironment component of the culture.

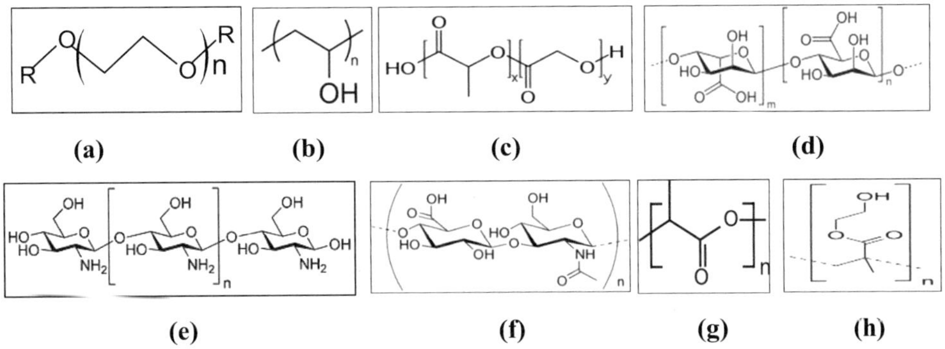

Figure 3: Some commonly used polymers' monomeric structures: **(a)**. PEG: monomeric unit, n stands for repeating monomeric units, R stands for chain continuation; **(b)**. Polyvinyl alcohol (PVA); **(c)** PLGA, x, y, n stands for repeating units; **(d)** Alginate or alginic acid, m & n stands for repeating units; **(e)** Chitosan, n stands for repeating units; **(f)** Hyaluronan; **(g)** Polylactic acid (PLA) or Polylactide; **(h)** Poly-2-Hydroxy-Ethyl-Methacrylate (PHEMA).

Consequently, hydrogels, in terms of generating better controls, can be formed from natural and synthetic materials as single polymeric entity and as blend and composite material from a number of polymeric and non-polymeric substances including polyethylene glycol (PEG), polyvinyl alcohol, poly-2-hydroxy-ethyl methacrylate, alginate, collagen, chitosan, hyaluronan and, hydroxyapatite through various methods from employing simple to complex gelation techniques and non-conventional/enhanced chemistries involving photo-polymerization [269-278]. Among other bio-compatible and biodegradable polymers that serve as scaffold material for preparing 3D-scaffolds are the polylactides and glycolides from polyester class, *e.g.*, polylactic acid (PLA), poly-L-lactic acid (PLLA), poly-D, L-lactic acid (PDLLA), polyglycolic acid (PGA), and poly-L-lactic-glycolic acid (PLGA) [279-282] (Fig. **3**). The polyester's structural backbone composed of ester linkage offers avenues for maneuvering of the macromolecular based properties according to specific scaffold's applications which includes mechanical strength as well as pore specifications based on the scaffold's raw material and its properties.

Among the synthetic polymers based hydrogels, the PEG-hydrogels maintain the functional feasibility of entrapped and/or encapsulated cells and, allows the extra-cellular materials deposition [283]. The PEG-hydrogels can function as 3D-cell culture platform without the availability of integrin-binding ligands in its material.

These types of inert gels are easily processed and are highly reproducible in nature (Fig. **4**). They freely allow facile adjustment(s) of physico-mechanical properties but lack endogenous factors to promote cell behavior [97].

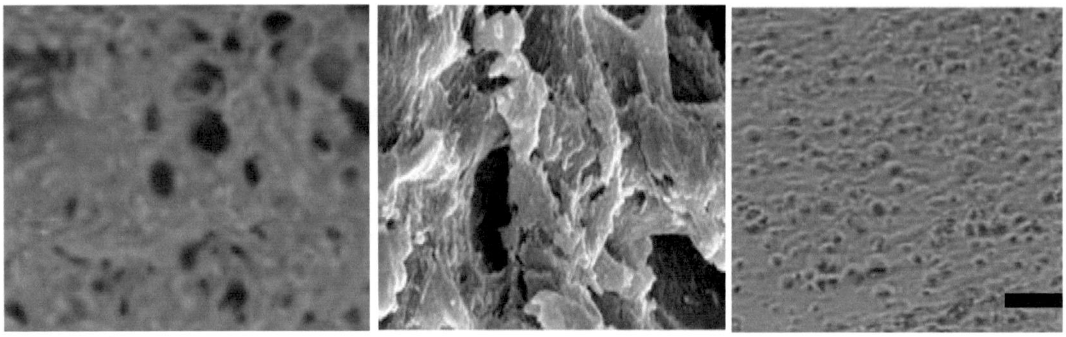

Figure 4: Typical Hydrogels: Porous and loaded, (scale bar 50 μm).

For synthetic/artificial hydrogels to mimic natural/native extracellular microenvironment, it is pertinent to understand the cell's natural environment, cell(s) interaction in the extracellular microenvironment, cells environmental modulation of the extracellular microenvironment, cells remodeling of the exo-environment and, the cell migration through the extracellular microenvironment. In order to properly design the scaffold for extracellular microenvironment. The necessary information on cell behavior patterns must be recreated and, recaptured in 3D-environment in presence of critical factors already embedded in the experimental gel matrix. It is advantageous to allow cells to self-dictate changes in their own environment. The user-defined controls of the physico-mechanical and, biochemical properties and, related functions can be advantageous to test complex and, inter-related assumptions about the effects of specific cell-extracellular microenvironment interactions in the 3D-scaffold models. The path to designing an ideal extracellular microenvironment mimic is dependent upon the intended culture and its outcome which requires multiple and orthogonal chemistries. For instance, photo-labile chemistries are useful to erode regions of an extracellular microenvironment gel and thus, expose surfaces for adhesion of bio and/or chemical natured ligands or, presence of linkers to encourage the cells adhesion and cells migration. The more complex scaffold formed of interpenetrating networks of cells with user-controlled chemistries are expected to arise eventually in the hydrogel series to the advantage of bioengineers. In fact, there is no single network type that can mimic the formation of complex extracellular microenvironment of every tissue types and which can incorporate rational, bio-inspired inputs into artificial gel to provide robust and, diverse scaffold for several cells types and their culture systems [284]. However, artificial hydrogels have been prepared to emulate typical and, natural disposition of growth-factors of the natural extracellular microenvironment. The heparin molecules were incorporated in scaffold's building blocks to entangle proteins within the gel-framework that subsequently was released. Recent improvements have been made by covalently-bound protein-specific ligands located in scaffold's backbone [285, 286] which certainly pin-points the need for presence of binding ligands onto the scaffold's matrix for selectively associating the desired growth-factors to the matrix and, its later release based on the factor(s) responsible for cellular uptake cycles. This is made feasible through the chemical un-bonding/detachment process. The local concentrations of binding ligand(s) provide avenues for protein biomolecules to bind at the site. Multiple soluble factors acting synergistically and/or antagonistically, within the environment, positively contribute towards development of signaling cascade, tissue development and homeostasis in the scaffold's medium [287]. To recreate these signaling domains, multiple chemical and, biochemical functionalities, process designs and, molecular entities needed to be incorporated within the scaffold's gel that does sequester and, present several orthogonal sites at desired time-points in the development of the intended cellular mass. Some of these approaches mimic strength and outcome of the extracellular microenvironment but the artificially created hydrogels, sometimes, fail to capitalize on physico-mechanical and biochemical/chemical structure of the extracellular microenvironment including the inbuilt functional processes as part of their design format. It is plausible to engineer chronological events' self- controlled bio-degradation capabilities into synthetic matrix so as the viable cells can deposit their own extracellular matrix in the degraded void and migrate and undergo desired morphogenesis in more efficient manner for tissue bioengineering to proceed [288-290] according to the intended plan for the tissue bioengineering product outcome. Thus, the artificial hydrogels are pre-designed to hydrolytically degrade and which can be achieved by incorporating polylactic acid and/or poly-caprolactone structural units in the scaffold's backbone's molecular structure network [291, 292]. The numbers of hydrolytic bondings dictate the rate of degradation of the scaffold which is a slower process in itself than the normal cellular processes occurring in the cells in the extracellular microenvironment and this may prove to be detrimental for the scaffold. However, Michael addition and, photo-initiated reactions of end tail-functionalized PEGs and, the thiol-labeled groups' cleavable peptide sequences are used to prepare enhanced hydrogels with cellular-action driven degradation on a shortened time-scale [293]. The poly-cationic hydrogels [291], antibiotic-loaded/based hydrogels with superior dual functions of antibacterial efficiency and better cell response platform [294, 295], as well as PEG-based hydrogels for guided cell growth and other naturally-sourced synthetic/semi-synthetic homo- and hetero-polymers like alginate, chitosan, chitosan-g-hematin, collagen-hyaluronan [294-296] have produced improved functional platforms serving at enhanced functionalization levels through pre-designed scaffold produced with inputs from known variables by incorporating molecular control switches/factors in the chemical structure of the scaffold preparation material [297]. The dynamical feedback between the extracellular microenvironment's structural components and, cell behavior is very helpful in studying the cell migration, cell differentiation and cell proliferation to properly design the requisite scaffold. The tumor morphogenesis and cancer metastasis studies in the hydrogel formats of the extracellular microenvironment have also led to designing of proper scaffold for various applications of tissue bioengineering and cell growth. This implies a better understanding of the tumorogenesis *at par* with the normal cell differentiation, the cells' adhesion, progression and outreach in the extracellular microenvironment.

TISSUE BIOENGINEERING: NANO-FIBROUS SCAFFOLDS

The extra-cellular matrix is an elastic, complex and, functional platform that provides specific advantages and arena for cell growth and their further sustainability in a specified micro-environment called extra-cellular micro-environment. It needs to be replicated for effective tissue regeneration capabilities after cells implantations. In this regard, electro-spinning is a straightforward technique for producing extracellular microenvironment-mimicking scaffolds [298-302]. Natural polymers are one of the best suited options due to their cellular affinity and bio-compatibility properties. However, these are affected by fast degradation and, low mechanical stability because of their molecular framework and back-bone infrastructure. These limitations can be overcome, to a large extent, by means of polymeric material's cross-linking. Majority of the biologically efficient scaffolds are meeting the challenges thrust upon them from tissue bioengineering and regenerative medicine related demands for performance as competitive cell-culture platforms. In context to extracellular microenvironment where tissue bioengineering outcomes are greatly influenced by the physico-chemical and, physico-mechanical manipulations of the bio and non- bio-origins polymeric materials and by the techniques and processing methodology employed in developing the substitutes for implantation platforms into the body and/or outside of the body to foster regeneration and/or remodeling of the tissue in active and functional forms, the fabrication of nano-fibrous scaffolds assumes significance. The advantages involve better quality of repairing, replacements, maintenance and enhancements of the functions of any particular tissue types and/or organ(s) through various tissue bio-engineering techniques' employment for the purpose. The methodological inputs come from cell technology, scaffold construct technology and, technology for *in vivo* integration of various techniques towards better tissue bioengineering outcomes. The scaffold construct technology focuses on designing, manufacturing and, characterization of the 3D-scaffolds for cell seeding and, *in vitro* or, *in vivo* culturing of the target biomass [303]. The nano-scaffold completely fits in this mold/role. These nano-scale substrate's based scaffolds have exhibited great performance with cell attachment, proliferation and, penetration both *in vitro* and *in vivo* observations in high and, satisfactory orders. The nanofibers have demonstrated improved applications in tissue bioengineering of blood vessels, bones, muscles, skin tissues and, neural fibers [304]. An intended nano-fibrous scaffold needs to have certain characteristics of non-antigenicity, bio-compatibility and biodegradation capabilities with degradation rate out-performing or, atleast matching the rate of new tissue formations with high degree of matrix porosity and large surface area in a 3D-structure format with adequate levels of wettability characteristics [305] to sustain any culturing in an effective and, fruitful manner. The prepared scaffold should also possess the required structural integrity and, design ingredients to prevent the pores of the scaffold from collapsing during tissue formations under mechanical and other duress factors. The scaffold should also be insignificantly shrinkable after swelling and stabilization, and be non-toxic to cells in the micro-[illegible] with extended scope and, functioning feasibility along with interactions supporting the tissue culture processes by performing and, promoting cell adhesion, proliferation, migration and, any differentiated cell functions as and when required in the scaffold. More details on engineering bone-tissues, using electrospun nanofibers (Fig. **5**), began with the work on muscle cells [306]. Cells growth on PLGA nanofibers indicated that nanofiber structures positively promote cell-matrix and cell-cell interactions to facilitate the desired tissue culturing. For bone-tissue bioengineering purposes, the nanofibers made from poly-ε-caprolactone (PCL) have been extensively evaluated [307]. It was observed that mesenchymal stem cells (MSC) feasibly penetrate into the PCL matrix accompanied by the (abundant) extracellular materials after 1 week of seeding and, mineralization while for type-I collagen, the penetration occurred at 4th week. An *in vivo* experiment was also conducted by implanting mesenchymal stem cells cultured PCL construct (4th week) in the omenta (omentum: a fold of peritoneum connecting the stomach with other abdominal organs) of rats. After 4 weeks, the constructs maintained the size and, shape of the original scaffolds and, had a bone-like appearance. In addition to the pure PCL nanofibers, nanofibers from gelatin/PCL blend and, PCL-composite with $CaCO_3$ nanoparticles as well as hydroxyapatite (HAp)-nanoparticles were also investigated successfully for bone scaffold purposes [308]. Various nano-fiber based scaffolds have demonstrated encouraging results for skin replacement tissue repair, cartilage repair and its regeneration. The synthetic and, natural polymer based scaffolds have shown to be providing adequate mechanical support, enough adhesion capacity and, proliferation feasibility for the seeded cells in the scaffold and, their support for the cells migrating from surrounding tissues. Gelatin, chitosan and, hyaluronic acid are some of the materials used in preparing scaffolds for various tissue bioengineering applications. The physico-chemical properties of these materials provided desired microenvironment for cell growth which has been used as regenerating system for skin and, cartilage tissue bioengineering [309]. Several nanofibers have been spun as extracellular matrix [310] for a variety of applications including endothelial cell proliferation [311-313], micro-environmental platform for osteogenic gene expression of marrow stromal cells [314], cardio-vascular system's related tissue bioengineering for blood vessels including heart-valves [315-317] and, hemostasis [318].

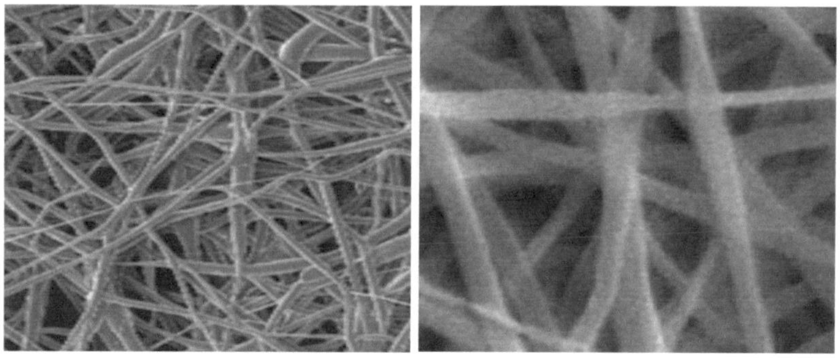

Figure 5: Typical spun nanofibers.

For blood vessels with different sizes and, individual mechanical strength along with their biochemical properties with various holdings/exchanges of different contents which is dependent on their contents, location and, specific functions; the vascular grafts tissue bio-engineering is more intricate, complex and, demanding. The blood vessel replacement, particularly a fine blood vessel sized <6 μm fiber diameter has remained a challenge. A comparison of cell growth on electrospun poly-(*L*-lactide-co)-ε-caprolactone (P(LLA-CL) nanofiber and on smooth film cast from P(LLA-CL) revealed that the function of vascular endothelial cells on the P(LLA-CL) cast film, rather than on the electrospun nanofiber mat is quite enhanced [319]. Because the electrospun nanofiber mats can provide good support during the initial growth of vascular smooth muscle cells, the bio-films combined with electrospun nanofiber mats could form a better 3D-scaffolding for blood vessel tissue bio-engineering [320]. In a separate set-up, the collagen nanofibers were used to assess the feasibility of culturing the smooth muscle cells. The cell growth on the collagen nanofiber was promoted and, the cells were well integrated into the nanofiber network after 7 days of seeding. The smooth muscle cells also adhered and proliferated well on other polymer-based nanofiber mats including a polymer blend of the collagen and, on P(LLA-CL) (poly-L-lactide-co-ε-caprolactone) [321, 322]. The incorporation of collagen as molecular entity in the structures of the nanofibers improved the fiber's elasticity and, tensile strength which effectively increased the cell adhesion capacity of the scaffold. The fiber surface wettability also influences the cell attachment. It was reported that when the poly-sulfone nanofibers were treated with argon plasma, the fiber wettability was increased significantly. As a result, cell attachment was increased by two folds. Also, the alignment of nanofibers can induce cell orientation and, promote skeletal muscle cell morphogenesis as well as aligned myotube formations in the medium. The cells cultured on the aligned nanofibers exhibited an alignment factor of 0.74, compared with 0.19 on the randomly orientated scaffolds substrate. The degradation rates of smooth muscle cells cultured on PGA, PLGA and, P(LLA-CL) nanofibers were investigated and, the results showed that the degradation rates were in the order of PGA >> PLGA > P(LLA-CL). The PGA based nanofibers degraded in 3 weeks duration while supporting the cell growth only for few days whereas PLGA nanofiber scaffolds facilitated the cell growth for about 30 days (4+ weeks) after the seeding which significantly slowed down thereafter the month while the P(LLA-CL) nanofibers facilitated the long-term (1-3 months) cell growth. Thus, the cell culturing significantly increased the degradation rate of PGA nanofibers, whereas PLGA and, P(LLA-CL) nanofibers were less affected by the degradation [323].

Skin is the largest body tissue, regulating heat loss and water retention. Cosmetic surgeries and, various other skin repairs involve nanofiber mats. Nanofibers from various types of homo- and, hetero-polymeric materials have been made and, evaluated for their cytocompatibility including compatibility to fibroblast and keratinocytes [324]. In most of the cases, the electrospun nanofiber mats exhibited good capability of supporting the cell attachment cell differentiation and cells proliferation. The cells seeded on the nanofibrous scaffolds tend to maintain their phenotypic shape and were guided on the scaffold to grow according to the nanofibers orientations. The cell culture on a 3D-nanofiber mat containing large pores was also evaluated. A 3D-silk-fibroin nanofiber matrix was prepared by directly depositing the newly-electrospun silk-fibroin nanofibers in a methanol solution. This 3D-nanofiber matrix supported the cell proliferation inside the fiber matrix very well. A 3D-nanofiber mat was also prepared by adding a chemical blowing agent into nanofibers *via* electrospinning and, a post-electrospinning heat-treatment leading to large pores within the nanofiber mat was prepared. These large pores provided good access to the cells to penetrate inside the nanofiber mat. The relationship between inter-fiber distance and, skin tissue regeneration has been investigated on electrospun gelatin scaffolds also. The cell infiltration was found only into the upper regions of the scaffold, provided the inter-fiber distance was smaller than 5.5 μm. The distances between 5 to 10 μm have yielded most favorable skin substitutes as the tissue end-product of the intended cultures. Electrospun nanofibers were also evaluated for the nervous system's tissue repair outcomes. In an *in vitro* cell culture study, the inducted neural stem cells on PLLA nanofiber scaffold were not only supported for the neural stem cells differentiation and, neurite outgrowth but the scaffold also promoted the neural stem cells adhesions. More work on aligned PLLA nanofibers and, the microfiber based scaffolds demonstrated that the directions of neural stem cells elongation and, its neurite outgrowth were parallel to the direction of the fiber alignment in the scaffold. The aligned fibers in the scaffold provided topographical guidance, enhanced cell proliferation and, the required neurite outgrowth [325]. The aligned fibers also have potential for bridging long peripheral nerve gaps during the neurite growth. So far, the effects of fiber diameter on cell culturing are far from clear. However, the PLLA nanofibers demonstrated higher levels of neural stem cell differentiation than the microfiber based scaffold with employment of similar cells orientations. Among the three kinds of poly-ether-sulfone based nanofibers having the diameters in the ranges of 283, 749 and, 1452 nanometers, the cell differentiation, cell proliferation and, cell spreading considerably increased with the decrease in the fiber's diameters [326].

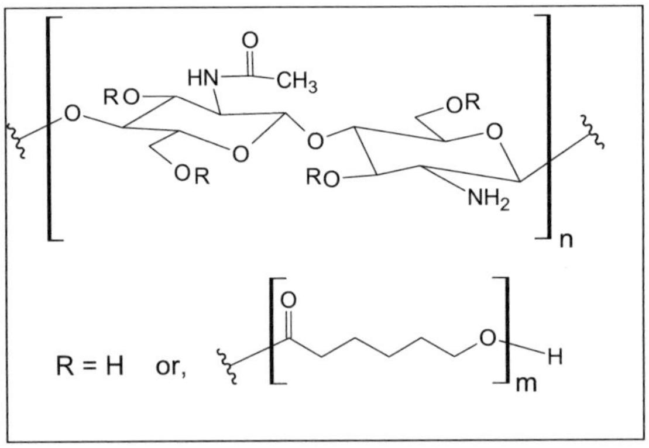

Figure 6: Amine-protected Poly-ε-caprolactone grafted chitosan monomeric structure.

In another investigation, PLLA fibers with different diameters were produced and, tested wherein shortest length electrospun fibers were the finest in action [327]. The neural growth promotion on nanofibrous scaffold can be regulated by extra-cellular matrix proteins [328], by the fiber's structure and, by the nano-fiber scaffold's surface properties [329, 330]. It can also be regulated by rat de-cellularized brain extracellular matrix [331], chitosan-PVA blend nanofiber [332] and fibrins [333] and, fibrin matrices containing affinity-based delivery systems [334] as well as some drug delivery systems [335]. Bio-responsive scaffolds and nanofiber platforms for tissue bioengineering related to spinal cord repair [336, 337], dermal substitutes [338], tracheal scaffolding [339], and stem cells differentiation on fibrin [340] and stem cells derived cardiomycetes' timed-notch activation, control of cell growth [341] are other advanced applications in the nano-fibers, and some of the normal gelated/non-gelated scaffoldings. Studies into tumor based biomaterial's requirements and, their behavior have been taken up with a view to understand the tumor model in general as well as their specific applications in the field [342, 343]. Several nano-fibers based scaffolds from modified [344] and, composite-scaffold of various natural and, synthetic polymers including chitosan, silk-fibrin and, nanoparticles-mix have been prepared, *e.g.*, PCL/Hydroxyapatite-chitosan/PVA, chitosan-PVA-PLGA, chitosan/HAp/PEO poly(ethylene oxide) with different polymers and, nanoparticles (NPs) in blend. The chitosan/gelatin/AgNPs, chitosan-Hap, chitosan-poly-*L*-lactide-co-ε-caprolactone, chitosan-PCL, (Fig. **6**) and, other nanofibers have also been prepared and, evaluated [345-353]. A comparatively less utilized techniques, soft-lithography, for scaffold preparation using poly-di-methyl-siloxane (PDMS) polymer with 3.5 mm diameter micro-bioreactors platform arranged in a 4 x 3 arrays provided means to study the growth behavior and, cellular differentiation activities of the human embryonic stem cells (hESCs) under controlled conditions [354]. The micro-bioreactor arrays provide the advantages of microarray and the bioreactor. The nanofiber based matrix is a niche for stem cells [355].

TISSUE BIOENGINEERING: NON-FIBROUS SCAFFOLDS

Tissue bioengineering through incorporation of new and, ingeniously produced biomaterials is replacing tissue-grafts in organ/cells repair owing to their advantages at cellular and, functional levels with improved and, newer methodologies to prepare the scaffolds (Fig. **7**) [356-359]. The enhanced bio-acceptability, system's integration at cellular levels, scaffold's degradability to pave way for generated biomass and, functional superiority of the scaffold based materials generated *versus* traditional tissue-grafts, have given way to new developments in the field [360, 361]. Scaffold have been used for initiating tissue development through peptide modifications for targeted cell-signaling [362] and for creating microenvironment platforms for stem cells' central nervous system (CNS) transplants [363]. The composite and non-composite materials including hydroxyapatite [364] and hyaluronic acid [365] based scaffold matrices have been employed. Again, the critically sized, 3D-biodegradable matrix for bone ingrowth [366] and rapid prototyping of tissue bio-engineering constructs using photo-polymerizable scaffolds with stereo-lithography techniques towards appropriate scaffolds generation have been produced, utilized and evaluated [367].

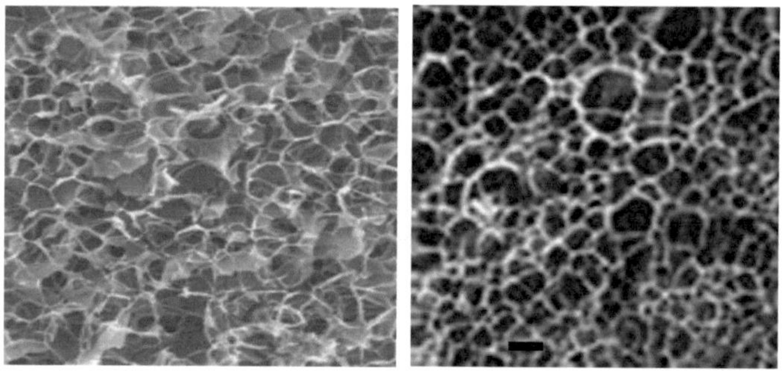

Figure 7: Typical polymeric scaffold, (scale bar 50 μm).

The solid 'free-form' fabrication technology of scaffolds preparation for holding and, continuously supporting of the tissue bioengineering processes along with ability to recreate scaffold-platforms with reproducible architecture and, compositional integrity each time with the liberty of introducing various structurally different starting raw materials containing new chemical groups/entities based on their chemical feasibilities for scaffold preparation have made this technology among the most accepted design formats. The functionalization of novel composite scaffolds as mixture of synthetic and, natural raw-materials together with combinations of growth factors and, immobilization techniques have further differentiated the scaffolds design to cater needs in specific conditions of tissue regeneration. The 3D-scaffolds have been bio-engineered to provide support for various types of cells survival, their adhesion, propagation and, proliferation within the scaffold including their over-all growth. The chronologically advanced biodegradation and, biocompatibility to cellular environment are well-considered and, important design criteria for promoting tissue regeneration. A wide range of biomaterials have been explored in this connection [368, 369]. Biomaterials modification strategies with chemical and, physical considerations and, aspects to bioengineer microenvironment with desired mechanical and, chemical properties to usher in with best physicochemical and, physico-mechanical feasibilities to stimulate cell growth and, adhesion are being developed currently. The scaffold's design dictating topographical requirements, physico-chemical characteristics and, ways to deliver growth-factors and, solubilized materials for tissue generation have potentially reiterated the microenvironment [370]. The extracellular matrix's mechanical microenvironment appropriately targets and, redirects different cells progression. The polymeric scaffold composites incorporating PVA, chitosan and, other synthetic/natural materials have attracted much attention owing to their processing feasibility and, characteristics' maneuverability [371-377]. Scaffolds catering the needs of adipose tissue regeneration [378], bone regeneration [379], generation of inter-vertebral matrix [380], cartilage tissue bioengineering [381, 382], neuronal growth [383] and, scaffolds fabricated from polyurethane-poly-(ethylene-glycol)-mono-methyl-ether methacrylate (PU/PEGMA) have been developed [384].

CURRENT MEDICAL APPLICATIONS: RESTRICTED UTILIZATIONS

Regenerative medicine, organ repair and, transplants have used the available classical tissue bioengineering techniques. The cells and/or, scaffolds induce tissues to grow from the incorporated biomass and/or supplied biochemical and, mechanical stimuli in a laboratory and/or natural settings for organ repair and, replacement purposes. Due to the inherent complexities in bringing out successful tissue-engineered organ(s) to commercial use, only a limited number of clinically competent devices are available today. Currently, (*ca* 2015), the extracellular microenvironment-mimic biomaterials for commercial medical applications are derived from processing human or, animal origin cell-free contents which are permanently integrated into the body of implant. The cells are induced in the extracellular microenvironment-mimicking device or, scaffold to grow in presence of biochemical and, physico-mechanical stimuli for organ repair/replacement which in its approach, technique and, practice is different than the allo- and, xeno-graft implants. A limited number of clinically available devices for artificial skin, venous leg and, diabetic foot ulcer healings are commercially available and are US-FDA approved [385]. Since implanting cell-free scaffolds for extracellular microenvironment generation, with or, without biochemical growth-factors induction, has become a favorable feature of current clinico-surgical practices, a range of medical applications like hernia repair [386], skin repair, wound healing [387], breast reconstruction [388], tendon repair [389], bone-healing [390], soft tissue repairs [391] and, demineralized bone-grafts [392-395] have become available to public.

FUTURE CHALLENGES: SCOPE & DIRECTIONS FOR FURTHER DEVELOPMENT

The ongoing advances in polymer chemistry, tissue bioengineering and, material sciences are driving the evolution of sophisticated, synthetic-biomimetic, artificial platforms wherein 3D-scaffold-based culturing of various cells in designed and, self-developed microenvironments is plausible but not without challenges. The oxygen availability is one of the issues requiring prime attention. The transitions in replacement of 2D-scaffolds to 3D have opened up the involved heterogeneities of natural 3D-extracellular microenvironment. In a 3D-scaffold, various structural and, functional gradients and, innate defects can precipitate through material's physico-mechanical-chemical properties and, playing out of biochemical and, other involved factors and, substances [396] may aggravate the situation. There might be situations where extracellular ingredients can be of limited-diffusion characteristics leading to non-equal distributions of them in the scaffold and, oxygen and, other nutrient raw materials ingredients may arise out of proportions as the diffusion goes through the scaffold, thereby disturbing the microenvironment's further creation and, maintenance. Moreover, the control over distribution of soluble growth-factors influencing the cellular differentiation along with the cell propagation and, scaffold's environment's homeostasis may also become increasingly complicated within the 3D-network which is solely dependent upon the distribution coefficients of the available bulk concentrations of the biochemical and, other essential ingredients in the surrounding medium. The diffusion of materials and, gases within the 3D gel-structure as well as final cellular uptake of nutrients dictate the status of the created extracellular microenvironment's functioning. It has been observed that an increase in the hydrostatic fluid pressure enhances matrix synthesis and, accumulation of supply in 3D-scaffold's culture, as reported in case of bovine chondrocytes, is observed [397]. More perturbations with respect to factors manipulating the formation, maintenance and degradation of different kinds of scaffolds is expected which will give rise to development of highly-evolved bio-responsive, fully functional and, completely controlled smart biomaterials capable of interacting with the surroundings based on the molecular, geometric and, physico-mechanical stimuli response's recognition [398]. However, currently on the analysis front, standard techniques employed for detecting cell-functions and, cells produce are facing complications in a 3D-matrix where cells are least accessible for immuno-staining and, material extractions. Headway in imaging is overdue. Moreover, as the produced proteins can be difficult to obtain from these scaffolds, a tissue culture's non-invasive methodology is also warranted for status and, functional checks of the scaffold along with the extracellular microenvironment. The cell-imaging is now tougher for optical microscopy utilizing refraction and, light scattering techniques. The analysis of/in the 3D-matrix which is laden with functional cells and, consumable supplies, is unapproachable for satisfactory analysis. These challenges clearly point to the need for sophisticated technique(s) for scaffold's environments checks to conjoin the ability on real-time basis with the biological performance analyses of the scaffold. Through this manipulation of the extracellular microenvironment or, direct manipulation of the scaffold matrix, functional characteristics and, controllable properties, processes can be managed for the tissue bioengineering set-up either *in vivo* or *in vitro* conditions. Since, the natural extracellular microenvironment is a dynamic entity and, to facilitate complex cellular behavior during the artificial tissue bioengineering processes, the extracellular microenvironment -mimics also need to be dynamic at cellular levels at par with the natural counterparts to achieve the desired levels of functioning along with feasibility for desired material outcomes for different cells/constructs types. Hence, it is important to develop a 3D-scaffold which is fully user-defined with exogenous control, with spatio-temporal controlled integrin-binding ligands, growth factor releases' known gradients, biochemical properties and, also with known physico-mechanical characteristics along with the fully compatible imaging and, analyses techniques with parameters well defined. Moreover, the living radical polymerization (LRP) technique as synthetic methodology is leading to total chemical synthesis of materials with complex chemical functionalities for further use in the scaffold. The 'Click Chemistries' has attracted attention for their automated way to chemically-contain the desired specifications as part of the 3D-network through chemical modifications which themselves are manipulated through understanding of the ligand chemistries in post-fabrication stages of the polymer based-scaffolds. The photo-labile chemistries are employed to spatially and, temporally regulate the scaffold-based extracellular microenvironment's physico-mechanical and, biochemical properties [399]. Coupling these advanced chemistries and, other cyto-compatible, orthogonal chemistries happening simultaneously in the scaffold, it seems, may provide a template for further development of cells and, cellular masses, the desired scaffold and, the extracellular microenvironment with increased interfacial/interphase interactions at the culture level. Hence, the physico-mechano-transduction of cells and, materials can be achieved. Further approaches to finally release the cellular mass, the proteinaceous materials/bio-macromolecules and, other molecular entities from the 3D-scaffolds need to be inbuilt as part of the design and, composition of these platforms to help in analytical-bio-assays which can be carried out easily to ascertain the role of matrix interactions on cell development and, scaffolds' functions in near future [400-402].

SUMMARY

The field of tissue bioengineering and, simpler cell culture techniques have significantly advanced from the old prototypical way of cell-seeding into the 2D-scaffolds based methodology to development of desired bio-constructs. The regeneration of various tissues types with intended physico-mechanical & biochemical complexities as well as functional superiority in an advanced 3D-platform based technique has evolved. As better and, in-depth understanding of the biological processes in its entirety of effect(s) and functional variability is being continuously sought; the fundamental structure-function relationship of cells & its surroundings as well as the factors that influence cell growth, morphogenesis, apoptosis along with other variables of gene expression are being conjoined into the design template & methodologies of fabrication, a more clear picture is emerging. The factors including signaling advances, loops of cellular and, extra-cellular microenvironment sensing & recognition, sensing in a 3D-context about cell-cell & cell-matrix interactions, role of the geometry of the surroundings, soluble factors and, physico-mechanical forces as well as inputs from stem cell technology are also being incorporated into the advancing tissue bioengineering techniques with certain fusion in tissue bio-engineering as well as aspects of stem cell technology which seems plausible and, far-reaching with initial lines appearing. The impetus to produce functional tissues at par with the natural counterparts, a better understanding of the relationship between the cells and, their microenvironment is essential at the primary level. To gain better insight of the cell behavior in and, outside of the bio-medium, a number of approaches have taken course. As building 3D-matrix that emulates the natural microenvironment for providing better alternatives for tissue repairs, organogenesis and, regenerative medicine; it is pertinent to know that the tissue development from cells is a complicated and, tedious process which involves changes in biochemical, morphological and, physico-mechanical specifications at every stage of development of the biomass in the scaffold. The generation of extracellular microenvironment to emulate the dynamical nature and, biochemical complexity of natural setting of tissue biogenesis or, cell culture biosynthesis is, currently, a significant challenge in tissue bioengineering domain. The aspects in fabrication of materials and, developmental techniques achieved till now is far from adequate where developmental work is still going on. Therefore, the ability to control the extracellular microenvironmental cues for shaping bulk properties at the macro and nano-scales by maneuvering the fabrication methods for 3D-scaffolds generation may further advance the tissue bioengineering field. Moreover, lacuna in smart materials availability with specific properties suitable for employing in various tissue bioengineering is required to be filled. The material's properties, functions and, ability to sustain and, reproduce physico-chemical-mechanical characteristics need to be further evolved. Hence, the goals of fabricating complex biomass is nonetheless an arduous task, whereby researchers have devised several micro/nano scale platforms that allow better control of various aspects of extracellular microenvironment's build-up and, properties. Some of these include polymeric scaffolds, synthetic hydrogels and, microfluidics platforms with micro/nanofabrication techniques. It is envisioned that these tools will allow to generate more complex and, controllable artificial microenvironment with the spatial and, temporal specifications for emulating the natural tissue bioengineering in an ethical way [403] to fill the lacunae in the technique and, material.

ACKNOWLEDGEMENTS

Author thanks colleagues for their support at initial stages of this work.

CONFLICT OF INTEREST

The author confirms that this chapter contents have no conflict of interest.

REFERENCES

[1] Frantz C, Stewart KM, Weaver VM. The extracellular matrix at a glance. J Cell Sci 2010; 123: 4195-00.
[2] Mikos AG, Lu L, Temenoff JS, Temmser JK. Synthetic Bioresorbable polymer scaffolds. In: An Introduction to Material in Medicine, Ed: Ratner BD, Hoffman AS, Schoen FJ, Lemons JE. Elsevier Academic Press. 2004; p 743.
[3] Bell E, Ehrlich HP, Buttle DJ, Nakatsuji T. Science, 1981; 211: 1052-54
[4] Kretlow JD, Mikos AG. From Material to Tissue: Biomaterial Development, Scaffold Fabrication, and Tissue Engineering. AIChE J 2008; 54: 3048-67.
[5] Farach-Carson MC, Wagner RC, Kiick KL. Extracellular matrix: Structure, function, and applications to tissue engineering. In: Tissue Engineering, Ed: Fisher JP, Mikos AG, Bronzino JD. CRC Press: Boca Raton, FL, USA, 2007.
[6] Martins-Green M. The dynamics of Cell-ECM interactions with implications for tissue engineering. In Principles of Tissue Engineering, Ed: Lanza R, Langer R, Chick W. RG Landes Company, Austin, TX, USA. 1997.
[7] Langer R, Vacanti JP. Tissue Engineering. Science 1993; 260: 920-26.
[8] Ciardelli G, Chiono V, Cristallini C, *et al.* Innovative tissue engineering structures through advanced manufacturing technologies. J Mater Sci Mater Med 2004; 15: 305-10.
[9] Vacanti JP, Langer R, Upton J, Marler JJ. Transplantation of cells in matrices for tissue regeneration. Adv Drug Deliv Rev 1998; 33: 165-82.
[10] Huang NF. Tissue Engineering and Regenerative Medicine: Role of Extracellular Matrix Microenvironment, In: Stem Cells and Cancer Stem Cells 2013; 9: 313-24; ISBN 940-075-6445.
[11] Wenk GE, Bowlin GL. Encyclopedia of Biomaterials & Biomedical Engineering, Marcel Dekker, New York, USA.

[12] Guangyuan L, Baiyang S, Gan W, *et al.* Controlling the degradation of covalently cross-linked carboxymethyl chitosan utilizing bimodal molecular weight distribution. J Biomater Appl 2009; 23: 435-51.

[13] Williams SF, Martin DP, Horowitz DM, Peoples OP. PHA applications: Addressing the price performance issue. I. Tissue Engineering. Int J Biol Macromol 1999; 25: 111-21.

[14] Meyers MA, Chen P-Y, Lin AY-M, Seki Y. Biological materials: Structure and mechanical properties. Prog Mat Sci 2008; 53: 1-206.

[15] Ma PX. Biomimetic materials for tissue engineering. Adv Drug Deliv Rev 2008; 60: 184-98.

[16] Deans TL, Elisseeff J. Mimicking Extracellular Matrix to Direct Stem Cell Differentiation. World Stem Cell Summit. Detroit, MI, USA. pp 97-100, October 4-6, 2010.

[17] Higuchi A, Ling Q-D, Hsu S-T, Umezawa A. Biomimetic Cell Culture Proteins as Extra Cellular Matrices for Stem Cell Differentiation. Chemical Reviews 2012; 112: 4507-40.

[18] Eberli D, Freitas FL, Atala A, Yoo JJ. Composite scaffolds for the engineering of hollow organs and tissues. Methods, 2009, 47: 109-15.

[19] O'Brien FJ. Biomaterials and scaffolds for tissue engineering. Mater Today 2011; 14: 88-95.

[20] Alsberg E, Anderson KW, Albeiruti A, Rowley JA, Mooney DJ. Engineering growing tissues. Proc Natl Acad Sci USA 2002; 99: 12025-30.

[21] Chen P-Y, Lin AYM, Lin Y-S, Meyers MA, J McKittrick J. Structure, function and mechanical properties of selected biological Materials. TSMH Med Nursing J 2010; 16: 135-66.

[22] Kasoju N, Bhonde RR, Bora U. Preparation and Characterization of Antheraea Assama Silk Fibroin Based Novel Non-Woven Scaffold for Tissue Engineering Applications. J Tissue Eng & Regen Med 2009; 3: 539-52.

[23] Gelinsky M, Welzel PB, Simon P, Bernhardt A, König U. Porous Three-Dimensional Scaffolds Made of Mineralized Collagen: Preparation and Properties of a Biomimetic Nanocomposite Material for Tissue Engineering of Bone. Chem Eng J 2008; 137: 84-96.

[24] Altman GH, Diaz F, Jakuba C, *et al.* Silk based biomaterials. Biomat 2003; 24: 401-16.

[25] Vepari C, Kaplan DL. Silk as a Biomaterial. Prog Polym Sci 2007; 32: 991-07.

[26] Zarkoob S, Eby RK, Reneker DH, Hudson SD, Ertley D, Adams WW. Structure and morphology of electrospun silk nanofibers. Polymer 2004; 45: 3973-77.

[27] Sukigara S, Gandhi M, Ayutsede J, Micklus M, Ko F. Regeneration of Bombyx mori silk by electrospinning- Part 1: processing parameters and geometric properties. Polym 2003; 44: 5721-27.

[28] Nazarov R, Jin HJ, Kaplan DL. Porous 3-D scaffolds from regenerated silk fibroin. Biomacromol 2004; 5: 718-26.

[29] Jin HJ, Chen J, Karageorgiou V, Altman GH, Kaplan DL. Human bone marrow stromal cell responses to electrospun silk fibroin mats. Biomat 2004; 25: 1039-47.

[30] Kundu SC, Dash BC, Dash R, Kaplan DL. Natural protective glue protein sericin, bioengineered by silkworms: otential for biomedical and biotechnological applications. Prog Polym Sci 2008; 33: 998-012.

[31] Yakimets I, Wellner N, Smith AC, Wilson RH, Farhat I, Mitchell J. Mechanical Properties with Respect to Water Content of Gelatin Films in Glassy State. Polymer 2005; 46: 12577-585.

[32] Villalobos P, Acevedo CA, Albornoz F, R *et al.* A Bod Monitoring Disposable Reactor With Alginate Entrapped Bacteria. Bioprocess & Biosys Eng 2010; 33: 961-70.

[33] Chin C, Khanna K, Sia SK. A Microfabricated Porous Collagen-Based Scaffold as Prototype for Skin Substitutes. Biomedical Micro-devices 2008; 10: 459-67.

[34] Chen G, Ushida T, Tateishi T. Scaffold Design for Tissue Engineering. Macromol Biosci 2002; 2: 67-77.

[35] Stitzel JD, Pawlowski KJ, Wnek GE, Simpson DG, Bowlin GL. Arterial smooth muscle cell proliferation on a novel biomimicking, biodegradable vascular graft scaffold. J Biomat Appl 2001; 16: 22-33.

[36] Ma Z, Kotaki M, Inai R, Ramakrishna S. Potential of nanofiber matrix as tissue engineering scaffolds, Tissue Eng, 2005; 11: 101-09.

[37] Cima G, Vacanti JP, Vacanti C, Ingber D, Mooney D, Langer R. Tissue engineering by cell transplantation using degradable polymer substrates. J Biomech Eng 1991; 113: 143–51.

[38] Kweona HY, Yoo MK, Park IK, *et al.* A novel degradable poly caprolactone networks for tissue engineering. Biomat 2003; 24: 801-08.

[39] Rydholm AE, Held NL, Benoit DSW, Bowman CN, Anseth KS. Modifying Network Chemistry in Thiol-Acrylate Photopolymers through Post-Polymerization Functionalization to Control Cell-Material Interactions. J Biomed Mater Res A 2008; 86: 23-30.

[40] Tan H, Wu J, Lao L, Gao C. Gelatin/Chitosan/Hyaluronan Scaffold Integrated With PLGA Microspheres for Cartilage Tissue Engineering. Acta Biomat 2009; 5: 328-37.

[41] Mandal BB, Kundu SC. Non-bioengineered silk gland fibroin protein: characterization and evaluation of matrices for potential tissue engineering applications. Biotech & Bioeng 2008; 100: 1237-50.

[42] Mandal BB, Kundu SC. Non-bioengineered high strength three-dimensional gland fibroin scaffolds from tropical non-mulberry silkworm for potential tissue engineering applications. Macromol Biosci 2008; 8: 807-18.

[43] Mandal BB, Kundu SC. Cell proliferation and migration in silk fibroin 3D-scaffolds. Biomat 2009, 30: 2956-65.

[44] Mandal BB, Kundu SC. Osteogenic and adipogenic differentiation of rat bone marrow cells on non-mulberry and mulberry silk gland fibroin 3D-scaffolds. Biomat 2009; 30: 5019-30.

[45] Liu H, Mao J, Yao K, Yang G, Cui L, Cao Y. A Study on a Chitosan-Gelatin-Hyaluronic Acid Scaffold as Artificial Skin *in vitro* and its Tissue Engineering Applications. J Biomat Sci, Polymer Edu 2004; 15: 25-40.

[46] Xia W, Liu W, Cui L, *et al.* Tissue Engineering of Cartilage with the Use of Chitosan-Gelatin Complex Scaffolds. J Biomed Mat Res, Part B: Applied Biomat 2004; 71 B: 373-80.

[47] Shi C, Zhu Y, Ran X, Wang M, Su Y, Cheng T. Therapeutic Potential of Chitosan and its Derivatives in Regenerative Medicine. J Surgical Res 2006; 133: 185-92.

[48] Collombel C, Damour O, Gagnieu C, Poinsignon F, Echinard C, Marichy J. 1992. Biomaterials Based on Mixtures of Collagen, Chitosan and Glycosaminoglycans, Method for preparing them and their Use in Human Medicine. US Patent 5166187

[49] Lee CR, Grodzinsky AJ, Spector M. The effects of cross-linking of collagen glycosaminoglycan scaffolds on compressive stiffness, chondrocyte-mediated contraction, proliferation, and biosynthesis. Biomat 2001; 22, 3145-54.

[50] Chiara G. Porous Polymeric Bioresorbable Scaffolds for Tissue Engineering. PhD Dissertation, University of Bologna, 2010. http: //amsdottorato.unibo.it/2507/1/gualandi_chiara_tesi.pdf

[51] Chiu JJ, Chen LJ, Lee PL, *et al.* Shear stress inhibits adhesion molecule expression in vascular endothelial cells induced by co-culture with smooth muscles cells. J Blood 2003; 101: 2667-74.

[52] Nahzat SN. Thermal Analysis of Biomaterials, In: Principles and Applications of Thermal Analysis, Ed: Gabbot P. Blackwell Publishing, Oxford, UK. 2008, pp 257-85. ISBN 9780470697702.

[53] Kim BS, Baez CE, Atala A. Biomaterials for tissue engineering. World J Urol 2009; 1: 2-9.

[54] Hutmacher DW. Scaffold in tissue engineering bone and cartilage. Biomat 2000; 21: 2529-43.

[55] Marreco PR, Moreira PDL, Genari, Selma C, Moraes MA. Effects of Different Sterilization Methods on The Morphology, Mechanical Properties, and Cytotoxicity of Chitosan Membranes Used as Wound Dressings. J Biomed Mat Res, Part B: Applied Biomat 2004; 71 B: 268-77.

[56] Weinstein-Oppenheimer CR, Aceituno AR, Brown DI, *et al.* The Effect of an Autologous Cellular Gel-Matrix Integrated Implant System on Wound Healing. J Translational Med 2010; 8: 59-69.

[57] Schoof H, Apel J, Heschel I, Rau G. Control of pore structure and size in freeze dried collagen sponges. J Biomed Mater Res 2001; 58: 352-57.

[58] Yaszemski MJ. Quantitative analysis of interconnectivity of porous biodegradable scaffolds with micro computed tomography. J Biomed Mater Res Part A 2004; 71: 258-67.

[59] Shoichet MS. Polymer Scaffolds for Biomaterials Applications. Macromol 2010; 4: 581-91.

[60] Harley BAC, Kim HD, Zaman MH, Yannas IV, Lauffenburger DA, Gibson LJ. Microarchitecture of Three Dimensional Scaffolds Influences Cell Migration Behavior *via* Junction Interactions. Biophysical J 2008; 95: 4013-24.

[61] O'Brien FJ, Harley BA, Yannas IV, Gibson LJ. The effect of pore size on cell adhesion in collagen-GAG scaffolds. Biomat, 2005; 26: 433-41.

[62] Karageorgiou V, Kaplan D. Porosity of 3D-biomaterial scaffolds and, osteogenesis. Biomat 2005; 26: 5474-49.

[63] Peroglio M, Gremillard L, Gauthier C, *et al.* Mechanical properties and, cyto-compatibility of poly-(Ɛ-caprolactone)-infiltrated biphasic calcium phosphate scaffolds with bimodal pore distribution. Acta Biomat 2010; 6: 4369-79.

[64] Murphy CA, Haugh MG, O'Brien FJ. The effect of mean pore size on cell attachment, proliferation and, migration in collagen-glycosaminoglycan scaffolds for bone tissue engineering. Biomat 2010; 31: 461-66.
[65] Intranuovo F, Sardella E, Gristina R, *et al.* PE-CVD processes improve cell affinity of polymer scaffolds for tissue engineering. Surface & Coat Technol 2011; 205: S548-51.
[66] Schugens C, Maquet V, Grandfils R, Jerome R, Teyssie P. Biodegradable and, macroporous polylactide implants for cell transplantation. I. Preparation of macroporous polylactide supports by solid-liquid phase separation. Polymer 1996; 37: 1027-38.
[67] Elisseeff J, McIntosh W, Fu K, Blunk R, Langer R. Controlled-release of IGF-I and, TGF-b1 in a photo-polymerizing hydrogel for cartilage tissue engineering. J Orthop Res 2001; 19: 1098-04.
[68] Levenberg S, Huang NF, Lavik E, Rogers AB, Itskovitz-Eldor J, Langer R. Differentiation of human embryonic stem cells on three-dimensional polymer scaffolds. Proc Natl Acad Sci USA 2003; 100: 12741-46.
[69] Shea LD, Smiley E, Bonadio J, Mooney DJ. DNA delivery from polymer matrices for tissue engineering. Nat Biotechnol 1999; 17: 551-54.
[70] Yim EKF, Leong KW. Proliferation and, differentiation of human embryonic germ cell derivatives in bioactive polymeric fibrous scaffold. J Biomat Sci Polym Ed 2005; 16: 1193-17.
[71] Semino CE, Merok JR, Crane GG, Panagiotakos G, Zhang SG. Functional differentiation of hepatocyte-like spheroid structures from putative liver progenitor cells in three-dimensional peptide scaffolds. Differentiation 2003; 71: 262-70.
[72] Silva GA, Czeisler C, Niece KL, *et al.* Selective differentiation of neural progenitor cells by high epitope density nanofibers. Science 2004; 303: 1352-55.
[73] Peppas NA, Hilt JZ, Khademhosseini A, Langer R. Hydrogels in biology and, medicine: From molecular principles to bio-nanotechnology. Adv Mater 2006; 18: 1345-60.
[74] Schnaper HW, Grant DS, Stetlerstevenson WG, *et al.* Type IV collagenase(s) and, TIMPs modulate endothelial cell morphogenesis *in vitro*. J Cell Physiol 1993; 156: 235-46.
[75] Petersen OW, Ronnovjessen L, Howlett AR, Bissell MJ. Interaction with basement-membrane serves to rapidly distinguish growth and differentiation pattern of normal and, malignant human breast epithelial cells. Proc Natl Acad Sci USA 1992; 89: 9064-68.
[76] Tanaka H, Murphy CL, Murphy C, Kimura M, Kawai S, Polak JM. Chondrogenic differentiation of murine embryonic stem cells: Effects of culture conditions and, dexamethasone. J Cell Biochem 2004; 93: 454- 62.
[77] Ito A, Mase A, Takizawa Y, *et al.* Transglutaminase-mediated gelatin matrices incorporating cell adhesion factors as a biomaterial for tissue engineering. J Ortho 2003; 95: 196-99.
[78] Ma PX. Scaffolds for tissue fabrication. Mater Today 2004; 7: 30-40.
[79] Li J, Shi R. Fabrication of patterned multi-walled poly-l-lactic acid conduits for nerve regeneration. J Neurosci Methods 2007; 165: 257-64.
[80] Birgersdotter A, Sandberg R, Ernberg I. Gene expression perturbation *in vitro* - a growing case for three-dimensional (3D) culture systems. Semin Cancer Biol 2005; 15: 405-12.
[81] Rothan HA, Djordjevic I, Bahrani H, *et al.* Three-Dimensional Culture Environment Increases the Efficacy of Platelet Rich Plasma Releasate in Prompting Skin Fibroblast Differentiation and Extracellular Matrix Formation. Int J Med Sci 2014; 11: 1029-38.
[82] Gieni RS, Hendzel MJ. Mechano-transduction from the ECM to the genome: Are the pieces now in place? J Cell Biochem 2008; 104: 1964-87.
[83] Ashe HL, Briscoe J. The interpretation of morphogen gradients. Development 2006; 133: 385-94.
[84] Le Beyec J, Xu R, Lee SY, *et al.* Cell shape regulates global histone acetylation in human mammary epithelial cells. Exp Cell Res 2007; 313: 3066-75.
[85] Chen CS, Mrksich M, Huang S, Whitesides GM, Ingber DE. Geometric control of cell life and, death. Science 1997; 276: 1425-28.
[86] Hamill OP, McBride DW Jr. Induced membrane hypo/hyper-mechano-sensitivity: a limitation of patch-clamp recording. Annu Rev Physiol 1997; 59: 621-31.
[87] Martinac B. Mechano-sensitive ion channels: molecules of mechano-transduction. J. Cell Sci 2004; 117: 2449-60.
[88] Hayakawa K, Tatsumi H, Sokabe M. Actin stress fibers transmit and, focus force to activate mechano-sensitive channels. J Cell Sci 2008; 121: 496-03.

[89] Sukharev S, Corey DP. Mechano-sensitive channels: multiplicity of families and, gating paradigms. Science. STKE 2004; re.4, pp 1-24. DOI: 10.1126/stke.2192004re4
[90] Kobayashi T, Sokabe M. Sensing substrate rigidity by mechano-sensitive ion channels with stress fibers and, focal adhesions. Curr Opin Cell Biol 2010; 22: 669-76.
[91] Kim TJ, Seong JH, Ouyang MX, *et al.* Substrate rigidity regulates Ca2+ oscillation *via* RhoA pathway in stem cells. J Cell Physiol 2009; 218: 285-93.
[92] Daley WP, Peters SB, Larsen M. Extracellular matrix dynamics in development and, regenerative medicine. J Cell Sci 2008, 121: 255-64.
[93] Lee SH, Moon JJ, West JL. Three-dimensional micro-patterning of bioactive hydrogels *via* two-photon laser scanning photolithography for guided 3D-cell migration. Biomat 2008; 29: 2962-68.
[94] Salinas CN, Anseth KS. The enhancement of chondrogenic differentiation of human mesenchymal stem cells by enzymatically regulated RGD functionalities. Biomat 2008; 29: 2370-77.
[95] Mann BK, West JL. Cell adhesion peptides alter smooth muscle cell adhesion, proliferation, migration, and, matrix protein synthesis on modified surfaces and, in polymer scaffolds. J Biomed Mater Res 60: 86-93.
[96] West JL, Hubbell JA. 1999. Polymeric biomaterials with degradation sites for proteases involved in cell migration. Macromol 2002; 32: 241-44.
[97] Cushing MC, Anseth KS. Hydrogel Cell Cultures. Science 2007; 316: 1133-34.
[98] Wang N, Butler JP, Ingber DE. Mechano-transduction across the cell surface and, through the cytoskeleton. Science 1993; 260: 1124-27.
[99] Giancotti FG, Ruoslahti E. Integrin signaling. Science 1999; 285: 1028-33.
[100] Howe A, Aplin AE, Alahari SK, Juliano RL. Integrin signaling and, cell growth control. Curr Opin Cell Biol 1998; 10: 220-31.
[101] Weber LM, Hayda KN, Anseth KS. Cell-matrix interactions improve b-cell survival and, insulin secretion in three-dimensional culture. Tissue Eng A 2008; 14: 1959-68.
[102] Yan S, Leong KF, Du Z, Chua CK. The design of scaffolds for use in tissue engineering. Part II. Rapid prototyping techniques. Tissue Eng 2002; 8: 1-11.
[103] Sun W, Starly B, Darling A, Gomez C. Computer-aided tissue engineering: application to biomimetic modelling and design of tissue scaffolds. Biotechnol Appl Biochem 2004; 39: 49-58.
[104] Hutmacher DW, Sittinger M, Risbud MV. Scaffold-based tissue engineering: rationale for computer-aided design and solid free-form fabrication systems. Trends Biotechnol 2004; 22: 354-62.
[105] Fang Z, Starly B, Sun W. Computer-aided characterization for effective mechanical properties of porous tissue scaffolds. Comput Aid Design 2005; 37: 65-72.
[106] Cheah CM, Chua CK, Leong KF, Cheong CH, Naing M W. Automatic algorithm for generating complex polyhedral scaffold structures for tissue engineering. Tissue Eng 2004; 10: 595-10.
[107] Elisseeff J. Hydrogels, Structure Starts to Gel. Nat Mater 2008; 7: 271-73.
[108] Hwang NS, Varghese S, Elisseeff J. Cartilage Tissue Engineering: Directed differentiation of embryonic stem cells in three dimensional hydrogel cultures. Methods Mol Biol 2007; 407: 351-73.
[109] Hwang NS, Varghese S, Elisseeff J. Controlled differentiation of stem cells. Adv Drug Deliv Rev 2008; 60: 199-14.
[110] Hollister SJ, Levy RA, Chu TM, Halloran JW, Feinberg SE. An image-based approach for designing and manufacturing craniofacial scaffolds. Int J Oral Maxillofac Surg 2002; 29: 67-71.
[111] Lin CY, Kikuchi N, Hollister SJ. A novel method for biomaterial scaffold internal architecture design to match bone elastic properties with desired porosity. J Biomech 2004; 37: 623-36.
[112] Anyaejia CI, Bassera PJ, Horkay F. Cartilage: Biomimetic Study of the Extracellular Matrix, MRS Proceedings, Vol 1622, Symposium E - Fundamentals of Gels and Self-Assembled Polymer Systems, ©Materials Research Society, 2014. DOI: http: //dx.doi.org/10.1557/opl.2014.55
[113] Kuboki Y, Jin Q, Takita H. Geometry of carriers controlling phenotypic expression in BMP-induced osteogenesis and chondrogenesis. J Bone Joint Surg Am 2001; 83A, Suppl 1 (PII): S105-115.
[114] Kima B, Peppas NA. Analysis of molecular interactions in poly (methacrylic acid-g-ethylene glycol) hydrogels. Polymer 2003; 44: 3701-07.
[115] Deans TL, Elisseeff J. World Stem Cell Summit, Detroit, MI, USA, 2010; Oct.60 - Nov. 6, pp 312-101, www.World stemcellsummit.com, www.genpol.org
[116] Badylak SF. The extracellular matrix as a scaffold for tissue reconstruction. Semin Cell Dev Biol, 2002, 13: 377-83.

[117] Nehrer S. Matrix collagen type and, pore size influence behavior of seeded canine chondrocytes. Biomat 1997; 18: 769-76.
[118] DeLustro F, Dasch J, Keefe J, Ellingsworth L. Immune responses to allogeneic and xenogeneic implants of collagen and collagen derivatives. Clin Orthop Relat Res 1990; 260: 263-79.
[119] Hwang NS, Varghese S, Elisseeff J. Derivation of chondrogenically-committed cells from human embryonic cells for cartilage tissue regeneration. PLoS One 2008; 3: p. e2498.
[120] Lee HJ. Collagen mimetic peptide-conjugated photo-polymerizable PEG hydrogel. Biomat 2006; 27: 5268-76.
[121] Lee HJ. Enhanced chondrogenesis of mesenchymal stem cells in collagen mimetic peptide-mediated micro environment. Tissue Eng A 2008; 14: 1843-51.
[122] Mueller MB, Tuan RS. Functional characterization of hypertrophy in chondrogenesis of human mesenchymal stem cells. Arthritis Rheum 2008; 58: 1377-88.
[123] Ruoslahti E. RGD and, other recognition sequences for integrins. Annu Rev of Cell Dev Biol 1996; 12: 697-715.
[124] Lutolf MP, Hubbell JA. Synthetic biomaterials as instructive extracellular microenvironments for morphogenesis in tissue engineering. Nat Biotechnol 2005; 23: 47-55.
[125] Nuttelman CR, Tripodi MC, Anseth KS. Synthetic hydrogel niches that promote hMSC viability. Matrix Biol 2005; 24: 208-18.
[126] Park KH, Na K, Chung HM. Enhancement of the adhesion of fibroblasts by peptide containing an Arg-Gly-Asp sequence with poly (ethylene glycol) into a thermo-reversible hydrogel as a synthetic extracellular matrix. Biotechnol Lett 2005; 27: 227-31.
[127] Patel PN, Gobin AS, West JL, Patrick CW. Poly (ethylene glycol) hydrogel system supports pre-adipocyte viability, adhesion, and, proliferation. Tissue Eng 2005; 11: 1498-05.
[128] Massia SP, Hubbell JA. An RGD spacing of 440 nm is sufficient for integrin αVβ3-mediated fibroblast spreading and, 140 nm for focal contact and, stress fiber formation. J Cell Biol 1991; 114: 1089-00.
[129] Gobin AS, West JL. Cell migration through defined, synthetic extracellular matrix analogues. FASEB J 2002; 16: 751-60.
[130] Khire VS, Benoit DSW, Anseth KS, Bowman CN. Ultrathin gradient films using thiol-ene polymerizations. J Polym Sci Part A Polym Chem 2006; 44: 7027-39.
[131] Hoyle CE, Bowman CN. Thiol-ene Click Chemistry. Angew Chem Int Ed Engl 2010; 49: 1540-73.
[132] Salinas CN, Anseth KS. Mixed mode thiol-acrylate photopolymerizations for the synthesis of PEG-peptide hydrogels. Macromol 2008; 41: 6019-26.
[133] Rigogliuso S, Pavia FC, Brucato V, *et al.* Use of Modified 3D-Scaffolds to Improve Cell Adhesion and Drive Desired Cell Responses. Chemical Eng Trans 2012; 27: 1-6.
[134] Kim DH, Lipke EA, Kim P, *et al.* Nanoscale cues regulate the structure and function of macroscopic cardiac tissue constructs. Proc Natl Acad Sci, USA 2010; 107: 565-70.
[135] Stevens, MM, George JH. Exploring and engineering the cell surface interface. Science 2005; 310, 1135-38.
[136] Heino J, Kapyla J. Cellular receptors of extracellular matrix molecules. Curr Pharm Des 2009; 15: 1309-17.
[137] Teixeira AI, Abrams GA, Bertics PJ, Murphy CJ, Nealey PF. Epithelial contact guidance on well-defined micro- and nanostructured substrates. J Cell Sci 2003; 116: 1881-92.
[138] Bettinger CJ, Langer R, Borenstein JT. Engineering substrate topography at the micro- and nanoscale to control cell function. Angew Chem Int Ed 2009; 48: 5406-15.
[139] Lu P, Takai K, Weaver VM, Werb Z. Extracellular matrix degradation and remodeling in development and disease. Cold Spring Harb Perspect Biol. 2011 Dec 1; 3(12). pii: a005058. DOI: 10.1101/cshperspect.a005058.
[140] Freeman I, Kedem A, Cohen S. The effect of sulfation of alginate hydrogels on the specific binding and controlled release of heparin-binding proteins. Biomat 2008; 29: 3260-68.
[141] Re'em T, Tsur-Gang O, Cohen S. The effect of immobilized RGD peptide in macroporous alginate scaffolds on TGFbeta1-induced chondrogenesis of human mesenchymal stem cells. Biomat 2010; 31: 6746-55.
[142] Jeon O, Powell C, Ahmed SM, Alsberg E. Biodegradable, photocrosslinked alginate hydrogels with independently tailorable physical properties and cell adhesivity. Tissue Eng A 2010; 16: 2915-25.

[143] Schofer MD, Boudriot U, Bockelmann S, *et al.* Effect of direct RGD incorporation in PLLA nanofibres on growth and osteogenic differentiation of human mesenchymal stem cells. J Mater Sci Mater Med 2009; 20: 1535-40.
[144] Good MC, Zalatan JG, Lim WA. Scaffold Proteins: Hubs for Controlling the Flow of Cellular Information. Science 2011; 332: 680-86.
[145] Hao N, Nayak S, Behar M, *et al.* Regulation of Cell Signaling Dynamics by the Protein Kinase-Scaffold Ste5. Molecular Cell 2008; 30: 649-56.
[146] Basu K, Liu X. Mathematical modeling for multisite phosphorylation with scaffold binding in cell signaling. Math Meth Appl Sci 2013; 1-9. DOI: 10.1002/mma.2864.
[147] Burack WR, Shaw AS. Signal transduction: hanging on a scaffold. Cell Regulation. Current Opinion in Cell Biology 2000; 12: 211-16.
[148] Shaw AS, Filbert EL. Scaffold proteins and immune-cell signaling. Nature Reviews, Immunology 2009; 9: 47-56.
[149] Hollister SJ. Porous scaffold design for tissue engineering. Nature Materials 2005; 4: 518-24.
[150] Lima MJ, Correlo VM, Reis RL. Micro/nano replication and 3D-assembling techniques for scaffold fabrication. Materials Sci Eng C 2014; 42: 615-21.
[151] Izadifar Z, Chen X, Kulyk W. Strategic Design and Fabrication of Engineered Scaffolds for Articular Cartilage Repair. J Funct Biomater 2012; 3: 799-38.
[152] Pan Z, Ding J. Poly (lactide-co-glycolide) porous scaffolds for tissue engineering and regenerative medicine. Interface Focus 2012; 2: 366-77.
[153] Sell SA, Wolfe PS, Garg K, McCool JM, Rodriguez IA, Bowlin GL. The Use of Natural Polymers in Tissue Engineering: A Focus on Electrospun Extracellular Matrix Analogues. Polymers 2010; 2: 522-53.
[154] Mirzaei E, Faridi-Majidi R, Shokrgozar MA, Paskiabi FA. Genipin cross-linked electrospun chitosan-based nanofibrous mat as tissue engineering scaffold. Nanomed J 2014; 1: 137-46.
[155] Nguyen T-H, Lee B-T. Fabrication and characterization of cross-linked gelatin electro-spun nano-fibers. J Biomed Sci Eng 2010; 3: 1117-24.
[156] Ma PX, Langer R. Fabrication of biodegradable polymer foams for cell transplantation and tissue engineering. In: Tissue Engineering Methods and Protocols. Ed: Morgan J, Yarmush M. Humana Press, NJ. 1999; p 47.
[157] Lu L, Peter S, Lyman M, *et al. In vitro* and *in vivo* degradation of porous poly (DL-lactic-co-glycolic acid) foams. Biomat 2000; 21: 1595-05.
[158] Sachlos E, Czernuszka JT. Making tissue engineering scaffolds work. Review on the application of solid free from fabrication technology to the production of tissue engineering scaffolds. Euro Cells Mater 2003; 5: 29-40.
[159] Smith LA, Liu X, Ma PX. Tissue Engineering with Nano-Fibrous Scaffolds. Soft Matter 2008; 4(11): 2144–2149.
[160] Laurencin CT, Ambrosio AMA, Borden MD, Cooper JA. Tissue engineering: orthopedic applications. Annu Rev Biomed Eng 1999; 1: 19-46.
[161] Lavik E, Langer R. Tissue engineering: current state & perspective. Appl Microbiol Biotechnol 2004; 65: 1-8.
[162] Hutmacher DW. Scaffold design and fabrication technologies for engineering tissues- state of the art and future perspectives. J Biomat Sci Polym Edu 2001; 12: 107-24.
[163] Subia B, Kundu J, Kundu SC. In: Biomaterial Scaffold Fabrication Techniques for Potential Tissue Engineering Applications, Tissue Engineering, Ed: Eberli D. InTech 2010; ISBN: 978-953-307-079-7.
[164] Chen AA, Tsang VL, Albrecht DR, Bhatia SN. 3-D Fabrication Technology for Tissue Engineering. BioMEMS and Biomedical Nanotechnology, Vol III: Therapeutic Micro/Nanotechnology. Ed: Ferrari M, Desai T, Bhatia SN. 2007; XXIV, p 376. Springer. ISBN: 978-0-387-25565-1 (accessed 4/12/2014).
[165] Partap S, Lyons F, O'Brien FJ. IV.1. Scaffolds & Surfaces. Stud Health Technol Inform. 2010; 152:187-201.
[166] Lu T, Li Y, Chen T. Techniques for fabrication and construction of three-dimensional scaffolds for tissue engineering. International J Nanomed 2013; 8: 337-50.

[167] Jin GR, Prabhakaran MP, Ramakrishna S. Stem cell differentiation to epidermal lineages on electrospun nanofibrous substrates for skin tissue engineering. Acta Biomater 2011; 7: 3113-22.
[168] Chandrasekaran AR, Venugopal J, Sundarrajan S, Ramakrishna S. Fabrication of a nanofibrous scaffold with improved bioactivity for culture of human dermal fibroblasts for skin regeneration. Biomed Mater 2011; 6: 015001.
[169] Megelski S, Stephens JS, Chase DB, Rabolt JF. Micro- and nanostructured surface morphology on electrospun polymer fibers. Macromol 2002; 35: 8456-66.
[170] Shin HJ, Lee CH, Cho IH, *et al.* Electrospun PLGA nanofiber scaffolds for articular cartilage reconstruction: mechanical stability, degradation and cellular responses under mechanical stimulation *in vitro*. J Biomat Sci Polym Edu 2006; 17: 103-19.
[171] You Y, Youk JH, Lee SW, Min BM, Lee SJ, Park WH. Preparation of porous ultrafine PGA fibers *via* selective dissolution of electrospun PGA/PLA blend fibers. Mater Lett 2006; 60: 757-60.
[172] Tillman BW, Yazdani SK, Lee SJ, Geary RL, Atala A, Yoo JJ. The *in vivo* stability of electrospun polycaprolactone-collagen scaffolds in vascular reconstruction. Biomat 2009; 30: 583-88.
[173] Yang W, Fu J, Wang D, *et al.* Study on chitosan/polycaprolactone blending vascular scaffolds by electrospinning. J Biomed Nanotech 2010; 6: 254-59.
[174] Cooper A, Bhattarai N, Zhang MQ. Fabrication and cellular compatibility of aligned chitosan-PCL fibers for nerve tissue regeneration. Carbohydr Polym 2011; 85: 149-56.
[175] Wei K, Li Y, Kim KO, *et al.* Fabrication of nano-hydroxyapatite on electrospun silk fibroin nanofiber and their effects in osteoblastic behavior. J Biomed Mater Res A 2011; 97A: 272-80.
[176] Ma PX, Zhang RY. Synthetic nano-scale fibrous extracellular matrix. J Biomed Mater Res 1999; 46: 60-72.
[177] Ma PX, Zhang RY, Xiao GZ, Franceschi R. Engineering new bone tissue *in vitro* on highly porous poly(alpha-hydroxyl acids)/hydroxyapatite composite scaffolds. J Biomed Mater Res 2001; 54: 284-93.
[178] Zhao J, Han W, Chen H, *et al.* Preparation, structure and crystallinity of chitosan nano-fibers by a solid-liquid phase separation technique. Carbohyd Poly 2011; 83: 1541-46.
[179] Nishi T, Wang TT, Kwei TK. Thermally induced phase separation behavior of compatible polymer mixtures. Macromol 1975; 8: 227-34.
[180] Niu XF, Li X, Liu H, et *al.* Homogeneous chitosan/poly-(L-lactide) composite scaffolds prepared by emulsion freeze-drying. J Biomater Sci Polym Edu 2012; 23: 391-404.
[181] Boyde A, Franc F. Freeze-drying shrinkage of glutaraldehyde fixed liver. J Microsc 1981; 122: 75-86.
[182] Haugh MG, Murphy CM, O'Brien FJ. Novel freeze-drying methods to produce a range of collagen-glycosaminoglycan scaffolds with tailored mean pore sizes. Tissue Eng Part C Methods 2010; 16: 887-94.
[183] Dragusin D-M, Van Vlierberghe S, Dubruel P, *et al.* Novel gelatin-PHEMA porous scaffolds for tissue engineering applications. Soft Matter 2012; 8: 9589-02.
[184] Liu JP, Song H, Zhang LL, Xu HY, Zhao XJ. Self-assembly-peptide hydrogels as tissue-engineering scaffolds for three-dimensional culture of chondrocytes *in vitro*. Macromol Biosci 2010; 10: 1164-70.
[185] Castelletto V, Hamley IW, Adamcik J, Mezzenga R, Gummel J. Modulating self-assembly of a nanotape-forming peptide amphiphile with an oppositely charged surfactant. Soft Matter 2012; 8: 217-26.
[186] Stupp SI. Self-assembly and biomaterials. Nano Lett 2010; 10: 4783-86.
[187] Hamley IW. Self-assembly of amphiphilic peptides. Soft Matter 2011; 7: 4122-38.
[188] Kumaraswamy P, Lakshmanan R, Sethuraman S, Krishnan UM. Self-assembly of peptides: influence of substrate, pH and medium on the formation of supramolecular assemblies. Soft Matter 2011; 7: 2744-54.
[189] Zhang SG, Lockshin C, Herbert A, Winter E, Rich A. Zuotin, a putative Z-DNA binding-protein in Saccharomyces cerevisiae. EMBO J 1992; 11: 3787-96.
[190] Holmes TC. Novel peptide-based biomaterial scaffolds for tissue engineering. Trends Biotechnol 2002; 20: 16-21.
[191] Mikos AG, Sarakinos G, Vacanti JP, Langar RS, Cima LG.1996. Biocompatible polymer membranes and methods of preparation of three dimensional membrane structures. US Patent 5,514,378.
[192] Mikos AG, Temenoff J. Formation of highly porous biodegradable scaffolds for tissue engineering. Electron J Biotechnol 2000; 3: 1-6.

[193] Plikk P, Målberg S, Albertsson AC. Design of resorbable porous tubular copolyester scaffolds for use in nerve regeneration. Biomacromol 2009; 10: 1259-64.

[194] Mano JF, Silva GA, Azevedo HS, *et al.* Natural origin biodegradable systems in tissue engineering and regenerative medicine: present status and some moving trends. J R Soc Interface 2007; 4: 999-30.

[195] Hou QP, Grijpma DW, Feijen J. Porous polymeric structures for tissue engineering prepared by a coagulation, compression moulding and salt leaching technique. Biomat 2003; 24: 1937-47.

[196] Mikos AG, Sarakinos G, Leite SM, Vacanti JP, Langer R. Laminated three dimensional bio degradable foams for use in tissue engineering. Biomat 1993; 14: 323-30.

[197] Ikada Y. Scope of Tissue Engineering, In: Tissue engineering: Fundamental and Applications. 2006; pp 29, Academic Press, USA.

[198] Huang YC, Mooney DJ. Gas foaming to fabricate polymer scaffolds in tissue engineering. In: Scaffoldings in tissue engineering, Ed: Ma X P, Elisseeff J. Taylor & Francis. CRC Press, 2005; p 159.

[199] Zhang S. Fabrication of novel biomaterial through molecular self assembly. J Nat Biotechnol 2003; 21: 1171-78.

[200] Zhang S, Zhao X, Spirio L. PuraMatrix: Self assembling peptide nanofiber scaffolds, In: Scaffolding in Tissue Engineering, Ed: Ma PX, Elisseeff J. 2006; pp 217-236. CRC Press, Boca Raton, FL, USA.

[201] Tambralli A, Blakeney B, Anderson J, *et al.* A hybrid biomimetic scaffold composed of electrospun poly-caprolactone nanofibers and self-assembled peptide amphiphile nanofibers. Biofab 2009; 1. 025001, 11.

[202] Matson JB, Stupp SI. Self-assembling peptide scaffolds for regenerative medicine. Chem Comm (Camb) 2012; 48: 26-33.

[203] Liu GJ, Ding JF, Qiao LJ, *et al.* Polystyrene block- poly (2-cinnamoylethyl methacrylate) nanofibers: Preparation, characterization, and liquid crystalline properties. Chem Eur J 1999; 5: 2740-49.

[204] Leong MF, Rasheed MZ, Lim TC, Chian KS. *In vitro* cell infiltration and *in vivo* cell in filtration and vascularization in fibrous highly porous poly (D, L-Lactic acid) scaffold fabrication by electro-spining technique. J Biomed Res A 2008; 91: 231-40.

[205] Liang D, Hsiao BS, Chu B. Functional electrospun nanofibrous scaffolds for biomedical applications. Adv. Drug Deliv Rev 2007; 59: 1392-12.

[206] Chen H, Truckenmüller R, Van Blitterswijk C, Moroni L. Fabrication of nanofibrous scaffolds for tissue engineering applications. University of Twente, The Netherlands. DOI: 10.1533/9780857097231.1.158; Woodhead Online Publishing 2013; 158-183. Accessed at *doc.utwente.nl/88370/1/fulltext.pdf*

[207] Hua FJ, Kim GE, Lee JD, Son YK, Lee DS. Macroporous poly (l- lactide) scaffold 1. Preparation of a macroporous scaffold by liquid-liquid phase separation of a PLLA- dioxane- water system. J Biomed Mater Res 2002; 63: 61-67.

[208] Kai H, Wang X, Madhukar KS, *et al.* Fabrication of two-level tumor bone repair biomaterial based on rapid prototyping technique. Biofab 2009; 1(2) 025003. DOI: 10.1088/1758-5082/1/2/025003.

[209] Woodfield TB, Guggenheim M, von Rechenberg B, Riesle J, van Blitterswijk CA, Wedler V. Rapid prototyping of anatomically shaped, tissue engineered implants for restoring congruent articulating surfaces in small joints. Cell Prolif 2009; 4: 485-97.

[210] Chen G, Ushida T, Tateishi T. Development of biodegradable porous scaffolds for tissue engineering. J. Mater Sci Eng C 2002; 17: 63-69.

[211] Martins AM, Pham QP, Malafaya PB, *et al.* The Role of Lipase and alpha-Amylase in the Degradation of Starch/Poly(varepsilon-Caprolactone) Fiber Meshes and the Osteogenic Differentiation of Cultured Marrow Stromal Cells. Tissue Eng Part A 2009; 15: 295-05.

[212] Mooney DJ, McNamara K, Hern D, Vacanti JP, Langer R. Stabilized polyglycolic acid fibre-based tubes for tissue engineering. Biomat 1996; 17(2): 115-24.

[213] Moroni L, Hamann D, Paoluzzi L, Pieper J, de Wijn JR, van Blitterswijk CA. Regenerating articular tissue by converging technologies. PLoS One 3: e3032.

[214] Mikos AG, Bao Y, Cima LG, Ingber DE, Vacanti JP, Langer R. Preparation of Poly (glycolic acid) bonded fiber structures for cell attachment and transplantation. J Biomed Mat Res 1993; 27: 183-89.

[215] Thompson RC, Wake MC, Yaszemski MJ, Mikos AG. Biodegradable polymer scaffolds to regenerate organs. Biopolymers II. Adv Polymer Sci 1995; 122: 245-74.

[216] Lee SH, Kim BS, Kim SH, Kang SW, Kim YH. Thermally produced biodegradable scaffolds for cartilage tissue engineering. Macromol Biosci 2004; 4: 802-10.

[217] Maquet V, Jerome R. Design of macroporous biodegradable polymer scaffolds for cell transplantations, Mater Sci Forum 1997; 250: 5-24.

[218] Boland ED, Espy PG, Bowlin GL. Tissue engineering scaffolds. In: Encyclopedia of Biomaterials and biomedical engineering. Ed: Wenk GE, Bowlin GL. 2004; pp 1633-1635. Richmond, Virginia, USA.

[219] Lin L, Ju S, Cen L, Zhang H, Hu Q. Fabrication of porous, TCP scaffolds by combination of rapid prototyping and freeze drying technology. Ed: Peng Y, Weng X. 2008; pp 88-91, APCMBE 2008, IFMBE Proceedings 19, Springer-Verlag, Berlin Heidelberg, Germany.

[220] Egea AMC, Vieu C. Microcontact printing of biomolecular gratings from SU-8 masters duplicated by Thermal Soft UV NIL, Microelectron Eng 2011; 88: 1935-38.

[221] Giannitelli SM, Abbruzzese F, Mozetic P, *et al.* Surface decoration of electrospun scaffolds by micro-contact printing. Asia Pac J Chem Eng 2014; 401-06.

[222] Poon JKS, Huang Y, Paloczi GT, Yariv A. Soft lithography replica molding of critically coupled polymer microring resonators, IEEE Photonics Technol Lett 2004; 16: 2496-98.

[223] Suh K-Y, Park MC, Kim P. Capillary force lithography: a versatile tool for structured biomaterials interfaces towards cell and tissue engineering. Adv Funct Mater 2009; 19: 2699-12.

[224] Wang G-J, Hsueh C-C, Hsu S-H, Hung S-H. Fabrication of PLGA micro-vessel scaffolds with circular microchannels using soft lithography. J Micromech Microeng 2007; 17(10): 2000-05.

[225] Chou SY, Krauss PR. Imprint lithography with sub-10 nm feature size and high throughput, Microelectron Eng 1997; 35: 237-40.

[226] Schvartzman M, Wind SJ. Robust pattern transfer of nano-imprinted features for sub-5-nm fabrication, Nano Lett 2009; 9: 3629-34.

[227] Peng L, Deng Y, Yi P, Lai X. Micro hot embossing of thermoplastic polymers: a review, J Micromech Microeng 2014; 24: 013001.

[228] Ifty A. Biomedical Materials Science. Hot Topics in Biomaterials, Future Science Book Series, Future Science Ltd 2014; pp 2-4.

[229] Elsayed M, Merkel OM. Nano-imprinting of topographical and 3D-cell culture scaffolds. Nanomed 2014; 9: 349-66.

[230] Borenstein JT, Terai H, King KR, Weinberg EJ, Kaazempur-Mofrad MR, Vacanti JP. Micro-fabrication technology for vascularized tissue engineering. Biomed Microdev 2002; 4: 167-75.

[231] Yang S, Leong KF, Du Z, Chua CK. The design of scaffolds for use in tissue engineering. Part II. Rapid prototyping techniques. Tissue Eng 2002; 8: 1-11.

[232] Leong KF, Cheah CM, Chua CK. Solid freeform fabrication of three-dimensional scaffolds forengineering replacement tissues and organs. Biomat 2003; 24: 2363-78.

[233] Lee G. Biocompatibility of SLS-formed Calcium Phosphate Implants. Proceedings of Solid Free form Fabrication Symposium, Austin, TX: University of Texas, TX, USA. 1996; pp 15-22.

[234] Zein I, Hutmacher DW, Tan KC, Teoh SH. Fused deposition modeling of novel scaffold architectures for tissue engineering applications. Biomat 2002; 23: 1169-85.

[235] Landers R, Hübner U, Schmelzeisen R, Mülhaupt R. Rapid prototyping of scaffolds derived from thermos-reversible hydrogels and tailored for applications in tissue engineering. Biomat 2002; 23: 4437-47.

[236] Jacobs PF. Fundamental Processes, In: Rapid Prototyping & Manufacturing: Fundamentals of Stereolithography, Jacobs PF & Reid DT (Eds). Dearborn, Michigan, Society of Manufacturing Engineers, 1992, pp 79-110. Melchels FPW, Bertoldi K, Gabbreilli R, Velders AH, Feijan J, Grijpma DW. Mathematically defined tissue engineering scaffold architectures prepared by stereolithography. Biomaterials 2010; 31: 6909-6916.

[237] Park A, Wu B, Griffith LG. Integration of surface modification and 3D-fabrication techniques to prepare patterned poly (L-lactide) substrates allowing regionally selective cell adhesion. J Biomater Sci Polym Edu 1998; 9: 89-110.

[238] Kim SS, Utsunomiya H, Koski JA, et *al.* Survival and function of hepatocytes on a novel three-dimensional synthetic biodegradable polymer scaffold with an intrinsic network of channels. Ann Surg 1998; 228: 8-13.

[239] Zeltinger J, Sherwood JK, Graham DA, Müeller R, Griffith LG. Effect of pore size and void fraction on cellular adhesion, proliferation, and matrix deposition. Tissue Eng 2001; 7: 557-72.
[240] Sachlos E, Reis N, Ainsley C, Derby B, Czernuszka JT. Novel collagen scaffolds with predefined internal morphology made by solid free form fabrication. Biomat 2003; 24: 1487-97.
[241] Mironov V, Boland T, Trusk T, Forgacs G, Markwald RR. Organ printing: Computer-aided jet-based 3D-tissue engineering. Trends Biotechnol 2003; 21: 157-61.
[242] Xu T, Jin J, Gregory C, Hickman JJ, Boland T. Inkjet printing of viable mammalian cells. Biomat, 2005, 26: 93-99.
[243] Odde DJ, Renn MJ. Laser-guided direct writing of living cells. Biotech Bioeng 2000; 67: 312-18.
[244] Liu VA, Bhatia SN. Three-dimensional photo patterning of hydrogels containing living cells. Biomed Microdev 2002; 4: 257-66.
[245] Liu E, Treiser MD, Johnson PA, *et al.* Quantitative Biorelevant Profiling of Material Microstructure within 3D-Porous Scaffolds *via* Multi-photon Fluorescence Microscopy. J Biomed Mat Res Part B: App Biomat 2007; 82B: 284-97.
[246] Nair A, Shen J, Thevenot P, *et al.* Enhanced intra-tumoral uptake of quantum dots concealed within hydrogel nanoparticle. Nanotech 2008; 19: 485102. DOI: 10.1088/0957-4484/19/48/485102.
[247] England CG, Priest T, Zhang G, et *al.* Enhanced penetration into 3D-cell culture using two and three layered gold nanoparticles. Int J Nanomed 2013; 8: 3603-17.
[248] van de Ven AL, Wu M, Lowengrub J, *et al.* Integrated intravital microscopy and mathematical modeling to optimize nanotherapeutics delivery to tumors. AIP Adv 2012; 2: 11208. DOI 10.1063/1.3699060.
[249] Foster B. Focus on Microscopy: A Technique for Imaging Live Cell Interactions and Mechanisms. Am Lab Nov 2004, pp 21-27.
[250] Acosta MA, Ymele-Leki P, Kostov YV, Leach JB. Fluorescent micro-particles for sensing cell microenvironment oxygen levels within 3D-scaffolds. Biomat 2009; 30: 3068-74.
[251] Liu Y, Wang S. 3D-inverted opal hydrogel scaffolds with oxygen sensing capability. Colloids & Surfaces B: Biointerfaces 2007; 58: 8-13.
[252] Saha K, Pollock JF, Schaffer DV, Healy KE. Designing synthetic materials to control stem cell phenotype. Curr Opin Chem Biol 2007; 11: 381-87.
[253] Shu XZ, Ahmad S, Liu YC, Prestwich GD. Synthesis and, evaluation of injectable, *in situ* cross-linkable synthetic extracellular matrices for tissue engineering. J Biomed Mater Res A 2006; 79: 902-12.
[254] Drury JL, Mooney DJ. Hydrogels for tissue engineering: scaffold design variables and applications. Biomaterials. 2003; 24:4337-4351.
[255] Ehrbar M, Sala A, Lienemann P, *et al.* Elucidating the role of matrix stiffness in 3D-cell migration and remodeling. Biophys J 2011; 100: 284-93.
[256] Griffith LG, Naughton G. Tissue engineering- current challenges and, expanding opportunities. Science 2002; 295: 1009-14.
[257] Hartgerink JD, Beniash E, Stupp SI. Self-assembly and, mineralization of peptide-ampiphile nanofibers. Science 2001; 294: 1684-88.
[258] Butcher JT, Nerem RM. Porcine aortic valve interstitial cells in three dimensional cultures: Comparison of phenotype with aortic smooth muscle cells. J Heart Valve Dis 2004; 13: 478-85.
[259] Eyrich D, Brandl F, Appel B, *et al.* Long-term stable fibrin gels for cartilage engineering. Biomat 2007; 28: 55-65.
[260] Masters KS, Shah DN, Walker G, Leinwand LA, Anseth KS. Designing scaffolds for valvular interstitial cells: Cell adhesion and, function on naturally derived materials. J Biomed Mater Res A 2004; 71: 172-80.
[261] Gu Y, Zhu J, Xue C, *et al.* Chitosan/silk fibroin-based, Schwann cell-derived extracellular matrix-modified scaffolds for bridging rat sciatic nerve gaps. Biomat 2014; 35: 2253-63.
[262] Azab AK, Orkin B, Doviner V, *et al.* Cross-linked chitosan implants as potential degradable devices for brachytherapy: *In vitro* and, *in vivo* analysis. J Control Release 2006; 111: 281-89.
[263] Barralet JE, Wang L, Lawson M, Triffitt JT, Cooper PR, Shelton RM. Comparison of bone marrow cell growth on 2D and, 3D-alginate hydrogels. J Mater Sci Mater Med 2005; 16: 515-19.
[264] Dawson E, Mapili G, Erickson K, Taqvi S, Roy K. Biomaterials for stem cell differentiation. Adv Drug Deliv Rev 2008; 60: 215-28.

[265] Behravesh E, Zygourakis K, Mikos AG. Adhesion and migration of marrow-derived osteoblasts on injectable *in situ* crosslinkable poly(propylene fumarate-co-ethylene glycol)-based hydrogels with a covalently linked RGDS peptide. J Biomed Mater Res 2003; 65A: 260-70.

[266] Beebe DJ, Moore JS, Bauer JM, *et al.* Functional hydrogel structures for autonomous flow control inside microfluidic channels. Nature 2000; 404: 588-90.

[267] Peppas NA, Bures P, Leobandung W, Ichikawa H. Hydrogels in pharmaceutical formulations. Eur J Pharm Biopharm 2000; 50: 27-46.

[268] Halstenberg S, Panitch A, Rizzi S, Hall H, Hubbell JA. Biologically engineered protein-graft-poly(ethylene glycol) hydrogels: a cell adhesive and plasmin-degradable biosynthetic material for tissue repair. Biomacromol 2002; 3: 710-23.

[269] Lutolf MP, Lauer-Fields JL, Schmoekel HG, *et al.* Synthetic matrix metalloproteinase-sensitive hydrogels for the conduction of tissue regeneration: engineering cell-invasion characteristics. Proc Natl Acad Sci USA 2003; 100: 5413-18.

[270] Koh WG, Revzin A, Pishko MV. Poly(ethylene glycol) hydrogel microstructures encapsulating living cells. Langmuir 2002; 18: 2459-62.

[271] Yu T, Ober CK. Methods for the topographical patterning and patterned surface modification of hydrogels based on hydroxyethyl methacrylate. Biomacromol 2003; 4: 1126-31.

[272] Nguyen KT, West JL. Photo-polymerizable hydrogels for tissue engineering applications. Biomat 2002; 23: 4307-14.

[273] Schmedlen RH, Masters KS, Westn JL. Photo-crosslink-able polyvinyl alcohol hydrogels that can be modified with cell adhesion peptides for use in tissue engineering. Biomat 2002; 23: 4325-32.

[274] Sawhney AS. Optimization of photo-polymerized bio-erodible hydrogel properties for adhesion prevention. J Biomed Mat Res 1994; 28: 831-38.

[275] Mann BK, Gobin AS, Tsai AT, Schmedlen RH, West JL. Smooth muscle cell growth in photo-polymerized hydrogels with cell adhesive and proteolytically degradable domains: synthetic ECM analogs for tissue engineering. Biomat 2001; 22: 3045-51.

[276] Sawhney AS, Pathak CP, Hubbell JA. Bioerodible hydrogels based on photo-polymerized poly (ethylene glycol)-co-poly (a-hydroxy acid) diacrylate macromers. Macromol 1993; 26: 581-87.

[277] Martens P, Anseth KS. Characterization of hydrogels formed from acrylate modified poly (vinyl alcohol) macromers. Polymer 2000; 41: 7715-22.

[278] Chirila TV, Constable IJ, Crawford GJ, *et al.* Poly (2-hydroxyethel methacrylate) sponges as implant materials: *In vivo* and, *in vitro* evaluation of cellular invasion. Biomat 1993; 14: 26-38.

[279] Papavasiliou G, Sokic S, Turturro M. Synthetic PEG Hydrogels as Extracellular Matrix Mimics for Tissue Engineering Applications, Biotechnology - Molecular Studies and Novel Applications for Improved Quality of Human Life, Ed:Sammour R. 2012. ISBN: 978-953-51-0151-2, InTech. http: //www.intechopen.com/books/biotechnology.

[280] Nam YS, Park TG. Biodegradable polymeric microcellular foams by modified thermally induced phase separation method. Biomat 1999; 20: 1783-90.

[281] Bryant SJ, Durand KL, Anseth KS. Manipulations in hydrogel chemistry control photo-encapsulated chondrocyte behavior and their extracellular matrix production. J Biomed Mater Res A 2003; 67: 1430-36.

[282] Tibbitt MW, Anseth KS. Hydrogels as Extracellular Matrix Mimics for 3D-Cell Culture. Biotechnol Bioeng 2009; 103: 655-63.

[283] Yamaguchi N, Kiick KL. Polysaccharide-poly (ethylene glycol) star copolymer as a scaffold for the production of bioactive hydrogels. Biomacromol 2005; 6: 1921-30.

[284] Willerth SM, Johnson PJ, Maxwell DJ, Parsons SR, Doukas ME, Sakiyama-Elbert SE. Rationally designed peptides for controlled release of nerve growth factor from fibrin matrices. J Biomed Mater Res A 2007; 80: 13-23.

[285] Gurdon JB, Bourillot PY.Morphogen gradient interpretation. Nature 2001; 413(6858):797-803.

[286] Bryant SJ, Anseth KS. Hydrogel properties influence ECM production by chondrocytes photo-encapsulated in poly (ethylene glycol) hydrogels. J Biomed Mater Res 2002; 59: 63-72.

[287] Raeber GP, Lutolf MP, Hubbell JA. Mechanisms of 3-D migration and, matrix remodeling of fibroblasts within artificial ECMs. Acta Biomat 2007; 3: 615-29.

[288] Mahoney MJ, Anseth KS. Three-dimensional growth and, function of neural tissue in degradable polyethylene glycol hydrogels. Biomat 2006; 27: 2265-74.

[289] Metters AT, Anseth KS, Bowman CN. Fundamental studies of a novel, biodegradable PEG-b-PLA hydrogel. Polymer 2000; 41: 3993-04.
[290] Nuttelman CR, Kloxin AM, Anseth KS. Temporal changes in PEG hydrogel structure influence human mesenchymal stem cell proliferation and, matrix mineralization. In: Tissue Engineering, Ed: Fisher JP, Springer-Verlag, Berlin 2006; pp 135-149.
[291] Li P, Poon YF, Li W, *et al.* A polycationic antimicrobial and biocompatible hydrogel with microbe membrane suctioning ability, Nature Materials 2011; 10: 149-56.
[292] Wu F, Meng G, He J, Wu Y, Gu Z. Antibiotic-loaded chitosan hydrogel with superior dual functions: antibacterial efficacy and osteoblastic cell responses. ACS Appl Mater Interfaces 2014; 9(6): 10005-13.
[293] Grandhi TSP, Potta T, Faust J, Regel K. Design of Bone Microenvironment Mimicking Antibiotic-based Hydrogels for Generation of 3-D Tumor Models of Dormancy and Relapse. Conference of the Society for Biomaterials. 2014; Abstract #191.
[294] Morais DS, Rodrigues MA, Lopes MA, *et al.* Biological evaluation of alginate-based hydrogels with antimicrobial features by Ce (III) incorporation as a vehicle for bone substitute. J Mater Sci Mater Med 2013; 24: 2145-55.
[295] Ryu JH, Lee Y, Do MJ, *et al.* Chitosan-g-hematin: Enzyme-mimicking polymeric catalyst for adhesive hydrogels. Acta Biomat 2014; 10: 224-33.
[296] Calderon L, Collin E, Velasco-Bayon D, Murphy M, O'Halloran D, Pandit A. Type II Collagen-Hyaluronan Hydrogel - A Step Towards a Scaffold for Intervertebral Disc Tissue Engineering. Euro Cells Mat 2010; 20: 134-48.
[297] Schulte VA, Alves DF, Dalton PP, Moeller M, Lensen MC, Mela P. Micro-engineered PEG hydrogels: 3D-scaffolds for guided cell growth. Macromol Biosci 2013; 13: 562-72.
[298] Baker BM, Gee AO, Metter RB, *et al.* The potential to improve cell infiltration in composite fiber-aligned electrospun scaffolds by the selective removal of sacrificial fibers. Biomat 2008; 29: 2348-58.
[299] Reneker DH, Chun I. Nanometer diameter fibres of polymer, produced by electrospinning Nanotech, 1996, 7: 216-223.
[300] Hartgerink JD, Beniash E, Stupp SI. Peptide-amphiphile nanofibers: a versatile scaffold for the preparation of self-assembling materials. Proc Natl Acad Sci USA 2002; 99: 5133-18.
[301] Li MY, Mondrinos MJ, Gandhi MR, Ko FK, Weiss AS, Lelkes PI. Electrospun protein fibers as matrices for tissue engineering. Biomat 2005; 26: 5999-08.
[302] Doshi J, Reneker DH. Electro-spinning process and application of electrospun fibers. J Electrostat 1995; 35: 151-60.
[303] Jia C, Mahjour SB, Chan L, Hou Da-wei, Wang H. Chapter 6: Biomimetic Design of Extra-Cellular Matrix likes Substrate for Tissue Regeneration, Eds: Zhang LG, Khademhosseini A, Webster TJ. Pan Stanford Publishing, 2014. ISBN 978-981-4411-67-4.
[304] Ma Z, He W, Yong T, Ramakrishna S. Grafting of gelatin on electrospun poly (capro-lactone) nanofibers to improve endothelial cells spreading, and proliferation to control cell orientation. Tissue Eng 2005; 11: 1149-58.
[305] Dainiak MB, Allan IU, Savina IN, *et al.* Gelatin-Fibrinogen Cryogel Dermal Matrices for Wound Repair: Preparation, Optimisation and *In vitro* Study. Biomat 2010; 31: 67-76.
[306] Li WJ, Laurencin CT, Caterson EJ, Tuan RS, Ko FK. Electrospun nanofibrous structure: A novel scaffold for tissue engineering. J Biomed Mat Res 2002; 60: 613-21.
[307] Ekaputra AK, Zhou Y, Cool SM, Hutmacher DW. Composite electrospun scaffolds for engineering tubular bone grafts. Tissue Eng, Part A 2009; 15: 3779-88.
[308] Fujihara K, Kotaki M, Ramakrishna S. Guided bone regeneration membrane made of poly caprolactone/calcium carbonates composite nano-fibers. Biomat 2005; 26: 4139-47.
[309] Yoshimotoa H, Shina YM, Teraia H, Vacantia JP. A biodegradable nanofiber scaffold by electrospinning and its potential for bone tissue engineering. Biomat 2003; 24: 2077-82.
[310] Chen F, Li X, Mo X, He C, Wang H, Ikada Y. Electrospun chitosan-P(LLA-CL) nanofibers for biomimetic extracellular matrix. J Biomat Sci Polymer Edu 2008;19: 677-91.
[311] Zhong S, Zhang Y, Lim CT. Fabrication of large pores in electrospun nanofibrous scaffolds for cellular infiltration: a review. Tissue Eng B 2012; 18: 77-87.

[312] Mo XM, Xub CY, Kotakib M, Ramakrishna S. Electrospun P (LLA-CL) nanofiber: a biomimetic extracellular matrix for smooth muscle cell and endothelial cell proliferation. Biomat 2004; 25: 1883-90.

[313] Hea W, Mab ZW, Yonga T, Teoc WE, Ramakrishna S. Fabrication of collagen-coated biodegradable polymer nanofiber mesh and its potential for endothelial cells growth. Biomat 2005; 26: 7606-15.

[314] Ravindran S, Gao Q, Kotecha M, *et al.* Biomimetic Extracellular Matrix-Incorporated Scaffold Induces Osteogenic Gene Expression in Human Marrow Stromal Cells. Tissue Eng A 2012; 18: 295-09.

[315] Xua CY, Inaic R, Kotakib M, Ramakrishna S. Aligned biodegradable nanofibrous structure: a potential scaffold for blood vessel engineering. Biomat 2004; 25: 877-86.

[316] Flanagan TC, Wilkins B, Black A, Jockenhoevel S, Smith TJ, Pandit AS. A collagen-glycosaminoglycan co-culture model for heart valve tissue engineering applications. Biomat 2006; 27: 2233-46.

[317] Hao H, Nian G, Jia S, Chao G, Ping H. Fabrication of biomatrix/polymer hybrid scaffold for heart valve tissue engineering *in vitro*. ASAIO J 2008; 54: 627-32.

[318] Luo ZL, Wang SK, Zhang SG. Fabrication of self-assembling D-form peptide nanofiber scaffold d-EAK16 for rapid hemostasis. Biomat 2011; 32: 2013-20.

[319] Xu C, Yang F, Wang S, Ramakrishna S. *In vitro* study of human vascular endothelial cell functions on materials with various surface roughnesses. J Biomed Mat Res Part A 2004; 71: 154-61.

[320] Hanjaya-Putra D, Wanjare M, Gerecht S. Chapter: 16: Biomaterials Approaches in Vascular Engineering: A Review of Past and Future Trends, In: Biomaterials for Tissue Engineering Applications, Eds: Burdick JA & Mauck RL. 2011; pp 457-488. Springer-Verlag, Wien. DOI 10.1007/978-3-7091-0385-2_16

[321] Chen ZG, Wang PW, Wei B, Mo XM, Cui FZ. Electrospun collagen chitosan nanofiber: A biomimetic extracellular matrix for endothelial cell and, smooth muscle cells. Acta Biomat 2010; 6: 372-82.

[322] Kumbar SG, James R, Nukavarapu SP, Laurencin CT. Electrospun nanofiber scaffolds: engineering soft tissues. Biomed Mater 2008; 3(3): 034002. DOI: 10.1088/1748-6041/3/3/034002.

[323] Dong Y,Yong T, Liao S, Chan CK, Stevens MM, Ramakrishna S. Distinctive degradation behaviors of electrospun poly glycolide, poly (L-lactide-co-glycolide), and poly (L-lactide-co-ε-caprolactone) nanofibers cultured with/without porcine smooth muscle cells. Tissue Eng A 2010; 16: 283-98.

[324] Ki CS, Kim, JW, Hyun JH, *et al.* Electrospun three-dimensional silk fibroin nanofibrous scaffold. J App Polymer Sci 2007; 106: 3922-28. Wang X, Ding B, Li B. Biomimetic electrospun nanofibrous structures for tissue engineering. Materials Today 2013; 16: 229-241. Pereira RF, Barrias CC, Granja PL *et al.* Advanced Biofabrication Strategies for Skin Regeneration and Repair. Nanomed 2013; 8(4): 603-621. Guadalupe E, Ramos D, Shelke NB, *et al.* Bioactive polymeric nanofiber matrices for skin regeneration. J App Polym Sci 2015; 132: 41879. Venugopal J, Ramakrishna S. Biocompatible nanofiber matrices for the engineering of a dermal substitute for skin regeneration. Tissue Engg 2005; 11(5-6): 847-854.

[325] Mukhatyar VJ, Salmeron-Sanchez M, Rudra S, *et al.* Role of fibronectin in topographical guidance of neurite extension on electrospun fibers. Biomat 2011; 32: 3958-68.

[326] Christopherson GT, Song H, Mao H-Q. The influence of fiber diameter of electrospun substrates on neural stem cell differentiation and proliferation. Biomat 2008; 30: 556-64.

[327] Wang HB, Mullins ME, Cregg JM, McCarthy CW, Gilbert RJ. Varying the diameter of aligned electrospun fibers alters neurite outgrowth and Schwann cell migration. Acta Biomat 2010; 6: 2970-78.

[328] Cooke MJ, Zahir T, Phillips SR, *et al.* Neural differentiation regulated by biomimetic surfaces presenting motifs of extracellular matrix proteins. J Biomed Mater Res 2010, 93A: 824-32.

[329] Xie J, MacEwan MR, Li X, Sakiyama-Elbert SE, Xia Y. Neurite Outgrowth on Nanofiber Scaffolds with Different Orders, Structures, and Surface Properties. ACS Nano 2009; 3: 1151-59.

[330] Xie J, Macewan MR, Willerth SM, *et al.* Conductive Core-Sheath Nanofibers and Their Potential Application in Neural Tissue Engineering. Adv Functional Mat 2009; 19: 2312-18.

[331] Baiguera S, Del-Gaudio C, Lucatelli E, *et al.* Electrospun gelatin scaffolds incorporating rat de-cellularized brain extracellular matrix for neural tissue engineering. Biomat 2014; 35: 1205-14.

[332] Alhosseini SN, Moztarzadeh F, Mozafari M, *et al.* Synthesis and characterization of electrospun polyvinyl alcohol nanofibrous scaffolds modified by blending with chitosan for neural tissue engineering. Int J Nanomed 2012; 7: 25-34.
[333] Wood MD, Moore AM, Hunter DA, *et al.* Affinity-based Release of Glial-Derived Neurotrophic Factor from Fibrin Matrices Enhances Sciatic Nerve Regeneration. Acta Biomat 2009; 5: 959-68.
[334] Wood MD, Sakiyama-Elbert SE. Release Rate Controls Biological Activity of Nerve Growth Factor Released from Fibrin Matrices Containing Affinity-Based Delivery Systems. J Biomed Mat Res Part A 2008; 84: 300-12.
[335] Willerth SM, Sakiyama-Elbert SE. Approaches to neural tissue engineering using scaffolds for drug delivery. Adv Drug Delivery Rev 2007; 59: 325-38.
[336] Johnson PJ, Parker SR, Sakiyama-Elbert SE. Fibrin-based tissue engineering scaffolds enhance neural fiber sprouting and delay the accumulation of reactive astrocytes at the lesion in a sub-acute model of spinal-cord injury. J Biomed Mat Res 2010; 92: 152-63.
[337] McCreedy D, Wilems T, Xu H, *et al.* Survival, Differentiation, and Migration of High-Purity Mouse Embryonic Stem Cell-derived Progenitor Motor Neurons in Fibrin Scaffolds after Sub-Acute Spinal-Cord Injury. Biomat Sci 2014; 2: 1672-82.
[338] Sundaramurthi D, Vasanthan KS, Kuppan P, Krishnan UM, Sethuraman S. Electrospun nanostructured chitosan-poly (vinyl alcohol) scaffolds: a biomimetic extracellular matrix as dermal substitute. Biomed Mater 2012; 7: 045005.
[339] Ajalloueian F, Lim ML, Lemon G, *et al.* Biomechanical and biocompatibility characteristics of electrospun polymeric tracheal scaffolds. Biomat 2014; 35: 5307-15.
[340] Willerth SM, Faxel TE, Gottlieb D, Sakiyama-Elbert SE. The Effects of Soluble Growth Factors on Embryonic Stem Cell Differentiation Inside of Fibrin Scaffolds. Stem Cells 2007; 25: 2235-44.
[341] Jason CT, Sharon LP, Buddy DR, Charles EM, Cecilia MG. Engineered Biomaterials Control Differentiation and Proliferation of Human-Embryonic-Stem-Cell-Derived Cardiomyocytes *via* Timed Notch Activation. Stem Cell Reports 2014; 2: 271-81.
[342] Qiu P, Qu X, Brackett DJ, Lerner MR, Li D, Mao C. Silica-Based Branched Hollow Microfibers as a Biomimetic Extracellular Matrix for Promoting Tumor Cell Growth *in vitro* and *in vivo*. Adv Mater 2013; 25: 2492-96.
[343] Shea LD, Woodruff TK, Shikanov A. Bioengineering the Ovarian Follicle Microenvironment. Ann Rev Biomed Eng 2014; 16: 29-52.
[344] Vasanthakumari BL, Nair APV, Hussain FSJ, Valiyaveettil S. Peptide modified polymer nanofibers as biomimetic extracellular matrices for optimized cell adhesion and differentiation. Division of Analytical Chemistry, The 234th ACS National Meeting, Boston, MA, August 19-23, 2007; Abstract: ANAYL 208.
[345] Zhang K, Wang H, Huang C, Su Y, Mo X, Ikada Y. Fabrication of Silk Fibroin Blended P(LLA-CL) Nanofibrous Scaffolds for Tissue Engineering. J Biomed Mat Res A 2010; 93: 984-93.
[346] Zargarian SS, Haddadi-Asl V. A Nanofibrous Composite Scaffold of PCL/Hydroxyapatite-chitosan/PVA Prepared by Electrospinning. Iranian Polymer J 2010; 19: 457-68.
[347] Sun K, Li ZH. Preparations, properties and applications of chitosan based nanofibers fabricated by electrospinning. Express Polymer Lett 2011; 5: 342-61.
[348] Jian F, Wang X, Lin T. Ed: Lin T: Functional Applications of Electrospun Nanofibers, Nanofibers - Production, Properties and Functional Applications, 2011. ISBN: 978-953-307-420-7, www.InTech.com.
[349] Hassan MI, Sultana N, Hamdan S. Bioactivity Assessment of Poly (ε-caprolactone)/Hydroxyapatite Electrospun Fibers for Bone Tissue Engineering Application. J Nanomat 2014; 2014: 1-6. http: //dx. doi. org/ 10. 1155 /2014/573238.
[350] Roozbahani F, Sultana N, Ismail AF, Nouparvar H. Effects of Chitosan Alkali Pretreatment on the Preparation of Electrospun PCL/Chitosan Blend Nanofibrous Scaffolds for Tissue Engineering Application. J Nanomat 2013; 2013: 1-6. http: //dx.doi.org/10.1155/2013/641502.
[351] Ding F, Deng H, Du Y, Shi X, Wang Q. Emerging chitin and chitosan nanofibrous materials for biomedical applications. Nanoscale 2014; 6: 9477-93.
[352] Figallo E, Cannizzaro C, Gerecht S, *et al.* Micro-bioreactor array for controlling cellular microenvironments. Lab on a Chip 2007; 7: 710 -19.

[353] Song W, Markel DC, Wang S, Shi T, Mao G, Ren W. Electrospun polyvinyl alcohol-collagen-hydroxyapatite nanofibers: a biomimetic extracellular matrix for osteoblastic cells. Nanotech 2012; 23: 115101. DOI: 10.1088/0957-4484/23/11/115101.

[354] Beachley V, Wen X. Polymer nanofibrous structures: Fabrication, bio-functionalization, and cell interactions. Prog Polym Sci 2010; 35: 868-92.

[355] Hai-Quan M, Shawn HL, Shuming Z, Christopherson G, Korey K, Stephen F. The Nanofiber Matrix as an Artificial Stem Cell Niche. Stud Mechanobiol Tissue Eng Biomat 2010; pp 89-118. Springer-Verlag Berlin, Heidelberg. DOI: 10.1007/8415_2010_2.

[356] Yang S, Leong KF, Du Z, Chua CK. The design of scaffolds for use in tissue engineering. Part I. Traditional factors. Tissue Eng 2001; 7: 679-89.

[357] Leong MF, Chan WY, Chian KS, Rasheed MZ, Anderson JM. Fabrication and *in vitro* and *in vivo* cell infiltration study of a bi-layered cryogenic electrospun poly(D,L-lactide) scaffold. J Biomed Mater Res A 2010; 94: 1141-49.

[358] Yeong WY, Chua CK, Leong KF, Chandrasekaran M.Rapid prototyping in tissue engineering: challenges and potential. Trends Biotechnol 2004; 22: 643-52.

[359] Guan J, Fujimoto KL, Sacks MS, Wagner WR. Preparation and characterization of highly porous, biodegradable polyurethane scaffolds for soft tissue applications. Biomat 2005; 26: 3961-71.

[360] Lee JW, Kim JY, Cho DW. Solid Free-form Fabrication Technology and Its Application to Bone Tissue Engineering. Int J Stem Cells 2010; 3: 85–95.

[361] Kim H, Wosnick JH, Yu LMY, Shoichet MS. Neural Tissue Engineering: Fundamental and Translational Approaches, In: Translational Approaches in Tissue Engineering and Regenerative Medicine, Eds: Mao J, Vunjak-Novakovic G, Mikos A, Atala A. Artech House, Norwood, MA, USA.

[362] Levesque S, Wylie R, Aizawa Y, Shoichet M. Chapter 9: Peptide Modification of Polysaccharide Scaffolds for Targeted Cell Signalling, In: Natural-based Polymers for Biomedical Applications, Eds: Reis RL, Neves NM, Mano JF, Gomes ME, Marques AP, Azevedo HS. Woodhead Publishing, Cambridge. 2007; pp 260-287.

[363] Kim H, Cooke MJ, Shoichet MS. Creating permissive microenvironments for stem cell transplantation into the central nervous system. Trends in Biotech 2012; 30: 55-63.

[364] Guobao W, Ma PX. Structure & Property of Nano-hydroxyapatite/Polymer Composite Scaffolds for Bone Tissue Eng Biomat 2004; 25: 4749-57.

[365] Collins MN, Birkinshaw C. Hyaluronic acid based scaffolds for tissue engineering-A review, Carbohydrate Polymers 2013; 92:1262-79.

[366] Cooke MN, Fisher JP, Dean D, Rimnac C, Mikos AG. Use of stereo-lithography to manufacture critical-sized 3D-biodegradable scaffolds for bone ingrowth. J Biomed Mater Res 2003; 64B: 65-69.

[367] Dhariwala B, Hunt E, Boland T. 2004. Rapid prototyping of tissue-engineering constructs, using photo-polymerizable hydrogels and stereo-lithography. Tissue Eng 2004; 10: 1316-22.

[368] Vrana NE. PhD Thesis: Use of Poly Vinyl Alcohol (PVA) Cryogelation for Tissue Engineering: Composites, Scaffold Formation and Cell Encapsulation. Dublin City University, Ireland, 2009.

[369] Stella JA, D'Amore A, Wagner WR, Sacks MS. On the biomechanical function of scaffolds for engineering load-bearing soft tissues. Acta Biomat 2010; 6: 2365-81.

[370] Schmidt CE, Leach JB. Neural tissue engineering: strategies for repair and regeneration. Annul Rev Biomed Eng 2003; 5: 293-47.

[371] Shuai C, Mao Z, Lu H, Nie Y, Hu H, Peng S. Fabrication of porous polyvinyl alcohol scaffold for bone tissue engineering *via* selective laser sintering. Biofab 2013; 5: 015014.

[372] Hossan MY, Molla MAI, Islam MS, Rana AA, Gafur MA, Karim MM. Fabrication and, Characterization of Polyvinyl Alcohol-Hydroxy Apatite Biomimetic Scaffold by Freeze Thawing *in situ* Synthesized Hybrid Suspension for Bone Tissue Engineering. Int J Emerg Tech & Adv Eng 2012; 2: 696-01.

[373] Abbas AA, Lee SY, Selvaratnam L, Yusof N, Kamarul T. Porous PVA-Chitosan Based Hydrogel as an Extracellular Matrix Scaffold for Cartilage Regeneration. Euro Cells & Mater 2008; 16(2): 1-50.

[374] Cho SH, Oh SH, Lee JH. Fabrication and characterization of porous alginate/polyvinyl alcohol hybrid scaffolds for 3D-cell culture. J Biomat Sci Polym Edu 2005; 16: 933-47.

[375] Stampella A, Papi A, Rizzitelli G, *et al*. Synthesis and characterization of a novel poly (vinyl alcohol) 3D-platform for the evaluation of hepatocytes' response to drug administration. J Mater Chem B 2013; 1: 3083-98.

[376] Nanda S, Sood N, Reddy BVK, Markandeywar TS. Preparation and Characterization of Poly (vinyl alcohol)-chondroitin Sulphate Hydrogel as Scaffolds for Articular Cartilage Regeneration, Indian J Mater Sci 2013; 2013: 1-8.

[377] Laurienzo P. Marine Polysaccharides in Pharmaceutical Applications: An Overview. Mar Drugs, 2010, 8: 2435-2465.

[378] Davidenko N, Campbell JJ, Thian ES, Watson CJ, Cameron RE. Collagen-Hyaluronic Acid Scaffolds for Adipose Tissue Engineering. Acta Biomat 2010; 10: 3957-68.

[379] Forcino RG. Characterization of blended PLGA: PEG scaffolds for bone regeneration applications. PhD Dissertation, University of the Sciences in Philadelphia. Pennsylvania, PA, USA. 2007; ISBN: 9780549447542.

[380] Alini M, Li W, Markovic P, Aebi M, Spiro RC, Roughley PJ. The Potential and Limitations of a Cell-Seeded Collagen/Hyaluronan Scaffold to Engineer an Intervertebral Disc-Like Matrix. Spine 2003; 28: 446-53.

[381] Yan J, Li X, Liu L, Wang F, Zhu TW, Zhang Q. Potential Use of Collagen-Chitosan-Hyaluronan Tri-Copolymer Scaffold for Cartilage Tissue Engineering. Artif Cells, Blood Subs & Biotech 2006; 34: 27-39.

[382] Huang W, Liao H. Research progress of scaffold materials in skeletal muscle tissue engineering. Zhongguo Xiu Fu Chong Jian Wai Ke Za Zhi (Chinese). 2010, 24: 1386-91. http: //www.ncbi.nlm.nih.gov/pubmed/21226367.

[383] Yu LMY, Leipzig ND, Shoichet MS. Promoting neuron adhesion and growth. Materials Today 2008; 11: 36-43.

[384] Wang H, Feng Y, An B, *et al.* Fabrication of PU/PEGMA cross-linked hybrid scaffolds by *in situ* UV photo-polymerization favoring human endothelial cells growth for vascular tissue engineering. J Mater Sci Mater Med 2012; 23: 1499-10.

[385] Espinosa HD, Rim JE, Barthelat F, Buehler MJ. Merger of Structure and Material in Nacre and Bone - Perspectives on *de novo* Biomimetic Materials. Prog Mat Sci 2009; 54: 1059-00.

[386] Eisenbud D, Huang NF, Luke S, Silberklang M. Skin Substitutes and Wound Healing: Current Status and Challenges. Wounds, 2004: 16: Part 1 & 2, http: //www.woundsresearch.com/article/2269 & 2270.

[387] Boutros C, Somasundar P, Espat NJ. Early results on the use of biomaterials as adjuvant to abdominal wall closure following cytoreduction and hyperthermic intraperitoneal chemotherapy. World J Surgical Oncol 2010; 8: 72-78. DOI: 10.1186/1477-7819-8-72.

[388] Jones I, Currie L, Martin R. A guide to biological skin substitutes. British J Plastic Surg: JPRAS 2002; 55: 185-93.

[389] Breast reconstruction advances fix distortions left by lumpectomy. 2008. http: // www. sciencedaily.com/ releases /2008 /04/080423121425.htm (accessed on 2014/11/25).

[390] Coons, DA, Barber AF. Tendon graft substitutes-rotator cuff patches. Sports Med Arthrosc 2006; 14: 185-90.

[391] Boden SD, Martin GJ Jr, Morone M, Ugbo JL, Titus L, Hutton WC. The use of coralline hydroxyapatite with bone marrow, autogenous bone graft, or osteo-inductive bone protein extract for posterolateral lumbar spine fusion. Spine 1999; 24: 320-27.

[392] Cornwell KG, Landsman A, James KS. 2009. Extracellular Matrix Biomaterials for Soft Tissue Repair. Clin Podiatr Med Surg 2009; 26: 507-23.

[393] Peri-Guard Repair, 2014, Jan. 10. http: //www.accessdata.fda.gov/cdrh_docs/pdf13/k132660.pdf (accessed on 2014/11/25).

[394] Dental Implant: Bone Graft, http: //www.biohorizons.com/grafton.aspx (accessed on 2014/11/25).

[395] Jeronimidis G. Design and function of structural biological materials, In: Structural biological materials, Ed: Elices M, 2001, Pergamon Press, New York, USA.

[396] DeForest CA, Anseth KS. Advances in Bioactive Hydrogels to Probe and Direct Cell Fate. Annu Rev Chem Biomol Eng 2012; 3:421–44.

[397] Mizuno S, Tateishi T, Ushida T, Glowacki J. Hydrostatic fluid pressure enhances matrix synthesis and accumulation by bovine chondrocytes in three-dimensional culture. J Cellular Physiol 2002; 193: 319-27.

[398] Ebara M, Kotsuchibashi Y, Narain R, *et al.* Smart Biomaterials, 2015. Springer, New York, USA. ISBN 978-4-431-54400-5.

[399] Palsson BO, Bhatia SN. Tissue Eng, 2004, Upper Saddle River, Pearson Prentice Hall, NJ, USA.
[400] Kloxin AM, Kasko AK, Salinas CN, Anseth KS. Photo-labile hydrogels for dynamic tuning of physical and, chemical properties. Science 2009; 324: 59-63.
[401] Porter MM, Imperio R, Wen M, Meyers MA, McKittrick J. Bio-inspired Scaffolds with Varying Pore Architectures and Mechanical Properties. Adv Funct Mater 2014; 24: 1978-87.
[402] Madurantakam PA, Cost CP, Simpson DG, Bowlin GL. Science of nanofibrous scaffold fabrication: strategies for next generation tissue-engineering scaffolds. Nanomed 2009; 4: 193-06
[403] Taylor DA, Baiguera, Caplan AL, Macchiarini P. Ethics of bioengineering organs and tissues. Expert Opin Biol Ther 2014; 14: 879-82.

CHAPTER 11

Biopolymers in Non-Aqueous Emulsions for Food and Pharmaceutical Applications

Marcelo R. Ceballos[1], Dante M. Beltramo[1,2,3] and Ismael D. Bianco[1,2,4,*]

[1]Centro de Excelencia en Productos y Procesos de Córdoba (CEPROCOR), Argentina; [2]Consejo Nacional de Investigaciones Científicas y Técnicas (CONICET), Argentina; [3]Facultad de Ciencias Químicas – Universidad Católica de Córdoba – Argentina and [4]Departamento de Ciencias Exactas, Físicas y Naturales - Universidad Nacional de La Rioja – Argentina

Abstract: Emulsions and suspensions are systems formed by at least two phases; one of them dispersed into another. For their preparation these generally unstable structures require the addition of extra molecules that improve either their thermodynamic or kinetic stability. Traditionally small amphiphilic molecules have been used for this purpose. More recently, the surface and rheologic properties of biopolymers have been modified in order to "tune" their properties as to improve emulsions and suspensions stability.

Most of the research and technological developments related to emulsions performed so far imply water as one of the involved phases. Non-aqueous, oil-in-oil, anhydrous or waterless emulsions, are common names for the dispersed systems in which water is absent. Even though the first reports in this subject were published around 50 years ago the literature on this area is sparse. In the recent years, non-aqueous emulsions became attractive as potential vehicles for the development of drug delivery systems and healthier foods. This led to the design of novel and simpler methods for the dissolution of polymeric surfactants into edible vegetable oils and improved dispersion of non-aqueous phases into structured oils giving a result very stable non-aqueous emulsion with tunable rheological properties. These emulsions have potential applicability to improve nutritional qualities of foods as well as to design vehicles for hydrophobic active pharmaceutical ingredients.

In this chapter, we will summarize the recent advances in the basic principles involved in biopolymer stabilization of non-aqueous emulsions and their potential applications for the design of foods and medicines with improved performance.

Keywords: Biopolymers, emulsifier, ethylcellulose, graft copolymers, inulin, kinetic stability, non-aqueous emulsions, rheology, thermodynamic stability, viscosity.

***Corresponding author Ismael D. Bianco:** Centro de Excelencia en Productos y Procesos de Córdoba (CEPROCOR), Ministerio de Industria, Comercio, Minería y Desarrollo Científico Tecnológico, Pabellón CEPROCOR, CP 5164, Santa María de Punilla, Córdoba – Argentina; Tel: 54-3541-489651/53; Ext: 143; Fax: 54-3541-488181; E-mails: ibianco@ceprocor.uncor.edu; ibianco@hotmail.com

Viness Pillay, Yahya E. Choonara and Pradeep Kumar (Eds.)

INTRODUCTION

It is fair to say that the mixing of two different, immiscible fluids deep in the primitive oceans was a key event in the origin of life in our planet. No matter to which of the various theories of the origin of life we ascribe, there are at least two essential requirements that the primitive cells should had: a) they should had been able to compartmentalize chemical components and b) they should had been able to accumulate and use energy. These key processes both require the presence of water insoluble substances, probably iron-sulfur minerals at the beginning and lipids at present, that form bubbles, membranes, tubules or any of the numerous fantastic structures with which the presence of life in our planet has surprised us. From this, it is easy to see that nonhomogeneous systems are widely present and essential components in nature as colloids, emulsions and foams. For this reason, it is not surprising to find through history that the different inhomogeneous systems have been used for many different applications since ancient times [1-4].

When a two phase system is formed between two or more immiscible liquids in which one liquid is dispersed in the other in the form of microscopic droplets the system is known as an emulsion. There are in nature systems that are or contain emulsions that confer them special properties essential for their function. In the context of technological applications, the preparation of emulsions has been generally reduced to dispersing one oily phase into an aqueous one. Less common has been the inverse pathway: dispersing water into a non-polar oil phase. In this way, the resultant dispersed systems are represented by the o/w (oil-in-water) or w/o (water-in-oil) notation respectively [1, 2].

An important aspect that should be kept in mind when working with emulsions is that most of these systems are thermodynamically unstable and their existence in time (this is, their kinetic stabilization) will depend on the use of amphiphilic molecules that are frequently dissolved into one of the two involved immiscible liquids. This event, will generally define which type of emulsion is obtained; *i.e.*, which component will form the continuous phase (usually that in which the surfactant is dissolved) and which the dispersed one (this is known as Bancroft's rule, see below). Those amphipatic chemical substances are commonly known as *emulsifiers* or *surfactants*. Although the above is a very simplified description of the forming process of aqueous emulsions, it is on this basis that it is generally possible to predict the formation, stability, rheological behavior and transportation properties of these dispersed systems. The general aspects and the major properties of water containing emulsions have been exhaustively characterized [1-3]. However, there are other types of emulsions very less common but not less important, the water-in-water (w/w) and the oil-in-oil (o/o) emulsions. These last ones have been also designed as non-aqueous, anhydrous or waterless emulsions.

Non-aqueous emulsions are scarcely studied dispersed systems. The first results in this field were published around 50 years ago, with the work of Petersen and Hamill characterizing the preparation and stability for their use in cosmetics [4]. More recently, the efforts have been focused on their behavior as drug carriers [5, 6].

Related to foodstuffs, Patel, Schmid, and Lawrence replaced water with glycerol and propylene glycol (PG) as non-aqueous polar phases and used soy lecithin as emulsifier. With this experimental design, oil-in-PG emulsions with variable structural features were obtained, depending on lecithin solubility in the polyol-based continuous phase. However, it is worth mentioning at this point that all these emulsion systems require a co-surfactant to prepare relatively stable emulsions [7].

Almost invariably, the aqueous emulsions have been stabilized with relatively small amphiphilic molecules, where the stabilization mechanism generally implies the adsorption of surfactant at the interface between continuous and dispersed phases. This phenomenon is responsible of reducing the free energy in the interface, avoiding drop coalescence and phase separation. On the other hand, non-aqueous emulsions stability has been also achieved through the use of polymeric molecules. This has been mainly so because of the higher instability of non-aqueous emulsions related to aqueous ones. Thus, more complex mechanisms than interfacial surfactant adsorption are required in order to avoid or impair the coalescence velocity between droplets of the dispersed phase. In this connection, the possibilities that the drops of dispersed phase can travel, meet each other, fuse and coalesce, will be restricted by the increment of viscosity of the continuous phase. One of the easiest ways to accomplish this kinetic effect can be through the use of a diverse type of polymers. This was the basis of the original uses of polymers as emulsion stabilizers. With time, more and more polymers with surface activity have been developed and studied in order to be used in emulsion stability applications. Therefore, at least a two fold mechanism of emulsion stabilization can be attained, thermodynamic (through a reduction in surface free energy of dispersed droplets) and kinetic (through an increase in viscosity of the dispersing medium).

An interesting aspect of polymers is the enormous possibility of their structure to be chemically manipulated, *e.g.* adding specially selected chemical groups to the main chain. Consequently, it is possible to modulate or "tune" their polarity according to specific requirements. Hence, tailored polymers have been designed for the preparation of specific emulsions.

On the other hand, an important part of the research about non-aqueous emulsions is based in systems in which the dispersed phase is the less polar one, and consequently the most polar solvent ends forming the continuous phase. Generally, these results in a less viscous continuous phase, as a result of which it will have lower capacity to stabilize the emulsion through the kinetic mechanism described above.

Working in stabilization of anhydrous oil-based emulsions, a recent work of our group describes the use of modified cellulose, ethycellulose, as a successful stabilizer of a PG-in-oil emulsion. The oily continuous phase was in this case either a mixture of medium chain triglycerides or commercial vegetable oil. These systems show several advantages given by the use of edible materials in their composition, which allow them to be used in the formulation of healthier foods as well as water-free drug delivery systems [8].

This chapter aims to discuss the role that different polymers play in stabilization of non-aqueous emulsions and the underlying mechanisms. To achieve this objective, the most relevant research about non-aqueous emulsions must necessarily be reviewed first, because it is not a very common topic in the literature.

Emulsifier Trends - Overview

Foodstuffs represent the one of the most complex emulsions and their stabilization has been a hard challenge for manufacturers, scientists and food developers. In this context, the application of surfactants in some foods includes also their use in foam aeration in cakes, dispersion stabilization in peanut butter, dough strengthening in bread, clouding in citrus beverages, preventing of crystallization in salad oils, antisticking in candies, rheological tuning in chocolate, regulation of fat packing in ice creams and gloss enhancement in confectionery coatings, among others [9, 10].

In the pharmaceutical and cosmetic fields, emulsions are not less important: they are frequently used as drug delivery systems. These industries have developed emulsion/dispersion products which are stable for several years [11-13].

Emulsifiers and Emulsion Types – Historical Perspective

Beyond the above well described achievements, food and pharmaceutical industries have continuously renewed challenges when new products are developed, because the requirements can be different and will depend on several factors.

In the food industry, nutritional facts (e. g., for healthier foods), scientific aspects (e. g., for flavor modulation) and technological features (e. g., optimization of contact between raw materials to allow the use of lower processing temperatures) are just a few examples of the mentioned challenges [11].

In spite of this, many artificial surfactants have been widely used in the food industry. For over a century and a half of chemical handling of oils and fats, chemists have been developing non-metabolized surfactants [14]. Many of them are considered GRAS (generally recognized as safe) products and have more than 60 successful years in the food market [10].

On the other hand, the use of synthetic emulsifiers in the pharmaceutical industry has been associated directly or indirectly with toxic events. In some cases, these surfactants can be toxic by themselves or may affect the delivery of co-administered drugs [13, 15-17]. Despite these facts, pharmaceutical emulsions have been historically stabilized by synthetic surfactants [18].

All of these aspects, the properties, toxicity and efficiency of natural and synthetic surfactants used in aqueous based emulsions have been extensively reviewed and compared [10, 12, 19-23].

Understanding the basic principles involved in the microstructure of food colloids and food emulsions is a key point to choose the right molecule and procedure to improve their preparation, stability and long-term life. As mentioned above, either small-size amphiphiles or macromolecules have been shown to play important roles [10].

In the search for natural emulsifiers, surface active molecules have been extracted from natural sources or biotechnologically prepared and have also been modified (chemically or enzymatically) according to special requirements of novel food or pharmaceutical products.

Until the last 15 years, most of the work on emulsifiers was focused on those molecules that exert a stabilizing effect through interfacial adsorption, and excluded those that stabilize emulsions by a viscosity effect on the continuous phase. The review by Garti [10] starts with the low molecular weight emulsifiers to follow with a category named "high molecular weight amphiphiles, macromolecular emulsifiers, biopolymers as emulsifiers". Inside this group, the author includes proteins and polysaccharides extracted from natural sources and the biosurfactants. This term is used to define surface active compounds isolated from microorganisms. The author highlights several advantages of these biosurfactants over their chemically synthesized counterparts (lower toxicity, higher biodegradability, higher tolerance to extreme temperatures, pH and salinity and easier synthesis) [10].

Although known since long ago, surfactant association structures (*i.e.*: micelles, lyotropic liquid crystals) and food colloids and their effects on the stability of macrodispersed systems like emulsions has gained much more attention during the last few years [12].

The successful experience of food manufacturers using natural polymers (mainly proteins and polysaccharides) to stabilize aqueous emulsions moved promptly to the pharmaceutical industry. A complete analysis on aqueous emulsions stability from a physicochemical standpoint, modern methods to assess stability and even comparison between small-surfactant molecules and biopolymers has been recently reported [12].

Works briefly summarized above, are both exhaustive reviews and good literature sources about this section subject. However, it is worth mentioning that there is barely any reference to the potential application of the surfactants described as stabilizers of non-aqueous emulsions. As mentioned above, this type of emulsions are not commonly mentioned in the literature and their study has been generally confined to model systems rather than specific applications.

Since the pioneering work of Petersen and Hamill [4] there are only a few works until non-aqueous emulsions are mentioned as oil-in-oil (o/o) ones in a well renowned book on the subject [3]. However, they are not included in the posterior stability assessment studies described. It is clear from these reports that fundamental research on most types of emulsions is not simple owing to difficulties to produce reliable model systems. Moreover, general theoretical aspects related to stabilization of emulsions are inaccurate, and semi-empirical approaches should be used.

Emulsions are not easy because model systems are difficult to produce. In many cases, theories on emulsion stability are not exact and semiempirical approaches have to be used. Even so, the extensive studies performed using o/w and w/o emulsions has allowed to elaborate the theories that describe the energetics and dynamics of the formation, stability, and breakdown processes involved in these phenomena [3]. The exclusion of non-aqueous emulsions from this analysis is easy to understand: all theoretical and experimental models described so far, have been obtained from water-based systems. Despite these facts, special attention about how to prevent the different breakdown processes has been described, and prevention of creaming or sedimentation through the use of "thickeners" can also explain their use to improve the stability of some types of non-aqueous emulsions. (See below). As mentioned above, this was probably one of the initial reasons that led to the incorporation of polymers in non-aqueous emulsions. Due to their chemical versatility different copolymers ended being the most used polymers in non-aqueous emulsion stabilization [8, 24-26]. In the next sections we will discuss the fundamentals about non-aqueous emulsions and will attempt to introduce the generalizing approaches about their preparation, stability mechanisms and breakdown processes through the use of polymeric surfactants.

Polymeric Surfactants

Block and Graft Copolymers: Structure and Emulsifying Properties

Structural Features

A **homopolymer** is that whose monomers are all identical. On other hand, a **copolymer** is synthesized using two or more different monomers. In this way, copolymers have special properties that are achieved through the combination of the features of the individual monomers. Thus, the properties of the copolymer depend not only on the chemical nature of its constituent monomers but also in their configuration. For these reasons, they are frequently classified into four categories: alternating, random, block and graft (Fig. **1**).

Graft and block copolymers, have been the most used polymers in non-aqueous emulsion stabilization [8, 24-26]. Block copolymers are constituted by macromolecules in turn formed by blocks disposed in linear sequence. Each block is a defined portion that has some differential characteristic with respect to the adjacent blocks. In this way, if A and B are two different polymeric blocks, their disposition in the macromolecule could be AB (diblock copolymer) and ABA or BAB (triblock copolymers).

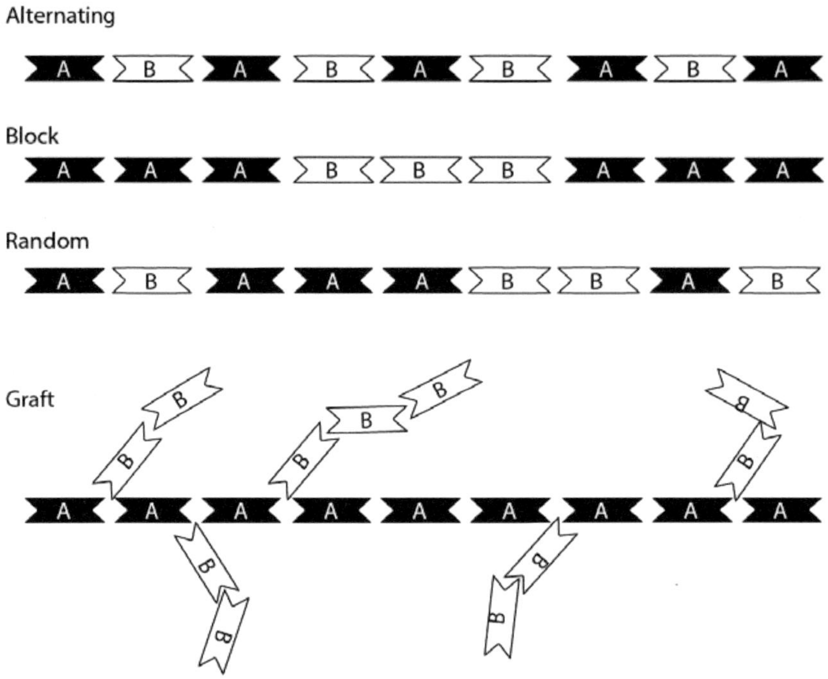

Figure 1: Schematic representation of molecular structure of main categories of copolymers.

Graft copolymers are polymers in which one or more polymeric blocks are joined to a principal macromolecule chain, forming a branched copolymer. As mentioned above, as these ramifications may have different properties than the main string these polymers are suitable candidates to be synthesized with a tunable amphipaticity.

Emulsifying Properties

Emulsion stability is a complex phenomenon, whereby its study involves several aspects not always observed in simplified model systems. The challenge is even greater in real products, such as foodstuffs and pharmaceutical formulations. In these cases, to the inherent thermodynamic instability of emulsions, other aspects must be added (steric and electrostatic repulsion forces, other surface active components, physical state of constituent phases, thermal treatments, *etc*. [13, 27, 28].

This instability can occur through some of the following well studied mechanisms (Fig. **2**):

A. Inversion

B. Creaming

C. Sedimentation

D. Flocculation

E. Coalescence

F. Ostwald ripening

It is beyond the scope of this chapter to give a detailed explanation of these mechanisms, but their broad picture will be enough to understanding the main aspects on how polymers act like emulsion stabilizers. If interested, the reader will find a good discussion of each of them by T. Tadros [3].

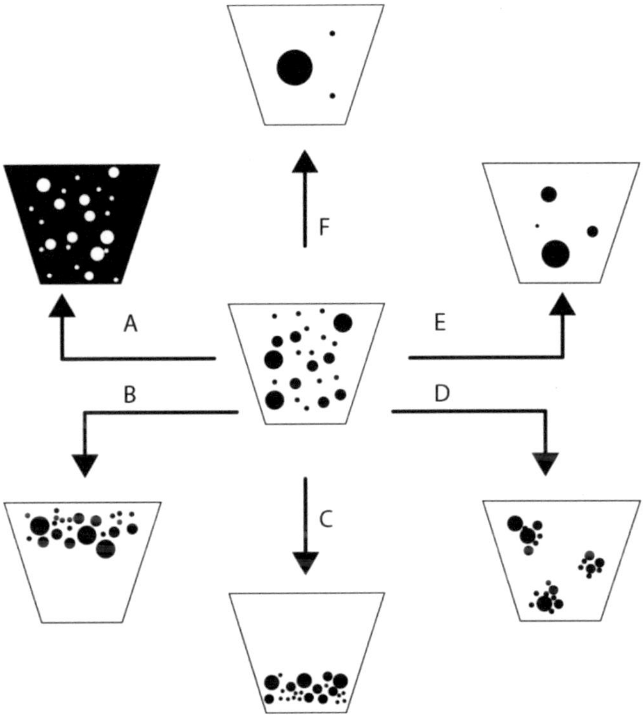

Figure 2: Schematic representation of different breakdown processes in emulsions.

IUPAC defines an emulsifier as "a surfactant which is positively adsorbed at interfaces and lowers the interfacial tension. It facilitates when present in small amounts the formation of an emulsion, or enhances its colloidal stability by decreasing either or both of the rates of aggregation and coalescence" [13].

Several polymers can exert emulsion stabilizing action through either or both mechanisms: surface adsorption or thickening of the continuous phase [9].

For the reasons mentioned above, homopolymers are not usually the most suitable emulsifiers; although slight structural modifications can improve their surface activity. On the other hand, the most convenient polymeric surfactants to stabilize emulsions and suspensions are either block or graft copolymers (see above).

Among block and graft synthetic polymeric surfactants, those with diverse combinations and different chain lengths of the following polymeric blocks are the most studied:

- poly (ethylene oxide) (PEO)
- poly (propylene oxide) (PPO)
- polystyrene (PS)
- polyvinyl alcohol (PVA)
- poly (methyl methacrylate) (PMMA)
- polybutadiene (PB)
- poly (ter butylstyrene) (PtBuS)

The first two in the list are the blocks most frequently used. However, in any case, the goal is usually to design specific molecules for stabilizing defined target structures. For this purpose, several groups have aimed to achieve it through the combination of individual properties of any polymeric block in the list [14, 24, 25, 29-32].

In case of PEO and PPO, triblock combinations of them are the most used polymeric surfactants. They can be used as emulsifying or dispersing agents; the latter when used in liquid/solid dispersions. In this way, it is assumed that the hydrophobic PPO is in contact with the hydrophobic solid surface or oil droplet, and the PEO portion remains in the aqueous/polar continuous phase, supplying steric repulsion.

Nevertheless, as there are still situations in which their effectiveness is not good enough there is still a permanent attention in the development of new graft copolymers [13, 26, 31]. In this connection, a series of graft copolymers have been obtained from inulin, a polyfructose with a glucosyl terminal group obtained from chicory roots [31]. For instance, by grafting C_{12} alkyl chains two main advantages are conferred to the modified inulin: (1) presence of C_{12} chains offers multi-point attachment residues to achieve a strong copolymer adsorption to a dispersed hydrophobic droplet or particle; (2) the polyfructose chains are able to be strongly hydrated both in water or highly concentrated electrolyte medium. As a result and in comparison with block copolymers features, a low level of desorption from the interface as well as effective steric stabilization are ensured [31].

This focus in graft copolymers includes a series of cellulose ethers which have been studied for quite some time. Within this group, both hydrophilic and hydrophobic polymeric surfactants are included, and the latter can be used in non-aqueous emulsions stabilization. For this reason, these cellulose-derivatives will be treated in more detail in the next sections.

Copolymers and Non-Aqueous Emulsions

Structural Aspects and Parameters for Choosing a Suitable Emulsifier

As stated above, general theories about emulsions stability are based on studies of water-based dispersed systems. Ionic strength, pH and electrolyte concentration together with kinetic or thermodynamic approaches are the main studied aspects. In fact, from these studies, few classifications of surface active molecules have been accomplished considering their physical and chemical features, *e.g.* HLB (hydrophile-liphophile balance) number, Bancroft's rule and molecular geometric disposition [33].

However, the selection of emulsifiers on the basis of their fundamental properties can be very difficult for non-aqueous emulsions. Nevertheless, by their empirical value, Bancroft's rule and HLB number can be used with an approximate predictive value to design water-free dispersed systems.

Briefly, Bancroft's rule states that the phase in which the surfactant is most soluble will end forming the continuous phase of the emulsion. Therefore, the solubility of the emulsifier would determine the type of emulsion that is formed. Thus, an oil-soluble emulsifier will generally give an emulsion with an oily continuous phase, and a water-soluble emulsifier will give an aqueous continuous one. This was established for low molecular weight surfactants, but it is also valid for polymeric molecules. However, there are always exceptions (see below).

On the other hand, HLB is an empirical scale based on the comparative proportion of hydrophilic and hydrophobic functional groups in the surface active molecule. Generally, surfactants with HLB values within 4-6 are able to form w/o emulsions, whereas molecules with HLB number in the range of 8-18 are suitable to form o/w emulsions.

Using these concepts, two approaches can be evaluated to obtain and stabilize non-aqueous emulsions [24, 34]:

1. Designing of surfactants whose molecule has two structurally incompatible portions, each selectively soluble in one of two immiscible liquids involved in the emulsion;

2. Finding polar liquids immiscible with oily phases able to replace water. In addition, they must show compatibility with well known polar solvent compatible emulsifiers.

Using the second approach, oil-in-formamide (FA) and oil-in-dymethyl sulfoxide (DMSO) emulsions were more effectively stabilized when polymeric surfactants (triblock copolymers) were used instead of six different nonionic surfactants derived from polyoxyethylene alkyls. For this purpose, three copolymers with the Polyoxyethylene – polyoxypropylene – Polyoxyethylene (POE-POP-POE) structure with different mean molecular weight (MMW) and POE content (%) were used [24]. Although the molecular properties involved in the successful emulsifying capacity are not clear in these cases, it was suggested that hydrogen bonding was more important than polarity itself [24]. Perhaps the most relevant finding was the fact that triblock copolymers were the best efficient surfactants. Their relative high molecular size could be decisive to stabilizing FA and DMSO emulsions, avoiding their breakdown in an almost indefinite way (after one month, there was still no breaking signals). These results agree with the fact that polymeric surfactants are more effective in stabilization of emulsions, whereas nonionic surfactants are generally more effective in the emulsification process (to produce small droplets) [3].

Considering the composition of the molecular blocks of surfactants, HLB scale can provide a useful guideline for the formulation of anhydrous dispersed systems. However, definition of HLB scale implies the concept of "hydrophilicity", and the possibility of using the HLB value to predict non-aqueous emulsions behavior must be carefully evaluated. In this sense, Imhof *et al.* proposed that for non-aqueous polar solvents, "solvophilic" concept and its appropriate scale should be defined [24].

In this sense, it is important to highlight that in all the systems mentioned above oil in polar solvent emulsions were generally obtained. This was so even when using lipophilic polyoxyethylene copolymers, with which it was expected that a w/o emulsion would be obtained (due to Bancroft's rule – see above).

Considering the first approach listed above, it is possible to obtain non-aqueous emulsions with controlled droplet size and good stability. This can be carried out through the controlled synthesis of polymeric surfactants as well as the complete characterization of the obtained emulsion.

The synthesis of the copolymer allows controlling its molecular weight and the content of different specific blocks [32, 35]. Thus, the selective solubility of each type of block in determined solvents can be controlled.

On the other hand, through the copolymer concentration it is possible to control the viscosity of either continuous or dispersed phases; thus droplet size of the latter can be adjusted to an optimum value.

On this basis, Riess *et al.* studied the efficiency of PB-PEO and PtBuS-PEO diblock copolymers with different molecular weights and PEO content as emulsifiers and stabilizers [25]. Continuous and dispersed phases were selected taking into account these copolymer structures. Thus, the continuous phase of the emulsion was prepared dissolving each copolymer at different concentrations in methylcyclohexane, whereas the dispersed phase was PEG 400. In this dispersed system and as mentioned above, PEO is selectively soluble in PEG 400, whereas PB and PtBuS are soluble in methylcyclohexane. In this way, the correct interfacial balance required for emulsion stability can be achieved. Thus, using block copolymers with PEO contents between 40 and 60 % and molecular weights between 5,000 and 15,000 Da, non-aqueous emulsions with controlled droplet size and excellent stability could be obtained [25].

Beyond surfactant properties and considering the composition of emulsions, the emulsifier activity of amphiphilic molecules is also dependent on the solvent. In this way, "a liquid capable of replacing water in an emulsion should have an appreciable polarity to make it immiscible with oils and to make it a good solvent for the solvophilic part of the surfactant molecules. Hydrogen bonding in the polar liquid is expected to play a role in solvating both ionic and non-ionic surfactants, and in the formation of a hydrogen-bonded network in the liquid itself" [34]. In this sense, measurement of critical micellar concentration (CMC) for a surfactant in both water and a non-aqueous solvent can be a useful tool to evaluate the capacity of the latter to form anhydrous emulsions. For instance, CMCs for polysorbate 20, 40, 60 and 80 are higher in formamide than water, showing that formamide is a good solvent for polysorbates. In addition, the hydrogen bonding capacity and dielectric constant of formamide are very similar to those of water. For these reasons, formamide has been one of the most used polar solvent to obtain non-aqueous emulsions [34].

Non-Aqueous Emulsions in Food Developments: A Challenging Area

Since foods are complex multi-colloidal systems where very different types of emulsions coexist, food processing as well as food developing area represents attractive fields of applications for anhydrous dispersed systems. This approach should be considered not only from structural aspects, but also considering foods as potential vehicles of bioactive components too.

Despite these facts, very few applied research on this field has been produced, perhaps because only the oily-based foods are suitable carriers of non-aqueous emulsions.

Ethylcellulose: A Versatile Polymeric Surfactant

Different hydrophilic cellulose ethers are proved o/w or w/o emulsifiers. These graft copolymers have been assessed in pharmaceutical field highlighting hydroxylpropyl methylcellulose (HPMC) (or hypromellose) graft copolymer as one of the most widespread use [36-39]. Among these ethers, the hydrophobic cellulose ether *ethylcellulose (EC)* shows interesting differential properties as a result of its particular chemical structure.

Like the others ethers, EC is derived from cellulose whose molecular backbone is composed of repeated anhydroglucose units linked *via* β (1-4) bonds which make it highly hydrophobic (Fig. **3a**). Each anhydroglucose ring has three hydroxyl sites that can react to form a "grafted" structure. In this way, if hydrophilic groups are inserted the resultant structure will have greater water absorption and swelling. On the other hand, if more lipophilic groups are introduced into the backbone chain, they will provide it with enhanced surface activity. Thus, when cellulose is treated with ethyl chloride under specific conditions, the reaction will yield ethylcellulose (Fig. **3b**). In this case, reaction conditions as well as reactants ratio will determine the extension of substitution [14].

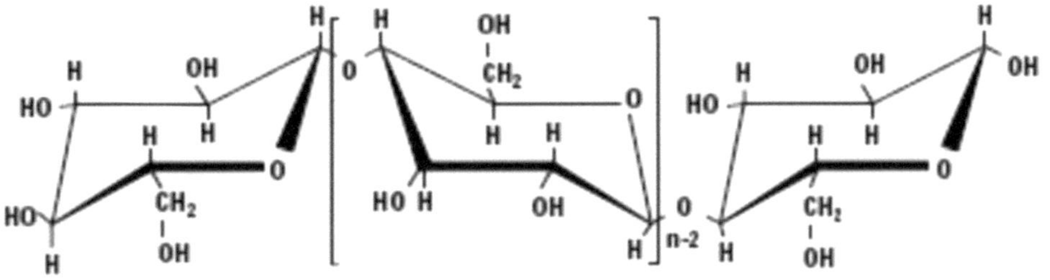

Figure 3(a): Chemical structure of cellulose.

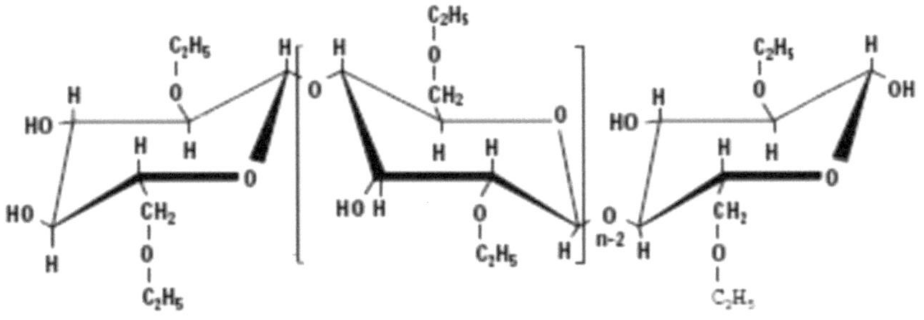

Figure 3(b): Chemical structure of ethylcellulose.

In an attempt to gain insight in the interfacial properties of EC, we evaluated the surface activity, molecular parameters and phase properties of EC using a Langmuir trough equipped with a Wilhelmy plate (Fig. **4**). The high collapse pressure is indicative of a great stability of this polymer at interfaces. It is clear that the oil solubility and good interfacial activity of this graft copolymer are a consequence of its structural features. Related to this, EC has been examined by different research groups, in order to find practical uses for its surfactant properties in food and pharmaceutical developments. These applications comprise its use as emulsifier of water based emulsions [13, 14], as a suspension stabilizer [26], for microparticle manufacturing [40] and as structuring agent in liquid oils from vegetal sources [41-44].

In spite of this proved interfacial activity, the stability of emulsions prepared with EC may not only be owing to the interfacial adsorption of the polymer. A second effect can also be due to the oil solubility of EC. In this way EC can be acting as a continuous phase "thickener" in an o/w emulsion, avoiding its breaking through steric stabilization [31]. A third stabilization mechanism could be participating since transmission electron microscopy (TEM) images suggest that colloidal EC in certain emulsions precipitate at the interface [13]. This type of systems can be classified as solid-stabilized emulsions or "Pickering emulsions".

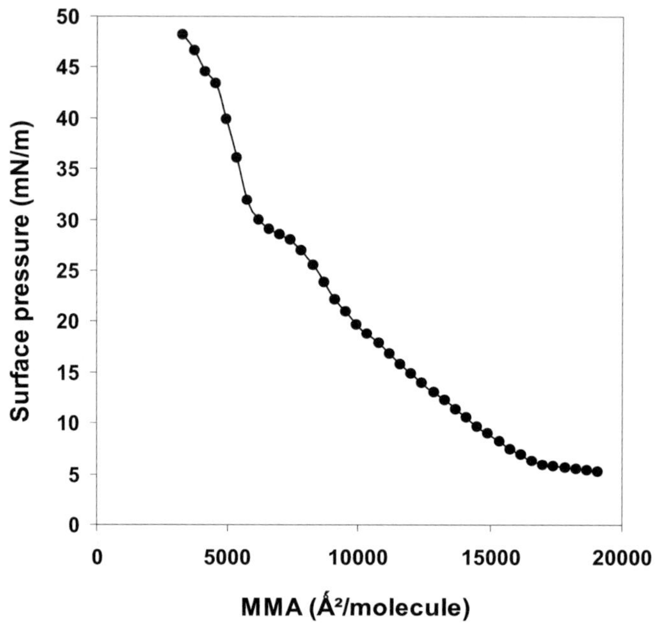

Figure 4: Surface pressure – MMA (Medium Molecular Area) compression isotherm of EC on a water subphase.

As mentioned above, it is difficult to disclose the contribution of each of these mechanisms in complex systems like foods or pharmaceutical products. Moreover, this problem becomes harder when dispersed non-aqueous systems must be characterized. In this context, chocolate represents an interesting water-free system on which emulsifiers play an irreplaceable role in the manufacturing process. Briefly, chocolate is a dispersion of sugar and cocoa particles into a fatty continuous phase (*i.e.*: cocoa butter or some approved replacers). In its production process two possible physical states are involved: in the early stages, chocolate is a fluid mass where its flow behavior must be closely controlled, whereas the late stages are a sum of perfectly traced crystallization steps that conclude in the solid final product.

In the fluid stage, the main role of surfactants is to coat the surface of sugar to favor the flowability of molten chocolate. Coating of dispersed particles avoids their aggregation by reducing both their interaction and friction. This controlled flow behavior leads to an important viscosity reduction which facilitates the next processing steps and reduces energy costs involved in this step [26].

At the last stages, a chocolate that have been badly crystallized results in the occurrence of "fat bloom", a grayish-white film on the product surface. Fat bloom is the main cause of economic losses in chocolate industry. This chocolate defect involves the existence of non-desired polymorphic forms of cocoa butter, and consequently a part of fat migrates from the core to the surface of chocolate [45, 46].

Emulsifiers influence both fluid and solid chocolate processing. Soybean lecithin is the most used chocolate emulsifier, and with other additives have been used to retard the development of fat bloom [47]. Interestingly, EC and soybean lecithin have shown synergic properties to achieve a good control of chocolate flow behavior. However, on its own EC was not effective for this purpose when compared with a reference sample containing only lecithin [26].

Considering that almost 100% of soy lecithin originates from genetically modified organisms (GMO) and that traces of DNA were found in soy by-products, the demand of non-GMO soybean lecithin has rising, especially in the European Union. Therefore, the suppliers of raw materials and consumers are expectant for soy lecithin replacers. With this in mind, in the search for alternatives to stabilize a lecithin-free chocolate using polymeric surfactants, we achieved both successful stabilization and flow behavior control using EC dissolved in medium chain triglycerides without a requirement of co-surfactants. Thus dissolved, EC was effective at very low concentration leading to a final product that was not only stable but was also absolutely free of off-flavors (Ceballos *et al.* unpublished results).

Nevertheless, and as mentioned above in different sections, it is very hard to disclose the mechanisms through which the emulsifier molecules exert the stabilizing effect in a non-aqueous dispersed system. Therefore, in order to gain insight into the mechanisms involved when using EC a simple water-free model can be used [8]. The model system chosen consists of PG-in-oil emulsions using EC as emulsifier. A mixture of medium chain triglycerides (MCT) and commercial soybean oil (SO) were used as continuous phase, and four ECs with different MMW were tested. The emulsions were prepared by dissolving EC in both oily continuous phases at concentrations between 2 and 23 % (w/w).

Thus, different proportions of PG (10-50 % by weight) were emulsified into above described EC solutions through vortexing or high shearing rotor-stator forces. Although EC cannot be dissolved in PG, homogeneous cream-like emulsions were obtained within a few seconds after PG addition to EC solutions in either oil, at room temperature.

Interestingly, although EC is insoluble in PG, cream-like emulsions were obtained at room temperature within a few seconds after PG addition to EC solutions in either oil. This almost instantaneous effect shows the good emulsifying effect of EC in addition to its stabilizing property. As mentioned above, this is an additional feature of EC as polymeric surfactant considering the fact that generally this type of surfactants are more effective in stabilization phenomena, whereas nonionic surfactants are generally more effective in the emulsification process [3].

Stability of obtained emulsions was approximately one month. After that, only a slight amount of exudate was macroscopically perceived, at room temperature. Moreover, ease in preparation is important to highlight, since diverse technical troubles are frequent in non-aqueous emulsions making.

Assessment of microstructure of PG-in-oil emulsions stabilized by EC was performed by means of optical microscopy. Results showed that microstructure of emulsions was similar whatever the polymer concentration as well as MMW was used (Fig. **5a** and **b**). Furthermore, destabilization is not observed if PGPR is used like co-surfactant (Fig. **5c**).

(a)

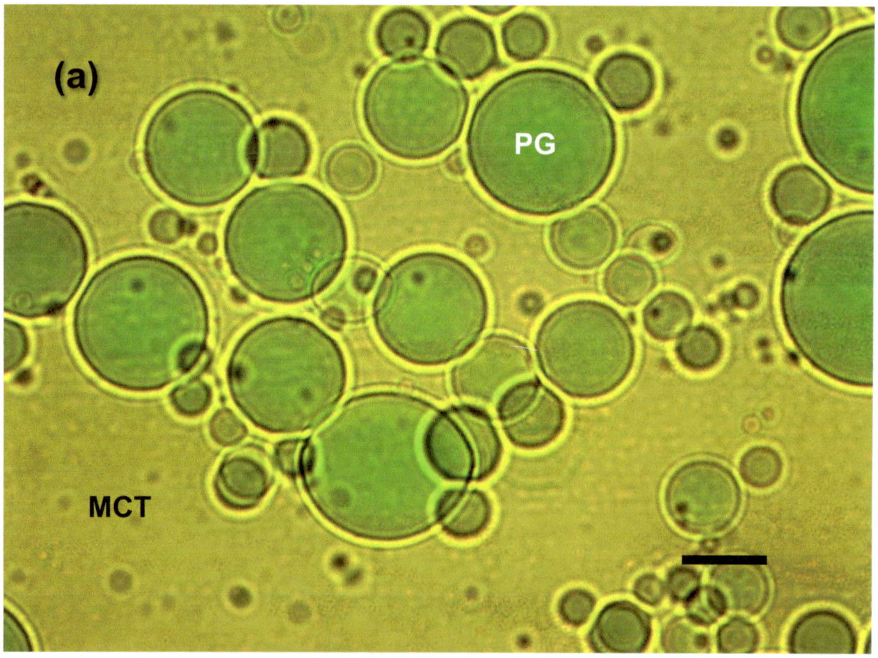

Fig. 5: contd…

(b)

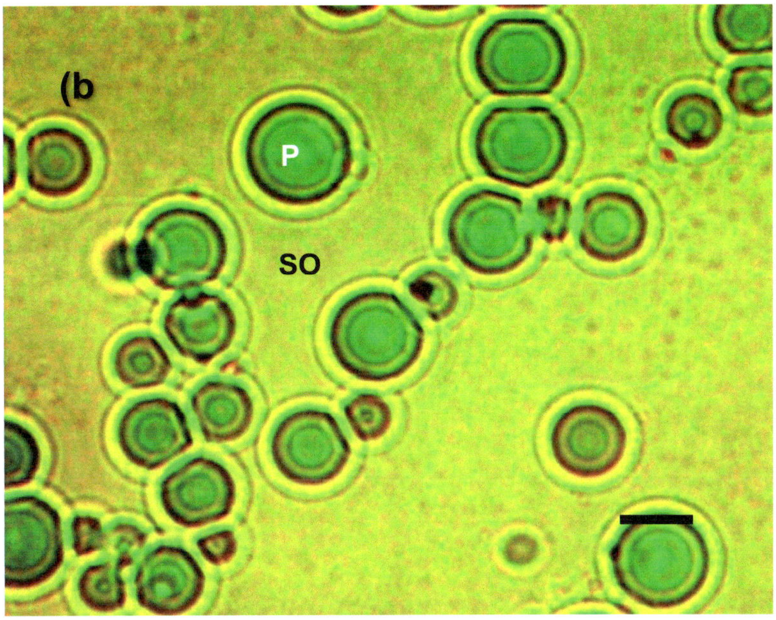

(c)

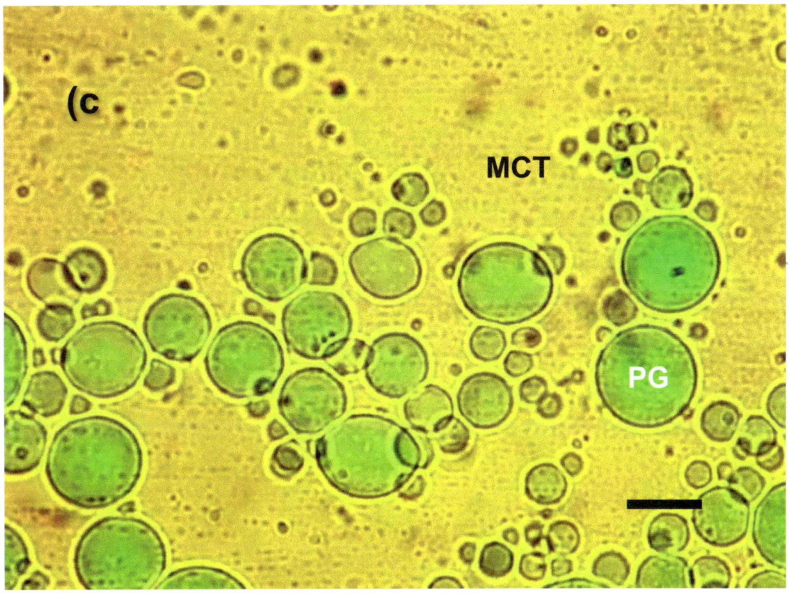

Figure 5: Photomicrograph of freshly prepared emulsions: **(a)** 2:1MCT:PG emulsion stabilized with 2% (w/w) EC-10 in the continuous phase, at 100×magnification (bar=100 μm), **(b)** 1:1 SO:PG emulsion stabilized with 1% (w/w) EC-20 in the continuous phase, at 100× magnification (bar =100 μm) and **(c)** 2:1 MCT:PG emulsion stabilized with 2% (w/w) EC-20 and 1% (w/w) PGPR in the continuous phase, at 100× magnification (bar=100 μm).

Among several important findings it is remarkable that at all EC concentrations PG-in-oil emulsions were obtained and phase inversion phenomena were never observed. The independence of flow behavior with the MMW of the polymer at high shear stress, as well as the inverse proportionality observed at low shear stress, was maybe the most surprising effect at the evaluation of PG-in-MCT emulsions (Fig. **6**). Increasing viscosity as function of MMW is a common trend of polymer solutions. Consistently, this behavior was observed when EC was added to oils. In this context, the independence of emulsion viscosity on the MMW of the polymer, or even an inverse dependence, indicates that the major component in the stabilization of PG-in-oil emulsions is thermodynamic rather than kinetic and is due to a reduction in surface free energy by EC.

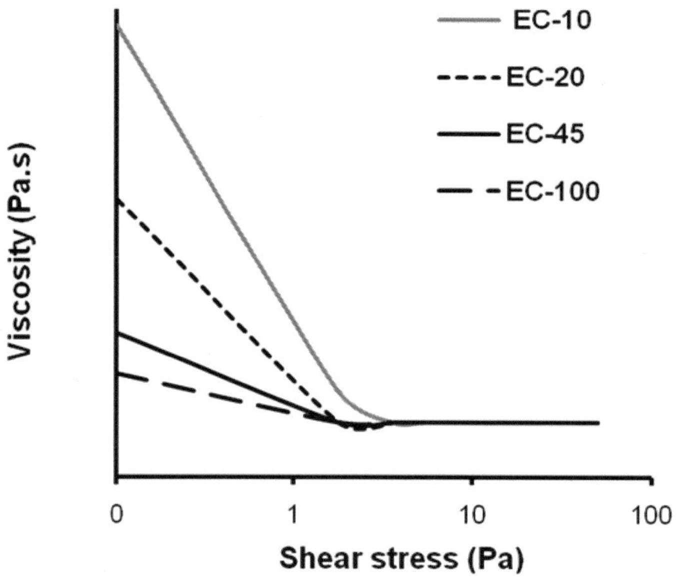

Figure 6: Flow behavior tendency at 25 °C for 50/50 PG/MCT emulsions stabilized with 2% EC of varying molecular weights, as a function of shear stress.

This effective stabilizing mechanism after interfacial adsorption of the polymer may be explained in a similar way to that observed with the inulin-based graft copolymer, but with inverse structuring of polymeric molecule (see above):

1. With EC, the strong anchoring of polymeric molecule to PG droplet surface should be produced through the glucose backbone chain. This

strong interaction could ensure very low desorption thus avoiding the eviction of the polymer from the interface.

2. Strong solvation occurring on the ethoxyl chains. Thus favoring an effective steric stabilization.

Non-Aqueous Emulsions and Pharmaceuticals Formulations

The rising field of drug design has generated an important number of molecules with specific therapeutic possibilities. However, in many cases the structural properties of these chemicals affect their water solubility [48]. In addition, a certain number of therapeutic drugs are more exposed to oxidative or hydrolytic phenomena in aqueous environments. In such cases, non-aqueous emulsions become an attractive option in the formulation of novel drug delivery systems [5, 49].

Research of Florence *et al.* is perhaps the most systematic work related to this topic. They have been working in formulation but with focus on comprehension of stability mechanisms of anhydrous dispersed systems, using dodecane - polyethylene glycol [50] and hydrocarbon – formamide systems [34]. Latest work has been revised in this chapter in relationship with the use of polymeric surfactants (see above), but it also includes a brief description of capacity of these systems to charge and release for dehydroepiandrosterone, a very liphopilic therapeutic drug [34].

In a posterior study and in the search of suitable components for pharmaceutical formulations, castor oil was dispersed in several types of silicone oils, and such systems were stabilized using polymeric surfactans [5, 51]. Release profiles for dexamethasone and dehydroepiandosterone were studied, previous solubilization of these lipophilic drugs into the dispersed phase [5, 51].

Silicone oils have also been used in polyol-in-silicone non-aqueous emulsions in order to protect the skin-whitening agent deoxyarbutin, which acts trough tyrosinase activity inhibition. This is a very effective drug in its class, but it is highly water-sensitive. However, this molecule has been protected when propylene glycol and glycerin were used to solubilize it instead water-based solvent, and then this solution was dispersed into silicone-oil. The resultant non-aqueous emulsions showed enhanced protection ability over deoxyarbutin when compared with water-based dispersed phase emulsions. An highlighting aspect of this work, is the fact that emulsions were stabilized through polyol-based emulsifiers, *i.e.* polymeric surfactants [49].

More recently, several advantages in either formulation as well as therapeutic efficacy have been evaluated using oil-in-oil emulsions as carriers for nonivamide, a pharmaceutical agent used in the treatment of different dermal diseases. Anew, the major stabilizing effect was achieved by the action of a polymeric emulsifier, a silicone surfactant. However, authors emphasize the need of previous experimental evaluation to define the optimal continuous-dispersed phases and surfactant combinations [52].

These are some examples of non-aqueous emulsions (in many cases, literally oil-in-oil emulsions) that have been increasingly used as vehicles of different therapeutic drugs. *In vitro* release and degradation profiles, as well as optimal concentration ratios for their formulation are well described. However, working with this type of emulsions becomes harder when lack of rules for their preparation and stabilization is considered. In addition, the well established rules to water-based emulsions preparation, only in a few cases are applicable to non-aqueous emulsions formulation (see previous sections).

CONCLUDING REMARKS AND FUTURE PERSPECTIVES

From the thermodynamic point of view, formation of an emulsion implies an increase in the surface area of the dispersed liquid in contact with the dispersion medium. This leads to a very important increase in surface free energy that ends being a major driving force leading to emulsion destabilization and breakdown. From this, any substance that reduces the surface free energy will have a stabilizing effect allowing that more surface area can be in contact without increasing the total free energy of the system. This is the most obvious use and definition of surfactants. The emulsions formed by this mechanism are generally microemulsions (size $\leq$ 100 nm).

On the other side, whenever a thermodynamically unstable system is formed kinetic barriers can be incorporated into the system to impair the movement, aggregation and fusion of the droplets of dispersed phase that would eventually lead to breaking of the emulsion. As an obvious consequence of their high molecular weight, polymers generally increase the viscosity of the solvent in which they are dissolved. Therefore, as a side effect, this increase in viscosity will necessarily reduce the movement of the dispersed droplets leading to a reduction in the kinetics of aggregation, fusion and phase separation.

A third aspect that is not frequently taken into account and that would lead to an increased stability of an emulsion is steric impairment of droplet aggregation and fusion due to particles associated to the surface of already formed droplets ("Pickering emulsion").

As mentioned above, there are two important aspects that should be taken into consideration when preparing an emulsion: a) the energy involved in the emulsification process and b) the stability of the product obtained.

In the particular case of non-aqueous emulsions, a rule of thumb would be that low molecular weight molecules will generally be useful for the emulsification process and high molecular weight amphiphiles will be better at contributing to the stability of the product. Therefore, combining different proportions of a specially selected surface active polymer dissolved in oil with low viscosity would generally allow modulating rheological parameters and stability of the product in order to optimize preparation procedure and final properties.

Although water is "the solvent", there are many active ingredients of foods or medicines that are either unstable or insoluble in water. Furthermore, under many circumstances water should be considered a reactive undesirable solvent. On the other side, there are nonhomogeneous systems that, like chocolate, are intrinsically free of water and therefore should be considered non aqueous. These facts, added to the advantages of biopolymers to be easily engineered for special applications, make it reasonable to assume that this field will surely expand to other disciplines and we will see greater strides in the near future.

As it has been discussed in this chapter, polymeric surfactants are the most preferable option to prepare and stabilize non-aqueous emulsions, because they can be manipulated in different ways (chemical structure, concentration, *etc*) to optimize kinetics and steric emulsions features. If a number of them are not strictly biopolymers, recent research works are encouraging related to obtaining totally biocompatible non-aqueous emulsions, mainly in pharmaceutical areas. Nevertheless, a lack of rules that allow to predict either its formulation as well as its stabilizing effect, are still an impediment that leads to an unavoidable loss of time at the moment of the experimental design. This represents the biggest challenge for the future research in this field.

ACKNOWLEDGEMENTS

DMB and IDB are staff members of CONICET. We wish to acknowledge Andrew Gee for his technical assistance with English grammar and spelling.

CONFLICT OF INTEREST

The authors declare no conflict of interest. All the work has been supported with grants from CEPROCOR and CONICET.

ABBREVIATIONS

DMSO	=	dymethyl sulfoxide
EC	=	ethycellulose
FA	=	formamide
HLB	=	hydrophile-liphophile balance
HPMC	=	hydroxylpropyl methylcellulose or hypromelose
MCT	=	medium chain triglycerides
MMA	=	mean molecular area
MMW	=	mean molecular weight
PB	=	polybutadiene
PEO	=	poly (ethylene oxide)
PG	=	propylene glycol
PMMA	=	poly (methyl methacrylate)
PPO	=	poly (propylene oxide)
PS	=	polystyrene
PtBuS	=	poly (ter butylstyrene)
PVA	=	polyvinyl alcohol
SO	=	soybean oil

REFERENCES

[1] Tadros TF, Vincent, B. Emulsion stability. In: Becher P, Ed. Encyclopedia of Emulsion Technology. New York, Marcel Dekker 1983; vol. *1*, pp. 129-285.

[2] Walstra P. Physical principles of emulsion science. In: Blanshard J, Lillford PJ, Eds. Food Structure and Behaviour. London, Academic Press 1987; pp. 87-106.

[3] Tadros TF. Emulsion Formation, Stability, and Rheology. In: Tadros TF, Ed. Emulsion Formation and Stability. 1st ed. KGaA, Weinheim, Wiley-VCH Verlag GmbH & Co 2013; pp. 1-75.

[4] Petersen RV, Hamill RD. Studies in nonaqueous emulsions. J Soc Cosmet Chem 1968; 19: 627–40.

[5] Suitthimeathegorn O, Jaitely V, Florence AT. Novel anhydrous emulsions: Formulation as controlled release vehicles. Int J Pharm 2005; 298: 367–71.

[6] Atanase L-I, Riess G. Block copolymer stabilized nonaqueous biocompatible sub-micron emulsions for topical applications. Int J Pharm 2013; 448: 339–45.

[7] Patel N, Schmid U, Lawrence MJ. Phospholipid-based microemulsions suitable for use in foods. J Agric Food Chem 2006; 54: 7817–24.

[8] Ceballos MR, Brailovsky V, Bierbrauer KL, Cuffini SL, Beltramo DM, Bianco ID. Effect of ethylcellulose on the structure and stability of non-aqueous oil based propylene glycol emulsions. Food Res Int 2014; 62: 416–23.

[9] Hasenhuettl GL. Overview of Food Emulsifiers. In: Hasenhuettl GL, Hartel RW, Eds. Food Emulsifiers and Their Applications. 2nd ed. New York, Springer 2010; pp. 1-9.

[10] Garti N. What can nature offer from an emulsifier point of view: trends and progress? Colloids Surf A Physicochem Eng Asp 1999; 152: 125–46.

[11] Hasenhuettl GL. Forecasting the Future of Food Emulsifiers. In: Hasenhuettl GL, Hartel RW, Eds. Food Emulsifiers and Their Applications. 2nd ed. New York, Springer 2010; pp. 395-402.

[12] Bouyer E, Mekhloufi G, Rosilio V, Grossiord J–L, Agnely F. Proteins, polysaccharides, and their complexes used as stabilizers for emulsions: Alternatives to synthetic surfactants in the pharmaceutical field? Int J Pharm 2012; 436: 359 – 78.

[13] Melzer E, Kreuter J, Daniels R. Ethylcellulose: A new type of emulsion stabilizer. Eur J Pharm Biopharm 2003; 56: 23–7.

[14] Hasenhuettl GL. Synthesis and Commercial Preparation of Food Emulsifiers. In: Hasenhuettl GL, Hartel RW, Eds. Food Emulsifiers and Their Applications. 2nd ed. New York, Springer 2010; pp. 11-37.

[15] Buggins TR, Dickinson PA, Taylor G. The effect of pharmaceutical excipients on drug disposition. Adv Drug Deliv Rev 2007; 59: 1482-503.

[16] Weiszhár Z, Czúcz J, Révész C, Rosivall L, Szebeni J, Rozsnyay Z. Complement activation by polyethoxylated pharmaceutical surfactants: Cremophor-EL, Tween-80 and Tween-20. Eur J Pharm Sci 2012; 45 (4): 492–8.

[17] Liwarska-Bizukoj E, Miksch K, Malachowska-Jutsz A, Kalka J. Acute toxicity and genotoxicity of five selected anionic and non-ionic surfactants. Chemosphere 2005; 58: 1249-523.

[18] Scherlunda M, Malmstenb M, Brodina A. Stabilization of a thermosetting emulsion system using ionic and nonionic surfactants. Int J Pharm 1998; 173: 103-16.

[19] Dickinson E. Towards more natural emulsifiers. Trends Food Sci Tech 1993; 4: 330–4.

[20] Benichou A, Aserin A, Garti N. Double emulsions stabilized with hybrids of natural polymers for entrapment and slow release of active matters. Adv Colloid Interface Sci 2004; 108–109: 29–41.

[21] Kralova I, Sjöblom J. Surfactants Used in Food Industry: A Review. J Dispers Sci Technol 2009; 30: 1363–83.

[22] Yang Y, Leser ME, Sher AA, McClements DJ. Formation and stability of emulsions using a natural small molecule surfactant: Quillaja saponin (Q-Naturale®). Food Hydrocoll 2013; 30 (2): 589–96.

[23] Bouyer E, Mekhloufi G, Huang N, Rosilio V, Agnely F. β-Lactoglobulin, gum arabic, and xanthan gum for emulsifying sweet almond oil: Formulation and stabilization mechanisms of pharmaceutical emulsions. Colloids Surf A Physicochem Eng Asp 2013; 433: 77–87.

[24] Imhof A, Pine DJ. Stability of Nonaqueous Emulsions. J Colloid Interface Sci 1997; 192: 368–74.

[25] Riess G, Cheymol A, Hoerner P, Krikorian R. Non-aqueous emulsions stabilized by block copolymers: application to liquid disinfectant-filled elastomeric films. Adv Colloid Interface Sci 2004; 108 –109: 43–8.

[26] Do T-A L, Mitchell JR, Wolf B, Vieira J. Use of ethylcellulose polymers as stabilizer in fat-based food suspensions examined on the example of model reduced-fat chocolate. React Funct Polym 2010; 70: 856–62.

[27] Ushikubo FY, Cunha RL. Stability mechanisms of liquid water-in-oil emulsions. Food Hydrocoll 2014; 34: 143–53.

[28] Nylander T, Arnebrant T, Bos M, Wilde P. Protein/Emulsifier Interactions. In: Hasenhuettl GL, Hartel RW, Eds. Food Emulsifiers and Their Applications. 2nd ed. New York, Springer 2010; pp. 89-171.

[29] Rutledge TF, Hughes FA, Galvin TJ, Zech JD. New Polymeric Surfactants from Butadiene, Urea and Sulfuric Acid. J Am Oil Chem Soc 1967; 44: 367-71.

[30] Kuwamura T, Takahashi H, Hatori T. Surface Active Block Copolymers: I. The Preparation and Some Surface Active Properties of Block Copolymers of Tetrahydrofuran and Ethylene Oxide. J Am Oil Chem Soc 1971; 48: 29-34.

[31] Tadros T. Polymeric surfactants in disperse systems. Adv Colloid Interface Sci 2009; 147–148: 281–99.

[32] Xie H-Q, Xie D. Molecular design, synthesis and properties of block and graft copolymers containing polyoxyethylene segments. Prog Polym Sci 1999; 24: 275–313.

[33] Narsimhan G, Wang Z. Guidelines for Processing Emulsion-Based Foods. In: Hasenhuettl GL, Hartel RW, Eds. Food Emulsifiers and Their Applications. 2nd ed. New York, Springer 2010; pp. 349-94.

[34] Sakthivel T, Jaitely V, Patel NV, Florence AT. Non-aqueous emulsions: hydrocarbon–formamide Systems. Int J Pharm 2001; 214: 43–8.

[35] Liu S, Armes SP. Recent advances in the synthesis of polymeric surfactants. Curr Opin Colloid Interface Sci 2001; 6: 249-56.

[36] Hayakawa K, Kawaguchi M, Kato T. Protective colloidal effects of hydroxypropyl methyl cellulose on the stability of silicone oil emulsions. Langmuir 1997; 13: 6069–73.

[37] Sun W, Sun D, Wei Y, Liu S, Zhang S. Oil-in-water emulsions stabilized by hydrophobically modified hydroxyethyl cellulose: Adsorption and thickening effect. J Colloid Interface Sci 2007; 311: 228–36.

[38] Schulz M, Daniels R. Hydroxypropylmethylcellulose (HPMC) as emulsifier for submicron emulsions: influence of molecular weight and substitution type on the droplet size after homogenization. Eur J Pharm Biopharm 2000; 49: 231–6.

[39] Mezdour S, Lepine A, Erazo-Majewicz P, Ducept F, Michon C. Oil/water surface rheological properties of hydroxypropyl cellulose (HPC) alone and mixed with lecithin: Contribution to emulsion stability. Colloids Surf A Physicochem Eng Asp 2008; 331: 76–83.

[40] Alhnan MA, Basit AW. Engineering polymer blend microparticles: An investigation into the influence of polymer blend distribution and interaction. Eur J Pharm Sci 2011; 42: 30–6.

[41] Aiache JM, Gauthier P, Aiache S. New gelification method for vegetable oils I: Cosmetic application. Int J Cosmetic Sci 1992; 14: 228–34.

[42] Gravelle AJ, Barbut S, Marangoni AG. Ethylcellulose oleogels: Manufacturing considerations and effects of oil oxidation. Food Res Int 2012; 48: 578–83.

[43] Zetzl AK, Marangoni AG, Barbut S. Mechanical properties of ethylcellulose oleogels and their potential for saturated fat reduction in frankfurters. Food Funct 2012; 3: 327–37.

[44] Almeida IF, Bahia MF. Evaluation of the physical stability of two oleogels. Int J Pharm 2006; 327: 73–7.

[45] Shenk H, Peschar R. Understanding the structure of chocolate. Radiat Phys Chem 2004; 71: 829–35.

[46] Smith KW, Zand I, Talbot G. Effect of Antibloom Fat Migration from a Nut Oil Filling on the Polymorphic Transformation of Cocoa Butter. J Agric Food Chem 2008; 56: 1602–5.

[47] Schantz B, Rohm H. Influence of lecithin–PGPR blends on the rheological properties of chocolate. Lebenson Wiss Technol 2005; 38: 41–5.

[48] Kalepun S, Manthina M, Padavala V. Oral lipid-based drug delivery systems – an overview. Acta Pharm Sin B 2013; 3 (6): 361–72.

[49] Lin CC, Yang CH, Chang NF, *et al.* Study on the stability of deoxyArbutin in an anhydrous emulsion system. Int J Mol Sci 2011; 12 (9): 5946-54.

[50] Sakthivel T, Wan KW, Florence AT. Formulation of non-aqueous emulsions. Pharm Sci Suppl 1999; 1: 685.

[51] Jaitely V, Sakthivel T, Magee G, Florence AT. Formulation of oil in oil emulsions: potential drug reservoirs for slow release. J Drug Deliv Sci Technol 2004; 14: 113–7.

[52] Rottke M, Lunter DJ, Daniels R. *In vitro* studies on release and skin permeation of nonivamide from novel oil-in-oil-emulsions. Eur J Pharm Biopharm 2014; 86: 260–6.

CHAPTER 12

Biopolymers for Neurotherapeutic Interventions

Valence M.K. Ndesendo*

Department of Pharmaceutics & Formulation Sciences, School of Pharmacy & Pharmaceutical Sciences, St. John's University of Tanzania, Dodoma, Tanzania

Abstract: Developing new neuropharmaceuticals presents formidable economic, scientific and medical challenges mainly due to the complex nature of the central nervous system (CNS), in particular the blood-brain barrier (BBB). The constituents of the BBB and its drug efflux system form some kind of a boundary which prevents the entry of potent neurotherapeutics from entering the brain thus leading to treatment inefficiency for various CNS diseases. Conventional neurotherapeutic interventions for major disease conditions include among others, ablative limbic system surgeries, vagus nerve stimulation, transcranial magnetic stimulation and deep brain stimulation. These interventions are however somehow invasive. The progress so far made in the area of interventional neurology has greatly widened the scope of viable pharmaceutical drug delivery systems for which biopolymers may play a substantial role. The major aim is to improve specificity in drug delivery (*e.g.* through drug targeting) and to avoid or reduce the adverse effects imposed by the conventional neurotherapeutic interventions. This Chapter therefore seeks to provide a detailed discussion of the potential application of biopolymers in neurotherapeutic interventions citing most the: i) strategies for neurotherapeutic interventions (including biopolymer gene-based drug delivery, bioactive release from nerve conduits, nonviral gene delivery vectors, molecular imaging as a novel technique for biopolymer-based neurotherapeutic intervention); and ii) biopolymers used in neurotherapeutic interventions (including biopolymeric vehicles, biopolymer scaffolds, multiple polymer-based layers, polysaccharide-based biopolymers, protein-mimicked polypeptides and protein-based biopolymers). Finally, this Chapter seeks to provide a detailed description about the future prospects for biopolymer-based neurotherapeutic interventions.

Keywords: Biopolymers, biopolymer scaffolds, biopolymeric vehicles, blood-brain barrier, drug delivery, drug targeting, gene delivery, molecular imaging, nerve conduits, neurotherapeutic interventions.

1. INTRODUCTION

Biopolymers are materials based on re-growing resources which have thus far found massive attraction in biomedical applications since they can inherently

***Corresponding author Valence M.K. Ndesendo:** Department of Pharmaceutics & Formulation Sciences, School of Pharmacy & Pharmaceutical Sciences, St. John's University of Tanzania, P.O. Box 47, Dodoma, TANZANIA; Tel: +255-26-2390044; Cell: +255-754-262263; Fax: +255-26-2390025; Fax/Email: +255-86-765-6130; E-mail: vndesendo@sjut.ac.tz/vndesendo@gmail.com

Viness Pillay, Yahya E. Choonara and Pradeep Kumar (Eds.)

combine degradability *in vivo,* can offer sites of adhesion for cells and proteins, often show good biocompatibility and may furthermore be used to release the embedded bioactive molecules in the most desirable manner [1, 2]. For proper management of neuronal infections and neurodegenerative diseases using neurotherapeutics, there has been various strategies in place for direct and indirect liberation of drugs to the brain including the employment of biopolymer-based therapeutic agent carriers [3-5]. However, due to the restrictions from the blood brain barrier (BBB), the delivery of neurotherapeutics to the brain has remained to be a challenging task to date [1, 6]. Traditionally, such methods like direct injections or convection-enhanced delivery of drug or cerebrospinal fluid or intranasal delivery have been used. These approaches have however proven to be not very safe, have high degree of invasiveness, and may just last for short periods of time [6, 7]. As an alternative, other molecular-based therapeutic modalities have been under development for the management of brain diseases. These therapeutic modalities have made it possible to target the neurotherapeutic agent to the designated locations thereby stopping the abnormal pathway signals in the brain. With this scenario therefore, therapeutic agents that may pass by the brain barriers, particularly the BBB, are better preferred so as to decrease as much as possible the neuronal injury-based risks, non-delivery, and treatment inefficiency. These drawbacks have as such called for the needful approaches in generating novel biopolmer-based drug carriers that are capable of passing through the blood cerebrospinal fluid barrier (BCB), cerebrospinal fluid (CSF) brain barrier, BBB and some specialized barriers such as blood tumor barrier (BTB) (in case of brain tumor) to deliver neurotherapeutics safely to the brain. These carriers must be able to passively pass through discontinuous tight junctions found in the BBB with perhaps the assistance from plasmalemmal vesicles and endothelial cells windows. BBB and the role that it plays in obstructing drug delivery to CNS is schematically illustrated in Fig. (**1**).

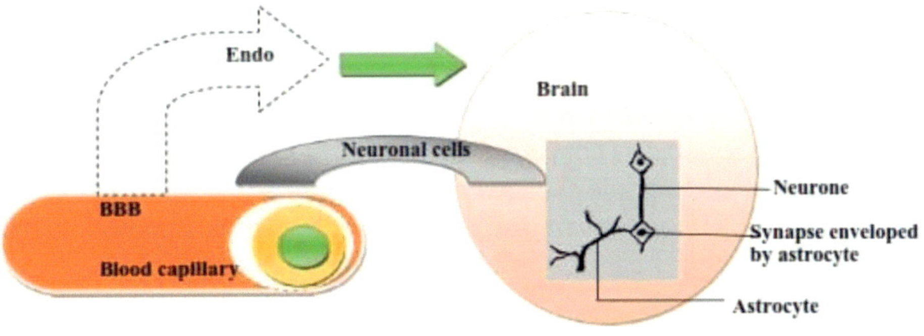

Figure 1: A schematic depicting the blood brain barrier at the blood capillary endothelium that restricts the delivery of drugs to the CNS [6].

Noteworthy is that special strategies must be designed so as to attain an improved safe delivery of various drugs to the CNS. Among the means used to achieve this strategy are the use of nanotechnology-based drug carriers such as colloids and nanoparticles and also the use of micelles and liposomes. More specifically, nano-enabled drug delivery systems have a bigger advantage in the sense that they often lead to improved uptake of neurotherapeutic agents to the brain and may also target these agents to the desired locations in the brain.

Gene therapy has found a great application in management of CNS conditions [8, 9]. In order to succeed in transferring genes to specific sites, they must be endowed with the capability of escaping the processes that interferes with the nature of the macromolecules. In addition, the gene degradation caused by serum nucleases should absolutely not be in place. For this matter, genes should be encapsulated in the process of their delivery so as to protect them up until when they reach their designated targets. The gene carries should ideally have the possible smallest size so as to be able to enter into the cells and finally cross over to the nucleus. These carries must also have the capability of escaping endosome–lysosome route following endocytosis [10].

Viral and non-viral vectors are among the delivery systems for genes. They have been found to have the desirable attributes for gene delivery including low immunogenicity as well as the ease with which their properties can be controlled. It has been shown that cationic polymers can undergo complexation with DNA [11]. Cationic polymer-DNA complexation is another form of non-viral vectors for gene therapy applications. The specificity onto the nanoparticle outermost layers may be introduced by conjugating cell-specific ligands to the outermost layers of nanoparticles which will thereby give the provision for targeted transgene expression [11]. In the recent years, nerve conduits have evolved to be considerably important when compared to their hollow polymer tube counterparts with regards to the significant modifications in scaffold design and selection of polymer materials [12]. Much as there has been major advances and various approaches in the design of nerve conduits and repair strategies, the issue of insufficient functional recovery after peripheral nerve injury has still remained to be a challenge that needs to be overcome [13-15]. Much research in this field of tissue regeneration, has concentrated most on the fabrication of nerve conduits that are able to provide multiple features that may possibly enhance nerve regrowth and restore functional recovery in a smaller time frame as opposed to previous strategies which were solely based on bridging nerve gap defects [16].

In the former years, research was primarily targeted at bridging nerve gaps and evaluation of the results thereafter with almost neglecting the determination of the extent of functional recovery achieved as well as the degree of similarity to physiological tissues [17]. At the moment, much focus is being placed on producing nerve conduits that are able to closely mimic the structure and function of native nerve tissues. This strategy will purportedly improve the promotion of nerve regeneration to a level equal to or above that of a nerve autograft [18-20]. In order to be able to attain such degree of resemblance to native nerve tissues, numerous physical, chemical and biological factors need to be well observed when incorporating multiple-functioning components into a nerve conduit [21, 22]. It has been shown that regeneration of an injured spinal cord may be hindered by the formation of glial scar tissue, inhibitory molecules contained within this scar, the failure in producing molecules that may promote axon growth, and the poor regenerative ability of most CNS neurons [23].

Several strategies have been investigated to promote spinal cord injury (SCI) repair, including cell transplantation, application of neurotrophic factors, administration of enzymes to remove growth inhibiting molecules, implantation of nerve guidance channels, and combinations of these treatments. Currently, there is no single available treatment that is capable of completely restoring function after SCI. It has been established that since there are various barriers to regeneration in an injured spinal cord, a combination of two approaches might be best for promoting axonal growth across a spinal cord lesion and promote functional recovery [4, 24]. Providing a growth permissive substrate in the form of a biomaterial scaffold while as well delivering molecules that can degrade the inhibitory glial scar and/or promote axon growth *via* the scaffold will enable the bridging of the lesion at the site of injury to the axons [4, 24].

Molecular imaging basically encompasses the techniques for enhancing the visualization and investigation of complex biochemical events. It may be defined as the noninvasive, real-time visualization of biochemical events at the cellular and molecular level within living entities [25-27]. Molecular imaging is as such emerging as a valuable technique for thoroughly studying and understanding the inner functions of the body both in disease and non-disease conditions. In this book chapter in particular, molecular imaging has therefore been cited as a novel technique for biopolymer-based neurotherapeutic interventions.

The aim of Chapter has therefore been to provide a detailed overview of the potential application of biopolymers in neurotherapeutic interventions, mainly citing at: i) strategies for neurotherapeutic interventions as well as biopolymers used in neurotherapeutic interventions. The strategies will include biopolymer gene-based drug delivery, bioactive release from nerve conduits, biopolymers as specific nonviral gene delivery vectors and molecular imaging as a novel technique for biopolymer-based neurotherapeutic intervention. Biopolymers used in neurotherapeutic interventions will include among others the biopolymeric vehicles and biopolymer scaffolds. Finally, this Chapter has sought to provide a detailed description about the future prospects for the use of biopolymers in neurotherapeutic interventions.

2. STRATEGIES FOR NEUROTHERAPEUTIC INTERVENTIONS

2.1. Biopolymer Gene-Based Drug Delivery for Neurotherapeutic Interventions

2.1.1 Gene-Based Liposomes for Neurotherapeutic Intervention

One of the most difficult endeavors in brain drug delivery is the passage of large anionic bioactive DNA across the brain cells [28]. Much as DNA becomes destroyed by circulating and intracellular deoxyribonucleases, it is still a needful requirement that it must be carried to the brain cells undestroyed. Thus far, liposomes have proved to be capable of delivering DNA intracellularly quite efficiently [29]. Fig. (**2**) is a schematic depicting a DNA-liposome complex.

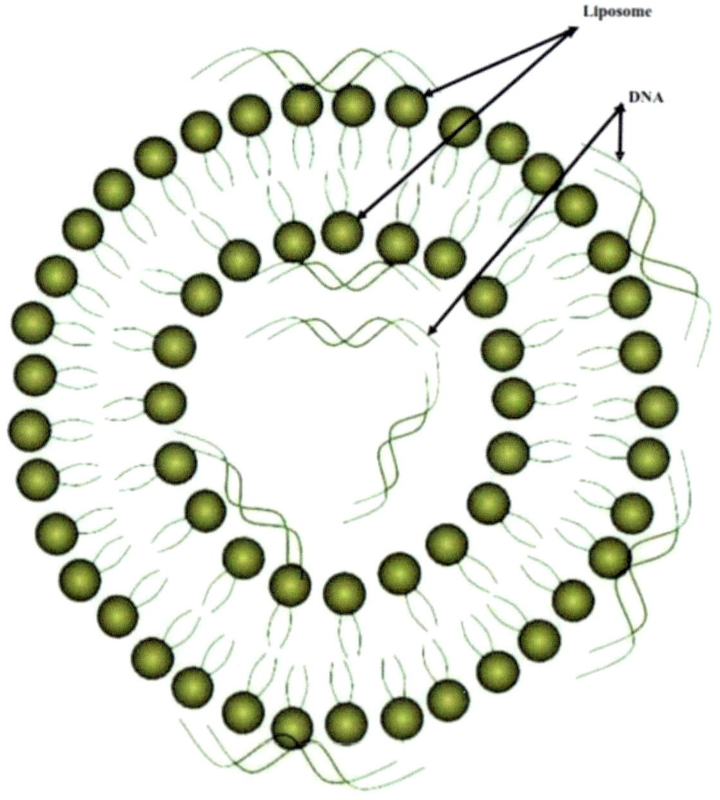

Figure 2: A Schematic depicting a DNA-Liposome complex.

2.1.2. Gene –Based Pegylated Immune Liposomes for Neurotherapeutic Intervention

There has been an attempt to treat Parkinson's disease (PD) using dual gene therapy with the purpose of replacing striatal TH gene expression with TH gene therapy, as well as halting or reversing nigro-striatal tract neurodegeneration with neurotrophin gene therapy using pegylated immune liposomes that are able to cross the BBB [30]. In this treatment strategy, liposomes are directed towards the BBB *via* an attachment to either insulin receptor or transferrin receptor (TfR) through which the pegylated immunoliposomes (PIL) becomes capable of reaching the neuronal nuclear compartment from the circulation [30].

2.2. Bioactive Release from Nerve Conduits as a Strategy for Neurotherapeutic Intervention

2.2.1. Optimal Peripheral Nerve Regeneration as a Strategy for Neurotherapeutic Intervention

An injured tissue requires four basic components in order to achieve successful peripheral nerve regeneration. These component are: i) scaffold; ii) the inclusion of cells; iii) growth factors; and iv) extracellular material (ECM) molecule for the neuronal survival, optimal growth, and ultimate regeneration potential (Fig. **3**). Essentially the design of artificial nerve conduits must be in such a way that they are ultimately able to provide a scaffold for the support of growing tissues in conjunction with at least one other component to create a basis for the model of a multifunctioning nerve conduit [31, 32]. Promotion of axonal regeneration and functional recovery can be significantly improved by employing such elements like scaffold design, growth factor delivery, supply of extracellular matrix proteins, and a substrate or matrix for the attachment of growing cells. These features are quite important since they contribute substantially in neural cell proliferation as well as the maintenance of cell shape in addition to guiding and providing mechanical strength to developing axons [33, 34].

Research has shown that integrating these components into a nerve conduit delivery system will enable to a significant degree the control over the length and processes of the Wallerian degeneration phase [35, 36]. Wallerian degeneration is a term that describes the rapid breakdown of axons and myelin sheaths following the injury of neural tissues (Fig. **4**) [37]. Much as it is a process triggered for the creation of a microenvironment supportive for nerve regeneration, it further initiates the onset of inflammation and intensification of pain [21, 35, 38].

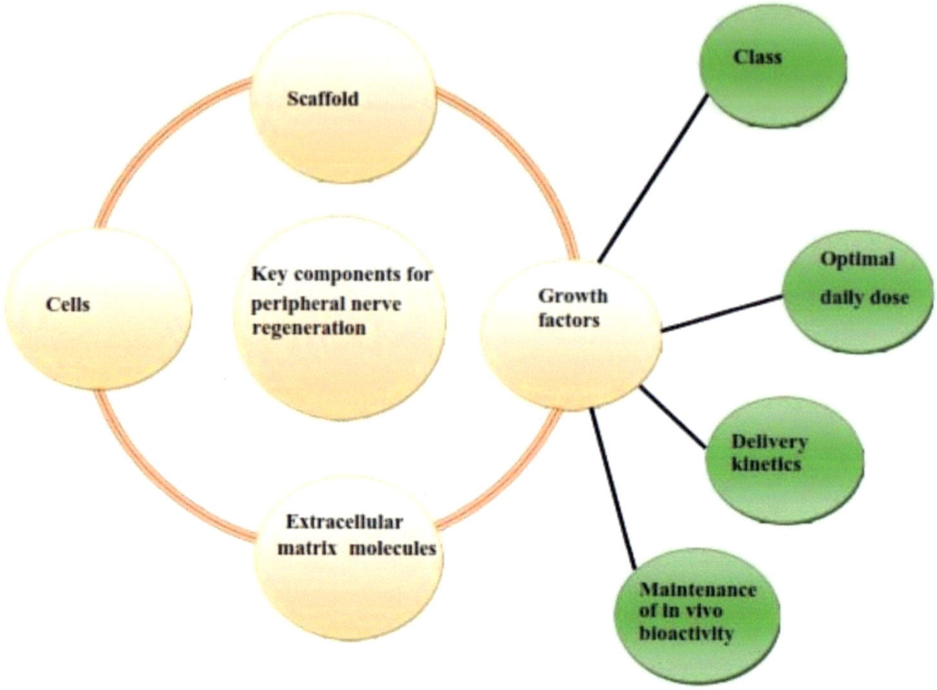

Figure 3: A schematic depicting the basic components required for peripheral nerve regeneration and the factors responsible for the delivery of growth factors {adapted from: [12]}.

2.3. Nonviral Gene Delivery Vectors for Neurotherapeutic Interventions

Viruses are essentially efficient gene vectors that have found a great application in the clinical trials of gene therapy (> 70%) [39]. However, there have been several concerns with regards to their use as they are said to be toxic and also they are capable of generating a strong immune response [39]. It has been established that non-viral delivery systems are devoid of most of these problems (*i.e.* they are inherently safer than viral carriers) and therefore are currently featuring as favorable alternatives to viral vectors [40-42]. The polymeric delivery systems that employ the genes and liposomes are considered to be non-viral gene vectors that are specifically based on nanotechnology [41].

Researchers have thus far established that the benefits of non-viral drug delivery system (NDDS) to patients are among others; fewer adverse effects with the active ingredient becoming more efficacious, and also the great chance for treating diseases for which it may currently be impossible to treat with conventional drugs [39, 41]. Much as the production of NDDS might seem to be

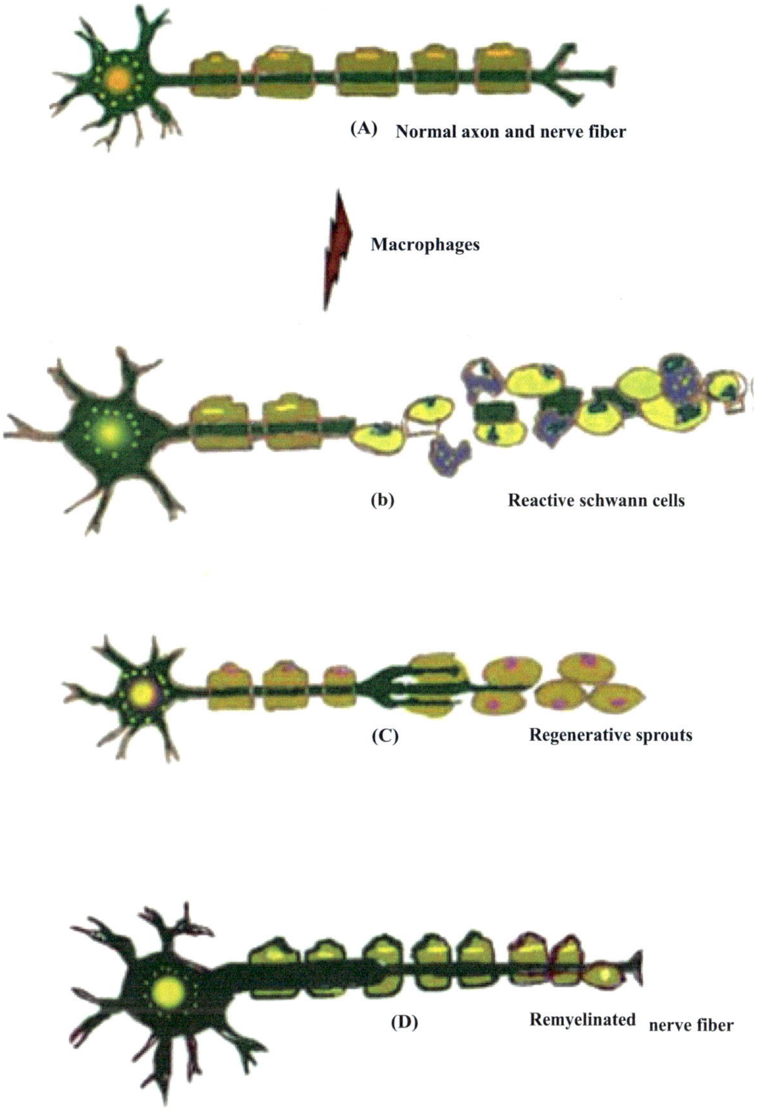

Figure 4: A schematic depicting degeneration and regeneration after peripheral nerve injury: **a**) normal neuron and nerve fiber; **b**) Wallerian degeneration; **c**) Schwann cells in the distal segment lining up in bands of Büngner; **d**) axonal reconnection with end organs as well as maturation and remyelination of the nerve fiber (adapted from: [12]).

an expensive exercise, the efficacy of this delivery system will obviously result into reduced disease treatment costs. It has been proved that non-viral vectors may overcome such safety problems like the risk of excessive immune response (*e.g.* by adenovirus) and insertion mutagenesis (*e.g.* by retroviruses) [3, 43]. Non-viral vectors are composed of either natural vectors (plasmid DNA or small nucleic acids, antisense oligonucleotides, small interfering RNAs) or synthetic vectors (liposomes, cationic polymers). The easiest way by which a NDDS can transfer a gene into a target cell is by the use of naked DNA, usually in plasmid form [44, 45]. The transferance of a bare plasmid do take place quite inefficiently. This has necessitated the employment of various physical (electroporation, ultrasound, gas-filled micro-bubbles) and chemical (liposomes) means to improve this process [46]. Non-viral vectors may as well be classified as organic (lipid complexes, conjugated polymers, cationic polymers *etc*) or inorganic (magnetic nanoparticles, quantum dots, carbon nanotubes, gold nanoparticles, *etc*) delivery systems [47]. Natural biomaterials are among the materials that have been recently employed to design non-viral vectors the main reason being the exclusive characteristics with which they are endowed with including biodegradability, biocompatibility and controlled release.

Nanoparticles have been found to be very useful in gene delivery by the virtue of the fact that they are capable of ensuring both targeted and sustained delivery since they can protect the genes from being destroyed by nuclease degradation thereby making them even more stable [48-51].

2.4. Molecular Imaging as a Novel Technique for Biopolymer-Based Neurotherapeutic Intervention

2.4.1. Application of Molecular Imaging in Biopolymer-Based Neurotherapeutic Intervention

Despite the remarkable achievements in neurological research in the recent past, brain is such a complicated entity that some of its basic functions and pathologies have remained to be enigmatic. Apparently the means by which the brain cells and synapses coordinate into forming intricate networks, and participate in cross-talk are still not well understood [25]. The meager understanding about the brain and its functionalities coupled with the molecular processes involved in neurological disease, contributes to the difficulties encountered in early diagnosis of brain diseases.

In the past, a CNS disease would be discovered when the disease condition has become worse and sometimes during postmortem analysis. However, this trend has been changing gradually, certainly due to the introduction of noninvasive imaging techniques such as magnetic resonance imaging (MRI), positron emission tomography (PET) and single photon emission computed tomography

(SPECT). Such techniques have made it possible to disclose major features of various neurological diseases, while keeping track of development and progression of these diseases as time goes on. Efficient and successful employment of molecular imaging has resulted into the provision of imminent overview on a multitude of neuroscience areas including among others; neurodegenerative diseases [52, 53], neuroinflammatory conditions [54, 55], stroke [56], brain tumors [57], neuropsychiatric conditions [58], as well as changes in brain metabolism. Various theories with regards to the etiology of this progression, have been put across including infection [59], free radical production [60], excitotoxic cell death [61], loss of neurotrophic factors [62], ingestion of a neurotoxin [59], mitochondrial dysfunction [63], and induction of apoptosis [64]; however, there has not been an establishment of a single etiological source of neurodegeneration to date. Neurodegeneration is commonly associated with disorders like Alzheimer's disease (AD), Parkinson's disease (PD), multiple sclerosis (MS), Huntington's disease (HD), human immunodeficiency virus (HIV)-associated dementia, and Creutzberg-Jakob disease.

3. BIOPOLYMERS USED IN NEUROTHERAPEUTIC INTERVENTIONS

Recently much interest has been realized in the application of biopolymers as neurotherapeutic delivery agents for drugs and genes to the CNS. Biopolymers have been found to be suitable materials in clinical neurotherapeutic use mainly because of their biocompatibility, biodegradability and low immunogenicity [65]. Biopolymers are natural products produced by living organisms, and may be divided into 3 major groups namely polysaccharides, proteins and nucleic acids. Control of particle size, charge, morphology of surface and release rate of the loaded molecules is absolutely important and this is where the biopolymer-based nanoparticles (as drug/gene delivery carriers) find their popularity. This Section seeks to provide an insight into the use of biopolymers in neurotherapeutic interventions (with appropriate clinical applications) citing most on biopolymeric vehicles, biopolymer scaffolds, multiple polymer-based layers, polysaccharide-based biopolymers (*e.g.* chitosan, alginate, starch, heparin), protein-mimicked polypeptides and protein-based biopolymers (*e.g.* collagen, gelatin, albumin).

3.1. Biopolymeric Vehicles for Neurotherapeutic Intervention

3.1.1. Nanoparticles as Drug Delivery Vehicles for Neurotherapeutic Intervention

Nanoparticles are ultra small entities with dimensions measured in nanometres (nm). They are composed of natural or artificial polymers with a size range of 10-

1000nm. Due to the properties that they are endowed with, they have been found to be highly useful in drug targeting and biodistribution of pharmaceuticals in a controlled manner. The drug may either be bound to the nanoparticle as a solid solution or dispersed within the nanoparticle or adsorbed to the surface or chemically attached onto the nanoparticle; all of which resulting into drug-loaded nanoparticle (Fig. **5**). Essentially nanoparticles have got a vast application in the medical field to the extent of emerging as highly useful entities in nanomedicine when compared to usual drug delivery techniques; attributes that have resulted into huge advantages with regards to targeted drug delivery as well as controlled drug release kinetics [66]. Furthermore, upon the tying together of these potential attributes, nanoparticles have ended up into finding a profound use in nanomedicine-based diagnosis and therapy [67].

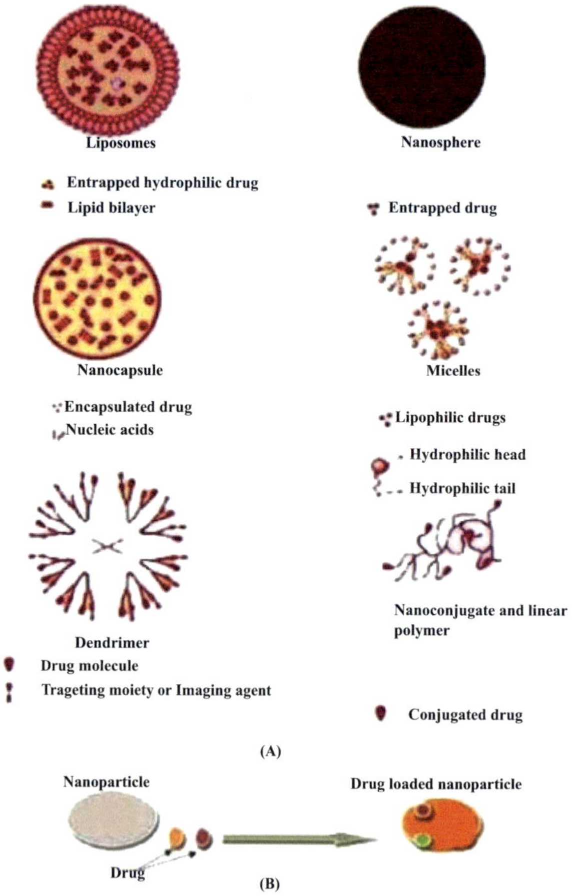

Figure 5: A schematic depicting: **a**) Various types of drug delivery vehicles; and **b**) a drug-loaded nanoparticle (adapted from [6]).

Nanoparticles have thus far proved to be novel delivery systems for carrying therapeutic agents to various body parts in the management of a variety of neurodegenerative and virus-based diseases. In addition, they are potentially useful in transferring of molecular and immunological agents to the biological systems. This potentiality is so much useful in gene delivery as well as in peptide delivery [6].

Nanoparticles may as be used as carrier agents for neurotrophic agents to treat neurodegenerative diseases. One other advantage is that, their desirable permeation properties give provision for safe and sustained release of drugs at targeted sites [68]. On a wider range, nanoparticles application have had a vast application in the management of cancer diseases (*e.g.* brain carcinoma) and neurodegenerative diseases *e.g.* Alzheimer's disease (AD) [69].

Thus far, novel nanotechnology-based drug delivery techniques (employing nanoparticles) are the best methods for transferring therapeutic agents to CNS [70]. Nanoparticles may pass across the BBB with ease, carrying with them the drug to the designated target sites in the brain and tumors. One of the best examples is human serum albumin (HSA) nanoparticles [6, 69, 71]. To make nanoparticles more effective they are normally coated with materials that can easily degrade in the body which in turn increases their permeability across the BBB. Typical examples of nanoparticles that have been made for targeting drugs across the BBB include lipid shelled and nonlipid shelled nanoparticles [72-74], biodegradable polymeric nanoparticles [75], transferrin-conjugated, fluorescein-loaded magnetic nanoparticles [76], solid lipid nanoparticles [74] and chitosan based nanoparticles [77]. Likewise, hydrogel-based ionotropic delivery devices have been developed for transdermal delivery of peptide/protein drugs [78].

For the purpose of improving the drug release, biodistribution as well as the improvement of treatment modalities and effectiveness, ester prodrugs are often included in the nanoparticle formulations [73] and coated using hydrophilic or hydrophobic drug materials [79]. In most of the cases, polysorbate-coated nanoparticles are employed to transfer drugs to the brain since they have proved to be more efficient when compared to uncoated nanoparticles [79]. In addition, solid nanolipid carriers are used as colloidal drug carriers for a variety of neurotherapeutics [80]. By the virtue of their small size, nanoparticles have been found to be able to penetrate into even small capillaries and they can easily be taken up within cells. More importantly is that the development of nanotechnology-based strategies (using nanoparticles) have made it possible to have an established equilibrium between cerebrovascular permeability outside and inside the nerve cells.

3.1.2. Polymersomes as Delivery Vehicles for Neurotherapeutic Intervention

Polymersomes is a sub-family of polymer-based liposomes used widely as delivering vehicles for carrying neurotherapeutics to designated locations in the human body [6]. Basically they are phospholipid-based vesicles that can form spontaneously in solutions and which can trap dissolved particles in solutions [6]. Due to the fact that most of the drugs cannot cross the BBB, polymersome technology has proven to be highly applicable since liposomes can cross BBB. Some recent advancement in liposomal drug delivery has resulted into the development of long circulating and potentially stable drug formulations. It has been established that synthesis of nanoparticles while emphasizing most on the employment of self-assembled systems such as micelles, microemulsions, nanoemulsions, and liposomes can enhance the drug distribution, bioavailability and the degree at which they can be more targeted [81]. However, in order to achieve more stable circulation, liposomes have to be conjugated with peptides and non-targeted liposomes so as to enhance the drug delivery to the respective sites [6].

The targetable properties for liposomes are easy to assess due to the ease with which their surface can be modified in comparison with other drug carrier agents such as nanoparticles and microemulsions [82, 83]. To date, different technicalities have been used in the bid to increase the amount of drug that may accumulate in polymer-based liposomes as well as their therapeutic efficacy employing a variety of biodegradable polymers [84]. For instance, a palmitic cationic liposomal *in situ* gel is said to be able to protect the drug from external compounds and can also ensure that the drug remains intact in its natural form throughout [85]. Furthermore, in order to develop liposomes with targetable carrier properties, cell-specific antibodies have been non-covalently bonded to liposomes to produce liposome-antibody complexes [86]. Noteworthy is that similar covalently-based complexation of poly- and monoclonal antibodies to the liposomes and coating of liposomes with heat aggregated immunoglobulins M (IgM) has been done successfully [87]. Likewise, natural and synthetic glycolipid, glycoprotein bearing liposomes, and transferrin coated paclitaxel loaded, lysozyme liposomes and neuroipilin-1-targeted liposomes have been developed to improve the delivery and bio-efficiency of drugs [84, 88, 89]. Fig. (**6**) is schematic depicting the various forms of liposomes employed for CNS drug delivery. Table **1** depicts a variety of some biopolymer-based drug delivery vehicles used for the management of CNS disorders.

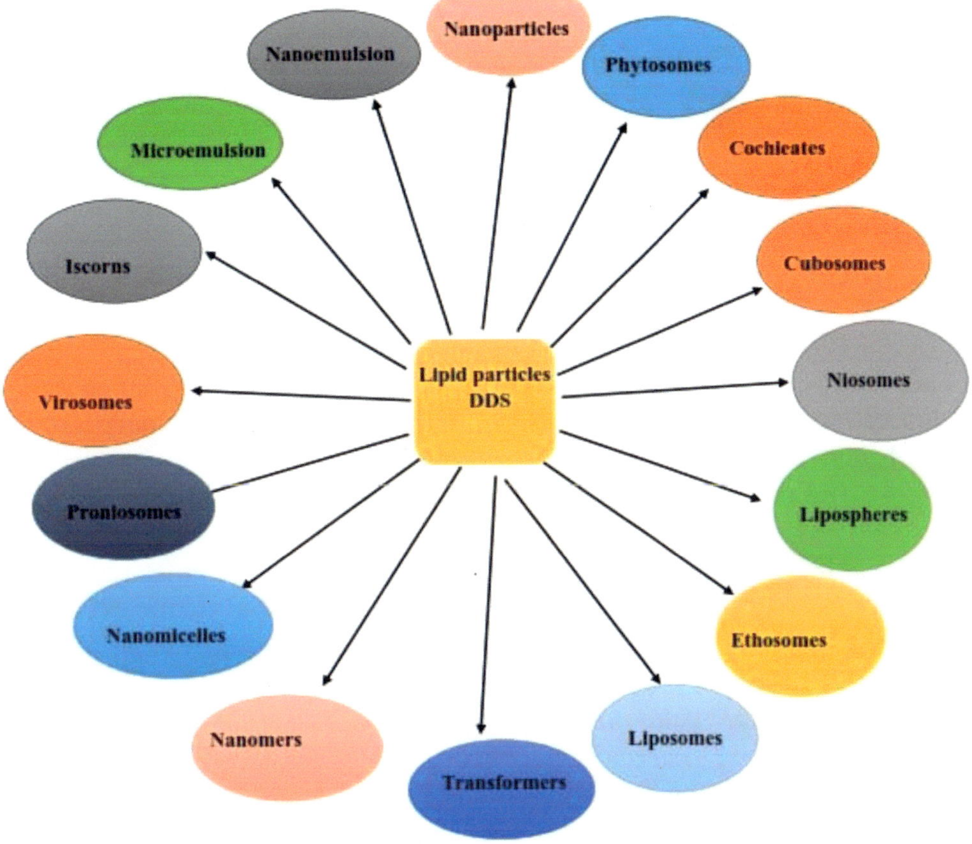

Figure 6: A schematic showing different types of liposomes used for drug delivery to CNS.

Table 1. Different types of biopolymer-based drug delivery vehicles used for neuroprotection and the treatment of brain diseases.

Delivery Vehicle/s	Route of Transfer	Method	Transport Mechanism	Advantages	Disadvantages	Applications	Refs.
Lipid nanoparticles	Intraventricular	Direct	DF	Enhance drug efficacy, neuroprotective	Toxic to membranes	Rhinitis and ischemic brain injury, for tumor, and global ischemia	[90]
Peptide masking	Injection	Direct	*DF/RMT**	Cholesteryl group traverse the drug through BBB	Poor biological activity	Multiple sclerosis, cancer and tumor	[91]
Chimeric peptides	Transcapillary	Indirect	RMT	Targeted drug delivery, stable during	Less permeable	Effective in treatment of various	[92]

Table 1: contd...

				transcytosis		neurodegenerative diseases	
Proline rich peptides	Transcapillary	Indirect	DF/RMT	Less toxic, safe and effective, and neuroprotective	Catalytically unstable	Use for treatment of cerebral infections neurocognitive disorders	[93]

*DF: diffusion and **RMT: receptor mediated transcytosis

3.2. Biopolymer Scaffolds for Neurotherapeutic Interventions

3.2.1. Spinal Cord Injury and the Mechanisms Thereof

There are two mechanisms of damage to the spinal cord following an acute spinal cord injury namely primary and secondary injury. The primary injury is associated with focal destruction of neural tissue as result of direct mechanical trauma [94]. The mechanical forces imposed at the primary injury site may cut off the neuronal and endothelial cell membranes, thereby resulting to necrosis and hemorrhage. This initial trauma then transforms to a progressive wave of secondary injury, an interdependent cascade of cellular and systemic events which can last for days to weeks and exacerbates injury to the spinal cord. The mechanisms for secondary injury include hemorrhage, ischemia and reperfusion injury, excitotoxicity, calcium-mediated secondary injury, disturbances in mitochondrial function, immunologic injury, apoptosis and axonal degeneration [95]. Several of these secondary injury mechanisms cause the formation of a large cystic cavity at the injury site, which often expands beyond the site of initial impact [95]. Damaging of the spinal cord produces an injury response known as reactive gliosis, or glial scarring. Finally, reactive astrocytes slowly penetrate into the injury site and proliferate, forming the bulk of the glial scar along with these other cells. The composition of the glial scar at its final form appears to be a meshwork of tightly interwoven astrocyte surrounding a large cystic cavity at the area of injury [96]. Fig. **(7)** is an illustration depicting spinal cord injury site (rounded circle in white color).

3.2.2. Biomaterial Design Strategies for Spinal Cord Repair in Spinal Cord Injury

Ideally an SCI repair scaffold should possess such properties like biocompatibility, biodegradability/porosity, the capability for the deliverance of bioactive factors, an internal-based matrix to cater for cell migration and mimic the structure of nerve fascicles, incorporation of support cells, and should also poses an electrical activity [98, 99]. Polymers as biomaterials have been found to be very useful in spinal cord repair spinal cord injury basically due to their high degree of biocompatibility and biodegradability potential. Biocompatibility may be defined as the ability of a biomaterial to be in contact with a living system without producing any undue adverse effects [100]. In the application of a scaffold for SCI repair, the biomaterial should elicit a minimal inflammatory response, minimize or at least not cause additional glial scar tissue formation, and should also allow neural cell adhesion and growth, and degrade in a nontoxic manner. Either, minimal swelling is required to maintain appropriate scaffold alignment and avoid compression of regenerating nerves [101]. It is important for the scaffold to maintain its structure *in vivo,* since structural collapse can lead to obstruction of axon regeneration through the length of the scaffold. The biomaterial should be sufficiently soft so that it does not physically damage the spinal cord as the patient moves.

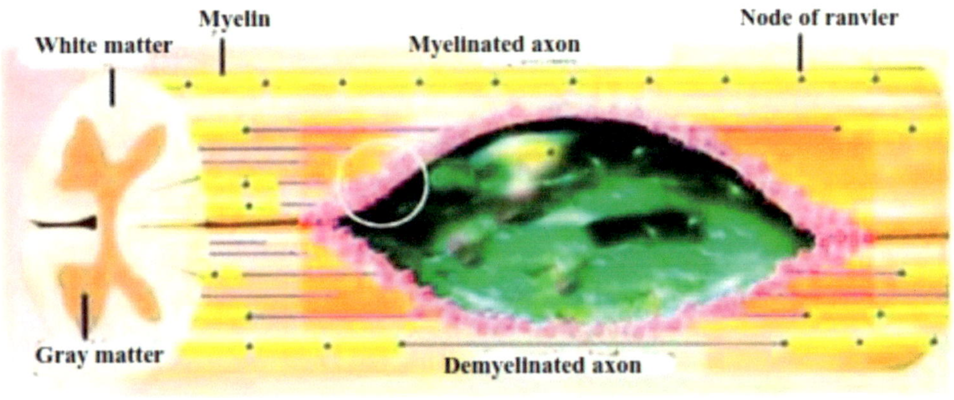

Figure 7: An illustration of spinal cord injury site {adapted from [97]}.

Cell attachment and axon regeneration can be aided by incorporating extracellular matrix (ECM) molecules, neurotrophic factors, and cell adhesion molecules into the scaffold. Neurotrophic factors promote survival, repair and recovery after CNS injury. Neurotrophic factors have a short half-life *in vivo* (hours) and factors delivered *via* injection diffuse away from the injury site. Implanting a scaffold to provide sustained delivery of these bioactive molecules can overcome these challenges. Incorporating cells into a scaffold is also an effective manner of supplying bioactive molecules to the injury site. Cells such as Schwann cells, olfactory ensheathing cells (OECs), bone marrow stromal cells (BMSCs), genetically engineered fibroblasts, and neural stem or progenitor cells have been loaded into scaffolds for delivery to the spinal cord [102-105]. Transplanted cells provide a permissive substrate for axon growth, while also producing cell adhesion molecules and neurotrophic factors that promote regeneration. Other bioactive molecules, such as glial scar-degrading chondroitinase ABC can be incorporated into scaffolds for sustained release [106].

Micro- and nanostructured topography of a scaffold, such as grooves, ridges, pores, and channels can provide physical cues to influence cell growth. Longitudinally aligned fibers or channels provide physical guidance for cell migration and oriented direction of axonal growth, leading to enhanced axon regeneration [107-109]. Compared to tubular scaffolds with a single lumen, multi-channel scaffolds provide more surface area for cell attachment and for release of incorporated bioactive molecules [72]. The internal matrix of a multi-channel scaffold also helps to prevent scaffold collapse [98].

3.3. Multiple Polymer-Based Layers and Polyelectrolyte Complexes for the Controlled Release of Neurotrophic Factors as a Neurotherapeutic Intervention

One of the strategies for controlling the release of neurotrophic factors (NTFs) is the deposition of bioactive molecules between layers of polymers varying in size, number, material selection, and their ionic charge [12]. In this way, release kinetics of NTFs may be regulated by changing the quantity of polymeric material *via* the following means: i) by delaying water penetration to hydrate the consequent polymer layers before reaching the NTF layer interface; ii) by increasing the diffusion pathway that must be crossed by the NTF molecules before becoming available to the target tissues; and iii) by extending polymer erosion or degradation before exposure of the NTF layer [110]. It has been shown that electrostatic properties may assist the enhanced entrapment of NTFs *via* ionic interaction between NTF and crosslinker or NTF and polymer [12]. Through layer to layer deposition or blending of oppositely charged polymers, various polyelectrolyte complexes (PECs) may be formed for immobilizing NTFs and controlling its release [111, 112]. The outcome on release profiles utilising PECs may be realized in a study that was conducted by Pfister and co-workers [113], using a multiple-layered PLGA-coated system. In this study, hollow nerve conduits were fabricated to control the release of nerve growth factor (NGF) by altering the position of NGF between various concentric layers of PLGA and an alginate-chitosan PEC (Fig. **8**).

Conduit A, which was made of NGF centrally placed within the alginate-chitosan PEC, was found to be capable of providing a single-phase near-perfect zero-order release kinetics with an indistinct initial burst release [12]. The firm entrapment of NGF within the PEC layer was purportedly due to electrostatic interaction between the positively charged NGF and the negatively charged alginate [114]. Different from what was observed in conduit A, a biphasic release pattern was reported for the 15-days study period with conduits B, C, and D [12]. An escalating rapid-release phase was noted during the first week followed by a plateaued effect achieving a near zero-order release over the final week. Conduit B which had the NGF layer deposited just outside the PEC layer exhibited the highest burst release, particularly during the rapid-release phase while Conduit D with the NGF layers interposed between additional layers of PLGA, was more efficient in minimising burst release and significantly prolonging the delivery of NGF at the end of 15 days [113].

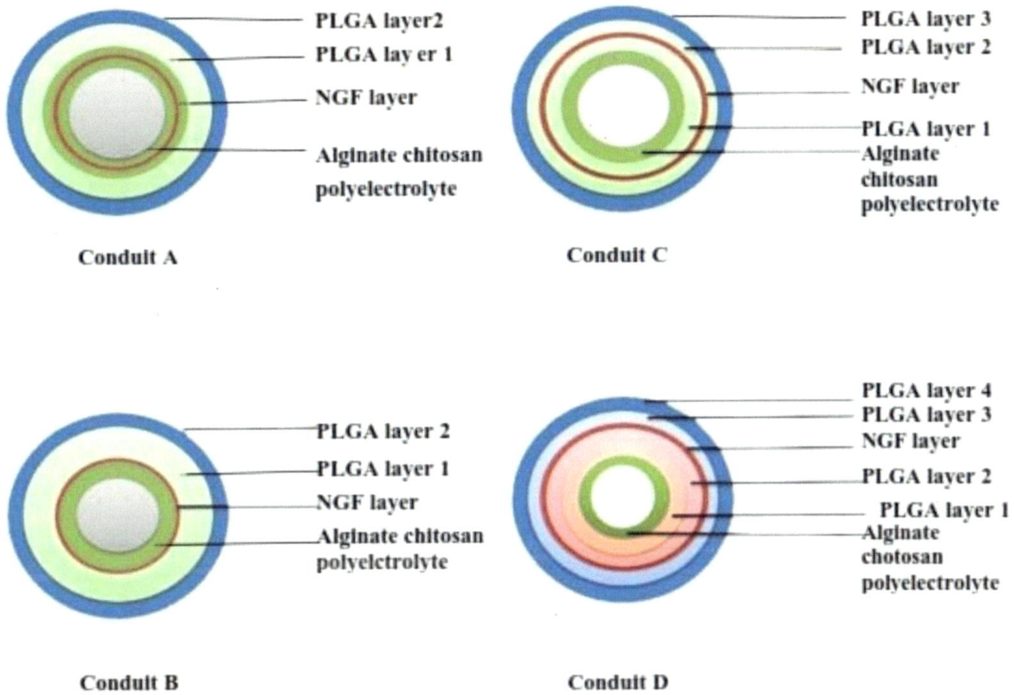

Figure 8: A schematic depicting the different conduits and various positioning of NGFs alginate-chitosan and PLGA layers {adapted from [12]}.

3.4. Polysaccharide-Based Biopolymers, Protein-Mimicked Polypeptides and Protein-Based Biopolymers

Polysaccharide-based biopolymers, protein-mimicked polypeptides and protein-based biopolymers that have thus far found a huge application in neurotherapeutics include among others: poly(alkylcyanoacrylates); polyacetates, ethylene-co-vinyl acetate (EVAc), PEGylated-based polymers, polylactic acid, polycaprolactone, poly(lactic-co-glycolic acid), poly(fumaric-co-sebacic) anhydride chitosan, modified chitosan, alginate, agarose, collagen, heparin and albumin. The application of some of these entities (*i.e.* polysaccharide-based biopolymers, protein-mimicked polypeptides and protein-based biopolymers) is summarized in Table **2** which depicts clearly the polymer/polymer delivery system, the bioactive employed, disease treated and finally the neurotherapeutic intervention outcome.

Table 2. Summary of various polymers/polymer delivery systems used in neurotherapeutic interventions.

Polymer/Polymer Delivery System	Bioactive Agent	Disease Treated	Neurotherapeutic Intervention Outcome	Reference
Alginate microspheres	NGF; BDNF	Brain injury; major depression	Prevented neuronal degeneration; antidepressant-like behavioral effects of BDNF	[115]
Agarose hydogels	BDNF	Spinal cord injury	Encouraged neurite growth into the channels; axonal regeneration;	[109, 116]
Protein bound to collagen in linearly ordered collagen scaffolds (LOCS)	BDNF	Spinal cord injury	Improved neuron survival and recovery of spinal cord injury	[117]
Hyaluronic-acid hydrogel scaffold	BDNF	Spinal cordinjury	Regeneration; improvement in locomotion	[118]
Agarose hydrogel coupled with laminin	NGF	Nerve injury		[119]
Collagen matrix implants	NT-3	Spinal cord Injury	Attraction of corticospinal tract fibers into the graft; recovery function	[120]
Fibrin matrix with heparin (or peptide) bound through electrostatic interactions to recombinant protein	BDNF; NGF; NT-3	Spinal cord injury; Sciatic nerve injury	Enhanced nerve regeneration across short nerve gaps; dose-dependent axonal regeneration; affinity-based delivery	[121, 122].

Table 2: cont….

			systems for neural tissue engineering	
PLGA-PLL-PEG biodegradable microspheres releasing recombinant protein	BDFN	CNS injury	Sustained release of bioactive human BDNF over 60 days	[123]
PLGA particles dispersed in a hydrogel of hyaluronic acid (HA) and methylcellulose	Chymotrypsin as a model of NT-3 and five other neuroprotectors	Spinal cord injury	Sustained release of NGF over 28 days from injectable composite hydrogels	[124]
PLGA biodegradable microspheres releasing recombinant protein	NGF	Huntington's disease	Improved protein stability; reduced striatal lesions	[125]
PLGA biodegradable microspheres releasing recombinant protein	GDNF	Parkinson'sdisease	Improved dopaminergic graftsurvival and function	[126, 127]
Ethylene-co-vinyl acetate (EVAc) discs for sustained release	NGF	Alzheimer's disease	Controlled release of NGF, high concentrations near the implant.	[128]
Implanted EVAc matrices	BDNF	Major depression	Dysregulation of BDNF-associated plasticity-related pathways upon sustained release; antidepressant-like effects upon short-term delivery	[115]

BDNF: Brain-derived neurotrophic factor; GDNF: Glial cell-derived neurotrophic factor; NGF: nerve growth factor; PLGA-PLL-PEG:PLGA-polylysine-polyethylene glycol; PLGA: poly(lactic-co-glycolic acid); EVAc: Ethylene-co-vinyl acetate

CONCLUSION

Recent advancement in the development of biopolymers and biopolymer-based drug carriers as brain drug delivery system have resulted in a great impact in neurotherapeutic interventions. As brain drug delivery systems, biopolymers are endowed with a multitude of favorable properties when compared to synthetic polymers including biocompatibility, biodegradability and abundant renewable sources. Needless to say, multi-functional biopolymer-based drug carries will have a tremendeous contribution in neurotherapeutics in the feature. Likewise, nanoscale drug delivery systems including liposomes seem to be capable of providing potential solutions in the area of neurodegeneration therapy. Polysaccharides have thus far proven to be among the most suitable biopolymers for developing delivery systems for neurotherapeutic intervention. This is mainly attributed to their biodegradability, biocompatibility and the ease with which molecular modification (both covalently and ionically) can be carried out onto them. One of the most useful features of protein-based biopolymers is the tendency and ability to form self-assembled α-helix, β-sheet or random structures, an attribute that makes it possible for them to easily form unique nanostructures such as nanoparticles, nanofibers and nanosheets that are highly applicable in neurotherapeutic delivery.

Biopolymer-based scaffolds have thus far proved to have the potential for rendering a controlled release of imbedded bioactives with their controlled release abilities proven to be very effective, leaving most of the bioactives entrapped within the walls. Obviously, the selection of scaffold type with differently incorporated bioactives will give a choice of drug release profile, potentially tailoring the desired release to the specific incorporated bioactive. The multi-functional scaffolds have shown to be able to promote the growth of DRG neurites thus demonstrating their potential for promoting axonal regeneration through the growth-inhibiting glial scar and release of chABC and/or neurotrophins. These multi-functional scaffolds have also proven to be suitable for use as a combination strategy for spinal cord repair due to their ability to: i) promote the directionally oriented growth of neuritis; and ii) release therapeutic biomolecules in a sustained manner.

It is a naked fact that molecular imaging has led into a remarkable revolution with regards to the studies of fundamental biological processes as well as the diagnosis, staging, and monitoring of various diseases. Much as molecular imaging has profoundly shown to have a great potential for a range of biomedical applications (in this particular case the biopolymer-based neurotherapeutic intervention), still there are some few challenges that need to be sorted out in order for a full potentiallty of this technique to be well realized.

Overall, the use of biopolymers and proper advancement in the field of biopolymer-based drug delivery as a strategy for neurotherapeutic intervention might in the future prove to be a potential non-invasive, non-toxic gold standard delivery method of neurotherapeutics for the treatment of neurodegenerative and neuropathological diseases of CNS.

5. FUTURE PROSPECTS FOR BIOPOLYMER-BASED NEUROTHERAPEUTIC INTERVENTIONS

For a successful delivery of neurotherapeutics to the CNS, proper carrier loading, targeting, and transport foolproof DDS has to be developed. Furthermore, there should be a desirable interaction between the developed DDS and the biological barriers in the brain particularly the BBB. Thus, more advanced methods must be developed for easy delivery of biopolymers and biopolymer-based materials to the brain for treatment of CNS diseases and disorders.

The improvement thus far obtained after say the implantation of biopolymer-based multimodal implants is self-effacing. There is however the need for future improvement in the following areas: i) the relationship between the implant and neuronal regeneration (including the functional outcome); ii) the effects of inflammation at the site of injury; iii) the prolonged survival of supporting cells; iv) the differentiation of stem cells; v) the effective delivery of neurotrophic factors; and vi) longer follow-up periods. Much emphasis should be directed at exploring how best the candidate polymers (as mentioned in Section 3.4) can be utilized in neurotherapeutic interventions.

When looking into the future of for instance molecular imaging, there are various areas with the potentiality of advancing substantially including: i) employment of modern imaging agents and imaging techniques as a standard operating procedure; ii) better screening and personalized medicine; iii) better dual-modality and multi-modality molecular imaging; iv) greater efficiency in the design and development of imaging agents; and v) development of multifunctional scaffolds for imaging agents.

The attachment of ligands to nanoparticle surfaces {*e.g.* transferrin (Tf) bound gold nanoparticles} so as to succeed in targeting specific sites will further potentiate the clinical application of biopolymer-based nanoparticles in neurotherapy. As such, the attention should not be paid to protein and/or polysaccharide-based nanoparticles only, but to biopolymer-synthetic polymer conjugates nanoparticles as well. Such multi-functional biopolymer-based and biopolymer-synthetic polymer conjugated nanoparticles will certainly become the potential candidates for targeted delivery of drugs and genes for neurotherapeutic intervention in the future.

Needless to say, future prospects in neurotherapeutic interventions should be based on putting more efforts towards the upgrading of pharmaceutical technologies through collaborative research in the attempt to explore new novel techniques for safely delivering neurotherapeutics to CNS using polymers and biopolymer-based materials. This will be possible through a substantial advancement in nanobiotechnology and biomaterial sciences which will stretch out the use of neurotherapeutics towards better and perfect means of curing neurological diseases and CNS impairments. Furthermore, biomedical researchers should work harder so as to increase the spectrum of neurotherapeutics as well as the methods for targeting them to the designated locations in the CNS.

ACKNOWLEDGEMENTS

The author would like to thank the initiators of this electronic Book (*i.e.* Wits Advanced Drug Delivery Platform-WADDP) for making him part of this important endeavour.

CONFLICT OF INTEREST

The author confirms that the chapter content has no conflict of interest.

REFERENCES

[1] Vidu R, Rahman M, Mahmoudi M, Enachescu M, Poteca TD, Opris I. Nanostructures: a platform for brain repair and augmentation. Front. Syst. Neurosci., doi: 10.3389/fnsys.2014.00091. 2014. Available from: http://journal.frontiersin.org/Journal/10.3389/fnsys.2014.00091/full [Cited: 13th Jan 2015].

[2] Neffe, AT, Wischke C, Racheva M, Lendlein, A. Progress in biopolymer-based biomaterials and their application in controlled drug delivery. Expert Rev Med Devic 2013; 6(8): 813-33.

[3] Dizaj SM, Jafari S, Khosroushahi AY. A sight on the current nanoparticle-based gene delivery vectors. Nanoscale Res Lett 2014; 9(1): 252.

[4] Hou S. Relay strategies combined with axon regeneration: a promising approach to restore spinal cord injury. Neural Regen Res 2014; 9(12): 1177-9.

[5] Martin-Banderas L, Holgado MA, Venero JL, Alvarez-Fuentes J, Fernandez-Arevalo M. Nanostrutures for Drug Delivery to the Brain. Curr Med Chem 2011; 18(34): 5303-21.

[6] Upadhyay RK. Drug Delivery Systems, CNS Protection, and the Blood Brain Barrier. BioMed Research International 2014; doi.http://dx.org/10.1155/2014/869269. Available from: http://www.hindawi.com/journals/bmri/2014/869269/ [cited: 8th Feb 2015].

[7] Schubert M, Breakefield X, Pedro R, Lowenstein PR. Gene delivery to the nervous system. Molecular therapy: The journal of the American Society of Gene Therapy 2008; 16(4): 640-8.

[8] Simonato M, Bennett J, Boulis NM, *et al.* Progress in gene therapy for neurological disorders. Nat Rev Neurol 2013; 9: 277-91.

[9] Naidoo J, Young D. Gene Regulation Systems for Gene Therapy Applications in the Central Nervous System. Neurol Res Int 2012; http://dx.doi.org/10.1155/2012/595410. Available from: http://www.hindawi.com/journals/nri/2012/595410/ [cited: 1st Feb 2015].

[10] Mansouri S, Lavigne P, Corsi K, Benderdour M, Beaumont E, Fernandes JC. Chitosan-DNA nanoparticles as non-viral vectors in gene therapy: strategies to improve transfection efficacy. Eur J Pharm Biopharm 2004; 57(1): 1–8.

[11] Dang JM, Leong KW. Natural polymers for gene delivery and tissue engineering. Adv Drug Deliv Rev 2006; 58(4): 487–99.

[12] Ramburrun P, Kumar P, Choonara YE, Bijukumar D, Toit LC and Pillay V. A Review of Bioactive Release from Nerve Conduits as a Neurotherapeutic Strategy for Neuronal Growth in Peripheral Nerve Injury. 2014; Doi: http://dx.doi.org/10.1155/2014/132350. Available from: http://www.hindawi.com/journals/bmri/2014/132350/ [cited: 21st Sept, 2014].

[13] Ma J, Liu J, Yu H, Wang Q, Chen Y, Xiang L. Curcumin promotes nerve regeneration and functional recovery in rat model of nerve crush injury. Neural Sci lett 2013; 547: 26-31.

[14] Dalton P, Harvey A, Oudega M, Plant G. Controlled release strategies in tissue engineering. In: van Blitterswijk C, Thomsen P, Lindahl A, Hubbell J, Williams DF, Cancedda R, de Bruijn JD, and Sohier J. (eds): Tissue Engineering, Burlington Academic Press 2008; pp. 611-47.

[15] Oh SH, Kim JH, Song KS *et al.* Peripheral nerve regeneration within an asymmetrically porous PLGA/Pluronic F127 nerve guide conduit. Biomaterials 2008; 29(11): 1601-9.

[16] Yamamoto K, Amako M, Yamamoto Y, *et al.* Therapeutic Effect of Exendin-4, a Long-Acting Analogue of Glucagon-Like Peptide-1 Receptor Agonist, on Nerve Regeneration after the Crush Nerve Injury. BioMed Research International 2013; doi: 10.1155/2013/315848. Available from: http://www.hindawi.com/journals/bmri/2013/315848/ [cited 22nd Feb 2015].

[17] Ichihara S, Inada Y, Nakamura T. Artificial nerve tubes and their application for the repair of peripheral nerve injury: an update of current concepts. Injury, Int. J. Care Injured 2008; 39(4): S29-S39.

[18] Marquardt LM, Sakiyama-Elbert SE. Engineering peripheral nerve repair. Curr Opin Biotech 2013; 24: 887-92.

[19] Tang X, Xue C, Wang Y, Ding F, Yang Y, Gu, X. Bridging peripheral nerve defects with a tissue engineered nerve graft composed of an *in vitro* cultured nerve equivalent and a silk fibroin-based scaffold. Biomaterials 2012; 33(15): 3860-67.

[20] Huang YC, Huang YY. Biomaterials and strategies for nerve regeneration. Artificial Organs 2006; 30: 514-22.

[21] Tang S, Zhu J, Xu Y, Xiang AP, Jiang MH, Quan D. The effects of gradients of nerve growth factor immobilized PCLA scaffolds on neurite outgrowth invitro and peripheral nerve regeneration in rats. Biomaterials 2013; 34(29): 7086–96.

[22] Quaglia F. Bioinspired tissue engineering: the great promise of protein delivery technologies. Int J Pharm 2008; 364(2): 281-97.

[23] Fawcett JW. Overcoming inhibition in the damaged spinal cord. J Neurotrauma 2006; 23(3-4): 371-83.

[24] Bradbury EJ, McMahon SB. Spinal cord repair strategies: why do they work? Nat Rev Neurosci 2006; 7: 644-53.

[25] James ML, Gambhir SS. Molecular Imaging Primer: Modalities, Imaging Agents, and Applications. Physiol Rev 2012; 92(2): 897-965.

[26] Osborn EA, Jaffer FA. Advances in molecular imaging of atherosclerotic vascular disease. Curr Opin Cardiol 2008; 23(6): 620-8.

[27] Weissleder R, Pittet MJ. Imaging in the era of molecular oncology. Nature. 2008; 452: 580-9.

[28] Spuch C, Navarro C. Liposomes for Targeted Delivery of Active Agents against Neurodegenerative Diseases (Alzheimer's Disease and Parkinson's Disease). Journal of Drug Delivery. 2011; Doi: http://dx.doi.org/10.1155/2011/469679. Available from: http://www.hindawi.com/journals/jdd/2011/469679/[cited: 11th Jan 2015].

[29] Kim BK, Doh KO, Nam JH *et al.* Synthesis of novel cholesterol-based cationic lipids for gene delivery. Bioorg Med Chem Lett 2009; 19(11): 2986–9.

[30] Pardridge, WM. Tyrosine hydroxylase replacement in experimental Parkinson's disease with transvascular gene therapy. NeuroRx. 2005; 2(1): 129-38.

[31] Bian YZ, Wang Y, Aibaidoula G, Chen GQ, Wu Q. Evaluation of poly (3-hydroxybutyrate-co-3-hydroxyhexanoate) conduits for peripheral nerve regeneration. Biomaterials 2009; 30(2): 217–25.

[32] Johnson EO, Soucacos PN. Nerve repair: experimental and clinical evaluation of biodegradable artificial nerve guides. Injury 2008; 39 Suppl 3:S30-6.

[33] Sohier J, Vlugt TJH, Cabrol N, Van Blitterswijk C, De Groot K, Bezemer JM. Dual release of proteins from porous polymeric scaffolds. J Control Rel 2006; 111(1-2): 95–106.

[34] Platt CI, Krekoski CA, Ward RV, Edwards DR, Gavrilovic J. Extracellular matrix and matrix metalloproteinases in sciatic nerve. J Neurosci Res 2003; 74(3): 417–29.

[35] Kehoe S, Zhang XF, Boyd D. FDA approved guidance conduits and wraps for peripheral nerve injury: a review of materials and efficacy. Injury 2012; 43(5):553–72.

[36] Shim S, Ming, G Roles of channels and receptors in the growth cone during PNS axonal regeneration. Exp Neurol 2010; 223(1): 38–44.

[37] Yong N, Guoping C. The role and mechanism of the up-regulation of fibrinolytic activity in painful peripheral nerve injury. J Neurosci Res 2009; 34(3): 587–92.

[38] Li M, Guo W, Zhang P, Li H, Gu X, Yao D. Signal flow and pathways in response to early Wallerian degeneration after rat sciatic nerve injury. Neurosci Lett 2013: 536: 56–63.

[39] Wagner V, Hüsing B, Bock A. Nanomedicine: Drivers for development and possible impacts. 2008; Available from: http://ftp.jrc.es/EURdoc/JRC46744.pdf [cited: 20th Feb, 2015].

[40] Higashi T, Khalil IA, Maiti KK, *et al.* Novel lipidated sorbitol-based molecular transporters for non-viral gene delivery. J Control Release 2009; 136(2): 140–7.

[41] Mastrobattista E, van der Aa MA, Hennink WE, Crommelin DJA. Artifical virsues: a nanotechnological approach to gene delivery, Nat Rev Drug Discov 2006; 5: 115-21.

[42] Kostarelos K, Miller AD. Synthetic, self-assembly ABCD nanoparticles; a structural paradigm for viable synthetic non-viral vectors. Chem Soc Rev 2005; 34(11): 970–94.

[43] Bharali DJ, Klejbor I, Stachowiak EK, *et al.* Organically modified silica nanoparticles: a nonviral vector for *in vivo* gene delivery and expression in the brain. Proc Natl Acad Sci USA 2005; 102(32): 11539–44.

[44] Mayrhofer P, Schleef M, Jechlinger W. Use of minicircle plasmids for gene therapy. Methods Mol Biol 2009; 542: 87–104.

[45] Conwell CC, Huang L. Recent advances in non-viral gene delivery. Adv Genet 2005; 53: 1–18.

[46] Deelman L, Sharma K. Mechanisms of kidney fibrosis and the role of antifibrotic therapies. Curr Opin Nephrol Hypertens 2009; 18(1): 85–90.

[47] Lee JM, Yoon TJ, Cho YS. Recent developments in nanoparticle-based siRNA delivery for cancer therapy. Biomed Res Int 2013; 1-10. Available from: http://www.hindawi.com/journals/bmri/2013/782041/ [cited: 15th March 2015].

[48] Morachis JM, Mahmoud EA, Sankaranarayanan J, Almutairi A. Triggered rapid degradation of nanoparticles for gene delivery. J Drug Deliv 2012; 2012: 1–7.

[49] Liu C, Zhang N. Nanoparticles in gene therapy: principles, prospects, and challenges. Prog Mol Biol Transl Sci (2011); 104: 509–62.

[50] Dobson, J. Gene therapy progress and prospects: magnetic nanoparticle-based gene delivery. Gene Ther 2006; 13(4): 283–7.

[51] Sinha R, Kim GJ, Nie S, Shin DM. Nanotechnology in cancer therapeutics: bioconjugated nanoparticles for drug delivery. Mol Cancer Ther 2006; 5(8): 1909–17.

[52] Drzezga A. Basic pathologies of neurodegenerative dementias and their relevance for state-of-the-art molecular imaging studies. Eur J Nucl Med Mol 2008; 35 (Suppl 1):S4-11.

[53] Felicio AC, Shih MC, Godeiro-Junior C, Andrade LA, Bressan RA, Ferraz HB. Molecular imaging studies in Parkinson disease: reducing diagnostic uncertainty. Neurologist 2009; 15(1): 6–16.

[54] Stoll G, Bendszus M. New approaches to neuroimaging of central nervous system inflammation. Curr Opin Neurol 2010; 23(3): 282–6.

[55] Venneti S, Lopresti BJ, Wiley CA. The peripheral benzodiazepine receptor (Translocator protein 18kDa) in microglia: from pathology to imaging. Prog Neurobiol 2006; 80(6): 308–22.

[56] Weissleder R. Molecular imaging in cancer. Science 2006; 312: 1168–71.

[57] Ullrich RT, Kracht LW, Jacobs AH. Neuroimaging in patients with gliomas. Semin Neurol 2008; 28(4): 484–94.

[58] Valotassiou V, Wozniak G, Sifakis N, Demakopoulos N, Georgoulias P. Radiopharmaceuticals in neurological and psychiatric disorders. Curr Clin Pharmacol 2008; 3(2): 99–107.

[59] Jones GJ, Barsby NL, Cohen EA, *et al.* HIV-1 Vpr causes neuronal apoptosis and *in vivo* neurodegeneration. J Neurosci 2007; 27(14): 3703–11.

[60] Prehn JH. Mitochondrial transmembrane potential and free radical production in excitotoxic neurodegeneration. Naunyn-Schmiedebergs Arch Pharmacol 1998; 357(3); 316–22.

[61] Park DS, Obeidat A, Giovanni A, Greene LA. Cell cycle regulators in neuronal death evoked by excitotoxic stress: implications for neurodegeneration and its treatment. Neurobiol Aging 2000; (21(6): 771–81.

[62] Hennigan A, O'Callaghan RM, Kelly AM. Neurotrophins and their receptors: roles in plasticity, neurodegeneration and neuroprotection. Biochem Soc Trans 2007; 35(2): 424–7.

[63] Beal MF. Mitochondria and neurodegeneration. Novartis Found Symp 2007; 287: 183–92.

[64] Okouchi M, Ekshyyan O, Maracine M, Aw TY. Neuronal apoptosis in neurodegeneration. Antioxid Redox Signal 2007; 9(8): 1059–96.

[65] Nitta SK, Keiji N. Biopolymer-Based Nanoparticles for Drug/Gene Delivery and Tissue Engineering. Int J Mol Sci 2013; 14(1): 1629-54.

[66] Liu Y, Tan J, Thomas A, Ou-Yang D, Muzykantov VR. The shape of things to come: Importance of design in nanotechnology for drug delivery. Ther Deliv 2012; 3(2): 181–94.

[67] Barratt G. Betbeder D. Lebleu B. Challenges for nanotechnology in delivery imaging. Preface. Int J Pharm 2009; 379(2): 199–200.

[68] Vinogradov SV, Bronich TK, Kabanov AV. Nanosized cationic hydrogels for drug delivery: preparation, properties and interactions with cells. Adv Drug Deliver Rev 2002; 54(1): 135–47.

[69] Sahni JK, Doggui S, Ali J, Baboota S, Dao L, Ramassamy C. Neurotherapeutic applications of nanoparticles in Alzheimer's disease. J Control Release 2011; 152(2): 208–31.

[70] Jain KK, Nanobiotechnology-based strategies for crossing the blood-brain barrier. Nanomedicine 2012; 7(8): 1225–33.

[71] Kreuter J. Drug delivery to the central nervous system by polymeric nanoparticles: what do we know ? Adv Drug Deliver Rev 2014; 71: 2–14.

[72] Wang X, Sykes DB, Miller DS. Constitutive androstane receptor-mediated up-regulation of ATP-driven xenobiotic efflux transporters at the blood-brain barrier. Mol pharmacol 2010; 78(3): 376–83.

[73] Wang J, Liu K, Sung K, Tsai C, Fang C. Lipid nanoparticles with different oil/fatty ester ratios as carriers of buprenorphine and its prodrugs for injection. Eur J Pharm Sci 2009; 38(2): 138–46.

[74] Üner M. Preparation, characterization and physico-chemical properties of solid lipid nanoparticles (SLN) and nanostructured lipid carriers (NLC): their benefits as colloidal drug carrier systems. Pharmazie 2006; 61(5): 375–86.

[75] Soppimath KS, Aminabhavi TM, Kulkarni AR, Rudzinski WE. Biodegradable polymeric nanoparticles as drug delivery devices. J Control Release 2001; 70(1-2): 1–20.

[76] Yan F, Wang Y, He S, Ku S, Gu W, Ye L. Transferrin-conjugated, fluorescein-loaded magnetic nanoparticles for targeted delivery across the blood-brain barrier. J Mater Sci-Mater M 2013; 24(10): 2371–79.

[77] Agnihotri SA, Mallikarjuna NN, Aminabhavi TM. Recent advances on chitosan-based micro- and nanoparticles in drug delivery. J Control Release 2004; 100(1): 5–28.

[78] Banga AK, Chien YW. Hydrogel-based iontotherapeutic delivery devices for transdermal delivery of peptide/protein drugs. Pharmaceut Res 1993; 10(5): 697–702.

[79] Wohlfart S. Khalansky AS, Gelperina S, Begle D, Kreuter J. Kinetics of transport of doxorubicin bound to nanoparticles across the blood-brain barrier. J Control Release 11; 154(1): 103–07.

[80] Müller-Goymann CC. Physicochemical characterization of colloidal drug delivery systems such as reverse micelles, vesicles, liquid crystals and nanoparticles for topical administration. Eur J Pharm Biopharm 2004; 58(2): 343–56.

[81] Sanchez-Dominguez, CN, Gallardo-Blanco HL, Rodriguez-Rodriguez AA, Vela-Gonzalez AV, Sanchez-Dominguez, M. Nanoparticles vs cancer: a multifuncional tool. Curr Top Med Chem 2014; 14(5):.664–75.

[82] El-Badry M, Fetih G, Shakeel F. Comparative topical delivery of antifungal drug croconazole using liposome and micro-emulsion-based gel formulations. Drug Deliv 2014; 21(1): 34–43.

[83] Li L, Hou J, Liu X, *et al.* Nucleolin-targeting liposomes guided by aptamer AS1411 for the delivery of siRNA for the treatment of malignant melanomas. Biomaterials 2014; 35(12): 3840–50.

[84] Paoli EE, Ingham ES, Zhang H, *et al.* Accumulation, internalization and therapeutic efficacy of neuropilin-1-targeted liposomes. J Control Release 2014; 178: 108–17.

[85] He W, Guo X, Feng M, Mao N. *In vitro* and *in vivo* studies on ocular vitamin A palmitate cationic liposomal *in situ* gels. Int J Pharm 2013; 458(2): 305–14.

[86] Abu-Lila AS, Uehara Y, Ishida T, Kiwada H. Application of polyglycerol coating to plasmid DNA lipoplex for the evasion of the accelerated blood clearance phenomenon in nucleic acid delivery. J Pharm Sci, 2013; 103(2): 557–66.

[87] Sarker SR, Hokama R,Takeoka S. Intracellular delivery of universal proteins using a lysine headgroup containing cationic liposomes: deciphering the uptake mechanism. Mol Pharmacol 2014; 11(1): 164–74.

[88] Gao J, Chen H, Song H, *et al.* Antibody-targeted immunoliposomes for cancer treatment. J Med Chem 2013; 13(14): 2026–35.

[89] Biswas S, Dodwadkar NS, Sawant RR, Torchilin VP. Development of the novel PEG-PE-based polymer for the reversible attachment of specific ligands to liposomes: synthesis and *in vitro* characterization. Bioconjugate Chem 2011; 22(10): 2005–13.
[90] Rautio J, Kumpulainen H, Heimbach T, *et al.* Prodrugs: design and clinical applications. Nat Rev Drug Discov 2008; 7(3): 255–70.
[91] Fross RD, Warnke PC, Groothuis DR. Blood flow and blood-to-tissue transport in 9L gliosarcomas: the role of the brain tumor model in drug delivery research. J Neurooncol 1991; 11(3): 185–97.
[92] Cornford EM, Cornford ME. New systems for delivery of drugs to the brain in neurological disease. Lancet Neurol 2002; 1(5): 306–15.
[93] Zhang JM, Zhao XM, Wang SJ, *et al.* Evaluation of 99mTcpeptide-ZHER2:342 Affidy molecule for *in vivo* molecular imaging. Br J Radiol 2014; 87(1033): 20130484
[94] Profyris C, Cheema SS, Zang D, *et al.* Degenerative and regenerative mechanisms governing spinal cord injury. Neurobiol Dis 2004; 15(3): 415-36.
[95] Dumont RJ, Okonkwo DO, Verma S, *et al.* Acute spinal cord injury, part I: pathophysiologic mechanisms. *Clin Neuropharmacol*, 2001; 24(5): 254-64.
[96] Fawcett JW, Asher RA. The glial scar and central nervous system repair. Brain Res Bull 1999; 49(6): 377-91.
[97] Macaya D, Spector M. Injectable hydrogel materials for spinal cord regeneration: a review. Biomed Mater 2012; 7(1): 012001.
[98] Straley KS, Foo CW, Heilshorn SC. (2010). Biomaterial design strategies for the treatment of spinal cord injuries. J Neurotrauma 2010; 27(1): 1-19.
[99] Ubramanian A, Krishnan UM, Sethuraman S. Development of biomaterial scaffold for nerve tissue engineering: Biomaterial mediated neural regeneration. J Biomed Sci 2009; 16; 108.
[100] Ratner BD. Biomaterials science: an introduction to materials in medicine. 2nd ed2004, Amsterdam; Boston: Elsevier Academic Press. 2004; 851.
[101] Madigan NN, McMahon S, O'Brien T, *et al.* Current tissue engineering and novel therapeutic approaches to axonal regeneration following spinal cord injury using polymer scaffolds. Respir Physiol Neurobiol 2009; 169(2): 183-99.
[102] Nomura H, Zahir T, Kim H, *et al.* Extramedullary chitosan channels promote survival of transplanted neural stem and progenitor cells and create a tissue bridge after complete spinal cord transection. Tissue Eng Part A, 2008; 14(5): 649-65.
[103] Fouad K, Schnell L, Bunge MB, Schwab ME, Liebscher T, Pearse DD. Combining Schwann cell bridges and olfactory-ensheathing glia grafts with chondroitinase promotes locomotor recovery after complete transection of the spinal cord. J Neurosci 2005; 25(5): 1169-78.
[104] Itosaka H, Kuroda S, Shichinohe H, *et al.* Fibrin matrix provides a suitable scaffold for bone marrow stromal cells transplanted into injured spinal cord: a novel material for CNS tissue engineering. Neuropathology 2009; 29(3): 248-57.
[105] Tobias CA, Han SSW, Shumsky JS, *et al.* Alginate encapsulated BDNF producing fibroblast grafts permit recovery of function after spinal cord injury in the absence of immune suppression. J Neurotrauma 2005; 22(1): 138-56.
[106] Hyatt AJ, Wang D, Kwok JC *et al.* Controlled release of chondroitinase ABC from fibrin gel reduces the level of inhibitory glycosaminoglycan chains in lesioned spinal cord. J Control Release 2010; 147(1) 24-9.
[107] Scott JB, Afshari M, Kotek R, Saul JM. The promotion of axon extension *in vitro* using polymer-templated fibrin scaffolds. Biomaterials 2011; 32(21): 4830-9.
[108] Gros T, Sakamoto JS, Blesch A, *et al.* Regeneration of long-tract axons through sites of spinal cord injury using templated agarose scaffolds. Biomaterials 2010; 31(26): 6719-29.
[109] Stokols S, Tuszynski MH. Freeze-dried agarose scaffolds with uniaxial channels stimulate and guide linear axonal growth following spinal cord injury. Biomaterials 2006; 27(3): 443–451.
[110] Kokai LE, Ghaznavi AM, Marra KG. Incorporation of double-walled microspheres into polymer nerve guides for the sustained delivery of glial cell line-derived neurotrophic factor. Biomaterials 2010; 31(8): 2313–22.
[111] Luo Y, Wang Q. Recent development of chitosan-based polyelectrolyte complexes with natural polysaccharides for drug delivery. Int J Biol Macromol 2014; 64: 353–67.

[112] Tsai RY, Chen PW, Kuo TY, *et al.* Chitosan/pectin/gum Arabic polyelectrolyte complex: Process-dependent appearance, microstructure analysis and its application. Carbohydrate Polymers 2014; 101: 752–9.

[113] Pfister LA, Alther E, Papaloïzos M, Merkle HP, Gander B. Controlled nerve growth factor release from multi-ply alginate/chitosan-based nerve conduits. European Journal of Pharmaceutics and Biopharmaceutics 2008; 69(2): 563–72.

[114] Chaudhari A, Vanmellaert L, Bauwens M, *et al.* *In vitro* and *in vivo* investigation of the potential of amorphous microporous silica as a protein delivery vehicle. BioMed Research International 2013; Doi: http://dx.doi.org/10.1155/2013/306418. Available from: http://www.hindawi.com/journals/bmri/2013/306418/ [cited 22nd Sept 2014].

[115] Sirianni RW, Olausson P, Chiu AS, Taylor JR, Saltzman WM. The behavioural and biochemical effects of BDNF containing polymers implanted in the hippocampus of rats. Brain Res 2010; 1321: 40–50.

[116] Jain A, Kim YT, McKeon RJ, Bellamkonda RV. *In situ* gelling hydrogels for conformal repair of spinal cord defects, and local delivery of BDNF after spinal cord injury. Biomaterials 206; 27(3): 497–504.

[117] Han QQ, Sun WJ, Lin H, *et al.* Linear ordered collagen scaffolds loaded with collagen-binding brain-derived neurotrophic factor improve the recovery of spinal cord injury in rats. Tissue Eng. Part A 2009; 15(10): 2927 35.

[118] Park J, Lim E, Back S, Na H, Park Y, Sun K. Nerve regeneration following spinal cord injury using matrix metalloproteinase-sensitive, hyaluronic acid-based biomimetic hydrogel scaffold containing brain-derived neurotrophic factor. J. Biomed. Mater. Res. A 2010; 93(3): 1091–9.

[119] Yu XJ, Dillon GP, Bellamkonda RV. A laminin and nerve growth factor-laden three-dimensional scaffold for enhanced neurite extension. Tissue Engen 1999; 5(4): 291–304.

[120] Houweling DA, Lankhorst AJ, Gispen WH, Bar PR, Joosten EAJ. Collagen containing neurotrophin-3 (NT-3) attracts regrowing injured corticospinal axons in the adult rat spinal cord and promotes partial functional recovery. Exp Neurol 1998; 153(1): 49–59.

[121] Maxwell DJ, Hicks BC, Parsons S, Sakiyama-Elbert SE. Development of rationally designed affinity-based drug delivery systems. Acta Biomater 2005; 1(1): 101–13.

[122] Taylor SJ, Rosenzweig ES, McDonald JW, Sakiyama-Elbert SE. Delivery of neurotrophin-3 from fibrin enhances neuronal fiber sprouting after spinal cord injury. J Control Release 2006; 113(3): 226–35.

[123] Bertram JP, Rauch MF, Chang K, Lavik E.B. Using polymer chemistry to modulate the delivery of neurotrophic factors from degradable microspheres: Delivery of BDNF. Pharm Res 2010; 27(1): 82–91.

[124] Bauman MD, Kang CE, Stanwick JC, *et al.* An injectable dtug delivery platform for sustained combinatory therapy. J Control Release 2009; 138(3): 205–13.

[125] Menei P, Pean JM, Nerriere-Daguin V, Jollivet C, Brachet P, Benoit J.P. Intracerebral implantation of NGF-releasing biodegradable microspheres protects striatum against excitotoxic damage. Exp Neurol 2000; 161(1): 259–72.

[126] Garbayo E, Ansorena E, Lanciego J.L, Blanco-Prieto M.J, Aymerich M.S. Long-term neuroprotection and neurorestoration by glial cell-derived neurotrophic factor microspheres for the treatment of Parkinson's disease. Mov Disord 2011; 26(10): 1943–7.

[127] Clavreul A, Sindji L, Aubert-Pouessel A, Benoi J.P, Menei P. Montero-Menei, C.N. Effect of GDNF-releasing biodegradable microspheres on the function and the survival of intrastriatal fetal ventral mesencephalic cell grafts. Eur J Pharm Biopharm 2006; 63(2): 221–8.

[128] Krewson CE, Klarman ML, Saltzman WM. Distribution of nerve growth factor following direct delivery to brain interstitium. Brain Res 1995; 680(1-2): 196–206.

CHAPTER 13

Porous and Non-Porous Metallic Biomaterials

Oluwatoyin Ayotomilola Adeleke*

Council for Scientific and Industrial Research, Meraka Institute, Centre for High Performance Computing, 627 Meiring Naudé Road, Pretoria 0001, South Africa

Abstract: Metallic biomaterials can be described as substances that do not contain drugs or active pharmaceutical ingredients, are natural or artificial in nature and are based on metallic elements or metallic alloys which are usually applied as a whole or part of biological systems and perform, amplify or substitute a function that has been lost through injury or illness. They can be used for an undefined duration in different spheres of patient health care such as medicine, surgery, dentistry and veterinary medicine. This chapter reviews the porous and non-porous structured class of metallic biomaterials, their desirable physicochemical and physicomechanical properties as well as their various types, fabrication methods and biomedical applications.

Keywords: Alloy, biomaterials, cobalt-chromium, implants, metals, non-porous, porous, prosthesis, stainless steel, titanium.

INTRODUCTION

Centuries ago, surgeons had little or no available alternative other than to remove diseased or damaged tissues which resulted in the progressive reduction in quality of tissues and only a slightly improved quality of life but in recent times (about 30 years ago), a significant revolution directed towards the successful replacement of impaired body tissues and organs which concurred with the significant increase in generally observed human longevity recurrently occur in medical care [1-3]. Such useful tissue replacement processes have been made possible through the invention or more frequently modification, of distinctive, biocompatible man-made materials otherwise known as biomaterials which are usually capable of interfacing with living, host tissues [1].

Typically, a biomaterial can be described as a natural or synthetic substance, matter, surface or construct (other than drugs) employed as a single or combined entity which can be introduced into the body as part of an implanted medical

***Corresponding author Oluwatoyin Ayotomilola Adeleke:** Council for Scientific and Industrial Research, Meraka Institute, Centre for High Performance Computing, 627 Meiring Naudé Road, Pretoria 0001, South Africa; Tel: +27 12 841 2962; Fax: +27 12 841 4720; E-mail: oluwatoyin.adeleke@gmail.com

Viness Pillay, Yahya E. Choonara and Pradeep Kumar (Eds.)

device or used to replace an organ, bodily function which interacts with biological systems and can be applied for an undefined period to amplify functions or replace partially or totally any tissue or organ within the body in order to conserve or improve the quality of life of the affected creature [4]. A biomaterial has been further described by Williams [2] as "a substance that has been engineered to take a form which, alone or as part of a complex system, is used to direct, by control of interactions with components of living systems, the course of any therapeutic or diagnostic procedure, in human or veterinary medicine". The biomedical application of biomaterials have been on since ancient times and have rapidly evolved from the use of natural materials such as wood into the use of better performing synthetic polymers, ceramics and metal alloys with improved functionality and reproducibility. These major improvements have resulted in the wide application of biomaterials built into devices like dental implants, hip and knee prosthesis, vascular stents and contact lenses which have improved and prolonged of human lifestyle and lifespan respectively [4]. Interestingly, the usefulness of biomaterials are not restricted to humans only but many constructs in which biomaterials form key parts are of great use in veterinary medicine [2]. Biomaterials were majorly employed to perform mechanical functions and preferred to be inert, not interacting with the biology of the host systems negatively, to prevent rejection by the biological environment [4]. In other words, it is necessary that biomaterials should function by focusing on the sequence of medical diagnosis or treatment by regulating undue interactions within the biological environments of the specific individual being managed [2]. While this chapter is focused on compiling important information on porous and non-porous metallic biomaterials, a concise overview of metallic biomaterials in general is presented in the ensuing section.

METALLIC BIOMATERIALS: A GENERAL OVERVIEW

In a nutshell, metallic biomaterials can be described as biocompatible materials (other than a drug) which can originate from natural or artificial sources, constructed based on metallic elements or alloys and usually applied as a whole or part of biological systems to perform, enhance or replace specific roles that have been lost through illness or injury [5, 6]. This class of materials can be utilized for either defined or undefined periods of patient health care duration and within varying medical spheres such as surgery, dentistry and general medicine. In comparison to ceramic and/or polymer based biomaterials, metallic biomaterials and their alloys possess excellent mechanical strength for enduring tensile stresses, resistance to fracture as well as good thermal and electrical conductivity properties, assuming proper production steps were maintained, which provides long term implant performance within implantation sites [5, 6]. These abovementioned qualities, the use of comparatively easy fabrication processes (*e.g.* casting, machining, forging) that produce simple and multifaceted forms of metallic biomaterials have promoted their extensive application in the biomedical arena such as orthopaedics, cardiology, neurology, dentistry [6]. The abovementioned attractive qualities of metallic biomaterials are attributable to the presence of inter-atomic bonding and mobile independent electric charge and thermal energy transferring electrons. These portable electrons are responsible for bonding positive metallic electrons together resulting in a characteristically strong and non-directional frame which is responsible for the closely packed, stable atomic arrangements within most metallic lattices. This is responsible for the high specific gravity, tensile strength, plastic behaviour and elevated melting point values commonly exhibited by most metals [5-7]. The processing technique selected for designing a metallic biomaterial determines its internal structure as well as the exhibited physicochemical and physicomechanical properties of the material. The elastic constants related to plastic deformation are an exception to this trend because these are practically structure independent properties dependent on inter-atomic bonding and equilibrium atom packing. The inter-atom equilibrium distance outlining the elemental cells of the metallic crystal and the intensity of inter-atomic interaction are dependent on material internal configuration (*e.g.* atom size, charge) and external conditions (*e.g.* temperature, pressure) [6]. Besides the malleable nature of metallic materials, the capability to deform plastically with the application of high loads results in the reduction of localized stress concentration leading to relatively high fracture toughness [6].

Metallic biomaterials are employed in the fabrication of various implantable biomedical devices which are applied at various parts of the body and this trend is expected to continue [8]. They have diverse applications which encompass use as inert alternatives for hard tissue replacement/high load skeletal reconstructions such as hip and knee partial/total endoprostheses, fracture restorative aids (bone plates, screws), dental implants, ankle prosthesis and spinal fixation devices (Fig. **1**) anchored upon their exceptionally good mechanical properties and ability to resist corrosion. Metallic biomaterials can also perform more dynamic, functional and non-loaded roles as in biomedical devices such as cochlea implants, orthodontic arch wires, vascular stents, catheter guide wires, valves, pump cages, pacemakers and conducting wires (Fig. **1**) [5, 7, 8].

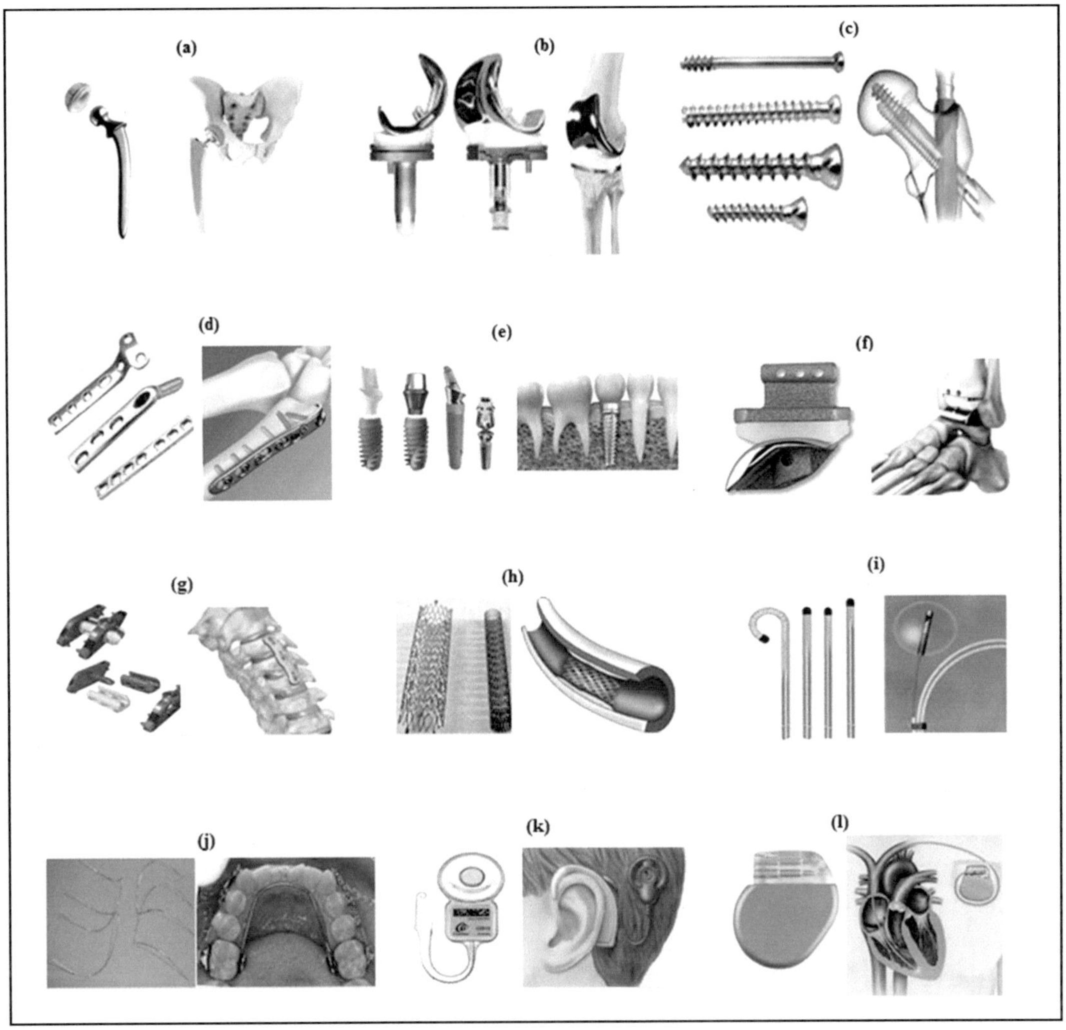

Figure 1: Images of some metallic biomaterial-based biomedical devices and some of their respective applications: (**a**) hip endoprostheses [10, 11], (**b**) knee endoprostheses [12, 13] (**c**) bone screws [14, 15], (**d**) bone plates [16, 17], (**e**) dental implants [18, 19], (**f**) ankle prosthesis [20, 21], (**g**) spinal fixation devices [22, 23], (**h**) vascular stents [24, 25], (**i**) catheter guide wires [26, 27], (**j**) orthodontic arch wires [28, 29], (**k**) cochlea implants [30, 31] and (**l**) heart pacemakers [32, 33].

Focusing on the aforementioned respective applications of metallic biomaterials, they can be utilized either as a porous or non-porous/dense surfaced architecture, depending on the expected biomedical function/application. Particularly, metallic biomaterial-based implantable devices with dense/non-porous characteristics have several drawbacks such as mismatch on elastic modulus which can lead to stress-shielding, interfacial instability with host tissue, possible production of fibrous tissue on implant surface and low volumetric porosity in the area of bone support and replacement. These disadvantages have resulted in researchers over the past two decades conducting many studies on porous materials for *in vivo* implantation [3].

Commonly, all biomaterials (metallic or non-metallic) are expected to satisfy certain requirements such as corrosion resistance, availability/accessibility, biocompatibility, bioadhesion (bone ingrowth), bio-functionality (satisfactory mechanical strength and a Young's modulus as close to that of the bone as possible), and processability which make them suitable for biomedical uses. These requirements are more or less adequately satisfied by the various regularly available groups of biomaterials but when compared, each material presents varying behaviour associated with its demands [5]. On a general note, the biocompatibility of metallic implantable biomaterials is considerably important because these devices can actually corrode in a biological environment resulting in material disintegration which can lead to weakening/loss of mechanical and tensile strength as well as the effects of corrosion products upon surrounding cells, tissues and organs can be detrimental [7]. Metallic biomaterials biocompatibility challenges can also include general *in vivo* toxic reactions and allergic reactions distant from the site of implantation [9].

Furthermore, a comprehension of the material properties and processes which are utilized to accomplish desirable features during the production of metallic biomaterials are crucial for ensuring that the implant is within acceptable standard limits to perform within biological systems. Failure of most engineered structures to display the expected optimal mechanical performance is completely unacceptable and with metallic biomaterial-based implantable devices, it is particularly so because a substandard product can lead to patient discomfort and pain, be fatal or result in the need for repetition of complicated, life threatening surgical procedures [6, 7].

POROUS AND NON-POROUS METALLIC BIOMATERIALS

Metals are employed as manufacturing materials for implantable biomaterials which are intended to substitute or enhance body functions (*e.g.* load-bearing function of the bone, joint flexibility) as well as show good and acceptable biocompatibility for an extended duration [3, 8]. Some of the metals which have been widely employed in the design of implantable biomaterials include aluminium (Al), chromium (Cr), zirconium (Zr), magnesium (Mg), tungsten (W), titanium (Ti), tantalum (Ta), iron (Fe), platinum (Pt), cobalt (Co), nickel (Ni), niobium (Nb), molybdenum (Mo), palladium (Pd), vanadium (V); alloys made up of a combination of metals such as Fe-Ni-Cr, Ti-Mo-Pd, Co-Ni-Cr, Al-V-Ti, Co-Cr-Mo, as well as stainless steel. Most metals like Fe, Cr, Co, W, Ni, Ti, Ta, Nb and Mo have been used for producing alloys for implantable devices in the past but there major demerit is that they can only be tolerated by the body in small quantities [3, 8].

Currently approved and commonly explored metallic biomaterials for biomedical applications or clinical use either as porous or non-porous frameworks include titanium and its alloys, stainless steels and cobalt-chromium alloys [3, 6, 8, 34-37]. This chapter will focus on the review of these selected metallic biomaterials in terms of their physical properties, porous or non-porous natured geometry, fabrication methods and biomedical applications.

Titanium Metal and Alloys

de Viteri and Fuentes [38] stated that "elemental metal titanium was first discovered in England by William Gregor in 1790 while in 1795, Klaproth named it". The combination of physical qualities such as low density, high mechanical strength to weight ratio, good biocompatibility, corrosion resistance with good plasticity and mechanical properties have set the pace for the use of Ti and its alloys in such industries as automotive, power, ship building, aviation, medicine, sports and architecture industries [38].

The use of Ti and its alloys as biomaterials for implantable purposes (*e.g.* fracture fixation and joint replacement) dates back to the 1930s. This application is basically due to its unique low weight, high specific strength, excellent corrosion resistance and good biocompatible properties [39]. Ti is a widely utilized metallic biomaterial and it has being applied over many years as a biological implantable device because of its capability to get integrated into contiguous bone indicative of its bioactivity when in contact with tissue, facilitating the outward growth of bone directly through and onto its surface.

Some reasons why Ti metal is useful as an implantable device include its: (i) highly reactive nature, (ii) capability to form a dense, coherent, passive oxide (stoichiometrically similar to biocompatible TiO_2) film (close-bone-to-implant apposition) which grows continually *in vivo* preventing the penetration of fragments of corrosion products into the surrounding environment during every phase of implantation and (iii) ability to facilitate the reformation of the abovementioned surface film coating (important advantage for permanent bone-interfacing implantation) into a Ti-OOH based surface which entraps the super oxide (O_2^-) generated during inflammatory reactions thereby impeding the discharge of hydroxyl (OH^-) free radical [6, 36].

Ti and its alloys are considerably costly (*e.g.* stainless steel) and they wear out easily in comparison to other biomedical metals and alloys and are thus not selected as suitable for high load bearing interfaces [36]. Considering an example such as partial or total hip bone replacement, the mechanical strength of pure Ti is considerably inadequate; therefore Ti alloys which are mechanically stronger are preferable [36].

Srivastav [36] classified the grades of commercially available, unalloyed Ti into four categories which vary based on the level of interstitial impurities such as oxygen (O), nitrogen (N), iron (Fe), carbon (C), hydrogen (H) contained in them. The quantity of inherent oxygen specifically poses significant negative impacts on the ductility and strength of these Ti grades by impacting the interstitial solidification of the metal [36]. Grade 1 Ti is the most ductile of the Ti grades but has the lowest O content and yield strength while grade 4 is the least ductile with the highest O content and yield strength. The grade 3 and 4 commercially available Ti is employed for building implantable devices for osteosynthetic applications however their low mechanical strength hinders their application in the development of joint replacement endoprostheses [36]. Table **1** enlists the interstitial element limits of grades 1, 2, 3 and 4 commercial purity Ti.

Demerits of Titanium and Its Alloys

A major drawback of Ti and its alloys which impede a wider application of this class of metallic biomaterial consist of their low wear resistance and poor tribological properties associated with low Ti hardness [38, 40, 41]. Ti metals and alloys exhibit poor tribological behaviours based on their high friction factor, severe wear and tear of adhesive capacity, intense tendency to seize and minimal resistance towards abrasion. Ti-based surfaces have the tendency of having its moving parts rub against another surface and eventually seize and loosen which can cause more intensive wear consequent to the creation of adhesion connections and mechanical instability of the passive oxide layers, particularly in the presence of more than two bodies [38]. Based on this effect, 10%-20% of cases of Ti-based total joint replacements usually require to be changed within 15-20 years after implantation [38].

Ultimately, very poor wear resistance remains the major problem of Ti and it alloys which makes them inappropriate for load-bearing articulating joint surfaces without some level of surface modification to provide greater resistance to wear and tear [6, 42]. Surface modification of this class of metals can be accomplished through ion implantation with N^+ or Ti-N film application using a suitable physical vapour deposition process. The introduction of modular hip implant designs (involves combining a more wear resistant bearing component made of either Co alloy or a ceramic) has served as an effective solution to this challenge, nonetheless, this issue still poses some concerns since modular designs will inevitably possess inter-component junctions where practical problematic motion can occur [6, 42, 43].

Table 1. The limits of the interstitial elements for grades 1-4 commercial purity Ti (adapted from [6].

	Maximum Interstitial Element Limits		
Commercial Purity Titanium	**Oxygen**	**Nitrogen**	**Hydrogen**
Grade 1	0.18	0.03	0.015
Grade 2	0.25	0.03	0.015
Grade 3	0.35	0.05	0.015
Grade 4	0.40	0.05	0.015

Chemistry of Titanium/Titanium Alloys and Processing Techniques

The lattice structure transition of pure Ti is body-centered cubic (β-phase) at temperatures above 883°C and hexagonal close-packed (α-phase) at lower temperatures [3, 44]. Titanium alloys containing aluminium and vanadium as a stabilizer element (Ti6Al4V) have been widely employed as implant biomaterials. Nevertheless, some studies revealed that the release of the ions *i.e.*, Al and V can be harmful to human body [3, 44]. Therefore, other alloy stabilizers were explored to reduce the biological adverse impacts. Also the introduction of biocompatible alloying element into Ti alloys can improve the mechanical properties of the implant [3]. The inclusion of other elements stabilizes either of the two phases. Stabilizers of the α-phase include Al, O, N and C while the β-phase stabilizers are the β-isomorphous (Mo, V, Nb and Ta) and the β-eutectoid (Fe, W, Cr, Ni, Co, H, Silicon (Si) and Manganese (Mn) [6, 45]. Thus far, several α + β and β –type Ti alloys have been commonly developed by researchers employing alloying elements like Ta, Nb, Zr and tin (Sn) which are selected because of the non-cytotoxicity while Mo and Si are not so desirable because of their cytotoxicity especially at certain concentrations. These elements have shown good biocompatibility, high corrosion resistance and complete solubilisation in Ti [3,

46]. Elemental Zr and Sn found in some Ti alloys are classified as "neutral" as they have no noticeable impact on neither α- nor β-phase stabilization [6].

The (α + β) Ti alloy has the longest history of effective clinical use as major load-bearing biomedical implants. They display higher mechanical strength, excellent corrosion resistance and osseointegration tendencies than pure Ti. Common examples include Ti6Al4V, Ti6Al7Nb and Ti5Al2.5Fe [6]. Billets of these alloys are produced using mill-annealing, a thermo-mechanical process, resulting in the manufacture of high mechanical strength materials. Mill annealing involves multiple processing steps which includes alloy mechanical manipulation at temperature just below the β-transition temperature followed by swift cooling to room temperature and recrystallization of worked alloy structure at about 750°C by annealing [6]. The α and metastable α-phases are produced during the rapid cooling phase with the retention of some β-phase while the metastable α-phase is converted to the stable α-phase during recrystallization. The level, size and distribution of α and β phase components is dependent on the degree of mechanical distortion during component formation, set formation temperature recrystallization treatment [6]. Implantable devices based on Ti metal alloy usually possess superior corrosion resistance qualities. The size and distribution of α and β phase regions determine the magnitude of the mechanical quantities (especially fatigue resistance) of the (α + β) Ti alloys [6, 42].

A number of the β-Ti alloys containing a higher level of β-stabilizing elements such as Mo, Nb have been developed for use as orthopaedic implants. Many studies focus on the β-Ti alloys because of their desired elastic modulus. Appropriate processing of the β-Ti alloys can result in them having reduced elastic moduli, good formability, high hardness, exceptional corrosion resistance and improved notch sensitivity than the (α + β) alloys [3, 6]. They are also fabricated using the annealing process but they are characterized by retention of a total β-phase on cooling from the heated β transition phase. Their wear resistance can be enhanced by improving surface hardness which can be accomplished by aging (heat to about 500°C) in an oxygenated surrounding. During this process, a tough surface oxide deposit and interstitial solid solution are formed to amalgamate the subsurface region [6].

Surface Modification and Biomedical Applications of Titanium/Titanium Alloys

Surface re-construction can play a vital role in prolonging the applicability of Ti-based implantable devices many times more than its original capacity. Surface treatment is aimed at improving implant tribological behaviour, corrosion resistance and osseointegration. Furthermore surfaces can be coated with biocompatible materials to maximize wear and corrosion resistance by increasing surface hardness/toughness. The application of surface coating can be achieved by different techniques such as thermo-chemical treatments (*e.g.* nitriding, carburizing, boriding), physical deposition (*e.g.* ion implantation, plasma spray coatings) and surface oxidation [38, 47, 48]. However, the Diamond-Like Carbon coating and Plasma Electrolytic Oxidation surface treatment techniques remain the most promising methods applied to Ti surfaces [38].

Diamond-like Carbon (DLC) is commonly termed as amorphous carbon films and it is employed as a coating to protect the alloy surface to solve tribological problems. The DLC coating is usually put on the alloy surface by either Physical Vapour or Plasma-Enhanced Chemical Vapour Deposition. DLC coatings are mainly amorphous, metastable and basically contain a mix of diamond (sp^3) and graphite (sp^2) and the combination proportion of these two phases will to a very large extent determine the properties of the coat. DLC films possess are characterized by good qualities such as relatively high mechanical strength, low friction coefficient, chemical stability and bio-, cyto- and hemo-compatibility which make them useful as biomedical materials [38, 49, 50].

A common surface modification process employed for Ti and it alloys still remains oxidation which involves the formation of oxide layers by either heat treatment [38, 51, 52] or electrolytic anodizing [52]. Oxidation by heating leads to the formation of a 15-30 µm thick TiO_2 layer while the conventional anodic oxidation in various solutions results in Ti surface passivation generating thin films of amorphous hydrated oxide or crystalline TiO_2. Elements such as Calcium (Ca) and Phosphorous (P) can be introduced onto the surface oxide present on the Ti metal and surface geometry can be changed by controlling electrolyte and electrochemical conditions. Research has shown that the presence of Ca ions on Ti metal surfaces has been shown to improve cell growth while implant containing Ca and P enhanced bone apposition [38, 51, 52].

The use of materials like ceramic as coating has been shown to improve biocompatibility of Ti implants [38]. Some studies conducted include modifying Ti surfaces using phosphonic acid *in vitro* to improve the biocompatibility of dental implants [53]. Outcomes indicated that pre-treatment of the implant with phosphonic acid had no cytotoxicity impact on the osteoblasts. Also, coating materials such as hydroxyapatite have been fabricated to improve mechanical bonding, biocompatibility and osseointegration [54]. Titanium oxide (TiO_2)

(synthesized by Plasma Electrolytic Oxidation), calcium phosphate, titania/hydroxyapatite composite and silica coating by the sol-gel method can also be put on the surface of the Ti and its alloys [38] to expand their *in vivo* effectiveness.

Ti surfaces have being modified by making them porous to overcome stress shielding upon their dense topography which often leads to aseptic loosening [3]. The porous structured variants of Ti are usually referred to as "metallic foams" or "Ti foams", and can be used in bulk form or as coating. They usually have an overall porosity ranging between 60% and 80% with pore diameters of 100-600 μm. The Ti foams display properties similar to those of cancellous bone with high surface friction capacity; enhanced porosity intensity and relatively low elastic modulus which all together provides a surface that can facilitate and maintain long-lasting bonding and substantial extent of bone ingrowth [39]. An example by Wen *et al.* [55] detailed the development of porous Ti-Zr alloy foams with porous structure and mechanical properties close to that of natural bone for tissue engineering purposes using the powder metallurgical mechanical alloying technique (Fig. **3**(**a**)). Dense/non-porous Ti-Zr alloy is potentially useful in the biomedical sphere because it is biocompatible, bioactive and mechanically stable. However, it has the problem of biomechanical mismatch of elastic modulus and internal structural inconsistencies with reference to bone tissue engineering applications which require interconnected porous surface structure to allow required space for cell ingrowth and vascularization [55]. Mg is a promising biomaterial for designing bone implants because of its bioresorbable and biodegradable properties. The fabrication of a porous foam composed of Ti-Mg alloy for bone substitution applications was also presented by Wen and co-workers [56] using a metallurgical process. Furthermore, the space-holder sintering process was used to manufacture an open cell porous structure Ti-Ni shape memory alloy foam. Generated Ti-Ni porous structure displayed mechanical properties that can be designed to match that of the bone and they displayed shape memory effect as well [57]. Calcium phosphate is a synthetic ceramic which can be employed for converting the surface geometry of Ti into a porous platform. Commonly employed calcium phosphate ceramics include tricalciumphosphate, octocalcium phosphate, hydroxyapatite, biphasic calcium phosphate [3].

Surface modification of Ti-based implantable devices to improve their capacity to function as biomimetic matrices appears to be a promising approach to enhance positive interaction between biological cellular sites and Ti-based implant interface [58]. The implantation of dental devices can be considered as a complete success if optimum osseointegration and development of biological tissue seal around the transmucosal site of the implant to prevent bacterial colonization, peri-implantitis is achieved [59]. The biomimetic modification of Ti surfaces with proteins or peptides with specific cell-binding moieties is reported as a potentially effective method for improving biological seal of dental implants [59]. Therefore, studies by Marín-Parejaaa and associates [59] looked into the process of immobilizing collagen on smooth and rough Ti surfaces and its impact on the human dermal fibroblast cell response. Ti samples were activated using either oxygen plasma or acid etching to produce smooth or rough surface and collagen grafting was achieved either physisorption or covalent bonding. Outcome showed improved fibroblast adhesion and activation [59]. Heller *et al.* [58] looked into the reaction of extracellular protein coated Ti surface with human umbilical cord and human osteoblast cells. It was evidenced by experimental results that the protein coats on the Ti surface led to a significant stimulation of cell adhesion in comparison to the control experiment. Additionally, Wei and co-workers [60] reported the use of a newly synthesized biocompatible, bio-inert hyperbranched polyglycerols functionalized with catechols as an ordered multi-layered coating on TiO_2 surfaces to decrease fouling which usually involves the speedy adsorption of proteins on biomaterial surfaces once they are in contact with biological fluids. Fouling usually results in undesirable host responses such as inflammation and irritation at the site of implantation as well as surrounding tissues, blood coagulation which can be detrimental to the wellbeing of the host body. The corrosion resistance and good blood compatibility of TiO_2 was a major motivation for the study conducted by Chen and colleagues [61]. TiO_2 was developed as a thin film with a modified surface with the potential to minimize thrombotic responses once in contact with blood. Basically, the antithrombotic capacity of TiO_2 was further enhanced for better blood compatibility by exposing the film surface to partial and full ultraviolet light irradiation.

A substantial amount of research is now being directed towards modifying the surface of Ti, its oxide and alloys into micro- or nano- configured topographies. These advancements have been reported to promote bone-to-implant integration by mechanical inter-locking and impart better biocompatibility respectively [62, 63] as well as enhance cell proliferation and corrosion resistance of Ti and its alloys [64]. Recently, scientists have delved into fabricating ordered micro/nano-topographies of the surface of Ti and its alloy (Ti6Al7Nb in this case) by etching with hydrofluoric acid followed by anodic oxidation. The effect of this surface modification was tested using bone mesenchymal stem cells and outcomes showed that both surface modified pure Ti metal and its alloy, Ti-6Al-7Nb, increased the spreading, adhesion and proliferation of the bone mesenchymal stem cells which are indicative of good biocompatibility [62, 63].

Furthermore, some other explored surface modifications for Ti pure metal and its alloys to improve their *in vivo* performance, integration and usefulness have been extended to the use of chitosan/hydroxyapatite composite as deposits or biological coating on the surface of Ti biomaterial [65-68], hydroxyapatite/carbon nanotube nanocomposite layer [69], calcium phosphate coatings of hydroxyapatite as surface coat [70], poly(vinylphosphonic acid), a hydrophilic polymer, as a superficial coat on the surface of a Ti alloy (Ti6Al4V) for improving friction interfaces [71], plasma electrolytic oxidation process utilizing tricalcium phosphate, wollastonite and silica as additives on the surface of Ti15Mo [72], progressive immobilization of oligo(ethylene glycol) and 2-(methacryloyloxyethyl phosphorylcholine) on Ti surface to improve its anticoagulation effect [73], a systematic immobilization of chitosan-lauric acid conjugate on a polydopamine film for surface functionalization to enhance osseointegration and reduce bacterial adhesion [74], polishing and sandblasting methods of surface preparation of Ti6Al4V and Ti6Al7Nb alloys to assess *in vivo* corrosion tendencies [75].

Porous metallic biomaterials have been employed as alternative to ceramics for orthopaedic prostheses [76]. Ti and its alloys are the most commonly utilized materials for this purpose due to their biocompatibility and osseointegration qualities. However, Ti's Young's modulus (100 GPa) noticeably exceeds that of bone (3-20 GPa) and this can lead to stress-shielding resulting into bone resorption [76]. Consequently, improving the inward growth of bone through implanted biomaterial by enhancing the surface porosity of the implant is very crucial to maintaining long-term stability of artificial joints [76, 77]. Hofman and co-workers [77] developed porous metallic scaffolds of NiTi alloy employing a selective laser melting method. The fabricated Ni-Ti matrices demonstrated good cytocompatibility and osteogenic potential. Porous Ti25Nb alloys fabricated using powder metallurgy was assessed by Xu and colleagues [77]. Alloy surface displayed interconnected pores and a pore diameter that ranged between 200 nm and 500 nm which was suitable for bone growth with compressive strength and elastic modulus similar to that of cortical bone and cancellous bone respectively. The porous Ti25Nb alloys displayed good biocompatibility, optimum mechanical strength and adequate pore diameter and porosity that allowed bone ingrowth. Xie *et al.* [78] prepared variants of porous Ti10Mo alloy from Ti, Mo and epoxy resin powders by selective laser sintering preforming, debinding and high heat sintering under argon. Produced alloys displayed flexible pore diameter, porosity and mechanical properties. Fig. (**2**) shows different images of surface modified Ti and its alloys.

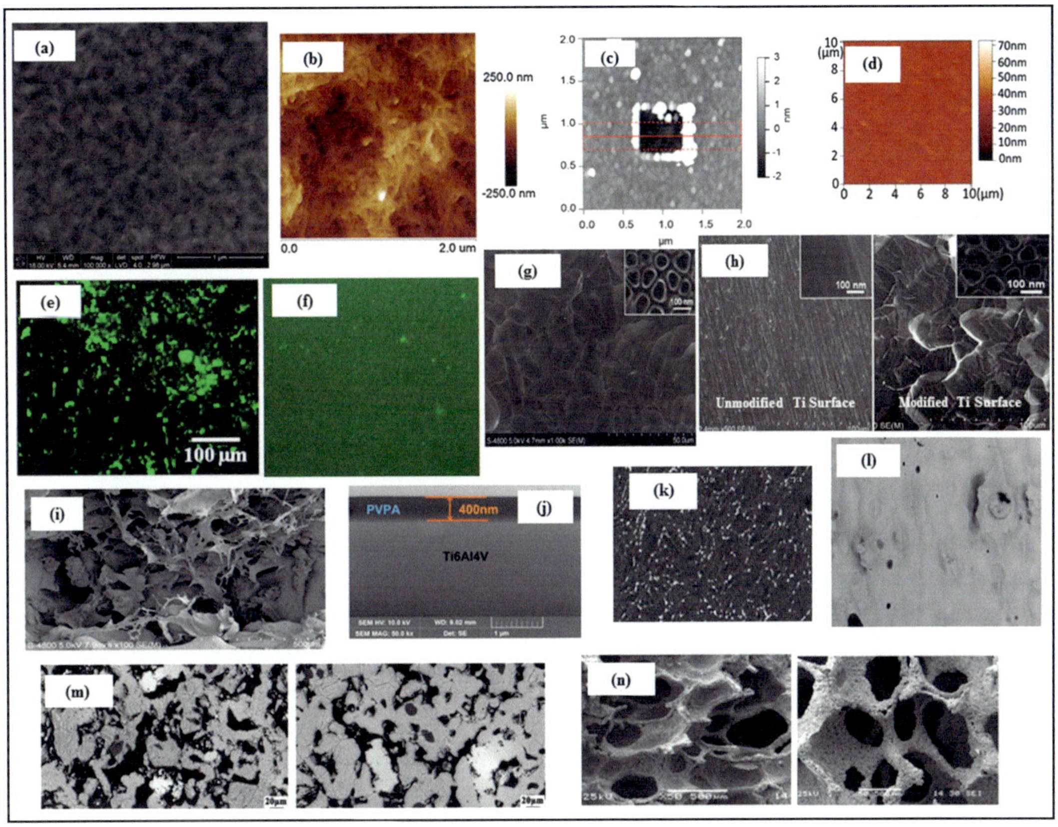

Figure 2: Typical illustrations of different surface structural modifications of Ti and its alloys: (**a**) and (**b**) carbon nanotube-hydroxyapatite deposit on Ti [69], (**c**) modified TiO_2 surfaces with hyperbranched polyglycerols functionalized with catechols [60], (**d**) ultraviolet light irradiated TiO_2 surface [61], (**e**) type 1 collagen functionalized Ti surface [59], (**f**) Ti surface treated with a primary amine [58], (**g**) modified Ti6Al7Nb surface [63], (**h**) acid etched and anodized Ti surface [62], (**i**) chitosan/hydroxyapatite Ti implant surface [67], (**j**) polyvinyl(vinylphosphonic acid) coatings on Ti [71], (**k**) polished Ti6Al7Nb alloy surface [75], (**l**) laser treated NiTi alloy porous scaffold surface [76], (**m**) microstructure of sintered porous Ti10Mo alloy [78], (**n**) porous Ti25Nb surface containing extended and interconnected pores [77].

The application of Ti and its alloys either as porous on non-porous implantable biomaterials continues to increase because of their unique physicochemical and physicomechanical properties mentioned earlier. Nevertheless, the mechanical strength of Ti is not sufficient enough for bone replacement so the Ti alloys are preferred for this application. Ti alloys may be applied as coatings on sintered beads, fibre metal mesh and plasma spray coatings. Porous Ti alloy foams are used extensively as coatings for primary and revision total hip and knee arthroplasty. Of course, not too many peer reviewed manuscripts are available on this category of Ti materials [39]. Also, Ti and its alloys are used almost exclusively for forming endosseous dental implants [38].

Stainless Steel

Stainless steels comprise the first set of reliable metals employed as prosthesis in orthopaedics. The key elements contained in stainless steels are Fe and C while other additional elements are usually Cr, V, Mo and Ni [36, 39]. Stainless steel needs to contain a certain amount of austenite stabilizing elements such as Ni or Mn for it to be austenitic at room temperature [79].

Overview of Stainless Steel Categories and Processing Technique

Vanadium steel was the first metal alloy manufactured for application in the human body for use as bone fracture plates and screws but it is not used for this purpose anymore because it corrodes within biological environment quite quickly. Later on, the austenitic stainless steel was developed for use in biological systems and up to date, it remains the most widely used for the development of implantable devices [36]. Some austenitic stainless steels (*e.g.* SUS 316 L) contain a significant amount of Ni which makes them capable of triggering Ni-related allergic reactions *in vivo*. Consequently, a number of researches have been directed towards preparing Ni-free austenitic stainless steel. Previously produced Ni-free stainless steels contained significant quantities of good austenite stabilizers similar to Ni namely N and Mn [80]. The biocompatibility of Mn has not been clearly established, therefore, the pressurized electron spin resonance method has been widely explored to produce high N-containing austenitic stainless steel which is reported to be more biocompatible [8]. Nevertheless, it is challenging to develop finished devices from high N-containing stainless steel because it's poorly plastic in nature which makes handling and processing quite problematic [8]. A method of manufacturing a Ni- and Mn-free, high N-austenitic stainless steel from ferritic stainless steel was develop in order to circumvent this problem of poor plasticity [81, 82]. This process involves refining ferritic stainless steel (Fe24Cr2Mo) crystal grains using a thermo-mechanical treatment followed by forging into the desired geometry of the final biomedical product which is then heated in a N_2 gas surrounding such that N penetrates into the product resulting in the stabilization of the austenite phase to produce a Ni- and Mn-free austenitic stainless steel which is cooled down systematically, shaped, ground and polished using respective procedures to generate the desired biomedical products [6, 8, 81, 82]. The overall exhibited properties (*e.g.* yield, ultimate strength, ductility, fatigue strength) of stainless steel are dependent on the intensity of the cooling phase. The uniformity and smoothness of the grain size of the stainless steel are also very important [6].

Corrosion of Stainless Steel

Generally, stainless steels are known to be corrosion resistant. This is due to the formation of a thin Cr- and Mo- containing inert, strongly adherent corrosion resistant surface oxide layer of Cr_2O_3 with the Mo specifically enabling stability in surroundings containing chlorine ion [6]. Most austenitic stainless steels are considered majorly single-phase, face-centered cubic alloys with the possibility of carbides forming on exposure to elevated temperatures (400°C-800°C) due to the presence of some C atoms within these lattices. The formation of the carbide deforms the internal structure of the stainless steel leading to the formation of a less stable structure that is more predisposed to inter-granular corrosion [6]. This process is described as alloy 'sensitization' and its occurrence can be minimized by preparing low carbon levels austenitic stainless steels especially for load-bearing applications [6].

Biomedical Applications of Stainless Steel

There are different types of stainless steels but the most widely utilized for the production of clinical implantable devices is the austenitic stainless steel, specifically the low carbon type, AISI 316 L which contains 0.03 %w/w C, 17–20 %w/w Cr, 12–14 %w/w Ni, 2–3 %w/w Mo and minute quantities of N, Mn, P, Si and S [79]. In comparison with other metal-based implants, stainless steels have lower mechanical strength, higher corrosion resistance, ductility and stiffness. Their highly stiff (low elasticity) nature makes them not as suitable for bone replacement applications like Ti [63]. However, austenitic stainless steel has been repeatedly accepted and applied over many decades as a biomaterial of choice for temporary fracture repair constructs such as hip nails, screws and fracture plates. Stainless steel's wide spread applicability is also attributable to its relatively low price, processing ease and accessibility [6, 79]. In addition, austenitic stainless steels have been employed for manufacturing joint replacement parts but this application has become more limited because alternative metallic alloys (*e.g.* Ti, Co-Crbased alloys) exist which demonstrate better corrosion and mechanical qualities. Nevertheless, austenitic stainless steel with Ni partially substituted with Mn and an elevated Cr (> 20%) and N (0.3-0.4%w/w) content is being used in joint prosthesis. N functions as an austenitic phase stabilizer as well as induces increase in both corrosion resistance and mechanical properties. Other grades of stainless steel such as 304 and 316, 316 L are applied (but not frequently) as surgical sutures, vascular stents, electrodes, pulse generator, conducting lead wires, microvascular clips for treatment of aneurysms, precipitation hardenable stainless steels (17-7PH) and martensitic stainless steels (420, 431 alloys) used for neurosurgical and microvascular clips. Austenitic stainless steels are also used for fabricating vascular stents as well as electrodes, conducting lead wires and pulse generator casings for cardiac pacing systems (304, 316, 316 L alloys) [6].

Surface Modification of Stainless Steel

Stainless steels (especially the AISI 316 L grade) remains a commonly employed biomaterial for internal fixation devices (*e.g.* screws, fracture plates) due to the fact that they are corrosion resistant, easy to process, mechanically strong, biocompatible and quite inexpensive compared to other metallic biomaterials [37]. Conventional stainless steel contains Ni which results in adverse effects under normal *in vivo* physiological conditions. The austenitic stabilizing property of N enables a reduction in stainless steel Ni content to nearly zero and this makes Ni-free, high N austenitic stainless steels more mechanically strong, resistant to corrosion, biocompatible and cost effective [37].

Various research activities are directed towards improving the quality and usefulness of stainless steels as metallic biomaterials. Choudhury and co-researchers [83] used an electro discharge machining oriented to construct a defined micro-dimple array on the head of a stainless steel hip prosthesis. The indented surface was further covered with hydrogenated amorphous carbon and tetrahedral amorphous carbon layers employing the magnetron sputtering technique. It was observed that this newly designed surface, when in contact with a ceramic ball under simulated hip joint conditions did not produce any wear debris and friction coefficient was significantly minimized. The carbon coatings amplified the hardness of the dimpled and non-dimpled surfaces. Zhang and colleagues [84] modified the surface properties of 316 L stainless steel wafers with SiCOH plasma nanocoatings (30-40 nm thickness) deposited by low temperature plasmas from a gas mixture of trimethysilane and oxygen at different ratios. The SiCOH plasma nanocoated 316 L stainless steel displayed good hemo- and cyto- compatibility compared to the uncoated 316 L stainless steel. Balla and co-workers [85] attempted to improve the osseointegration capacity of 316 L stainless steel by modifying it surface with bioactive hydroxyapatite using the laser-engineered net shaping manufacturing technique. Kenar *et al.* [86] reported the use of Femtosecond laser treatment technique to improve surface properties of

316 L stainless steel plates. The hydrophobicity and carbon content of these plates were improved which resulted in a significant improvement in osseointegration measured by an enhanced adhesion of human endothelial cells and human bone marrow mesenchymal cell respectively. Subramanian and associates [87] fabricated 316 L stainless steel substrates coated with nanoscale multi-layered TiN/VN employing the reactive dc magnetron sputtering method. The multi-layered stainless steel surface exhibited fewer platelet aggregation and lower corrosion rate than the plain substrates. Ni and colleagues [88] designed 316 L stainless steel with nano-pit surface using anodization technique in ethylene glycol electrolyte solution containing 5%v/v perchloric acid. The modified surface displayed better wettability properties compared to the untreated material. Generally, the nano-pitted surface intensified primary human dermal fibroblast attachment and growth. Fig. (**3**) highlights various images of surface modified medical grade stainless steel platforms.

Cobalt and Its Alloys

Co was discovered in 1735 by Brandt [89]. It is present in nature with cobaltite, smaltite, meteorites and erythrite minerals. It is also usually associated with nickel, silver, lead, copper and iron ores, which are usually obtained as by-products [89]. It is a brittle, hard metal which has a whitish with a bluish tinge appearance; it is diamagnetic and has magnetic permeability. It occurs as two allotropes over a wide temperature range namely the stable β-form (a closely-packed hexagonal crystal) which predominates below approximately 417°C and a stable, cubic crystal α-form which occurs mainly above 417°C until its melting temperature. Co is soluble in dilute sulphuric acid, nitric or hydrochloric acid and is slowly reactive with alkalis [89].

Common Types of Cobalt Alloys, Their Properties and Preparation

During the twentieth century, cobalt-chromium alloys were amongst the first group of metallic materials that were successfully used as orthopaedic implants [39]. Co-Cr alloys are attractive biomedical metals because of their properties like chemical inertness, biocompatibility and mechanical durability [39]. Of many Co-Cr alloys available, there are primarily, two types which are commonly applied as implantable metallic biomaterials. These two are: (i) the Cobalt-Chromium-Molybdenum (Co-Cr-Mo; Vitallium™) designated as ASTM F75 and F76 and (ii) Cobalt-Nickel-Chromium-Molybdenum (Co-Ni-Cr-Mo) designated as ASTM F562 alloys which are castable and wrought by hot forging respectively [36].

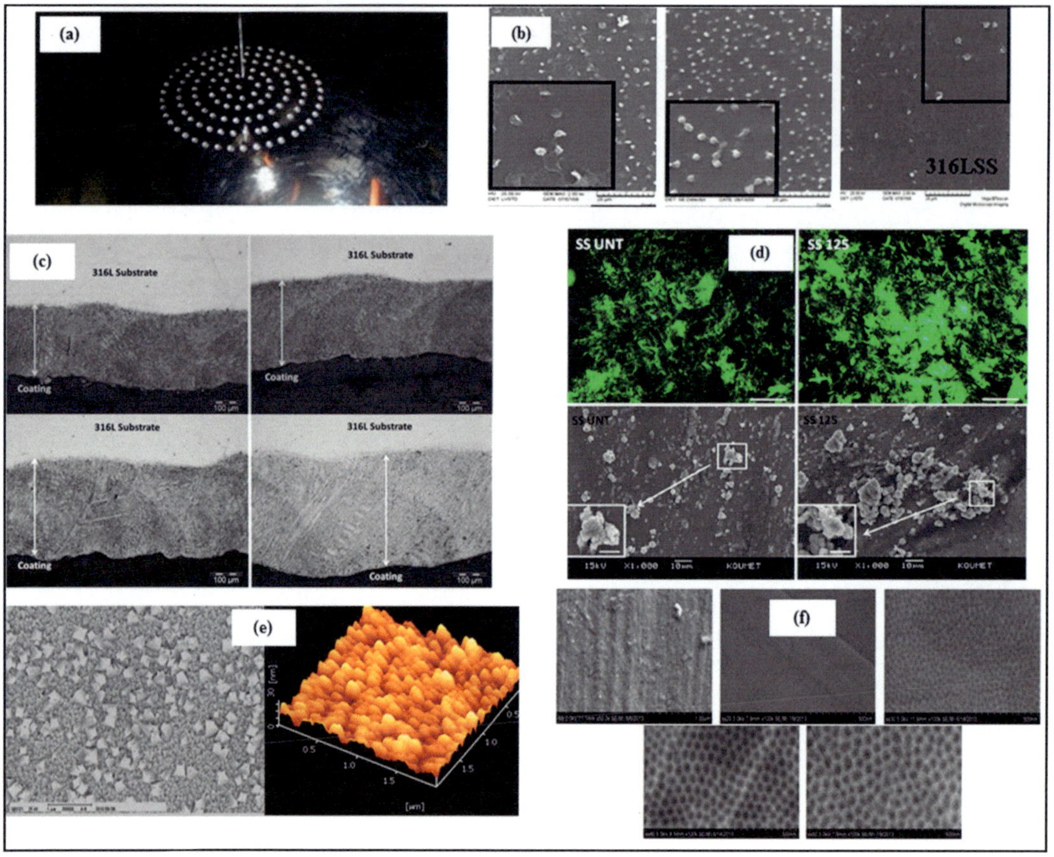

Figure 3: Images of different modified medical grade stainless steel surfaces: (**a**) micro-dimple array on the head of a stainless steel hip prosthesis [83], (**b**) platelet adhesion to SiCOH plasma nanocoated 316 L stainless steel and uncoated bare wafers [84], (**c**) bioactive hydroxyhapatite coated surface using Laser-engineered net shaping technique [85], (**d**) osteogenic differentiation of human endothelial cells and human bone marrow mesenchymal cell on untreated (SS UNT) and laser treated (SS 125) stainless steel samples [86], (**e**) TiN/VN coated stainless steel surface [87], (**f**) progression from non-porous untreated stainless steel surface to nano-porous surface structures after anodization [88].

Other Co alloys approved and recommended by the American Society for Testing and Materials for implant use are those amalgamated with tungsten (wrought Co-Cr-Ni-W also referred to as ASTM F90) and that amalgamated with iron (wrought Co-Ni-Cr-Mo-W-Fe also referred to as ASTM F563). Amidst these four earlier mentioned Co alloys, the Co-Ni-Cr-Mo is the most resistant to corrosion while its abrasive wear properties are similar to that of the Co-Cr-Mo alloy. It possesses good mechanical properties which make it suitable for long term use of implantable devices [36]. Co-Cr-based alloys are excellently resistant to corrosion even in chloride environments due to the spontaneous formation of a passive oxide layer within the *in vivo* environment [35, 79]. Their mechanical strength, especially fatigue strength is also superior and their elastic modulus (220-230 GPa) can be referred to as high and similar to that of stainless steel (approximately 200 GPa). They are not brittle because they have a minimum of 8% elongation and an order of magnitude higher than that of cortical bone (20–30 GPa) [35, 79]. Co alloys are also attractive because of their excellent creep resistance. They possess a matrix that is resistant to thermally-activated dislocated movements within the crystalline lattice as Co has a good tolerance for other elements in solid solution [89]. When Co-Cr alloy containing implantable devices are in contact with bone, they bear most of the weight due to their high modulus which enables stress shielding in the adjacent bone [79]. Toxic elements such as Ni, Cr and Co have been reported to be released from Co-Cr alloys due to the corrosion in the human body. These corrosion products of Co-Cr-Mo are more toxic than those of stainless steel 316 L [35]. There are Co alloys containing low Ni however a maximum Ni content of 1%w/w (because of Ni allergy) and C content of 0.35%w/w (because of destructive carbide formation if levels are too high) are concurrently allowed to enhance wear resistance. Research is now being directed towards fabricating Ni- and C-free Co alloys using thermomechanical processes conducted under elevated temperature conditions. A high mechanical strength and malleable Co-Cr-Mo alloy can be manufactured by moulding during the ductile phase region and rapidly cooling through the brittle region resulting in the formation of a stable wear resistant matrix [8].

Surface Modification of Cobalt and Its Alloys

Research has been directed towards making Co-Cr alloy surface porous because of their remarkable properties of inertness, biocompatibility and mechanical durability since the 1970's [39]. The Co-Cr beads manufactured from spherical Co-Cr alloy were sintered to the implant surface to produce a micro/macro-porous surface structure for bone ingrowth. This Co-Cr porous coating has been applied in acetabular cups, femoral stems and total knee arthroplasty components [39]. In a study of 72 hip implants with a Co-Cr beaded acetabular cup conducted by Matassi and co-workers [39], the occurrence of aseptic loosening was found to be 4% at an average of 8½ years follow up. Also, the Co-Cr beaded stems have been noted to facilitate about 95% stable ingrowth at an average of about 11 years. Nonetheless, ingrowth analysis showed minimal ingrowth and fibrous tissue formation in Co-Cr beaded implantable device which may be of significant concern as this may hinder the success of long term fixation [39].

Movement of the artificial femoral head into the pelvis region resulting from the degeneration of acetabular cartilage has come up as a serious concern in resurfacing or bipolar hemi-arthroplasty. Surface modification of Co-Cr-Mo alloy seems to be one of the very promising ways of tackling lubricant insufficiency thereby averting the relocation of the artificial femoral head [90]. On this basis, Kyomoto and co-researchers [90] attempted to modify the surface of Co-Cr-Mo alloy with biocompatible 2-methacryloyloxyethyl phosphorylcholine polymer employing the dip coating or grafting technique. The alloy-polymer grafted interface was hydro-stable and yielded reduced friction coefficient that can be compared to that of natural cartilage interface.

Various kinds of defects (such as delivery balloon expansion resulting into detachment) have been observed from the use of drug-eluting stents [91]. Thus Co-Cr alloy surface coupled with a layer of drug-in-polymer matrix using poly(caprolactone) brushes were designed by Bedair and colleague [91]. This combination was explored to prevent the detachment defect by increasing interfacial grip between the alloy and drug-in-polymer surfaces. The chemical alteration of the Co-Cr surface was achieved by embedding ricinoleic acid onto the alloy substrate surface followed by surface-initiated ring opening polymerization of ε-caprolactone. This approach enhanced both adhesion of drug-in-polymer matrix to the Co-Cr alloy as well as controlled drug release rate [91].

The survival of Co-Cr-Mo implants post a total knee arthroplasty was enhanced utilizing Titanium-nitride [92]. Overall experimental outputs obtained from testing this modified surface on osteoblast and cytokine proliferation and/or differentiation suggested that osteo-conduction on the titanium-nitride (TiN) surface were similar to that of the Co-Cr-Mo exterior [92]. Fig. (**4**) shows micrographs of modified Co alloy surface geometries.

Some Biomedical Applications of Cobalt and Its Alloys

Co–Cr alloys have also been efficaciously utilized in combination with polyethylene for the fabrication of artificial disc prostheses such as the SB Charite artificial disc prosthesis [79]. As mentioned earlier, the wrought and cast alloys comprise the range of Co–Cr alloys used clinically. The wrought alloys (*e.g.* ASTM F562, MP35N) are widely used in the production of orthopaedic biomedical devices [79]. The castable Co-Cr-Mo alloy is usually applied in medicine especially in dentistry and has also found good use in artificial joint construction. The wrought Co-Ni-Cr-Mo alloy is a relatively recent advancement and has been applied as an implantable material for making stems of high weight bearing joints such as the hip and knee prosthesis [36]. Table **2** highlights the general medical applications, merits and demerits of the metallic biomaterial discussed in this manuscript.

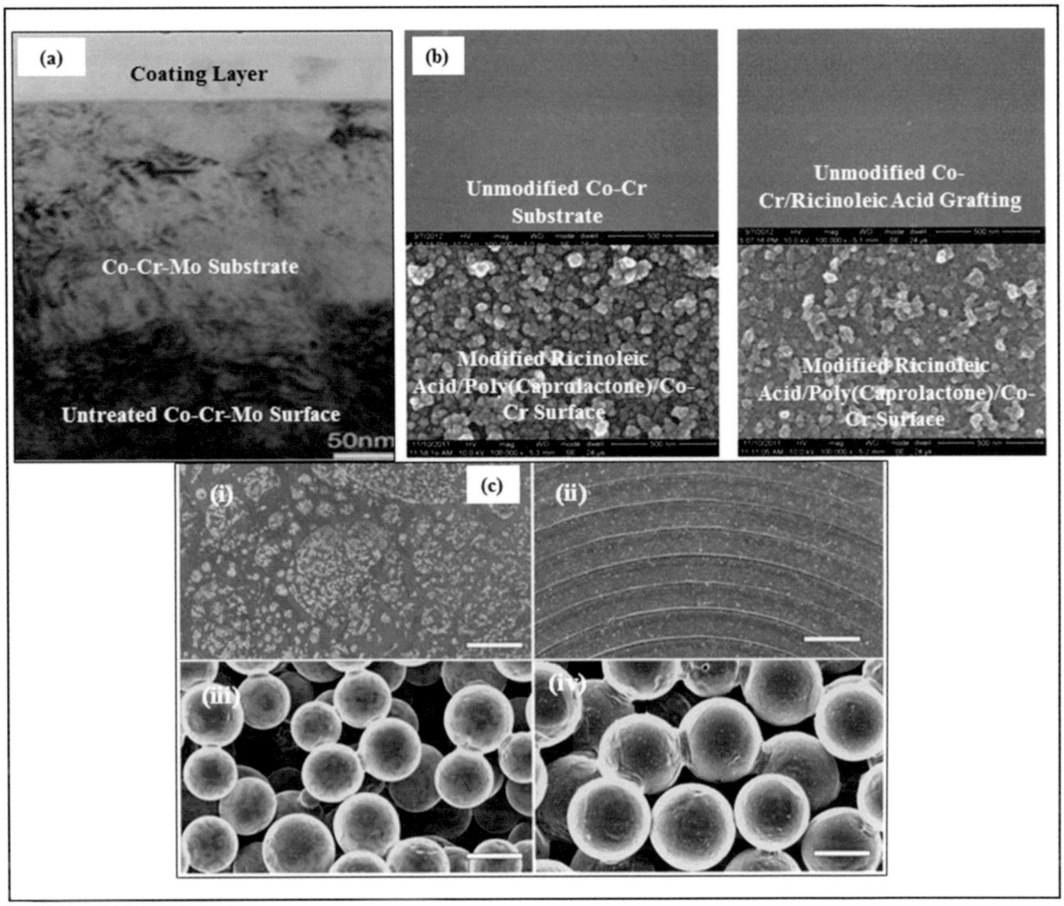

Figure 4: Cross-sectional images of: (**a**) modified surface of Co-Cr-Mo coated with 2-methacryloyloxyethyl phosphorylcholine polymer [90], (**b**) Co-Cr alloy surface coupled with a layer of drug-in-polymer matrix using poly(caprolactone) brushes [91], (**c**) (i) Co-Cr-Mo; (ii) TiN; (iii) porous coated Co-Cr-Mo; (iv) porous coated TiN surfaces [92].

CONCLUSION

In the last few decades, the utilization of metallic biomaterials have shifted from purely industrial applications to the development of materials that are potentially capable of interacting with the biological tissues/environment to generate desirable *in vivo* responses/effects. On a general note, metallic biomaterials (*e.g.*

Table 2. Medical applications, merits and demerits of Ti, Co, their alloys, and stainless steel.

Metallic Biomaterial	Common Bio-Applications	Advantages	Disadvantages
Titanium and its alloys	Endosseous dental implants [28], coatings on sintered beads, fibre metal mesh and plasma spray coatings and coatings for primary and revision hip and knee arthroplasty [39].	Adjoining bone integration, low weight, relatively high specific strength, corrosion resistance, biocompatibility, ability to form coherent passive oxide (TiO_2) [36, 39, 42].	Expensive, poor wear characteristics, destructive interfacial friction [36, 38], poor tribological properties [38, 40, 41].
Stainless Steel	Temporary fracture repair devices such as fracture plates, screws and hip nails [39], joint replacement components/joint prosthesis [79], surgical sutures, electrodes, conducting lead wires and pulse generator casings for cardiac pacing systems, microvascular clips for treatment of aneurysms, neurosurgical clips, vascular stents as well as electrodes, conducting lead wires and pulse generator casings for cardiac pacing systems [6].	Corrosion resistance, ductility and stiffness [39], relatively non-expensive and easy to process [6, 79].	Lower mechanical strength, low elasticity [39].
Cobalt and Its Alloys	Orthopaedic devices, artificial disc prostheses [79], dental implants and in high weight bearing joints applications such as the hip and knee prosthesis [36].	Superior mechanical strength [36], high elastic modulus, non-brittle [35], inert and biocompatible and durable [39] and have excellent creep resistance [89].	Release of toxic element due to corrosion, lubricant insufficiency [90].

Ti, Co, Cr, stainless steel, Ta, Sn *etc*.) are biocompatible materials which can originate from natural or artificial sources, but are based on metallic elements or alloys and usually applied as a whole or part of biological systems to perform, enhance or replace specific roles that have been lost through illness or injury. They can be utilized either for short-term or long-term purposes and most common applications are in surgery, dentistry and general medicine. The biomedical uses of metallic biomaterials are anchored upon their excellent mechanical properties, corrosion resistance, biocompatibility and durability. They have diverse applications for both high load bearing skeletal implantation as well as functional and unloaded roles in biological systems. Structural and surface modifications of implantable devices based on metallic biomaterials by introducing porous networks, coating with polymeric materials, ceramics *etc.* are crucial to making them versatile and highly suitable for their respective biomedical applications. Despite that there are still existing challenges with the use of metallic biomaterials; there are many more choices of possible sophisticated and highly effective solution options which therefore have led to the invention of more useful devices. It is recommended that future work should be more focused on the use biocompatible, but preferably non-biodegradable polymeric materials for the surface and interface modification of metallic biomaterials in such a way that wear resistance, mechanical strength, biocompatibility, osseointegration, flexibility/lubrication capacity and durability are reinforced especially for implants applied at high load-bearing *in vivo* regions. Besides, the fabrication of more sophisticated porous metallic biomaterial surfaces employing simple, effective and commercializable methods can be a giant stride towards developing implantable devices with the capacity of facilitating optimum osteointegration. Finally, future explorations can also concentrate more on the use of hybrid biomaterial-based templates (derived from *e.g.* metals, alloys, ceramics, polymers *etc.*), which conform to diverse universally set standards, as extended release drug/bioactive eluting stents/scaffolds for application at diseased peripheral or coronary arterial sites.

ACKNOWLEDGEMENTS

Declared None.

CONFLICT OF INTEREST

The author discloses that there are no commercial affiliations, consultancies, stock or equity interests, or patent licensing arrangements that could be considered to pose a conflict of interest regarding the submitted manuscript chapter.

REFERENCES

[1] Hench LL. Biomaterials: a forecast for the future. Biomaterials 1998; 19: 1419-23.
[2] Williams DF. Leading opinion on the nature of biomaterials. Biomaterials 2009; 30(2): 5897-909.
[3] Mediaswanti K, Wen C, Ivanova EP, *et al.* A review on bioactive porous metallic biomaterials. Biomim Biomater Tissue Eng 2013; 18(1): 1-8.
[4] Huebsch N, Mooney DJ. Inspiration and application in the evolution of biomaterials. Nature 2009; 462: 426-32.
[5] Breme J, Biehl V. Handbook of Biomaterials Properties. In: Black J, Hastings G, Eds. Chapman and Hall, London 1998; pp. 135-36.
[6] Pilliar RM. Biomedical Materials. In: Narayan R, Ed. Springer Science+Business Media, New York 2009; pp. 41-79.

[7] Park JB, Kim YK (2003) In: Park, JB and Bronzino, JD (Ed). Biomaterials principles and applications, CRC Press, Florida 1-2.

[8] Niinomi M. Metallic biomaterials. J Artif Organs 2008; 11(3): 105-10.

[9] White E, Shors EC. Biomaterial aspects of Interpore-200 porous hydroxyapatite. Dent Clin N Am 1986; 30(1): 49-67.

[10] Arcus Sportklinik Pforzheim, *Hip Endoprosthesis – Material and Fixation*. Available from: http://www.sportklinik-germany.com/operative-spectrum/endoprosthetics/hip-endoprosthesis/material-and-fixation.html (accessed 18 Nov 2014).

[11] B/Braun Sharing Expertise, *Total Hip Endoprostheis*. Available from: http://www.bbraun-vetcare.com/cps/rde/xchg/ae-vetcare-en-int/hs.xsl/products.html?prid=PRID00005501 (accessed 15 Nov 2014).

[12] Charité, *Julius Wolff Institute for Biomechanics and Musculoskeletal Regeneration*. Available from: http://jwi.charite.de/en/research/joint_loading/instrumented_implants/knee_joint/ (accessed 15 Nov 2014).

[13] Arcus Sportklinik Pforzheim, *Knee Endoprosthetics – Types of prostheses*. http://www.sportklinik-germany.com/operative-spectrum/endoprosthetics/knee-endoprosthetics/types-of-prostheses.html (accessed 16 Nov 2014).

[14] Komal Health Care, *Safe and Approved Medical Products*. Available from: http://www.gauzedswap.com/bone-screw-362973.html (accessed 18 Nov 2014).

[15] Smith and Nephew, *Hip Fracture Surgical Option*. Available from: http://www.smith-nephew.com/patient/treatments/hip-treatments/hip-fracture/ (accessed 15 Nov 2014).

[16] Beta Surgicare, *Bone Plates*. Available from: http://www.betasurgicare.com/bone-plates-1523183.html (accessed 18 Nov 2014).

[17] Medical Expo, *The Online Medical Devices Exposition*. Available from: http://www.medicalexpo.com/prod/depuy-synthes/ulna-compression-bone-plates-distal-human-79814-498418.html (accessed 18 Nov 2014).

[18] My Dental Implant, *Your Complete Resource About Dental Implants and Implantology*. Available from: http://mydentalimplants.org/cost/ (accessed 19 Nov 2014).

[19] Ran Family Dentistry, *Dental Implants*. Available from: http://randental.com/frisco-dental-services/dental-services/dental-implants/ (accessed 18 Nov 2014).

[20] Tornier, *Total Ankle Prosthesis*. Available from: http://www.tornier-us.com/lower/ankle/ankrec004/ (accessed 17 Nov 2014).

[21] Ortho-Praxis, *Arthrosis of the Ankle Joint*. Available from: http://www.ortho-praxis.ch/oberes-sprunggelenk-en.php (accessed 18 Nov 2014).

[22] X-Spine, *The Axle® Interspinous Fusion System* Available from: http://x-spine.com/surgeons/thoracolumbar/axle/ (accessed 18 Nov 2014).

[23] Medical Device Network, *Precision Spine Launches New Anterior Cervical Plate System*. Available from: http://www.medicaldevice-network.com/news/newsprecision-spine-launches-new-anterior-cervical-plate-system-us (accessed 17 Nov 2014).

[24] Wikipedia, *A document on Stent by Frank C. Müller*. Available from: http://en.wikipedia.org/wiki/stent (accessed 16 Nov 2014).

[25] Herbal Ignite, *Natural Virility*. Available from: http://herbalignite.com.au/erectile-dysfunction/treatments/invasive/vascular-surgery.html (accessed 16 Nov 2014).

[26] Bostonscientific, *Interventional Radiology Products*. Available from: http://www.bostonscientific.com/en-US/medical-specialities/interventional-radiology/products.html (accessed 10 Nov 2014).

[27] Medical Guide Wire, *the Online Medical Devices Exhibition*. Available from: http://www.medicalexpo.com/medical-manufacturer/catheter-guidewire-15311.html (accessed 18 Nov 2014).

[28] Shenzhen Rogin Dental Supply, *Orthodontic Niti Reverse Curved Arch Wires Dental Orthodontic Instruments*. Available from: http://www.china-dental-supply.com/sale-91239-orthodontic-niti-reverse-curved-arch-wires-round.html (accessed 18 Nov 2014).

[29] Wikipedia, *Lingual Arch*. Available from: http://en.wikipedia.org/wiki/Lingual_arch (accessed 14 Nov 2014).

[30] Sensorymedic, *The Cons of Cochlear Implantation.* Available from: http://www.sensorymedic.com/ear-and-hearing/111/ (accessed 10 Nov 2014).

[31] University of Southampton, *Auditory Implant Service*, Available from: http://ais.southampton.ac.uk/cochlear-implants/what-are-cochlear-implants/ (accessed 18 Nov 2014).

[32] MedCPS@Penn, *Heart Modeling for Closed-Loop Pacemaker Verification and Validation.* Available from: http://medcps.org/VHM.html (accessed 18 Nov 2014).

[33] St, Jude Medical, *Traditional Pacemakers*, Available from: http://www.sjm.com/leadlesspacing/intl/options/traditional-pacing (accessed 18 Nov 2014).

[34] Staigera MP, Pietaka AM, Huadmaia J, Dias G. Magnesium and its alloys as orthopaedic biomaterials: a review. Biomaterials 2006; 27(9): 1728-34.

[35] Nasab, MB, Hassan MR. Metallic biomaterials of knee and hip – A review. Trends Biomater Artif Organs 2010; 24(1): 69-82.

[36] Srivastav A. Biomedical Engineering, Trends in Materials Science. In: Laskovski AN, Ed. InTech, Croatia 2011; pp. 154-65.

[37] Talha M, Behera CK, Sinha OP. *In vitro* long term and electrochemical corrosion resistance of cold deformed nitrogen containing austenitic stainless steels in simulated body fluid. Mater Sci Eng C 2014; 40: 455-66.

[38] de Viteri VS, Fuentes E. Titanium and Titanium Alloys as Biomaterials. In: Gegner J, Ed. Tribology – Fundamentals and Advancements. InTech, Croatia 2013; pp. 155-81.

[39] Matassi F, Botti A, Sirleo L, Carulli C, Innocenti M. Porous metal for orthopaedics implants. Clin Cases Miner Bone Metab 2013; 10(2): 111-15.

[40] Yerokhin AL, Niea X, Leyland A, Matthews A. Characterization of oxide films produced by plasma electrolytic oxidation of a Ti-6AL-4V alloy. Surf Coat Tech 2000; 130: 195-206.

[41] Fraczek T, Olejnik M, Tokarz A. Evaluations of plasma nitriding efficiency of titanium alloys for medical applications. Metalurgija 2009; 48(2): 83-6.

[42] Pilliar RM, Weatherly, GC. Developments in implant alloys. Crit Rev Biocompat 1986; 1: 371-403.

[43] Gilbert JL, Buckley CA, Jacobs JJ. *In vivo* corrosion of modular hip prosthesis components in mixed and similar metal combinations - The effects of crevice, stress, motion, and alloy coupling. J Biomed Mater Res 2004; 27(12): 1533-44.

[44] Jafari SM, Benber B, Coyle C, Parvizi J, Sharkey PF, William, H. Do tantalum and titanium cups show similar results in revision hip arthroplasty. Clin Orthop Relat Res 2010; 468(2): 459-65.

[45] Long M, Rack HJ. Titanium alloys in total joint replacement – a material science perspective. Biomaterials 1998; 19: 1621-39.

[46] Lopez-Heredia MA, Sohier J, Gaillard C, Quillard S, Dorget M, Layrolle P. Rapid prototyped porous titanium coated with calcium phosphate as a scaffold for bone tissue engineering. Biomaterials 2008; 29: 2608-15.

[47] Carapeto AP, Serro AP, Nunes BMF, *et al.* Characterization of two DLC coatings for joint prosthesis: The role of albumin on the tribological behaviour. Surf Coat Technol 2010; 258(21-22): 3451-58.

[48] Ma G, Gong S, Lin G, Zhang L, Sun G. A study of structure and properties of Ti-doped DLC film by reactive magnetron sputtering with ion implantation. Appl Surf Sci 2012; 258(7): 3045-50.

[49] Dowling DP, Kola PV, Donnelly K, Kelly TC, Brumitt K, Lloyd L, Eloy R, Therin M, Weill N. Evaluation of diamond-like carbon coated orthopaedic implants. Diam Relat Mater 1997; 6(2-6): 390-93.

[50] Zhang ZY, Lu XC, Luo JB, Liu Y, Zhang, CH. Mechanical properties of La_2O_3 doped diamond-like carbon films. Surf Technol 2008; 202: 1621-27.

[51] Han Y, Hong SH, Xu KW. Porous nanocrystalline titania films by plasma electrolytic oxidation. Surf Coat Tech 2002; 154(2-3): 314-18.

[52] Huang P, Wang F, Xu K, Han Y. Mechanical properties of titania prepared by plasma electrolytic oxidation at different voltages. Surf Coat Tech 2007; 201(9-11): 5168-71.

[53] Viorney C, Guenther HL, Aronsson BO, Péchy P, Descouts P, Grätzel M. Osteoblast culture on polished titanium disks modified with phosphonic acids. J Biomed Mater Res 2002; 62(1): 149-55.

[54] Rashmir-Raven MA, Richardson DC, Aberman HM, De Young DJ. The response of cancellous and cortical canine bone to hydroxyapatite-coated and uncoated titanium rods. J App Biomater 1995; 6(4): 237-42.

[55] Wen CE, Yamada Y, Hodgson PD. Fabrication of novel TiZr alloy foams for biomedical applications. Mater Sci Eng C 2006; 26(6): 1439-44.

[56] Wen CE, Mabuchi M, Yamada Y, Shimojima K, Chino Y, Asahina T. Processing of biocompatible porous Ti and Mg. Scripta Mater 2001; 45(10): 1147-53.

[57] Xiong JY, Li YC, Wang XJ, Hodgson PD, Wen CE. Titanium-nickel shape memory alloy foams for bone tissue engineering. J Mech Behav Biomed Mater 2008; 1(3): 269-73.

[58] Heller M, Kämmerer PW, Al-Nawas B, Luszpinski M-A, Förch R, Brieger J. The effect of extracellular matrix proteins on the cellular response of HUVECS and HOBS after covalent immobilization onto titanium. J Biomed Mater Res A 2015; 103(6): 2035-44.

[59] Marín-Parejaa N, Salvagnia E, Guillem-Martia J, Apariciod C, Ginebraa M-P. Collagen-functionalised titanium surface for biological sealing of dental implants: Effect of immobilisation process on fibroblasts response. Colloid Surface B 2014; 122: 601-10.

[60] Wei Q, Krysiak S, Achazi K, *et al.* Multivalent anchored and crosslinked hyperbranched polyglycerolmonolayers as antifouling coating for titanium oxide surfaces. Colloid Surface B 2014; 122: 684-92.

[61] Chen J, Zhao A, Chen H, *et al.* The effect of full/partial UV-irradiation of TiO_2 films on altering the behaviour or fibrinogen and platelets. Colloid Surface B 2014; 122: 709-18.

[62] Zhao L, Liu L, Wu Z, Zhang Y, Chu PK. Effects of micropitted/nanotubular titania topographies on bone mesenchymal stem cell osteogenic differentiation. Biomaterials 2012; 33: 2629-41.

[63] Ren N, Zhang S, Li Y, *et al.* Bone mesenchymal stem cell functions on the hierarchical micro/nanotopographies of the Ti-6Al-7Nb alloy. Brit J Oral Max Surg 2014; 52(10): 907-12.

[64] Jindal S, Bansal R, Singh BP, *et al.* Enhanced osteoblast proliferation and corrosion resistance of commercially pure titanium through surface nanostructuring by ultrasonic shot peening and stress relieving. J Oral Implantol 2014; 40(S1): 347-55.

[65] Redepenning J, Venkataraman G, Chen J, Stafford N. Electrochemical preparation of chitosan/hydroxyapatite composite coatings on titanium substrates. J Biomed Mater Res A 2003; 66A(2): 411-6.

[66] Kashiwazaki H, Kishiya Y, Matsuda A, *et al.* Fabrication of porous chitosan/hydroxyapatite nanocomposites: their mechanical and biological properties. Bio-Med Mater Eng 2009; 19(2-3): 133-40.

[67] Ma X-Y, Feng Y-F, Ma Z-S, Li X, Wang J, Wang L. The promotion of osteointegration under diabetic conditions using chitosan/hydroxyapatite composite coating on porous titanium surfaces. Biomaterials 2014; 35(9): 7259-70.

[68] Yang J, Liu A, Han Y, Li Q, Tian J, Zhou C. Osteodifferentiation of mesenchymal stem cells on chitosan/hydroxyapatite composite film. J Biomed Mater Res A 2014; 102(4): 1202-09.

[69] Długoń E, Niemiec W, Frączek-Szczypta A, Jeleń P, Sitarz M, Błażewics M. Spectroscopic studies of electrophoretically deposited hybrid Hap/CNT coatings on titanium. Spectrochim Acta A 2014; 133: 872-75.

[70] Clèries L, Fernández-Pradas JM, Morenza JL. Bone growth on and resorption of calcium phosphate coating obtained by pulsed laser deposition. J Biomed Mater Res 2000; 49(1): 43-52.

[71] Zhang C, Liu Y, Wen S, Wang S. Poly(vinylphosphonic acid) (PVPA) on titanium alloy acting as effective cartilage-like superlubricity coatings. ACS Appl Mater and Inter 2014; 6(20): 17571-8.

[72] Kazek-Kęsik A, Krok-Borkowicz M, Pamuła E, Simka W. Electrochemical and biological characterization of coatings formed on Ti15Mo alloy by plasma electrolytic oxidation. Mater Sci Eng C 2014; 43: 172-81.

[73] Pan CJ, Hou YH, Liu HQ, Ding HY, Dong YX. Improved anticoagulant of titanium by sequential immobilization of oligo(ethyleneglycol) and 2-methacryloyloxyethyl phosphorylcholine. Colloid Surface B 2013; 112: 508-12.

[74] Zhao L, Hu Y, Xu D, Cai K. Surface functionalization of titanium substrates with chitosan-lauric acid conjugate to enhance osteoblasts function and inhibit bacterial adhesion. Colloid Surface B 2014; 119: 115-25.

[75] Burnat B, Walkowiak-Przybyło M, Błaszczyk T, Klimek L. Corrosion behaviour of polished and sandblasted titanium alloys in PBS solution. Acta Bioeng Biomech 2013; 15(1): 87-95.

[76] Hoffman W, Bormann T, Rossi A, *et al.* Rapid prototyped porous nickel-titanium scaffolds as bone substitutes. J Tissue Eng 2014; 5: 1-14.

[77] Xu J, Weng X-J, Wang X, *et al.* Potential use of porous titanium-niobium alloy in orthopedic implants: Preparation and experimental study of its biocompatibility *in vitro*. PloS ONE 2013; 8(11): e79289.

[78] Xie F, He X, Lu X, Cao S, Qu X. Preparation and properties of porous Ti-10Mo alloy by selective laser sintering. Mater Sci Eng C Mater Biol Appl 2013; 33(3): 1085-90.

[79] Navarro M, Michiardi A, Castaño O, Planell JA. Biomaterials in Orthopaedics. J R Soc Interface 2008; 5: 1137-58.

[80] Uggowitzer PJ, Bäher WF, Speidel MO. Metal injection molding of nickel-free stainless steels. Adv PM Part Mater 1997; 3: 113-21.

[81] Kuroda D, Hanawa T, Hibaru T, Kuroda S, Kobayashi M, Kobayashi T. New manufacturing process of nickel-free austenitic stainless steel with nitrogen absorption treatment. Mater Trans 2003; 44(3): 414-20.

[82] Yang K, Ren Y. Nickel-free austenitic stainless steels for medical applications. Sci Technol of Adv Mat 2010; 11(5): 1-13.

[83] Choudhury D, Ching HA, Mamat AB, *et al.* Fabrication and characterization of DLC coated microdimples on hip prosthesis heads. J Biomed Mater Res B 2015; 103(5): 1002-12.

[84] Zhang Q, Shen Y, Tang C, Wu X, Yu Q, Wang G. Surface modification of coronary stents with SiCOH plasma nanocoatings for improving endothelialisation and anticoagulation. J Biomed Mater Res B 2015; 103(2): 464-72.

[85] Balla VK, Das M, Bose S, Ram GD, Manna I. Laser surface modification of 316 L stainless steel with bioactive hydroxyapatite. Mater Sci Eng C 2013; 33(8): 4594-98.

[86] Kenar H, Akman E, Kacar E, *et al.* Femtosecond laser treatment of 316 L improves its surface nanoroughness and carbon content and promotes osseointegration: An *in vitro* evaluation. Colloid Surface B 2013; 108: 305-12.

[87] Subramanian B, Ananthakumar R, Kobayashi A, Jayachandran M. Surface modification of 316 L stainless steel with magnetron sputtered TiN/VN nanoscale multilayers for bio implant applications. J Mater Sci-Mater M 2012; 23(2): 329-38.

[88] Ni S, Sun L, Ercan B, Liu L, Ziemer K, Webster TJ. A mechanism for the enhanced attachment and proliferation of fibroblasts on anodized 316 L steel with nano-pit arrays. J Biomed Mater Res B 2014; 102B(6): 1297-303.

[89] Rodŕiguez B, Romero A, Soto O, de Varorna, O. Biomaterials for orthopedics. Review Article for Course on Mechanics of Materials – I at the University of Puerto Rico, Mayaguez (2004). Available from: http://jimboenk.heck.in/files/materi-biomaterial-2.pdf (accessed 10 Sept 2014).

[90] Kyomoto M, Moro T, Saiga K, *et al.* Lubricity and stability of poly(2-methacryloyloxyethyl phosphorylcholine) polymer layer on Co-Cr-Mo surface for hemi-arthroplasty to prevent degeneration of articular cartilage. Biomaterials 2010; 31(4): 658-68.

[91] Bedair TM, Cho Y, Kim TJ, *et al.* Reinforcement of interfacial adhesion of a coated polymer layer on a cobalt-chromium surface for drug-eluting stents. Langmuir 2014; 30(27): 8020-28.

[92] van Hove RP, Nolte PA, Semeins CM, Klein-Nulend J. Differences in proliferation, differentiation and cytokine production by bone cells seeded on titanium-nitride and cobalt-chromium-molybdenum surfaces. J Biomater Appl 2013; 28(2): 278-87.

CHAPTER 14

Derivatized Polysaccharides: A Potential in Micro/Nanoparticulate Based Drug Delivery

Vikas Rana*, **Sunil Kamboj and Kuldeep Singh**

Pharmaceutics Division, Department of Pharmaceutical Sciences and Drug Research, Punjabi University, Patiala, Punjab, India-147002

Abstract: The industrial demand of natural polysaccharides or their chemically modified derivatives have increasing due to their safety, biocompatibility, non-toxicity and biodegradability. Modified polysaccharides are industrially acceptable as disintegrating agent, thickening agent, gelling agent foam stabilizer, swelling agent, emulsifying agent, encapsulating agent, binding agent, *etc*. In addition, biopolymers are acceptable as materials for nanoparticles for clinical application. This is due to their versatile traits, including biodegradability, biocompatibility and low immunogenicity. These tailor-made materials are now commercially available and thus can give competition to already available synthetic excipients. Considering the current needs to produce stable nanoparticle systems, the chemical modification of these polysaccharides further improves their physicochemical as well as mechanical properties. The attempts have been made to obtain nano-carrier for therapeutic purposes and with their variation in materials and process preparation. This chapter is aimed at discussing the application of these modified polysaccharides with their physico-chemical characterization. Further, the factors influencing these process variables taken care off while making them are also enlightened in addition to influence of modifying polysaccharides on the drug release properties of pharmaceutical dosage forms and for other purposes in pharmaceutical industry.

Keywords: Polysaccharide, nanoparticle, microparticle, carboxmethylation, thiolation, grafting, crosslinking, ionotropic gelation, drug entarpment efficiency, nanoprecipitation, sonication, spraydrying.

1. INTRODUCTION

Polysaccharides chemically made up of multiple sugar units linked together to create long chain sugar. One such category of polysaccharide is natural gums that are frequently secreted by higher plants as a action against stress, physical injury and fungal attack. These gums are sometimes composed of heterogeneous units that upon hydrolysis breaks into simple sugar units like mannose, galactose, glucose, arabinose, uronic acids or xylose *etc*.

***Corresponding author Vikas Rana:** Pharmaceutics Division, Department of Pharmaceutical Sciences and Drug Research, Punjabi University, Patiala, Punjab, India-147002; E-mail: vikas_pbi@rediffmail.com

Now a days, these polysaccharides are becoming the most abundant and useful industrial raw materials. In addition, it have been the subject of intensive research due to their sustainability, biodegradability and biosafety [1]. Many natural gums possess 3-D structure of interconnected polymers strands and are also known as 'gels'. Natural polysaccharides are known for their swelling characteristics. The swelling is dependent upon their structure as the linear polymer showed higher swelling as compared to the branched one with comparable molecular weight. This behaviour of natural polysaccharides is due to entrapment of large amounts of water between their branches and chains of the polymer [2]. Natural polysaccharides are used in pharmaceutical and food for their wide range of applications. They are used as adhesive, binders, release modifiers, laxatives and disintegrants in pharmaceuticals [3]. In topical products and liquid oral natural polysaccharides are used as thickening and suspending and/or stabilizing agents. Thus, the choice of polysaccharides over synthetic materials is due to their, low cost, availability and non-toxicity.

The huge number of –OH moieties present in native form of natural polysaccharide hydrogen bonds with water molecules that made them highly water soluble. High water solubility limits their use for controlling drug release. Moreover, the researchers are forced to derivatize these natural polysaccharides due to their inability of –OH moieties to form strong ionic interactions with counter ions [4]. To modify drug release of dosage forms, various modifications strategies have been utilized by chemically altering the structure of natural polysaccharides with variety of functional groups. The extensively investigated derivatization are polyacrylamide, glutraldehyde grafting and phosphate crosslinking [5-9].

Natural polysaccharides are safe enough for human consumption as they are part of our food. This is probably the reason for their use as food additive or drug carrier in pharmaceuticals. However, the usage of natural polysaccharides associated with certain problems such as change in viscosity on storage, pH dependent solubility, uncontrolled rates of hydration, thickening, and microbial contamination. Chemical modification of gums is carried out with an aim to improve these drawbacks and enables their utilization for specific drug delivery purposes. Natural gums can also be modified in different ways to obtain tailor-made products that can compete with already available synthetic controlled release excipients. Various plant polysaccharides are still there that are not explored yet and studies on such sources can make significant contribution in this direction.

Polysaccharide-based micro/nano-carriers have received considerable attention in years due to their promising role in drug delivery. Polysaccharide-based micro/nano-carriers for drug delivery systems are defined as particulate dispersions or solid particles with smaller size range and with various morphologies, including micro/nanospheres, micro/nanocapsules, microbeads, nanomicelles, *etc*. Nanoparticulate drug delivery systems have advantages, like enhanced drug encapsulation efficiency, efficient drug stability against enzymatic or chemical degradation, ability to create a controlled release of drugs, cost effectiveness, cell internalization, heamo/cyto-compatibility [8-12]. Therefore, various attempts have been made to utilize natural polysaccharide for the preparation of drug loaded micro/nano-particles. In addition, the use of aqueous medium avoiding the use of organic solvents that are linked with various drawbacks like flammability, environmental pollution and induction of toxicity in chronic dosing even in traces amount of organic solvents in the final dosage forms [13].

Therefore, in future the interest in natural gums and their derivatization shall be to provide safe and better excipients for drug delivery systems. In light of the above, the present book chapter is aimed at providing a comprehensive review on the modified gums that were developed for the purpose of suitability as excipient in micro/nano particulate based drug delivery system.

2. SYNTHESIS OF DERIVATIZED POLYSACCHARIDES FOR MICRO/NANO DRUG DELIVERY

Gums were modified in order to enhance their physicochemical characteristics that could make it suitable as an excipient for development of micro/nano carriers. Different gums were modified by modifying their functional group at 6th or 4th position of the hexose unit present in the gums using chemical reactions like carboxymethylation, thiolation, using grafting techniques and crosslinking. However, there are other techniques like sulphation or phosphorylation that were efficiently used to derivitize polysaccharides, but they were not found suitable for micro/nano particulate based drug delivery system. Therefore, synthesis of derivatized polysaccharides for the purpose of micro/nano-particulate in drug delivery is detailed here.

2.1. Carboxymethylation

Carboxymethylation is the chemical reaction involving covalent attachment of –CH_2COOH moieties at 6th position of hexose replacing –OH moieties.

Carboxymethylation of gums was performed with an aim to increase aqueous solubility, hydrophilicity and solution clarity. In general, the polysaccharide is protonated with aqueous alkali hydroxide *e.g.* sodium hydroxide and reacted with monochloroacetic acid (Williamson ether synthesis) that yield carboxymethyl (CM) polysaccharide derivative. The carboxymethylation reaction procedure involved two steps. First, sodium hydroxide interacts with the hydroxyl moieties of the monosaccharide units to convert them into alkoxide moities. In second step, the attachment of carboxymethyl moieties occurs between the polymer alkoxide and monochloroacetic acid through a SN_2 reaction [14]. The general scheme of carboxymethylation is outlined in Fig. (**1**). In addition to polysaccharides, cellulose and starch from different sources are used as raw materials for this reaction. For example, carboxymethyl derivative of tamarind kernel powder, cassia tora gum, xantan gum, Auricularia auricula polysaccharides, gum kondagogu, chitosan and guar gum were prepared by Goyal *et al.* [14]; Sharma *et al.* [15, 16]; Yang *et al.* [17]; Silva *et al.* [18]; Kumar *et al.* [19]; Bidgoli *et al.* [20].

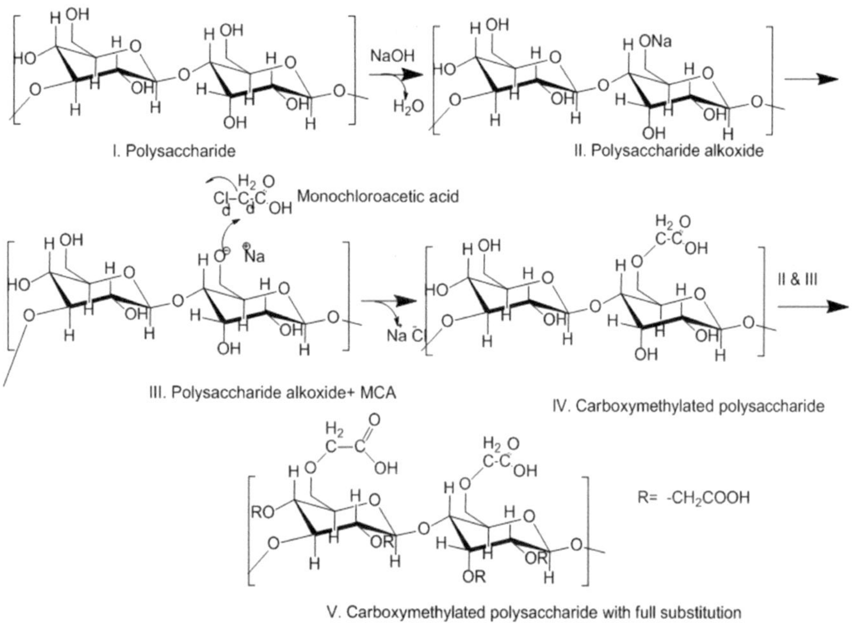

Figure 1: Chemistry involved in the synthesis of carboxymethyl derivative of natural polysaccharide. Steps involved in the carboxymethylation are **II**) alkoxide formation, **III**) polysaccharide and monochloroacetic acid conjugation, **IV**) carboxymethylated derivative with low substition and **V**) carboxymethylated derivative with full substitution.

Although, carboxymethylation is a simple and reproducible method, however concentration of sodium hydroxide and monochloroacetic acid are the key factors that vary from gum to gum as well as differ in degree of substitution within the same gum. For example amount of alkalie required to produce alkoxides was 80 mL of 45% w/v for gum kondagogu (1 g), 2.7 mL, 10 M for cashew gum and 0.158 mol for tamarind kernel polysaccharide (0.05 mol) to achieve degree of substitution 0.2, 2.21 and 0.649, respectively. Similarly, monochloroacetic acid content influenced DS of carboxymethylation of gum. For example addition of monochloroacetic acid 10 mL (75%w/v), 10 mL (26% w/v) and 0.09 moles, respectively for the carboxymethylation of tamarind kernel polysaccharide (TKP), cashew gum and gum kondagogu [14, 18, 19]. Further, it was observed that carboxymethylation of gum kondagogu increased its degree of crystallinity and surface roughness, reduced its viscosity and improved its mucoadhesive properties [19]. The carboxymethylation of Cashew tree gum was found to decrease in intrinsic viscosity and molecular weight with increasing DS but the solubility was improve [18]. Goyal *et al.* [14] synthesized carboxymethyl tamarind kernel powder and found enhancement in the cold water solubility along with its resistance to microbial attack. Rheological studies suggested the non-Newtonian pseudoplastic nature of CM-TKP solutions.

In another study the carboxymethylation reaction of GLP was optimized using orthogonal array design. The result indicated that concentration of NaOH in solution played a crucial role in enhancing the yield of carboxymethyl product. Carboxymethylation of GLP was conducted using 2.0 mL (volume of 4 mol/L monochloroacetic acid) and 7.0 mL (volume of 20 % NaOH) for 4 h (reaction time) at 60 °C reaction temperature. The results suggested increased reaction time ensured adequate swelling of GLP, the adsorption and diffusion of the reactants [21].

2.2. Thiolation

Thiolation is the chemical reaction which involves covalently attachment of sulfhydryl moieties like cysteine, thioglycolic acid and glutathione onto the backbone of natural polymer. Thiol bearing moieties in polysaccharide chain were developed through the amide or ester bonds. It was observed that the mucoadhesive as well as drug permeation ability of the thiolated polymer was higher when l-cysteine or thioglycolic acid were used as thiomers (Fig. **2**) [22-24].

Chitosan solution in acetic acid was reacted with 2-iminothiolane HCl with pH adjustment to 7. In some reactions 2-mercaptoethanol was added to prevent oxidation during coupling reaction. The total thiol content in conjugate was found to be 408.9μmol/g. The 2-iminothiolane chitosan derivative offered magnificent *in situ* gelling properties with additional strong mucoadhesive properties [25].

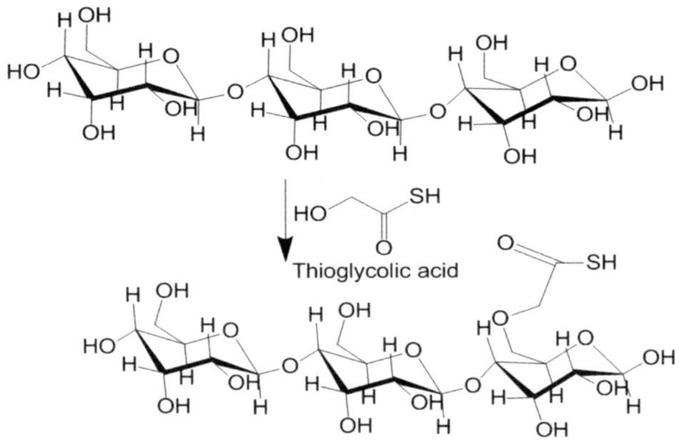

Figure 2: Chemistry behind the synthesis of thiolated natural polysaccharide derivative using thiolglycolic acid as thiolating agent *via* esterification reaction.

In another study, thiolated pectin was synthesized by esterifiaction between thioglycolic acid and pectin in acidic (HCl) environment. The optimum conditions for thiolation of pectin were: Pectin (16 g), thioglycolic acid (80%; 7.59 g) in acidic conditions under 80 ˚C (reaction temperature) for 150 min (reaction time). The thiolated pectin possess 0.60±0.04 mmol of thiol groups/g [26]. Yadav *et al.* [27] optimized thiolation procedure using different conditions for thiolation by reacting 1g of gellan gum with 0.3 mL of thioglycolic acid in acidic conditions (0.3 mL HCl, 7 N) for 3h at 80 ˚C. The thiol content in thiolated gellan gum was 13.92 mM of thiol groups/g of the conjugate. Thiolated gellan gum have slightly increased degree of crystallinity [27]. Thiolation of xyloglucan was also optimized by Mahajan *et al.* [28]. They found maximum thiolation when thioglycolic acid (0.223 mL), xyloglucan (1 g) and hydrochloride acid (1 mL, 7 N) at 80˚C for 150 min were reacted. Thiolated xyloglucan was determined to have 4 mmol of thiol groups/g of Polymer., Thiolation of xyloglucan was reported to have enhanced mucoadhesive property as compared to putative form [28].

2.3. Gums Grafted with Acrylic Acid or its Derivatives

Acrylic acid grafted gums have been used to improve the film forming nature, swelling characteristics and drug release properties. Fig. (**3**) describes the mechanism of grafting of natural gums using acrylamide and polyanailine.

Natural polysaccharides have been grafted, as modified polysaccharide offered enhanced higher mechanical strength, resistance to heat or abrasion, greater oil/water repellent qualities, or antibacterial activity. The pure polysaccharides are susceptible to degradation under shear and occur as useful flocculants [29], they also have poor shelf life as they are susceptible to biodegradation. The grafted natural gums can be easily converted into matrices with hybrid characteristics offering better applications [30, 31]. Grafting is an effective technique to improve the compatibility between synthetic polymers and natural polysaccharides to obtain novel materials with hybrid properties. In a grafting reaction, the

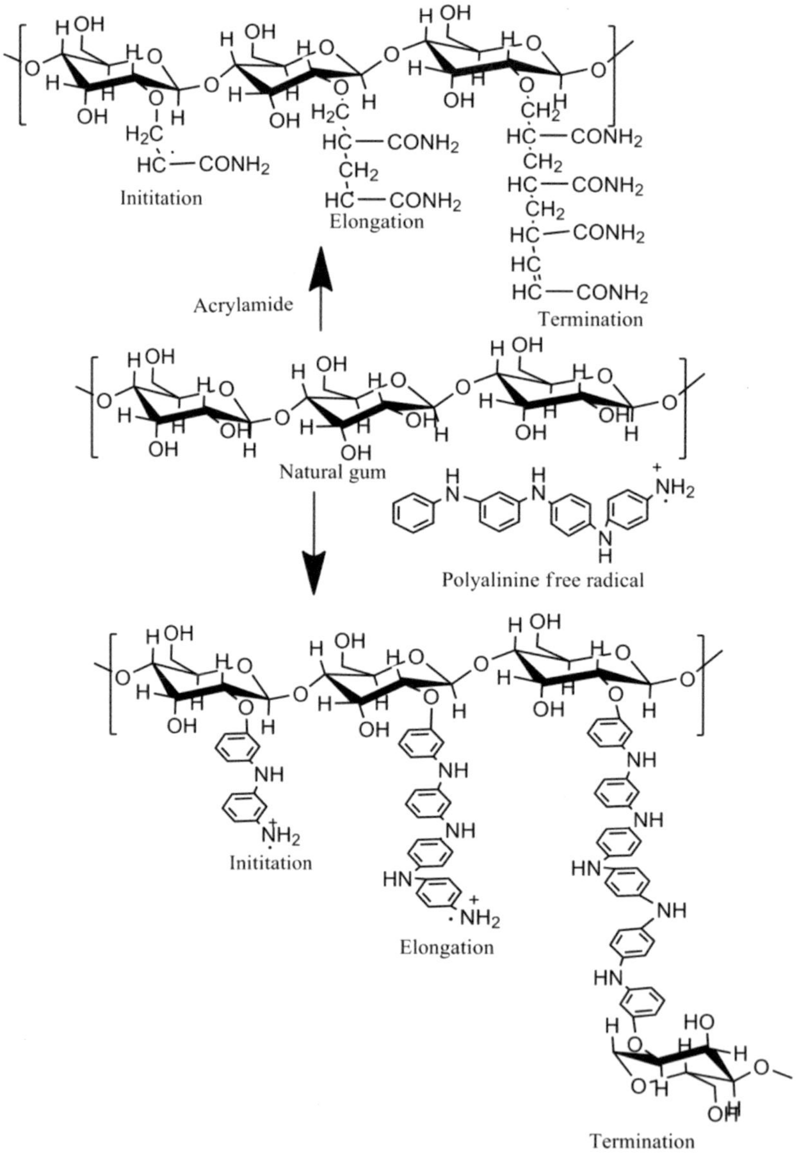

Figure 3: Mechanism of grafting of natural polysaccharide employing different copolymerizing agents **a**) acrylalmide and **b**) polyaniline free radical.

interaction between a vinyl monomer and $-COO^-$ and/or $-CH_2OH$ groups of natural gum was conducted. The grafting is advantageous as the resultant material can be designed to produce desired drug release profile. There are different types of vinyl monomers, such as acrylamide [32], N-tert-butylacrylamide [33], methylmethacrylate [34], acrylonitrile [35] and methacrylamide [36] have been grafted to many polysaccharides.

In another investigation, microwave assisted synthesis of grafted Gum acacia was conducted with PANI (polyaniline) [37] using initiator as ammonium peroxydisulfate (APS). The gum acacia-graft-polyaniline (GA-g-PANI) was found to have acceptable electrical conducting behaviour in the different pH conditions that was shown to be correlated with grafting characterization and on the pH of graft environment. The increase in ionic conductivity was found to be associated with increase in %G. The optimized GA-g-PANI grafted material offering controlled multifunctional electronic properties was synthesized. The optimum composition of the reaction mixture was 2×10^{-3} M HCl; 40×10^{-5} M $(NH_4)_2S_2O_8$; 0.1 M aniline and 8.0 g/L GA with microwave exposure of 500 W microwave power for 40 s.

In an similar experiment Poly(acrylamide)-graft-sodium alginate (SG-g-PAM) was synthesized using microwave [38]. The microwave power was used to initiate grafting. Different acrylamide concentrations (0.14 to 0.25 M), 900W microwave power with different exposure time (1-4 min) with fixed amount of sodium alginate was used to prepare various varieties of Poly(acrylamide)-graft-sodium alginate copolymers. The optimized batch showed 37%G using 0.21M acrylamide, 6.2 mM sodium alginate using 900 W microwave power with 3 min of exposure time.

Artemisia seed gum was grafted using acrylic acid to produce a superabsorbent grafted material [39]. The grafting was conducted using microwave radiation with APS as initiator. The reaction was optimized to prepare material with high water absorbency. Using 180W microwave power with 3 min exposure time, acrylic acid: Artemisia seed gum (10:1) and initiator (2%) produces a product that posses 400 times water absorbing capacity. In an another study a chitosan was grafted with acrylic acid, using microwave in presence of N,N-methylene-bis-acrylamide as a crosslinker and ceric ion as an initiator, to prepare a water sorbent material [40]. The %G was found to be affected by concentration of chitosan and microwave power. Using microwave irradiation the reaction rate increased 8 times as compared to conventional heating. Using microwave power 120W, 0.01 M crosslinker and 0.01 M initiator, the resultant material proced showed higher %G (0.865) with 704g/g of water absorbency.

The Guar gum was reported to be grafted with many monomers like acrylamide [32], acrylonitrile [41] and ethylacrylate [42] using microwave irradiation as the initiation source in the absence of radical initiator or catalyst. The synthesis of acrylamide grafted guar gum was carried out using conventional heating with initiator (APS/ascorbic acid)/catalyst (Ag^+) and microwave initiated grafting without initiator [32]. The results indicated guar-graft-poly(acrylamide) with 140 %G and 45% E was prepared in 90 min. Whereas, using 960 W microwave power with initiator (APS/ascorbic acid)/catalyst (Ag^+) for 1.6 min resulted in guar-graft-poly(acrylamide) with 182 %G and 63 %E. However, only microwave initiated manner without initiator and catalyst with low microwave power (840W) in only 0.33 min microwave time produces guar-graft-poly(acrylamide) with 120 %G and 42 %E. In a similar study, guar gum was grafted with poly(acrylonitrile) [41]. This reaction did not involve the use of Ag^+ as catalyst. Under the influence thermal conditions (±35 °C) 177 %G and 51 %E were obtained. However, microwave irradiated reaction produced 188%G and 54% E in 1.66 min, which do not statistically differ from conventional heating method. Poly(ethylacrylate) grafted guar gum was prepared using microwave initiated method [42]. The optimum conditions for synthesis of poly(ethylacrylate) grafted guar gum were 0.16 M ethylacrylate, 0.1 g guar gum, 1200 W microwave power and 15 s exposure time. The resultant copolymer had 295 %G and 73 % E.

In another study, microwave assisted synthesis of acrylonitrile (AN) graft-Xyloglucan (XG) was carried out [43]. Polymerization reaction was initiated by reacting XG with ceric ammonium nitrate in an aqueous medium with bubbling of N_2 to create inert atmosphere under microwave irradiation. The results indicated using microwave irradiation as reaction initiator, 92.5 %G was obtained in 120 s whereas, 76 %G was obtained in 24 h using ceric ions as an initiator at 40 °C.

Goyal *et al.* [44] developed acrylamide grafted TKP using ceric ammonium nitrate-nitric acid as initiator with enhanced % entrapment efficiency of 93.66 %. In another study, graft copolymerization of cassia tora gum with acrylamide was carried out using ceric ammonium nitrate–nitric acid as redox initiator [45].

3. TECHNIQUES FOR THE DESIGN OF MICRO/NANO-SPHERES OR PARTICLES

3.1. Ionotropic Gelation Technique

The technique is based on the capacity of polyelectrolytes to interact (crosslink) in the presence of counter ions to form hydrogel beads also called as microbeads/microspheres/nanospheres. Generally, the dropping of polymeric solution containing drug into the aqueous solution of polyvalent cations generate hydrogel beads. A three dimensional lattice of ionically crosslinked moieties were formed by the cationic diffusion of drug loaded polymeric drops. In this technique, biocompatibility and biodegradability is the main reason for increasing interest in the use of natural polysaccharides as drug carriers. By this technique, biopolymers like alginates, chitosan, gellan gum, pectin and carboxymethyl cellulose are widely acceptable for drug encapsulation [1, 5, 6, 46, 47]. The natural polysaccharides have anionic/cationic moieties on their chemical structure. These anionic/cationic moieties produces network like structure by interacting with the counter ions that induces gelation. In spite of having a property of release rate retardant, these polymers also bears property of coating of drug core. Fig. (**4**) explains the Ionotropic gelation of natural polymers.

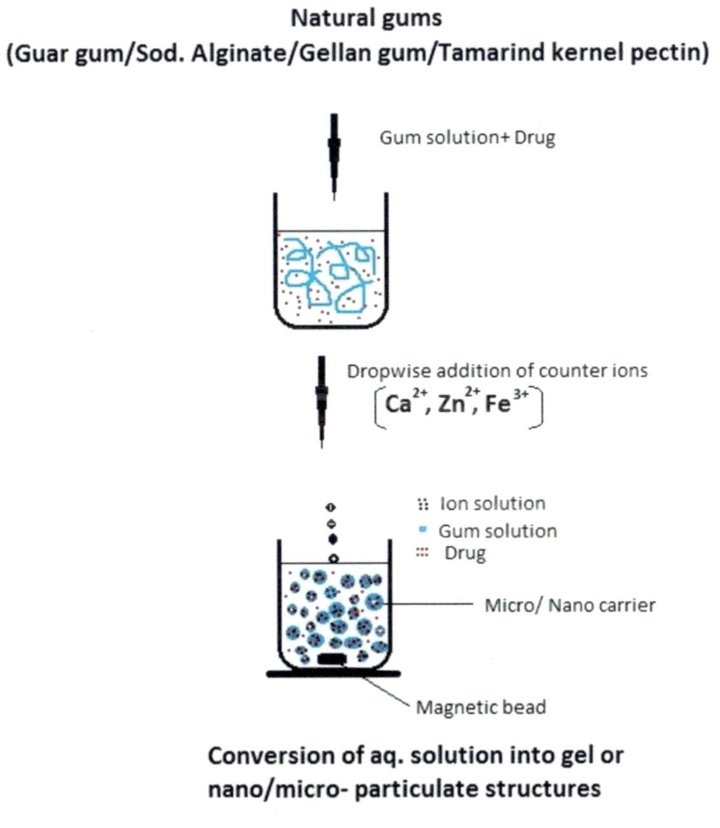

Figure 4: Schematic diagram showing formation of micro/nano carriers employing ionotropic gelation using counter ions.

3.1.1. Cross-Linking with Ions

Calcium chloride, barium chloride, aluminium chloride, *etc* are the ions that form bridges (Fig. **5**) with the polysaccharide chains to form microspheres, nanoparticles, beads *etc*. An ionotropic gelation technique was generally employed to cross linking polysaccharides with ions. For example sodium alginate and carboxymethylated guar gum chains were entangled with a bridge of Ba^{+2} ions to form beads [48].

Ionic crosslinking technique was used to prepare tranexamic acid (TA) containing Gellan gum (GG) microbeads by using aluminium chloride ($AlCl_3$) as cross-linking agent. Tranexamic acid was uniformLy dispersed in gellan gum solution; the solution was the added dropwise to $AlCl_3$ and tween 80 solution using 18-gauze flat-tipped needle. The results indicated the increase in water uptake with increasing gellan gum concentrations. The microbeads drug release following a non-fickian type release mechanism. The results showed that the microbeads were capable of minimizing the release of tranexamic acid in acidic environment and modulate release in alkaline environment and thus reduce the loss of drug due to first-pass metabolism and thus increased bioavailability [49].

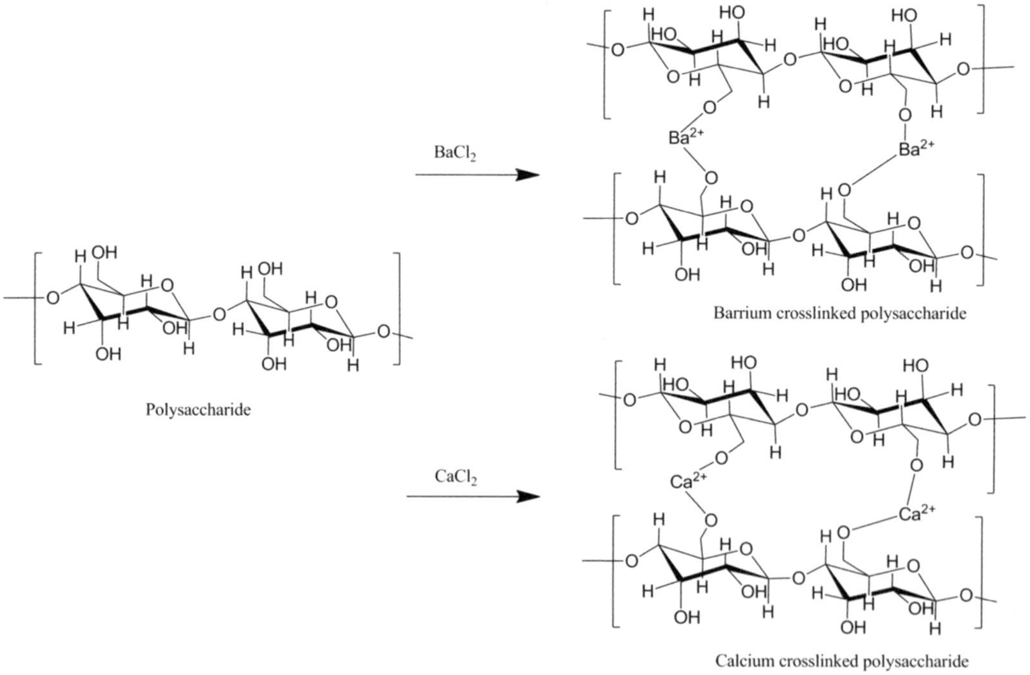

Figure 5: Crosslinking of natural polysaccharides using calcium and barium ions as crosslinking agents.

Anitha *et al.* [47]; Shi *et al.* [50], prepared O-carboxymethyl chitosan nanoparticles using ionic crosslinking technique. In this technique O-carboxymethyl chitosan solution was prepared in distilled water and to this solution $CaCl_2$ solution was added drop wise under constant stirring. Due to addition of $CaCl_2$ in O-carboxymethyl chitosan the clear solution converted into an turbid suspension. The nanoparticles obtained were centrifuged at 20,000 rpm for 45 min. The pellets were washed, centrifuged, redispersed in distilled water then lyophilized. Anitha *et al.* [47] observed the major factors like concentration of the polymer, cross linker and stirring speed had critical effects on nanoparticle formation.

A crosslinking bridge between alginate chains and guar gum were created using glutaraldehyde [51]. In this technique guar gum solution mixed with alginate solution in the ratio 2.75:1.25 and glutaraldehyde solution (0.5% w/v). This mixture was dropped into calcium chloride solution (0.5 M) to develop beads that were lyophilized after washing. It was observed that an increase in concentration of glutaraldehyde leads to enhancement in crosslinking density.

Polyacrylamide grafted carboxymethylxyloglucan (Pam-g-CMXG) microbeads containing acelofenac as model drug were prepared by ionic gelation method, using ferric ion (Fe^{3+}) as cross linking agent. The cross linking was induced when aqueous HPam-g-CMXG solution was slowly dropped into stagnant aqueous ferric ions solution using 25 mL hypodermic syringe (22G). Yellowish brown coloured microbeads were formed [1].

GG (guar gum) polysphers, ionotropically gelated with $ZnSO_4$ and covalently crosslinked with GA was prepared by Kulkarni *et al.* [52] to incorporate simivastatin. Drug–guar gum dispersion was added dropwise into an aqueous of $ZnSO_4$ solution under constant stirring. These ionotropically gelated spheres were further dipped into glutaraldehyde (GA) and 1 N HCl to obtain polyspheres of dual cross linking nature.

3.1.2. Crosslinker: Trisodium Metaphosphate (STMP)

Natural gums are not suitable for targeting drugs to distal parts of gut because of their high swelling characteristics. This drawback could be removed by phosphate cross-linking [53, 54]. Generally, phosphate cross-linked gums were synthesized by interacting gums with trisodium trimetaphosphate (STMP) solution in alkaline conditions (1 M sodium hydroxide solution, pH 11) (Fig. **6**) [4].

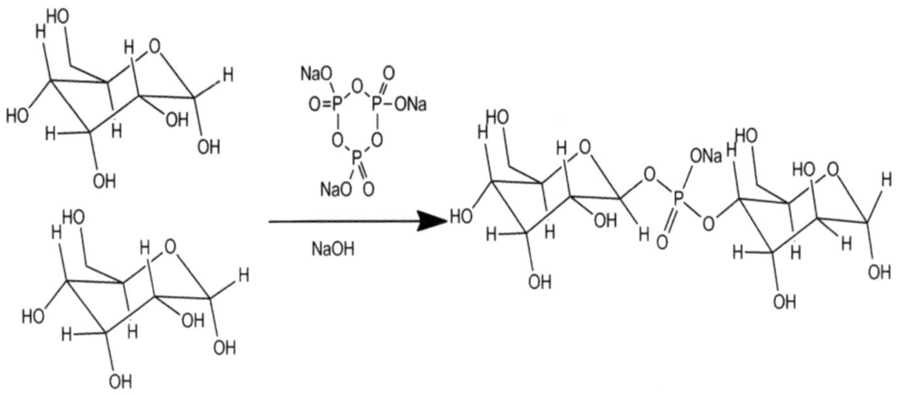

Figure 6: Scheme of phosphate crosslinking of gums.

Kabir *et al.* [55] crosslinked guar gum with increasing amount of aqueous solution of 30% w/v trisodium trimetaphosphate (STMP) assuming that each mole of STMP reacts with three pairs of hydroxyl groups of guar gum. Swelling of guar gum was reduced from 100-120 folds (native guar gum) to 10-35 folds after cross linking of guar gum with STMP. As a result of interaction the ionic charge of guar gum was precipitated in cross linking and this induces negatively charged moieties due to phosphate.

Kabir *et al.* [53] prepared hydrogel of crosslinked guar gum (GG) using hydrocortisone as model drug for colon-specific drug delivery and it was suggested that GG crosslinked with STMP was degraded by enzymes of gut wall. Therefore, it has a potential to retard the release of a low water-soluble drug. Thus, the polymer could be used as an excipient for colon-specific drug delivery.

In another study starch and xanthan gum was crosslinked using STMP. Starch and xanthan gum was dissolved by high shear mixer followed by boiling with addition of NaOH. The cross-linking was then introduced by including different concentrations of STMP and sodium sulfate. It was seen that the swelling behaviour of the films was enhanced at increased concentrations of xanthan gum and STMP. The gel mesh size was enhanced with increasing swelling ratio varying from 2.84 to 6.74 nm. Permeability of anionic drugs (caffeine, methylene blue, vitamin B12, ibuprofen (sodium salt), pyrogallol red, sodium salicylate) across the films was significantly reduces than their putative form. This may be due to occurrence of electrostatic repulsion between the polymer and the negatively charged drugs. The findings indicated starch–xanthan gum hydrogels have a potential of film-forming material that could be accepted in controlled release formulations [56].

3.1.3. Cross-Linking with Epichlorohydrin

In this crosslinked alkaline paste of cashew gum was kneaded with epichlorohydrin followed by heating at different temperature conditions provide cross linked gel. The reaction yield cashew gum (CG) with high crosslinking which enhances with enhancement of epichlorohydrin/gum molar ratio (E/G) up to 0.17 and then decreases. The swelling degree of CG gels was lowered with increasing E/G ratio due to the enhancement of crosslinking density [57]. Fig. (**7**) demonstrates the mechanism of epichlorohydrin crosslinking.

3.1.4. Covalent Crosslinking

Preparation of polyelectrolyte complex microspheres of chitosan and carboxymethyl cashew gum of different molar masses were prepared [58]. The CH concentrations were prepared containing 3% for high molar mass of chitosan and 5% for low molar mass chitosan. The CH solution was added dropwise to carboxymethyl cashew gum solution (1M NaOH/ethanol (2:1), air flow (10 L/h) with the help of syringe and a needle (21 mm length and 0.8 mm diameter). This results in the formation of polyelectrolyte complex microspheres of carboxymethyl cashew gum and chitosan.

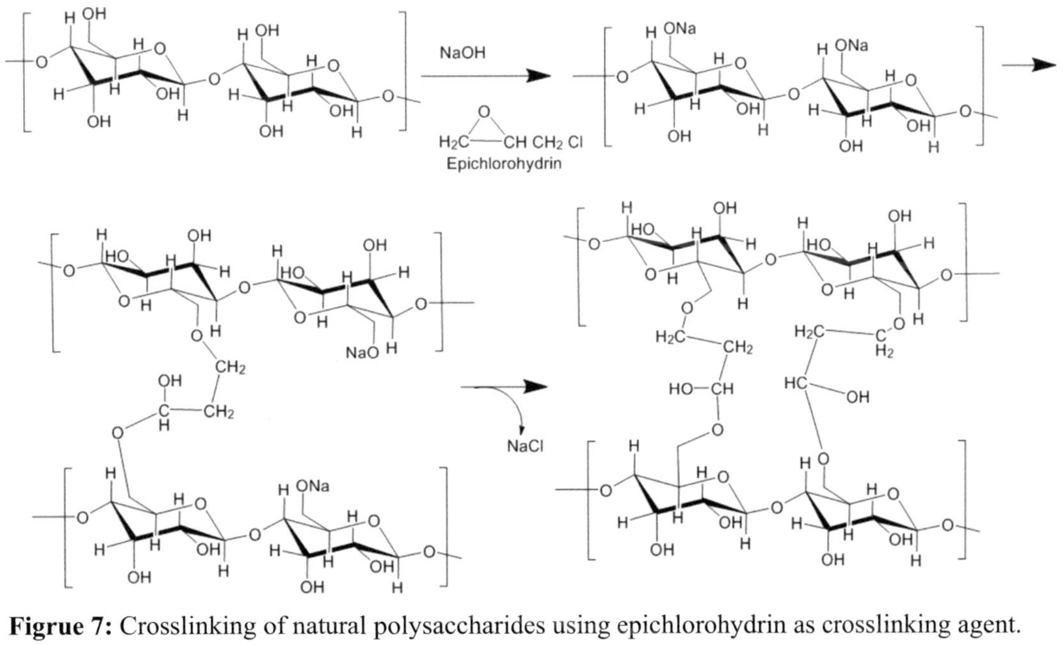

Figrue 7: Crosslinking of natural polysaccharides using epichlorohydrin as crosslinking agent.

In another study, Yeh *et al.* [9] prepared protein-loaded chondroitin (ChS) sulfate–chitosan (CS) and TPP-CS nanoparticles. The study utilizes the interaction between CS and ChS which was based on controlling of gelation between CS and its counter ion chondroitin sulfate (ChS). The occurrence of gelation is due to intermolecular interactions of carboxylate moieties of ChS with the amino moieties (positively charged) and sulphate (negatively charged) of CS. Protein-loaded chondroitin sulfate–chitosan (ChS–CS) nanoparticles and TPP–CS nanoparticles were formed immediately upon addition of TPP or ChS aqueous solution to a fixed volume of CS in acidic solution.

3.1.5. Nanoprecipitation, Sonication and Homogenization Method

The CMGG (carboxymethyl guar gum) nanoparticles were prepared using nanoprecipitation and sonication method [59]. A clear and viscous solution of guar gum in deionized water was prepared and kept overnight for complete solubilisation. The prepared solution was sonicated for 1 h. After sonication, the suspended colloidal solution was precipitated using acetone with continuous stirring to obtain whitish nanoprecipitates. It was found that the formation of nanoparticles might depend upon the sonication time, solvent, and stirring time.

Soumya *et al.* [60] prepared guar gum nanoparticles developed using nanoprecipitation and dialysis-cross-linking methods. Nanoprecipitation was carried out by dropping non-solvents (ethanol, ethylacetate, acetone, methanol and isopropyl alcohol) to the aqueous solutions of the depolymerized guar gum solution which was mixed with cross-linkers and surfactants. The surfactants include Tween 20 (non-ionic) and Triton X-100. Boric acid and Sodium tripolyphosphate (TPP) were used as ionic cross-linkers. In dialysis method dimethylsulphoxide (DMSO) was mixed with guar gum hydrolysate (5 mL) and 0.1×10^{-3} mM Triton X-100 (10μL), transferred inside a dialysis bag (12 kDa cut off) and dialyzed with respect to aqueous 0.1%TPP or borate for 24 h at 30±2 °C. Enzyme functionalized nanoparticles (EFNP) were prepared by sequentially mixing hydrolyzed GG (5 mL), Lipase PS (0.5 mL), Triton X-100 (0.1×10^{-3} mM, 5μL), 50% aqueous polyethyleneimine solution (10 μL), IPA (500l), 0.1% TPP (500μL) or boric acid and finally 25% glutaraldehyde (10μL) as crosslinker. The solution was vortexed and sonicated for 5 min at room temperature to get uniform Nanoparticles. The results suggested that minimum size was obtained when IPA was used as precipitating antisolvent. Boric acid produces stable Nanoparticles with ~32 nm average particle size (PDI = 0.491).

Rejinold *et al.* [7, 11] prepared nanoparticles of chitosan grafted polymer (chitosan grafted poly (N-isopropylacrylamide) and chitosan grafted poly(N-vinylcaprolactam)) by dissolving the graft polymer in acetic acid (1% solution) and then cross-linked with 0.5% TPP in 1:1 ratio which results in suspension formation. The resulting nanosuspension was centrifuged at 21,000 rpm for 30 min and resuspended in water. The suspension was sonicated with probe sonication at low amplitude of 35% till the pH become 7.4. This resulted in the blank nanoparticles with very smaller range100-150 nm. The loading of drugs *i.e.* 5-FU and curcumin was done using ionic crosslinking method by TPP *via* aqueous chemistry, this results in high drug loading and entrapment efficiency. This method is efficiently employed to prepare thermoresponsive, biocompatible, biodegradable nanocarriers.

3.2. Emulsion Cross-Linking Method

Fig. (**8**) explains the general scheme to prepare micro/nano DP carriers (derivatized polysaccharide based micro/nano carriers) employing emulsion crosslinking technique. The emulsion crosslinking method includes the emulsification of an organic solvent (usually methylene chloride) containing dissolved/dispersed drug in an excess amount of aqueous continuous phase and solubilized polymer. The cross linking was induced by means of chemical cross linkers (formaldehyde, di- acid chloride, glutaraldehyde *etc*) or heat. However, thermolabile substances are not suitable for heat denaturation. Upon drying the solution with appropriate method it resulted in the formation of dried drug loaded micro/nanocarriers.

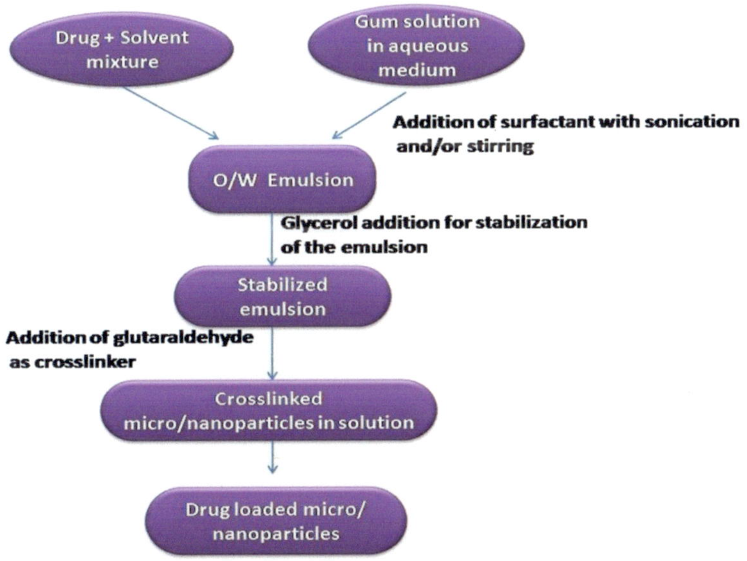

Figure 8: Schematic diagram showing preparation of micro/nano-carriers using emulsion crosslinking technique.

3.2.1. Oil in Water (o/w) Emulsion-polymer Cross-linking Method

Glutaraldehyde have used extensively for cross-linking polymers containing hydroxyl groups. Fig. (**9**) showed the mechanism of crosslinking of natural gums using glutaraldehyde as crosslinking agent.

Soppimath *et al.* [61] used polyvinyl alcohol and guar gum to develop interpenetrating network microspheres by crosslinking with glutraldehyde. In this technique, hydroxyl moieties of the polymers interact with the aldehyde moieties of glutaraldehyde to form acetal cross-links. The IR studies support crosslinking with glutraldehyde. The DSC studies showed enhancement in ΔH (*i.e.* enthalpy of transition) that could be attributed to the high level of energy required to break the highly cross-linked polymeric network structure. A dense macromolecular network arises due to increase in crosslinking, which was associated with decrease in molecular transport of liquid, resulting in reduced swelling. The *in vitro* release of nifedipine from these microspheres was observed to depend on the extent of cross-linking. This was due to the fact that the solvent uptake by the microspheres decreased with increased cross-linking. Thus, the release of the nifedipine from these microspheres increased initially due to the polymer relaxation process as water penetrated and converted the glassy polymer into a rubbery one. However, the latter part of the release profile from the fully swollen polymer was due to a diffusion process.

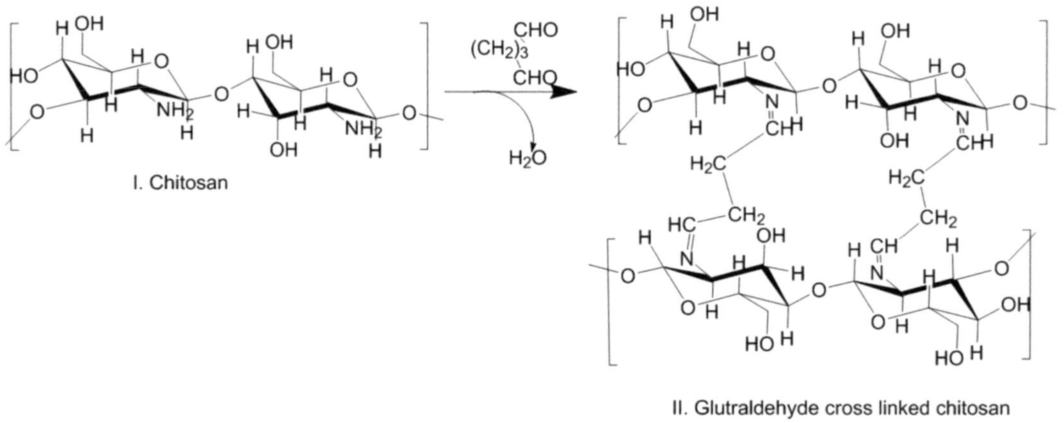

Figure 9: Chemistry of crosslinking of chitosan employing glutaraldehyde as crosslinking agent.

In another study, the interpenetrating polymer network (IPN) hydrogel microspheres were prepared by emulsion crosslinking method using gluteraldehyde as a crosslinker. Being a natural polysaccharide, locust bean gum

(LBG) has an inherent property to swell in presence of water. But this swelling behaviour was uncontrolled. To control this abrupt swelling property, LBG and polyvinyl alcohol was crosslinked with GA. Fig. (**10**) shows the mechanism of LBG and polyvinyl alcohol crosslinking with GA. The drug polymer mixture was transferred to light liquid paraffin containing Span 80 admixture. A milky white (w/o) emulsion was obtained. GA in 1 N HCl was transferred to this emulsion in order to produce hardened microspheres. GA belongs to a category of a bi-functional crosslinker and therefore forms the three dimensional polymer networks. This three dimensional network contribute in the entrapment of the drug and showed controlled drug release profile. The formation of microsphere rigid and insoluble matrix network was due to an acetal ring formation between the hydroxyl groups of LBG–polyvinyl alcohol polymer strands and aldehyde groups of GA to produce IPN [62].

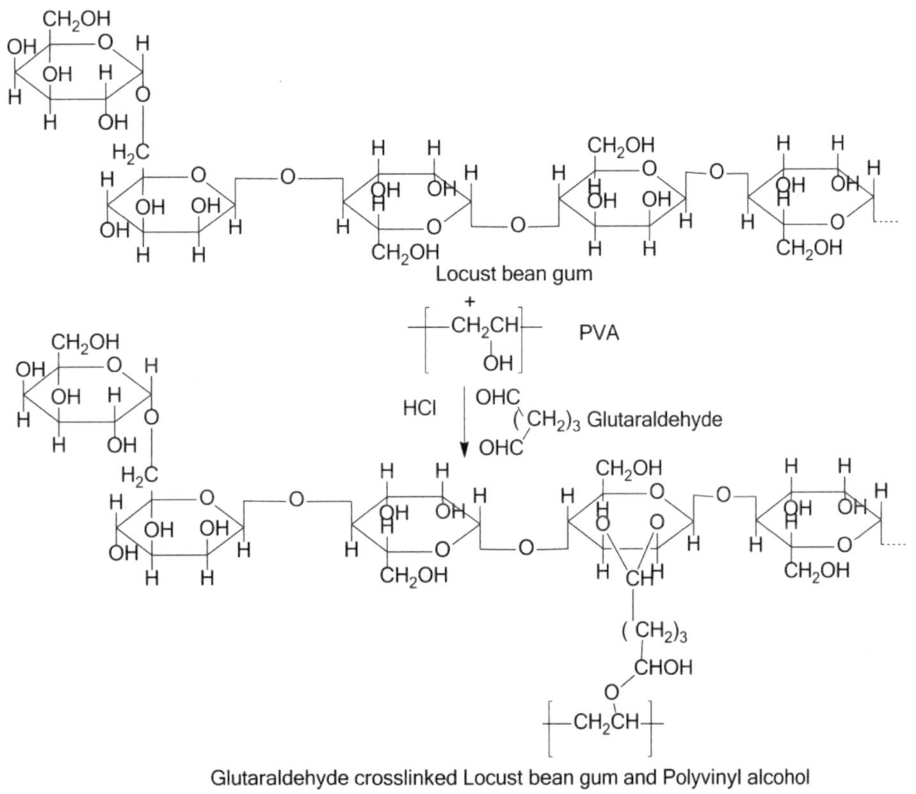

Figure 10: Mechanism behind the synthesis of microspheres of LBG and PVA interpenetrating network crosslinking with glutraldehyde

Cross-linked microspheres of polyacrylamide grafted guar gum (pAAm-g-GG) by water-in-oil (w/o) emulsification method was developed. A 5.0% (w/v) polymer solution was prepared and acidified with 5mL dilute sulphuric acid. Separately, glutaraldehyde solution 25% (w/v) was added to the polymer solution separately. These solutions were then emulsified into light liquid paraffin with 2 %w/v Tween 80. The hardened microspheres were filtered and washed with n-hexane followed by water to remove excess liquid paraffin, unreacted glutaraldehyde and any adhered Tween 80. The cross-linked pAAm-g-GG microspheres were loaded with nifedipine before cross-linking and by the soaking method; it was observed that higher crosslinked leads to lower encapsulation efficiency. Equilibrium uptake of the cross-linked microspheres exerts a profound influence on their release rates. The percent equilibrium water uptake data of the cross-linked empty microspheres indicate that, as the amount of GA in the matrices increases a significant decrease in equilibrium water uptake was observed and hence, the release of the nifedipine was modified [63].

Sarmah *et al.* [64] prepared tamoxifen citrate loaded guar gum nanoparticles (GG nanoparticles) by crosslinked with glutaraldehyde. Oil in water (o/w) emulsion polymer cross-linking method was used for preparation of sustained release nature biodegradable nanoparticles. The preparation of nanoparticles was found to be optimum when 0.5% guar gum concentration, 2M of 2% glutaraldehyde (GA), 4% Span 80 and 10 mL glycerol were used to formulate nanoparticles of relatively lower size (205 nm) with smooth and spherical texture.

Sun *et al.* [12] prepared DEX–ADM–PEI (Dextran–Adriamycin-graft–PEI) conjugate nanoparticles using slightly different procedure in three steps. The conjugation involves an aldehyde reaction between oxidized dextran and an amine group of PEI and ADM (Schiff base formation). Oxidation of dextran was performed using sodium periodate by dissolving in deionized water. After mixing the solutions, the reaction was performed for 24 h and stopped by adding ethylene glycol. This results in the formation of oxidized dextran white fluffy powder obtained by dialysis and subsequent lyophilization. In next step, Adriamycin (drug) was conjugated to the oxidized dextran with constant stirring at ambient temperature. In the last step, PEI 1800 (3 mmol) was reacted with the DEX–ADM solution for 48 h. Subsequently, the solution was treated with sodium borohydride (2 g $NaBH_4$/g dextran). The conjugate showed good coupling efficiency of ADM in the DEX–ADM–PEI conjugation which was 17.43±0.55 %.

The microparticle drug delivery system containing O-CMCS and Chelerythrine [65]; Insulin-loaded chitosan microspheres [66] were prepared by emulsion crosslinking methods. To achieve the optimized drug release property, chelerythrine or insulin was mixed with O-CMCS or chitosan and then crosslinked by glutaraldehyde.

3.2.2. Water-in-Oil (w/o) Emulsion-Polymer Crosslinking Method

Microspheres composed of Locust Bean Gum and Poly vinyl alcohol loaded with buflomedil hydrochloride (BH) were prepared by water-in-oil (w/o) emulsion crosslinking method [10]. The microspheres were prepared from IPN formed by reacting LBG and polyvinyl alcohol. The procedure involved preparation of 5% (w/v) aqueous polymeric solution of LBG and polyvinyl alcohol. First, polyvinyl alcohol was dissolved in water (80 °C) with continuous stirring to obtain transparent solution. LBG was then dispersed in hot polyvinyl alcohol solution with continuous stirring and kept for cooling at room temperature. Required amount of drug was added in the polymeric dispersion and stirred overnight. The drug polymer mixture was slowly added to light liquid paraffin containing 1% (w/w) Span80 under constant mechanical stirring to obtain milk white (w/o) emulsion. To this (w/o) emulsion, GA containing 1N HCl was added to produce hardened microspheres. The microspheres were then filtered and washed with acetone, 0.1 M glycine solution and water to remove excess amount of liquid paraffin, unreacted GA and surfactant, respectively.

Crosslinked guar gum (GG) microspheres were prepared by water-in-oil (w/o) emulsion crosslinking method. In this study a graft copolymer of GG was prepared by generating free radicals on the GG backbone and then allowing these radicals to serve as macro-initiators for the MAc (methacrylic acid) monomer. Ammonium persulfate (APS) was employed as initiator for graft copolymerization of the monomers onto GG. The initiator dissociates into free radicals *via* anion radical formation, which extract hydrogen from the C-6 on the galactose units of GG. These created free radical sites to initiate homolytical cleavage in MAc and this produces graft copolymer. Therefore, this condensation mechanism was utilized in the formation of covalent three-dimensional network [67, 68].

3.2.3. Water-in-Oil-in-Water ($W_1/O/W_2$) Double-Emulsion Method

Pullulan acetate microspheres (PAM, polysaccharide microspheres) were prepared *via* a double-emulsion-solvent evaporation technique (W/O/W, water in-oil-in-water) [69]. Exenatide dissolved in distilled water was mixed with methylene chloride and pullulan acetate, followed by of vortexing to form a W/O emulsion. This emulsion was added slowly into 0.5% (w/v) aqueous PVA solution using a 23G needle, the mixture was stirred. Methylene chloride was evaporated employing agitation for 2 h at room temperature, and the polysaccharide microspheres were solidified. Polysaccharides and their biocompatible modified forms have been only infrequently utilized to prepare microspheres *via* the double-emulsion method ($W_1/O/W_2$ technique), because they are not dissolved in volatile solvents. However, the solubility of PA in volatile solvent makes it possible to fabricate microspheres *via* a double-emulsion method. This method results in microspheres with regular and spherical shape with a broader range of 30–60 μm particle size.

3.3. Sol–Gel Polymerization

Sol–gel polymerization was used by Singh and Kumar, [70] to prepare carboxymethyl tamarind gum (CMTG) and silica nanohybrids for the incorporation of α-amylase. Carboxymethyl tamarind gum was dissolved in water and mixed with the separately prepared solution of tetramethoxysilane (TMOS) with constant stirring. The gel obtained was thoroughly washed with distilled water and vacuum dried under reduced pressure. A series of hybrids of CMTG were synthesized using different ratios of H_2O, TMOS and Methanol at fixed CMTG amount. The optimum hybrid sample was synthesized when 1.0 mL TMOS, 1.0 mL methanol with a gelation time of 20 min was used. This sample was calcined in N_2 atmosphere (at 200 °C, 400 °C and 600 °C for 2 h at each temperature) to obtain heat treated nanohybrids which were then evaluated for the immobilization of α-amylase. Nanohybrid (C4) impregnated with 40 μg of α-amylase was used for hydrolyzing soluble potato starch to glucose syrup at variable temperatures, pH values and enzyme concentrations as these parameters affect the kinetics of the studied reaction. The optimum pH and temperature for the hydrolysis reaction were pH 5 and 40 °C, respectively.

3.4. Spray Drying

Micro/nanoparticles formulation by spray drying method was prepared by dispersing a core material in a coating solution. The coating suspension was then atomizing into hot air stream. The heated air removes solvent from the coating suspension thus leads to formation of the micro/nano carrier.

Singh *et al.* [71] developed covalently crosslinked chitosan with ethylene diamine tetraacetic acid microparticles for the successfully delivery of nanoemulsion of amphotericin B. The complex spray drying parameters (concentration of CH:EDTA and inlet temperature of spray drier) and other synthesis conditions were optimized using 4^2 full factorial design. The particle size, shape and zeta potential were the dependent variables. The percentage yield of microparticles was found to be influenced by the proportion of CH:EDTA-diNa when it varies from lower level (40:60) to highest level (70:30) indicated the highest amount of CH was required as compared to EDTA. In addition, 170 °C inlet temperature was optimum to obtain highest yield of microparticles. Further, particle size of microparticles was found to be higher at 50 °C (inlet temperature) and the particle size was decreased to 1–5 μm when microparticles prepared at 170 °C.

Desai and Park, [72] prepared TPP cross-linked chitosan (chitosan–TPP) microspheres of acetaminophen as model drug using spray drying technique. A mixture of chitosan and model drug was prepared and aqueous solution of TPP (1% w/v) was added drop wise to the chitosan–drug suspension under constant stirring. The chitosan–TPP–drug suspension was then spray dried to obtain the acetaminophen loaded chitosan-TPP microspheres. Spray drying was performed using a liquid flow, compressed spray air flow, and inlet temperature at 2 mL/min, 1.2 m^3/min, at 175°C, respectively. The microspheres prepared by spray drying technique involves flow of acetaminophen-chitosan-TPP suspension from spray nozzle that atomized the liquid by force of compressed air, disrupting the liquid into small droplets. The droplets, together with hot air, were blown into a chamber where the solvent in the droplets was evaporated, discharged out through an exhaust tube and collected in a collection bottle. Chitosan–TPP microspheres prepared using a higher concentration of chitosan; higher volume of TPP solution, a higher molecular weight chitosan and/or a higher drug loading are spherical in shape and having slower release rate. Most importantly, the drug release rate was mainly controlled by the and thus, by the degree of swelling of the hydrogel matrix that further affected by chitosan–TPP crosslinking density.

4. A PHARMACEUTICAL POTENTIAL: AN OVERVIEW

4.1. Drug Entrapment Efficiency of Micro/Nano DP Carriers

The drug entrapment efficiency of the microcapsule or microsphere mainly affected by formulation parameters as the high polymer, low polymer to drug ratio, high crosslinker, high drug: polymer interaction, stirring speed, solvent used for drug dissolving, *etc*. Various researchers have optimized these formulation parameters to obtain higher loading/entrapment of the drug into the micro/nanoparticulate system is explained in the current section. The entrapment efficiency of various formulations is shown in Table **1**.

Table 1. Physical, electrical properties and drug encapsulating efficiency of micro/nano- carriers prep polysaccharides

S. No	Gum	Drug	Modification	Dosage Form	Method of Preparation	Particle Size	Zeta (mV)	Morpholo
1.	Xyloglucan	Aceclofenac	Polyacrylamide grafted carboxymethylxyloglucan	Mb	Ionic cross linking Fe^{3+}	1235 nm	-	Spherical in with slight fla the botto
2.	Locust bean gum	Buflomedil hydrochloride	Carboxymethyl locust bean gum	Ms	Water-in-oil (w/o) emulsion-cross-linking Method	755μm 0.19 PDI	-	Spherical in with Rough surf exhibiting po the surfa
3.	Chitosan–edta	Amphotericin B nano emulsion	Chitosan–edta complexation	Mp	Spray drying	1-2 μm	-	Spheric
4.	Chitosan	Insulin	Gluteraldehyde crosslinking	Ms	Emulsion polymer cross-linking method	32.6 nm	-	No aggrega
5.	Gellan gum and Carrageenan	Simivastatin	$ZnSO_4$ and glutraldehyde	Ps	Ionotropic gelation/covalent crosslinking method	1044 μm	-	Rough and surface
6.	Chitosan	Tetracycline	Carboxymethyl chitosan	Np	Ionic cross-linking using $CaCl_2$	200 nm	-	Spheric
7.	Carboxymethyl tamarind pectin	Tropicamide	Carboxymethyl Ca^{2+}	Np	Ionic Cross-linking	321 nm	-	Ovoid morph
8.	Guar gum	Tamoxifene	Glutraldehyde crosslinking guar gum	Np	Oil in water (o/w) emulsion polymer cross-linking method	204-205 nm	-	Spherical smooth Surface
9.	Chitosan	Curcumin	Carboxymethyl chitosan	Np	Ionic cross linking $CaCl_2$	150 nm	-30 to -35	Spheric
10.	Chitosan	Curcumin	Chitosan-g-poly (n-isopropylacrylamide)	Np	Ionic Cross-linking	180–220 nm	+28 to +25	Spherical s
11.	Chitosan	5-FU	Poly(n-vinylcaprolactam)	Np	Ionic Cross-linking	180-220 nm	-	Spherical s

Table 1: contd…

12.	Chitosan	Chelerythrine	Carboxymethyl chitosan	Ms	Oil in water (o/w) emulsion polymer cross-linking method	2-2.5μm	-	Spherical with rough surface	54
13.	Chondroitin sulfate /chitosan	Fluorescein Isothiocyanate conjugate bovine serum albumin	Crosslinking TPP	Np	Ionic cross linking	213nm	+32 to +50	Spherical with compact core surrounded by a thick but fluffy coat	
14.	Chitosan	Oxaliplatin	HA-coupled chitosan	Np	Ionotropic gelation technique	135nm	+10 to +40	Spherical and smooth shape	
15.	Guar gum	Ketoprofen	Grafted with poly(ε-caprolactone)	Ns	Emulsification method	60-100 nm	-	Spherical micelles	
16.	Carboxymethyl cashew gum/chitosan	Bovine serum albumin	Carboxymethyl cashew gum/chitosan	Ms	Polyelectrolye complex	500 μm	-	Spherical in nature	
17.	Acetylated pullulan	Exenatide	Acetylation	Ms	(w_1/o/w_2) double emulsion method	35-105 nm	-	Regular and spherical	
18.	Pectin	Theophylline	Ca-pectinate	Ms	Ionotropic gelation technique	1.90mm	-	Spherical and smooth shape	
19.	Xanthan	Bovine serum albumin	Carboxymethyl xanthan	Mp	Ionotropic gelation technique	1.09mm	-	Spherical and smooth shape	
20.	Guar gum	Methotrexate	Gluteraldehyde crosslinking	Ms	Emulsification method	10-22 μm		Spherical in shape	
21.	Chitosan	Acetaminophen	TPP	Mp	Spray drying	4.7-10μm	+18-+31	-	

Np= Naoparticles; Mp= Microparticles; Ms= Microspheres; Mb=Microbeads; Ps=Polyspheres; DEE= Drug entrapment efficiency; EE=Entrapment efficie[illegible] LE= Loading efficiency

The thermoresponsive nanoparticles were prepared by Rejinold *et al.* [7, 11] and confirmation of curcumin and 5-FU drug loading was done using FTIR by analyzing shift in amide peak. Due to curcumin and 5-FU loading on the nanoparticles, the amine peak of grafted polymer was shifted from 1629 to 1627cm^{-1} and the amide peak was merged at the same peak.

The tamoxifene loaded guar gum nanoparticles was optimized using the drug:polymer ratio. The optimized formulation contains 1:6 drug:polymer ratio for a batch with the highest drug loading. To evaluate tamoxifene loading in nanoparticles, a series of batches were prepared with 1:6 drug:polymer ratio and different drug loading solvents (1,2-dichloroethane, dichloromethane (DCM), chloroform and n-hexane). Dichloromethane as drug loading solvent provided the best loading for tamoxifen citrate (15.0%) because of its lowest boiling point and the highest rate of evaporation of dichloromethane which leads to rapid transfer of the drug into the polymer. The higher loading of tamoxifene onto guar gum nanoparticles was due to fast solidification of the nanoparticles and due to high solubility of the drug in dichloromethane [64].

Tetracycline loading on carboxymethyl chitosan nanoparticles was optimized by Maya *et al.* [8] to the ratio of 10:1 (carboxymethyl chitosan:Tet) to get maximum encapsulation efficiency, which was found to be 70%. The concentration of the polymers (carboxymethyl chitosan) showed marked effect on the drug loading, with increase in polymer solution concentration (0.05–0.5 wt.%) drug loading increased, with increase in polymer concentration and drug loading the particle size showed significant increase.

In another investigation, Anitha *et al.* [47] developed a nanocarrier for curcumin delivery, which facilitates maximum loading and entrapment curcumin. Entrapment and loading efficiency of curcumin in carboxymethyl chitosan nanoparticles was found to be 87% and 48%, respectively. The results suggested that carboxymethyl chitosan and curcumin concentration had critical influence on the drug loading. Higher amount of carboxymethyl chitosan promotes loading and entrapment, but resulted in large sized particles. At higher drug concentration entrapment efficiency was reduced due to precipitation of drug.

Setty *et al.* [1] investigated the effect of drug (Aceclofenac) loading on particle size of the Fe^{3+}crosslinked-polyacrylamide grafted carboxymethylaxyloglucan nanoparticles. The results indicated with increase in loading of drug onto microbeads, it was observed that the size of microbeads was increased. The reason for this significant increase in the particle size was explained by the fact that the shrinkage of microbeads during their cross linking process was restricted due to high concentration of drug that might have entrapped within fine matrix microstructure. Drug entrapment efficiency was found to be between 85 and 89%.

Kaity and Ghosh, [10] studied effect of formulation parameters on the drug entrapment efficiency. The results indicated with increase in the amount of polyvinyl alcohol in the blend results in higher drug entrapment efficiency. However, the drug entrapment efficiency was decrease with increase in carboxymethyl locust bean gum content in the formulations. In addition, the enhancement in the concentration of cross-linking agent increased the drug entrapment efficiency. This probably was due to the formation of rigid network that entrapped more drug particles and prevented drug leaching. The percentage drug entrapment efficiency was dependent on the percent drug loading. Formulations containing high drug loading exhibit slightly higher drug entrapment efficiency, this might be associated with accumulation of more drug particles inside the free void spaces of the matrixes.

Kulkarni *et al.* [52] prepared Simvastatin loaded composite polyspheres of $ZnSO_4$ crosslinked carrageenan and gellan gum, exhibiting the drug entrapment efficiency in the range of 77.37–84.31%. Results demonstrated that the drug entrapment efficiency of the polyspheres prepared by ionic crosslinking was higher than those prepared by dual crosslinking method. The drug entrapment efficiency of the polyspheres was mainly affected by concentration of $ZnSO_4$. The enhancement in the $ZnSO_4$concentration caused increase in drug entrapment efficiency as higher concentration of crosslinking agent results in stiff matrix and leakage of drug from matrix was low resulting in high drug entrapment efficiency.

In another study Li *et al.* [65] indicated that as the O/W phase ratio increased, the drug loading content and entrapment efficiency increased. The drug loading content and entrapment efficiency were also mainly affected by the polymer:drug ratio (O-carboxymethyl chitosan: Chelerythrine). The increase in the O-carboxymethyl chitosan: chelerythrine ratio lowering of chelerythrine content, the higher entrapment efficiency. The optimized formulation was prepared using optimum conditions *i.e.* O-carboxymethyl chitosan: Chelerythrine ratio 20:1 and the O/W phase ratio 3.5:1. Under these conditions, the entrapment efficiency and drug loading content were 54.78% (w/w) and 4.08% (w/w), respectively.

Bovine serum albumin (BSA) entrapment efficiency was found to be affected by gelation time. The drug entrapment decreased with enhancement in gelation time due to diffusion out of the water soluble protein in the aqueous gelation medium.

It was also explained that variation in the pH of SCMX (sodium carboxymethyl xanthan) gum solution did not affect the entrapment efficiency significantly. Further, the release of BSA was found to decrease as the gelation time to prepare the microparticles was increased [5].

Yeha *et al.* [9] explained that Protein loading in micro/nanoparticulate systems that was performed, either during the preparation of particles (incorporation) or after the formation of particles (incubation). Further, to avoid aggregation of nanoparticles during the process, the study was carried out by the incorporation method. The results indicated that while using the incorporation method the increase in entrapment efficiency was observed. Apparently, the results indicated the increase in entrapment efficiency with the increase in the FITC–BSA amount.

Peptide drug Exenatide (a drug for type II diabetes) was used to demonstrate the loading potential of acetylated pullulan microspherse. The exenatide loading efficiency of Polysaccharide microspheres was dependent on the degree of acetylation, *i.e.* higher acetylation increased drug entrapment. Loading efficiency was improved with higher concentration of Polysaccharide in the microspheres [69]. The encapsulation efficiency of acetaminophen loaded chitosan–TPP microspheres was in range of 48.9% to 99.5%. The encapsulation efficiency of chitosan–TPP microspheres was influenced by volume of TPP solution, molecular weight of chitosan and concentration of polymer. The rise concentration and molecular weight of chitosan enhanced encapsulation of acetaminophen in the microsphere [72].

Overall, the derivatized polysaccharides could be effectively utilized for the delivery of drugs with higher drug loading potential.

4.2. Physical Performance of Micro/Nano DP Carriers

The chemical modifications of natural polysaccharides drastically influenced physical properties. These derivatized polysaccharides when utilized in the preparation of micro/nano carriers; their physical, electrical performance was also modified. The changes appeared in physico-electrical performance are summarized in Table **1**.

Anitha *et al.* [47] observed zeta potential and particle size using DLS, the size distribution of O-carboxymethylnanoparticles lies within a size range of 70–100 nm and curucmin-O-carboxymethyl chitosan nanoparticles lies within 120–180 nm. The size and morphology of the nanoparticles were confirmed by AFM and SEM. The AFM images of O-carboxymethyl chitosan and curcumin-O-carboxymethyl chitosan nanoparticles showed uniformly spherical sized particles with a diameter of 100±20 and 150±30 nm, respectively. For O-carboxymethyl chitosan and curcumin-O-carboxymethyl chitosan Nanoparticles, the zeta potential value was found to be −35.12 and −30 mV indicating stability of nanoparticles. The negative surface charge of both O-carboxymethyl chitosan and curcumin-O-carboxymethyl chitosan nanoparticles was because of COO− groups.

The morphology of thermoresponsive chitosan-g-poly(N-isopropylacrylamide) co-polymeric nanoparticles loaded with curcumin was observed by Rejinold *et al.* [47]. Zeta potential value for the thermo-sensitive chitosan-g-poly-nanoparticles system was found to be +28 mV, whereas the curcumin loaded thermo-sensitive chitosan-g-poly-nanoparticles showed a value of +25 mV indicating that the nanoparticles system were stable.

Rejinold *et al.*[11] explained that the SEM of the 5-fluorouracil loaded thermo-sensitive chitosan-g-poly(N-vinylcaprolactam) nanoparticles and were spherical shape with an average diameter of 150 nm and 200 nm respectively. Whereas, the 5-FU–thermo-sensitive chitosan-g-poly-nanoparticles showed an increase in particle size could be due to the high loading of 5-FU on thermo-sensitive chitosan-g-poly-nanoparticles.

Sarmha *et al.* [64] found that as the concentration of span 80 increased from 1% to 6% (w/w) during the preparation procedure, the size of tamoxifen citrate loaded guar gum nanoparticles decreased from 354 to 205 nm. 4% w/w of span 80 was optimum to prepare nanoparticles, below that, nanoparticles tend to fuse and produce larger size particles.

Nanoparticles made from carboxymethyl tamarind kernel polysaccharide (CMTKP) and loaded with tropicamide were developed and evaluated for the purpose of ocular delivery [6]. An ionotropic gellation method was adopted to prepare nanoparticles employing optimization techniques using three-levels, two-factor central composite design. Concentration of polymer and crosslinker was found to be significantly and synergistically effect the particle size and % encapsulation efficiency of nanoparticles. Thus, indicated concentration of CMTKP bears more pronounced influence than the concentration of calcium chloride on particle size. Further, increase in the concentration of CMTKP was found to enhance the particle size of the nanoparticles. The optimized batch of tropicamide loaded CMTKP nanoparticulate formulation showed 339 nm particle size and 15.57 % encapsulation efficiency when prepared with CMTKP 0.10% (w/v) and calcium chloride 0.11% (w/v). The TEM image of the optimized batch of nanoparticles revealed the ovoid morphology with 20-40 nm size of the nanoparticles. The particle size and shape of nanoparticles could be utilized to estimate the ocular irritation potential of the particles in the cul-de-sac. Since, lack of any sharp edges and angles of the prepared tropicamide loaded CMTKP nanoparticles, indicated not to cause any irritation.

Ionotropic gelation technique was used to prepare L-OHP–loaded chitosan nanoparticles. Scanning electron micrographs and transmission electron micrographs showed the spherical shape of hyaluronic acid–coupled chitosan nanoparticles (HACT nanoparticles). The mean diameters of HA-coupled (HACTNPOPs) and uncoupled (CTNPOPs) chitosan nanoparticles formulations were 152 ± 5.2 and 136 ± 6.0 nm, respectively. Low PDI (0.155 and 0.110) was obtained indicating narrow size distribution. Nanoparticles of CTNPOP and HACTNPOP showed positive zeta potential *i.e.* 40.3±1.4 and 10.0±0.5mV, respectively [73].

Maestrelli *et al.* [46] showed irregular spherical shapes and a rugged surface, of calcium pectinate microspheres. No crystals of drug were visible on the microspheres surface, indicating their complete incorporation of drug within the polymeric matrix. In particular, microspheres prepared with AU701 (Pectin of low methoxylated grade, 36% > DE > 44%) were not stable and deliquescate also lost their integrity after few days of preparation. This was because of the higher hygroscopicity of calcium pectinate. The SEM images of CF020 (low methoxylated grade, 25% > DE > 31%) microspheres showed progressive morphologic modifications and the loss of integrity of these morphological changes might be due to calcium pectinate depolymerization phenomena.

The DEX–ADM–PEI/DNA (Dextran -g–PEI–adriamycin/DNA) nanoparticles were prepared using various N/P ratios (the ratios of moles of the amine groups of cationic polymers to those of the phosphate ones of DNA).the results indicated that with higher N/P ratio, the size of the nanoparticles decreased and the zeta potentials increased. At N/P ratio 0.5 the nanoparticle showed a size of 225±17 nm. However, at N/P ratio of 2, the nanoparticles size was decreased to 200 nm. The zeta potentials tendency of the nanoparticles with increased N/P ratios. At N/P ratio of 0.5 the zeta potential was +9.28 mV. With higher N/P ratios, the zeta potential increased (*i.e.* +24.71 mV, at N/P ratio 2). The morphological evaluation showed spherical shape and compact structure with a size of 166±15 nm [12].

The effect of gelation time and of gum solution concentration on the size was evaluated by Maiti *et al.* [5]. They found increase in pH from 6 to 7 and gelation time from 5 min to 15 min affects particle size of BSA-loaded microparticles from

1.18×10^3 µm to 1.05×10^3 µm. Increase in concentration of SCMX gum (from 1 % to 2 % w/v at pH 6) showed increase in particle size from 1.02×10^3 µm to 1.09×10^3 µm. Generally, the microparticles were of spherical morphology. However, the usage of 1 % SCMX gum to prepare microparticles produced ellipsoidal morphology.

As reported by Maya *et al.* [8], O-carboxymethyl chitosan Nanoparticles were prepared by the ionic crosslinking between carboxyl group of O-carboxymethyl chitosan with Ca^{2+} showed the size distribution for blank nanoparticles and drug loaded nanoparticles were found to be 60 ± 30 and 200 ± 20 nm. SEM confirmed the size of the spherical particles to be in the range of 200 nm.

The O-carboxymethyl chitosan (O-CMCS) microspheres prepared by Li *et al.* [65] had spherical morphology with rough surface having a diameter of about 2–2.5 µm (placebo) and 2.5- 3.0 µm (drug loaded). The SEM micrographs showed the presence of surface adhered drug. The particle size and distribution of the microspheres showed about 50 % of the measured particles were of 12.18 µm, about 10 % of the microspheres were below 3.59 µm and 90 % were below 26.56 µm average size range.

The polyelectrolyte complex formed using carboxymethyl cashew gum (CMCG) with different degree of substitution (DS) and chitosan with different molar mass but similar degree of deacetylation (DD) values were investigated for particle size and morphology. Chitosan with different molar masses in complex with CMCG with different DS produced spherical particles. The DS of CMC has affected the size of microsphere, higher DS produces larger particles regardless of molar mass of chitosan. For microspheres formed with the same DS of CMCG, the beads formed with high molecular mass chitosan sample had a larger sphere diameter. The increase in sphere diameter is likely to be mainly due to the increase in the chitosan chain length and also due to high CMCG incorporation [58].

Chitosan-chondroitin sulfate (CS-ChS) nanoparticles were prepared by Yeh *et al.* [9]. The results of the study indicated a linear relation with the concentration of chitosan at a fixed ChS/CS ratio. The particle size and zeta potential of the nanoparticle showed a zeta potential of +32 to +50 mV and particle size 77-213 nm. The SEM indicates well organized, dense, spherical morphology of nanoparticles.

The size of carboxymethyl guar gum particles in suspension have been found in the range 12–30 nm. The TEM images indicate the spherical morphology with the size of 12–30 nm along with few larger particles and clusters in the size range 50–70 nm. These clusters might be caused by drying induced aggregation during the TEM grid preparation. It was concluded that such type of nanoparticles may be used in pharmaceutical and drug delivery[59].

Chaurasia *et al.* [68] explained effect of various formulation parameters on the particle size. The particle size of glutaraldehyde cross-linked guar gum microspheres was found to be affected by the polymer concentration. With increase in polymer concentration from 1% w/w to 4% w/w, the particle size increased from 12.4 ± 1.02 µm to 16.5 ± 1.22 µm. 2% w/v of guar gum concentration produced uniform sized microspheres with increasing mixer rotational speed, from 2000 rpm to 5000 rpm. The particle size of microspheres decreased from 29.7 ± 1.27 µm to 14.6 ± 0.95 µm. The results indicated rotational speed as main parameter to control particle size of microsphere. The variation in Span 80 concentration from 1% w/w to 4% w/w, decreased the particle size from 17.1 ± 1.33 µm to 10.4 ± 0.93 µm. Overall, the results indicated rotational speed as important process parameter controlling particle size.

In another similar study by Desai and Park, [72], identified parameter which affect the particle size of the microsphere. They found concentration of chitosan, molecular weight of chitosan as the parameters affecting particle size and morphology of microspheres. With increase in concentration of chitosan from 0.5 to 1.0% w/v the particle size increased from 4.7 to 6.8 µm. This is due to the greater amount of chitosan contained in the small volume of a liquid droplet. In addition, the low molecular weight of chitosan showed lower particle size (3.1 µm) as compared to medium (6.8 µm) and higher (10.1 µm) molecular weight of chitosan. Non-cross-linked spray dried chitosan microspheres (control) were spherical with wrinkles on their surface. However, in case of drug-loaded chitosan-TPP microspheres, the formulation parameters such as drug loading, and concentration and molecular weight of the chitosan, induced remarkable change in the surface morphology of the microspheres.

Overall, the morphological performance of various micro/nano carriers showed a marked improvement in the formulation characteristics. Thus, derivatization could be considered as an important technique for the purpose of inclusion and exclusion properties of excipients as per our benefits in drug delivery.

4.3. Biological Performance

4.3.1. Toxicity

Toxicological and irritant effects of tropicamide-loaded carboxymethyl tamarind pectin nanoparticle formulation were measured by HET-CAM studies. From the potential irritation scores of tested formulations it was concluded that the optimized tropicamide-loaded CMTKP nanoparticle formulation was found to be non-irritant with no signs of irritation [6].

The oral toxicity study was carried out for carboxymethyl locust bean gum microparticles and polyacrylamide grafted carboxymethylxyloglucan based microbeads. The results for both the nanocarriers showed no mortality within the observation period of 14 days after dosing. The different hematological and blood chemistry data showed no abnormalities. Moreover, the histological sections of lung, kidney, liver, and stomach did not show any abnormality on 15th day of toxicity study. The LD_{50} value is greater than the 2000 mg/kg dose of carboxymethyl locust bean gum and polyacrylamide grafted carboxymethyl xyloglucan based microbeads, and both the product fall under "category 5" and the toxicity rating will be "zero" [1, 10]. The biodegradability studies were carried out to explain biodegradable nature of carboxymethyl locust bean gum microparticles the results reveled apparent fungi growth on 0, 3, 7, 14, and 21 days in all the Petri dishes which proves that the carbon present in the carboxymethyl locust bean gum had been utilized by the fungi for its growth [10].

O-carboxymethyl chitosan (O-CMC), Curcumin and curcumin-O-CMC nanoparticles toxicity was observed by Anitha *et al.* [47], in the concentration range of 1–5 mg/mL. The results indicated 80% of the cells were viable suggesting the non-toxicity of curcumin, O-CMC nanoparticles and curcumin-O-CMC nanoparticles towards normal cells. However, studies conducted with MCF-7 cell line showed O-CMC nanoparticles were non-toxic whereas curcumin and curcumin-O-CMC nanoparticles were considerably toxic. Similarly, for PC-3 cell line curcumin-O-CMC nanoparticles pointed towards reduced cell viability to 30% confirming its anticancer effects. Curcumin-O-CMC nanoparticles and curucmin showed comparable effect on cancer cells which reflected curcumin retains its anticancer activity even after being loaded into polymer matrix.

4.3.2. Cell Line Studies

Rejinold *et al.* [11] have shown curcumin loaded thermoresponsive nanoparticles to be more up taken by the cancer cell line compared the normal cell line. The study revealed direct relation of uptake and the time. A concentration dependent increase in apoptosis was also observed in the case of prostate cancer cells. Curcumin is a lipophilic molecule that could permeate across cellular membrane and is subsequently transported inside the cells. Difference in membrane structure, protein composition and size are the responsible factor for preferential uptake in tumour cells against normal cells. Flowcytometric analysis of curcumin loaded thermo-sensitive chitosan-g-poly-nanoparticles showed increased apoptosis on PC3 cells.

Maya *et al.* [8] tested the efficacy of tetracycline (Tet) encapsulated O-carboxymethyl chitosannanoparticles against intracellular survival of S. *aureus.* Encapsulation of Tet in o-carboxymethyl chitosan nanoparticles did not negate its anti-microbial action. Specific binding and aggregation of O-carboxymethyl chitosan nanoparticles with S. *aureus* indicated its potential use for targeted delivery against several S. *aureus* infections. The cytocompatibility (NIH-3T3, L-929 and HEK-293 epithelial cell lines and THP-1 monocytic cells using MTT assay showed no toxicity) and hemocompatibility (no change in PT and APTT) of Tet-O-carboxymethyl chitosan nanoparticles reflects its safety for IV administration as compared to normal cells. The confocal microscopic studies revealed the uptake of the nanoparticles by the macrophages (THP-1) and epithelial cells (HEK- 293) which helps to kill the bacteria within the macrophages. This helps Tet-O-carboxymethyl chitosan nanoparticles of delivering Tet intracellularly and showed a six fold enhanced antibacterial activity over that of free Tet against intracellular S. *aureus.*

4.3.3. Biocompatibility

The biocompatibility of DEX–ADM/DNA–PEI (dextran-g-PEI-adriamycin/DNA) nanoparticles was demonstrated as with the conjugation of oxidized dextran to PEI, oxidised dextran imparts macromolecular nature to the PEI, because of which the nanoparticles selectively penetrated across highly porous mass of tumour cells at the same time precluding normal tissues. Thus, it was hypothesized that the adriamycin coupled DEX–ADM–PEI conjugation would selectively positioned in the tumour mass. This increases the therapeutic margin of safety and decrease the side effects associated with adriamycin. Transfection efficiency of cationic biodegradable DEX–ADM–PEI/pEGFP–nanoparticles based on low molecular weight PEI and dextran to mediate efficient *in vitro* gene transfection was observed. The results indicated that transfection efficiency of DEX–ADM–PEI/pEGFP–nanoparticles was tended to decrease when N/P was increased above 8. The fluorescence-inverted microscopic and flow cytometric studies revealed that the DEX–ADM–PEI/pEGFP–N1 nanoparticles successfully transfected the nanoparticles into osteosarcoma cells [12].

4.4. Release Kinetics from Micro/Nano DP Carriers

The derivatization not only affects physicochemical, electrical and mechanical performance but also effect drug release profile from micro/nano-carrier based drug delivery system. Table **2** summarizes drug release performance when derivatized polysaccharides are used to prepare micro/nano-carriers.

Table 2. Drug Release performance from micro/nano delivery systems prepared by derivatized polysaccharides.

S. No.	Modification	Model Drug	Dosage Form	Remarks
1.	Polyacrylamide grafted carboxymethylxyloglucan	Aceclofenac	Microbeads	Showed a triphasic drug release pattern. Both the drug release and sw microbeads were significantly higher in simulated intestinal (alkaline) pl to simulated gastric (acidic) pH and this nature is desirable for targeted d
2.	GG_H-Cl-poly(MAc-)-GA	Bovine serum albumin (BSA)	Microspheres	The drug release from microspheres was extended beyond 12h with n release from 60% to 70% in alkaline conditions.
3.	Glutraldehyde crosslinked chitosan	Insulin	Microspheres	*In vitro* release showed a controlled profile up to 12 h and the formulatio 0.15% (w/v) of bacitracin revealed a maximum biological activity of a ±4.1%.
4.	Locust bean gum crosslinked with poly(vinyl) alcohol using glutaraldehyde as crosslinker	Buflomedil HCl	Interpenetrating polymer network (IPN) microspheres	The microspheres showed controlled release of drug without showi incompatibility in IPN device.
5.	Carboxymethyl gellan gum	Metformin HCl	Ionotropically gelled beads	Studies showed *ex vivo* bioadhesion of 100% over 24 h and it was obs carboxymethyl gellan gum beads released metformin at a rate faster than
6.	Ca-Pectinate	Rutin	Microbeads	Calcium pectinate microbeads loaded with rutin showed controlled rele
7.	Gellan gum grafted with carrageenan by gelation/covalent crosslinking method	Simvastatin	Polyspheres	The polyspheres were produced with entrapment efficiency as high as
8.	Carboxymethylated gum kondagogu-chitosan complex	ofloxacin	Nanopartcles	Nanoparticles provided 50% sustained release of drug in 24h following square root kinetics.
9.	Carboxymethyaltion of Gum kondagogu with monochloroacetic acid	Metformin	Ionotropically gelled beads	*Ex vivo* bioadhesion studies of CMGK beads revealed bioadhesion great over a period of 24h and drug release rate from the beads followed Ficki mechanism with Zero-order release kinetics.
10.	Carboxymethylated tamarind kernel (CMTKP)	Tropicamide	Nanoparticles	Tropicamide-loaded CMTKP formulation showed good ex-vivo corneal The mucoadhesive and non-irritant nature of CMTKP nanoparticles indic suitability as ocular delivery system
11.	O-carboxymethyl chitosan (O-CMC)	Tetracycline (Tet)	Nanoparticles	Nanoparticles showed sustained release of tetracycline. Tet-O-CMC na were six fold More effective in killing intracellular S. Aureus compared in HEK-293
12.	Guar gum crosslinked with glutraldehyde	Tamoxifene citrate	Nanoparticle	The drug release from the nanoparticles was time dependent release anc

Table 2: contd…

13.	Graft copolymer of Guar gum with acrylamide by crosslinking with glutaraldehyde	Ornidazole	Microsphere	At acidic ph, only a small fraction of drug was released; how was found to be higher in the presence of rat caecal conter susceptibility of guar gum matrix to colonic enzymes rele contents.
14.	Guar gum cross linked with glutaraldehyde	Tamoxifen citrate (TMX)	nanoparticles	*In vivo* studies demonstrated maximum uptake of TMX by albino female mice after 24h of administration with drug loa form of drug.
15.	Chitosan-g-poly (N-isopropylacrylamide) co-polymeric	Curcumin	Nanoparticles	The *in vitro* drug release was prominent at above Lower criti (LCST), cytocompatible and the nanoparticles showed spe cells
16.	Chondroitin sulfate–chitosan nanoparticles	Bovine serum albumin (BSA)	Nanoparticles	The release profile of FITC–BSA fro Chs–CS nanoparticles indicated a small burst release follov controlled release.
17.	O-carboxymethyl chitosan	Curcumin	Nanoparticles	*In vitro* drug release profile indicated slow, controlled and s from the nanoparticles matrix and also demonstrated enzym and release of the drug in the presence of lys
18.	chitosan-g-poly(N-vinylcaprolactam)	5-Fluorouracil	Nanoparticles	5-FU released above LCST of the system while at below L released. The drug release mechanism is based on the LCST system.
19.	Carboxymethyl Chitosan	Chelerythrine (CHE)	Microspheres	Mean particle size, drug loading content, entrapment Effici release of chelerythrine loaded microspheres were found to 54.78% and 35.30% at pH 7.4 in 20 h, respectively.
20.	DEX–ADM–PEI conjugate	Adriamycin and DNA	Nanoparticles	Nanoparticles were capable of high efficiency in deliver osteosarcoma cells, resulting in an enhanced drug efficacy adriamycin.
21.	Guar gum grafted with poly(ε-caprolactone) (GG-g-PCL)	Ketoprofen	Nanoparticle	Initial burst release followed by a sustained release of the en a period of 10–68 h.
22.	Ionic gelation between Arabic gum and chitosan	Insulin	Nanoparticles	Enhanced solubility of chitosan in acidic medium and better chains at ph >6.5 resulted in lower insulin release at ph 6.5 of the other ph mediums. Drug release profile followed n mechanism.

Table 2: contd…

23.	Hyaluronic acid–coupled chitosan	Oxaliplatin (L-OHP)	Nanoparticles	hyaluronic acid–coupled chitosan nanoparticles delivered 1.99 ± 0.82μg 1.10 μg of L-OHP/g of tissue in the colon and tumor, respectively after
24.	Calcium Carbonate/Carboxymethyl Chitosan Hybrid	Doxorubicin hydrochloride	Microspheres and Nanospheres	The drug release was effectively sustained. Compared with hybrid micro release from hybrid nanospheres was faster because of their higher surfa ratio and lower $CaCO_3$ content.
25.	Polyethyleneimine-Treated Xanthan	Diltiazem	Microparticles	40% and 80% drug was released in 2 hour in pH 1.2 and in 5 to 6 h in respectively. The prolonged release was attributed to decreased swelling due to PEI treatment. The drug release follows non-Fickian transport ph
26.	Chitosan/carboxymethyl cashew gum (CH/CMCG):	Bovine serum albumin (BSA)	Microspheres	Faster BSA release rates were observed for beads prepared with the hi Microsphere system showed non-Fickian BSA release mechanis
27.	Ca-Pectinate	Theophylline	Microspheres	Coated microspheres were able to adequately modulate drug release thro approach of pH and transit time control and reaching the colonic target 100% release was achieved within less than 24 h.
28.	Carboxymethyl Xanthan gum	Bovine serum albumin (BSA)	Microparticles	The release of BSA in acidic medium was almost equal to that in alkalin as compared up to 2 hr, the release in alkaline medium was found to be p a different extent depending upon the formulation variables.
29.	Guar gum crosslinking with glutaraldehyde	Methotrexate	Microsphere	The drug release in PBS (ph 7.4) and simulated gastric fluids followe release rate, while there was a significant increase in percent cumulative (91.0%) in the medium containing rat ceacal content
30.	Carboxymethyl chitosan	Doxorubicin (DOX)	Nanoparticles	Release studies showed an initial burst followed by an extended Slow re DOX release rate was hindered by CM-chitosan with high MW and DS.
31.	Crosslined chitosan microspheres	Metronidazole	Microsphere	Without rat caecal medium *in vitro* drug release was 31.23%, which wa with rat caecal medium (2% and 4%) upto 47.72% and 61.65%, resp
32.	Graft copolymer of Guar gum with acrylamide by crosslinking with glutaraldehyde	Verapamil and Nifedipine	Hydrogel Microspheres	Drug release followed non-Fickian mechanism.
33.	Locust bean gum:i- carrageenan microparticles	Gentamicin	Gel having gentamicin encapsulating by gum microparticles	*In vitro* experiments in isotonic phosphate buffer solution (ph 7-4) at 37° that the release of gentamicin sulphate was dependent on concentration

Rejinold *et al.* [7] investigated release kinetics of curcumin from thermo-sensitive chitosan-g-poly (N-isopropylacrylamide) nanocarriers. Initial swelling-controlled release mechanism followed by diffusion-controlled mechanism was seen. Penetration of water into the matrix during the initial period was due to hydration along with polymer chain reaction. During this process, enhanced release rate was induced due to the presence of the drug at and near the surface. After forming the gel structure, the release was shifted to the diffusion mechanism through the matrix by a much faster release rate.

Nanoformulation of thermo-sensitive chitosan-g-poly-nanoparticles showed very controlled and sustained release of 5-FU. In thermo-sensitive chitosan-g-poly-nanoparticles, it was observed that drug release from the nanoparticles was prominent above lower critical solution temperature (LCST) in the early phase as compared to that below LCST. About 10–20% of the 5-FU was released within 10 h at above LCST. After that, a slower and constant release rate was observed. The release was initially by swelling mechanism than by diffusion-controlled mechanism. However, the drug release was very slow and almost less than 5% at initial phase [11].

In another study, the time dependent cumulative percentage release profile of tamoxifene loaded guar gum nanoparticles was recorded. A sustained drug release profile was observed over a period of 12 h. The crosslinking effect might be the reason for significantly sustained the release of the drug from the nanoparticles. As in phosphate buffer pH 7.4 the crosslinking of nanoparticles broke, resulting in water penetration into the polymer matrix which initiates swelling that caused the drug to diffuse out at a faster rate. Release kinetics was evaluated by fitting the data to four kinetic models and mechanism of drug release was explored on the basis of regression coefficient analysis. It was observed that the nanoparticles follow time dependent Zero order kinetic model [64].

The *in vitro* release performance of Ca-Pectinate microbeads was evaluated by Maestrelli *et al.* [46]. An inverse relationship between calcium ions concentration and drug release rate was observed. Higher calcium concentration leads to greater degree of cross-linking and aggregation of pectin molecules which in turn increases the higher gel strength and decreases the swelling which further results in a slow drug release rate. On the contrary, type of pectin (amidated or nonamidated) and variations of crosslinking time (from 20 to 5 min) were not crucial factors to influence the drug release.

O-carboxymethyl chitosan particles showed 24% and 26% release of curcumin in first 12 h of study and reaches to 53% and 59% at pH 7.4 and 4.5, respectively after 6 days. O-CMC particles should pH dependent drug release which was due to swelling mechanism. Swelling in basic medium was low which in turn decrease the diffusion of the drug. However, lysozyme help to increase the release as it acts on chitosan and hydrolyse glucosamine-glucosamine, glucosamine-N-acetyl glucosamine and N-acetyl glucosamine-N-acetyl glucosamine linkages. Thus, in the presence of lysozyme release of the drug takes place by diffusion followed by enzymatic degradation of the polymer matrix [47].

The release kinetics of polyacrylamide grafted carboxymethylxyloglucan based micro beads were studied by Setty *et al.* [1]. The study revealed pH dependent drug release, the micro beads showed higher release in alkaline [pH-7.4] and moderate release as compared to acidic medium [pH-1.2]. The percentage cumulative drug release was directly proportional to the loading of drug and inversely proportional to the cross linking density. A triphasic release pattern was observed which can be attributed to the property of both the drugand matrix configuration of microbeads. The release kinetics showed Fickian or diffusion controlled release and anomalous or non-Fickian release for different formulations.

Drug diffusion, polymer swelling, polymer erosion or degradation can be the factors affecting *in vitro* drug release of tetracycline from O-carboxymethyl chitosan nanoparticles [8]. The drug release pattern showed that 42% of tetracycline was released within 24 h followed by almost sustained release of 62% after 60 h.

Kaity and Ghosh, [10] observed effect of different formulation parameters on the drug release. They observed that with higher amount of cross linker in fixed blend ratio, the drug release decrease due to higher cross-linker concentration that causes the rigidity of polymer network, which prevents the penetration of buffer into the polymer matrix. Therefore, this leads to decrease in erosion of the polymer network and leading to reduction in drug release. When the IPN blend ratio was changed from 1:1 to 1:2 (carboxymethyl locust bean gum:polyvinyl alcohol), the particles showed less drug release. Higher drug loading showed higher drug release.

Li *et al.* [65] demonstrates the release of chelerythrine from O-carboxymethyl chitosan microspheres. The swelling of microspheres was higher at pH-1.0 and lower at pH-7.4. The chelerythrine release profiles in pH 7.4 and pH 1.0 media were found to occur in a biphasic manner, with an initial fast release phase followed by a slower release phase. The optimized O-carboxymethyl chitosans microspheres followed the Higuchi matrix release model.

The exenatide release behaviors of polysaccharide microspheres and PLGA microspheres under *in vitro* conditions were investigated and compared by Yang *et al.* [69]. The poly lactic-co-glycolic acid (PLGA) microspheres showed a high initial burst 40% on the 1st day and then exenatide was released for 5 days. However, in the case of polysaccharide microspheres, the initial burst decreased markedly to less than 30% on day 1, and release for 20 days. The results showed polysaccharide microspheres formulation with the highest degree of acetylation, showed a lower level of burst, at 10 wt% for 1 day, and almost nearly zero-order release kinetics.

Soppimath *et al.* [74] explained the release of Diltiazem and Nifedipine release from polyacrylamide-g- guar gum nanoparticles. The presence of ionizable carboxylic functional groups makes the microgels sensitive to pH and ionic strength. The solvent front velocity and diffusion coefficients calculations showed that the rate of solvent diffusion was pronounced considerably at higher pH, *i.e.* above the pKa of the anionic microgels. The swelling mechanism was explained on the basis of the relaxation-controlled phenomenon in acidic pH, basic pH conditions and in ionic solution (1.0M NaCl). The release of both diltiazem hydrochloride and nifedipine followed the relaxation controlled process, which was found to be closely related to the swelling of the microgels in response to the pH changes.

Chaurasia *et al.* [68] enumerated the release of methotrexate from crossed linked guar gum microsphere matrix. A 15.3% of methotrexate was released in upper GIT which might be due to the dissolution of drug particles present on the surface of microspheres. The crosslinking of guar gum with glutaraldehyde induces rigidity in the polymer network. As the glutaraldehyde reacts with hydroxyl moieties of galactose and mannose, it prevents the engagement of -OH group with water, thus leading to reduce water uptake. This provides a slow release of methotrexate during the dissolution studies.

Thus, it could be concluded that the derivatization of polysaccharides improved controlled release characteristics of delivery systems. In addition, the crosslinking of derivatized polysaccharides either insitu or directly significantly shifts drug release performance to non-Fickian behavior, suggesting derivatized polysaccharide are advantageous in micro/nano carrier based drug delivery systems and their role in novel delivery of drugs could not be ruled out.

CONCLUSION

The chapter enlightened the recent approaches and of modifying gums and their subsequent use in the preparation of micro/nanocarrier based drug delivery system. The derivatization of gums was found to be advantageous technique to alter physicochemical properties of the native gum. The gums available in nature bear properties like low solubility, low swelling or high swelling, uncontrolled formulation characteristics, instability, *etc*. The derivatization of natural gums has helped various researchers in tailoring their physiochemical, rheological, mechanical, release kinetic behaviors *etc*. thus, tailor made products generated after derivatization was successfully used to generate micro/nano-carriers. In addition, derivatization of polysaccharide and incorporation into formulation had a significant impact on the physical properties and release kinetics. The advantages of derivatized polysaccharide based micro/nano delivery system over synthetic ones are biodegradability, high impact on the release profile, non toxicity, easily degraded by intestinal flora. Derivatized polymer based micro/nano-carrier microencapsulation technology has developed significantly and is applied in various fields such as pharmaceuticals, cosmetics, and food. This allows for controlled release of various drugs, enhanced stability of formulations and taste masking. In conclusion, this book chapter demonstrated that modified natural polysaccharides utilized micro/nano-carrier based drug delivery systems are promising matrix material for the development of intestinal/targeted pH responsive/controlled release drug delivery.

ACKNOWLEDGEMENTS

The authors would like to acknowledge the financial assistance provided by CSIR (Project no. 02/(0064)/12/EMR-II) and AICTE (F no. 8-188/RIFD/RPS/Policy-1/2014-15) New Delhi and Punjabi University, Patiala for providing necessary facilities.

CONFLICT OF INTEREST

The authors confirm that this chapter content has no conflict of interest.

REFERENCE

[1] Setty CM, Deshmukh AS, Badiger AM. Hydrolyzed polyacrylamide grafted carboxymethylxyloglucan based microbeads for pH responsive drug delivery. Int J Biol Macromol 2014; 67: 28-36.

[2] Jani GK, Shah DP, Prajapati VD, Jain VC. Gums and mucilages: versatile excipients for pharmaceutical formulations. Asian J Pharm Sci 2009; 4(5): 309-23.

[3] Rai PR, Tiwary A, Rana V. Superior disintegrating properties of calcium cross-linked *Cassia fistula* gum derivatives for fast dissolving tablets. Carbohydr Polym 2012; 87(2): 1098-104.

[4] Rana V, Rai P, Tiwary AK, Singh RS, Kennedy JF, Knill CJ. Modified gums: Approaches and applications in drug delivery. Carbohydr Polym 2011; 83(3): 1031-47.

[5] Maiti S, Ray S, Sa B. Effect of formulation variables on entrapment efficiency and release characteristics of bovine serum albumin from carboxymethyl xanthan microparticles. Polym Adv Technol 2008; 19(7): 922-7.

[6] Kaur H, Ahuja M, Kumar S, Dilbaghi N. Carboxymethyl tamarind kernel polysaccharide nanoparticles for ophthalmic drug delivery. Int J Biol Macromol 2012; 50(3): 833-9.

[7] Sanoj Rejinold N, Sreerekha P, Chennazhi K, Nair S, Jayakumar R. Biocompatible, biodegradable and thermo-sensitive chitosan-*g*-poly (N-isopropylacrylamide) nanocarrier for curcumin drug delivery. Int J Biol Macromol 2011; 49(2): 161-72.

[8] Maya S, Indulekha S, Sukhithasri V, Smitha K, Nair SV, Jayakumar R, *et al*. Efficacy of tetracycline encapsulated O-carboxymethyl chitosan nanoparticles against intracellular infections of *Staphylococcus aureus*. Int J Biol Macromol 2012; 51(4): 392-9.

[9] Yeh M-k, Cheng K-m, Hu C-s, Huang Y-c, Young J-j. Novel protein-loaded chondroitin sulfate–chitosan nanoparticles: preparation and characterization. Acta Biomater 2011; 7(10): 3804-12.

[10] Kaity S, Ghosh A. Carboxymethylation of locust bean gum: Application in interpenetrating polymer network microspheres for controlled drug delivery. Ind Eng Chem Res 2013; 52(30): 10033-45.

[11] Rejinold NS, Chennazhi K, Nair S, Tamura H, Jayakumar R. Biodegradable and thermo-sensitive chitosan-*g*-poly (N-vinylcaprolactam) nanoparticles as a 5-fluorouracil carrier. Carbohydr Polym 2011; 83(2): 776-86.

[12] Sun K, Wang J, Zhang J, Hua M, Liu C, Chen T. Dextran–*g*–PEI nanoparticles as a carrier for co-delivery of adriamycin and plasmid into osteosarcoma cells. Int J Biol Macromol 2011; 49(2): 173-80.

[13] Ray S, Maiti S, Sa B. Polyethyleneimine-treated xanthan beads for prolonged release of diltiazem: *in vitro* and *in vivo* evaluation. Arch Pharmacal Res 2010; 33(4): 575-83.

[14] Goyal P, Kumar V, Sharma P. Carboxymethylation of tamarind kernel powder. Carbohydr Polym 2007; 69(2): 251-5.

[15] Sharma BR, Kumar V, Soni P. Carbamoylethylation of guar gum. Carbohydr Polym 2004; 58(4): 449-53.

[16] Sharma BR, Kumar V, Soni P. Carbamoylethylation of *Cassia tora* gum. Carbohydr Polym 2003; 54(2): 143-7.

[17] Yang L, Zhao T, Wei H, Zhang M, Zou Y, Mao G, *et al*. Carboxymethylation of polysaccharides from *Auricularia auricula* and their antioxidant activities *in vitro*. Int J Biol Macromol 2011; 49(5): 1124-30.

[18] Silva DA, de Paula R, Feitosa J, de Brito AC, Maciel JS, Paula HC. Carboxymethylation of cashew tree exudate polysaccharide. Carbohydr Polym 2004; 58(2): 163-71.

[19] Kumar A, Ahuja M. Carboxymethyl gum kondagogu: Synthesis, characterization and evaluation as mucoadhesive polymer. Carbohydr Polym 2012; 90(1): 637-43.

[20] Bidgoli H, Zamani A, Taherzadeh MJ. Effect of carboxymethylation conditions on the water-binding capacity of chitosan-based superabsorbents. Carbohydr Res 2010; 345(18): 2683-9.

[21] Xu J, Liu W, Yao W, Pang X, Yin D, Gao X. Carboxymethylation of a polysaccharide extracted from *Ganoderma lucidum* enhances its antioxidant activities *in vitro*. Carbohydr Polym 2009; 78(2): 227-34.

[22] Bernkop-Schnürch A, Brandt U, Clausen A. Synthesis and *in vitro* evaluation of chitosan-cysteine conjugates. Sci Pharm 1999; 67: 196-208.

[23] Bernkop-Schnurch A, Hopf TE. Synthesis and *in vitro* evaluation of chitosan-thioglycolic acid conjugates. Sci Pharm 2001; 69(2): 109-18.

[24] Bernkop-Schnürch A, Kast CE, Richter MF. Improvement in the mucoadhesive properties of alginate by the covalent attachment of cysteine. J. Control Release 2001; 71(3): 277-85.

[25] Bernkop-Schnürch A, Hornof M, Zoidl T. Thiolated polymers—thiomers: synthesis and *in vitro* evaluation of chitosan–2-iminothiolane conjugates. Int J Pharm 2003; 260(2): 229-37.

[26] Sharma R, Ahuja M. Thiolated pectin: Synthesis, characterization and evaluation as a mucoadhesive polymer. Carbohydr Polym 2011; 85(3): 658-63.

[27] Yadav S, Ahuja M, Kumar A, Kaur H. Gellan–thioglycolic acid conjugate: Synthesis, characterization and evaluation as mucoadhesive polymer. Carbohydr Polym 2014; 99: 601-7.

[28] Mahajan HS, Tyagi VK, Patil RR, Dusunge SB. Thiolated xyloglucan: Synthesis, characterization and evaluation as mucoadhesive *in situ* gelling agent. Carbohydr Polym 2013; 91(2): 618-25.

[29] Singh R, Tripathy T, Karmakar G, Rath S, Karmakar N, Pandey S, *et al.* Novel biodegradable flocculants based on polysaccharides. Curr Sci 2000; 78(7): 798-803.

[30] Gref R, Rodrigues J, Couvreur P. Polysaccharides grafted with polyesters: novel amphiphilic copolymers for biomedical applications. Macromolecules 2002; 35(27): 9861-7.

[31] Ohya Y, Maruhashi S, Hirano T, Ouchi T. Preparation of poly (lactic acid)-grafted polysaccharides as biodegradable amphiphilic materials. Biomed Polym Polym Therap 2002: 139-48.

[32] Singh V, Tiwari A, Tripathi DN, Sanghi R. Microwave assisted synthesis of guar-g-polyacrylamide. Carbohydr Polym 2004; 58(1): 1-6.

[33] Fares MM, El-faqeeh AS, Osman ME. Graft copolymerization onto starch–I. Synthesis and optimization of starch grafted with N-tert-butylacrylamide copolymer and its hydrogels. J Polm Res 2003; 10(2): 119-25.

[34] Sharma BR, Kumar V, Soni P. Ce (IV)-ion initiated graft copolymerization of methyl methacrylate onto guar gum. J Macromol Sci Part A 2003; 40(1): 49-60.

[35] Ikhuoria E, Folayan A, Okieimen F. Studies in the graft copolymerization of acrylonitrile onto cassava starch by ceric ion induced initiation. Int J Biotechnol Mol Biol Res 2010; 1(1): 10-4.

[36] Celik M. Preparation and characterization of starch-g-polymethacrylamide copolymers. J Polm Res 2006; 13(5): 427-32.

[37] Tiwari A, Singh V. Microwave-induced synthesis of electrical conducting gum acacia- *graft*-polyaniline. Carbohydr Polym 2008; 74(3): 427-34.

[38] Sen G, Singh RP, Pal S. Microwave-initiated synthesis of polyacrylamide grafted sodium alginate: Synthesis and characterization. J Appl Polym Sci 2010; 115(1): 63-71.

[39] Zhang J, Zhang S, Yuan K, Wang Y. Graft copolymerization of artemisia seed gum with acrylic acid under microwave and its water absorbency. J Macromol Sci Pure Appl Chem 2007; 44(8): 881-5.

[40] Huacai G, Wan P, Dengke L. Graft copolymerization of chitosan with acrylic acid under microwave irradiation and its water absorbency. Carbohydr Polym 2006; 66(3): 372-8.

[41] Singh V, Tiwari A, Tripathi DN, Sanghi R. Grafting of polyacrylonitrile onto guar gum under microwave irradiation. J Appl Polym Sci 2004; 92(3): 1569-75.

[42] Singh V, Sharma AK, Maurya S. Efficient Cadmium(II) Removal from Aqueous Solution Using Microwave Synthesized Guar Gum-Graft-Poly(ethylacrylate). Ind Eng Chem Res 2009; 48(10): 4688-96.

[43] Mishra A, Clark JH, Vij A, Daswal S. Synthesis of graft copolymers of xyloglucan and acrylonitrile. Polym Adv Technol 2008; 19(2): 99-104.

[44] Goyal P, Kumar V, Sharma P. Graft copolymerization of acrylamide onto tamarind kernel powder in the presence of ceric ion. J Appl Polym Sci 2008; 108(6): 3696-701.

[45] Raj Sharma B, Kumar V, Soni P. Ceric ammonium nitrate-initiated graft copolymerization of acrylamide onto Cassia tora gum. J Appl Polym Sci 2002; 86(13): 3250-5.

[46] Maestrelli F, Cirri M, Corti G, Mennini N, Mura P. Development of enteric-coated calcium pectinate microspheres intended for colonic drug delivery. Eur J Pharm Biopharm 2008; 69(2): 508-18.

[47] Anitha A, Maya S, Deepa N, Chennazhi K, Nair S, Tamura H, *et al.* Efficient water soluble O-carboxymethyl chitosan nanocarrier for the delivery of curcumin to cancer cells. Carbohydr Polym 2011; 83(2): 452-61.

[48] Bajpai S, Saxena SK, Sharma S. Swelling behavior of barium ions-crosslinked bipolymeric sodium alginate–carboxymethyl guar gum blend beads. React Funct Polym 2006; 66(6): 659-66.

[49] Bhattacharya SS, Banerjee S, Chowdhury P, Ghosh A, Hegde RR, Mondal R. Tranexamic acid loaded gellan gum-based polymeric microbeads for controlled release: *In vitro* and *in vivo* assessment. Colloids Surf B 2013; 112: 483-91.

[50] Shi X, Du Y, Yang J, Zhang B, Sun L. Effect of degree of substitution and molecular weight of carboxymethyl chitosan nanoparticles on doxorubicin delivery. J Appl Polym Sci 2006; 100(6): 4689-96.

[51] George M, Abraham T. pH sensitive alginate–guar gum hydrogel for the controlled delivery of protein drugs. Int J Pharm 2007; 335(1): 123-9.

[52] Kulkarni RV, Nagathan VV, Biradar PR, Naikawadi AA. Simvastatin loaded composite polyspheres of gellan gum and carrageenan: *In vitro* and *in vivo* evaluation. Int J Biol Macromol 2013; 57: 238-44.

[53] Gliko-Kabir I, Yagen B, Baluom M, Rubinstein A. Phosphated crosslinked guar for colon-specific drug delivery: II. *In vitro* and *in vivo* evaluation in the rat. J Control Release 2000; 63(1): 129-34.
[54] Dulong V, Lack S, Le Cerf D, Picton L, Vannier J, Muller G. Hyaluronan-based hydrogels particles prepared by crosslinking with trisodium trimetaphosphate. Synthesis and characterization. Carbohydr Polym 2004; 57(1): 1-6.
[55] Gliko-Kabir I, Yagen B, Penhasi A, Rubinstein A. Phosphated crosslinked guar for colon-specific drug delivery: I. Preparation and physicochemical characterization. J Control Release 2000; 63(1): 121-7.
[56] Shalviri A, Liu Q, Abdekhodaie MJ, Wu XY. Novel modified starch–xanthan gum hydrogels for controlled drug delivery: Synthesis and characterization. Carbohydr Polym 2010; 79(4): 898-907.
[57] Silva DA, Feitosa J, Maciel JS, Paula HC, de Paula R. Characterization of crosslinked cashew gum derivatives. Carbohydr Polym 2006; 66(1): 16-26.
[58] Magalhães Jr GA, Santos CM, Silva DA, Maciel JS, Feitosa J, Paula HC, *et al.* Microspheres of chitosan/carboxymethyl cashew gum (CH/CMCG): Effect of chitosan molar mass and CMCG degree of substitution on the swelling and BSA release. Carbohydr Polym 2009; 77(2): 217-22.
[59] Gupta AP, Verma DK. Preparation and characterization of carboxymethyl guar gum nanoparticles. Int J Biol Macromol 2014; 68: 247-50.
[60] Soumya RS, Ghosh S, Abraham ET. Preparation and characterization of guar gum nanoparticles. Int J Biol Macromol 2010; 46(2): 267-9.
[61] Soppimath KS, Kulkarni AR, Aminabhavi TM. Controlled release of antihypertensive drug from the interpenetrating network poly (vinyl alcohol)–guar gum hydrogel microspheres. J Biomater Sci Polym Ed 2000; 11(1): 27-43.
[62] Kaity S, Isaac J, Ghosh A. Interpenetrating polymer network of locust bean gum-poly (vinyl alcohol) for controlled release drug delivery. Carbohydr Polym 2013; 94(1): 456-67.
[63] Soppirnath KS, Aminabhavi TM. Water transport and drug release study from cross-linked polyacrylamide grafted guar gum hydrogel microspheres for the controlled release application. Eur J Pharm Biopharm 2002; 53(1): 87-98.
[64] Sarmah JK, Mahanta R, Bhattacharjee SK, Mahanta R, Biswas A. Controlled release of tamoxifen citrate encapsulated in cross-linked guar gum nanoparticles. Int J Biol Macromol 2011; 49(3): 390-6.
[65] Li G-Y, Zhong M, Zhou Z-D, Ding P, Li Y-J. Novel carboxymethyl chitosan microspheres for controlled delivery of chelerythrine. J Macromol Sci Part A 2011; 48(11): 904-11.
[66] Jose S, Fangueiro J, Smitha J, *et al.* Predictive modeling of insulin release profile from cross-linked chitosan microspheres. Eur J Med Chem 2013; 60: 249-53.
[67] Sharma S, Kaur J, Sharma G, Thakur KK, Chauhan GS, Chauhan K. Preparation and characterization of pH-responsive guar gum microspheres. Int J Biol Macromol 2013; 62: 636-41.
[68] Chaurasia M, Chourasia MK, Jain NK, Jain A, Soni V, Gupta Y, *et al.* Cross-linked guar gum microspheres: A viable approach for improved delivery of anticancer drugs for the treatment of colorectal cancer. AAPS PharmSciTech 2006; 7(3): E143-E51.
[69] Yang HJ, Park IS, Na K. Biocompatible microspheres based on acetylated polysaccharide prepared from water-in-oil-in-water ($W_1/O/W_2$) double-emulsion method for delivery of type II diabetic drug (exenatide). Colloids Surf A 2009; 340(1): 115-20.
[70] Singh V, Kumar P. Carboxymethyl tamarind gum–silica nanohybrids for effective immobilization of amylase. J Mol Catal B: Enzym 2011; 70(1): 67-73.
[71] Singh K, Tiwary A, Rana V. Spray dried chitosan–EDTA superior microparticles as solid substrate for the oral delivery of amphotericin B. Int J Biol Macromol 2013; 58: 310-9.
[72] Desai KGH, Park HJ. Preparation and characterization of drug-loaded chitosan–tripolyphosphate microspheres by spray drying. Drug Dev Res 2005; 64(2): 114-28.
[73] Jain A, Jain SK, Ganesh N, Barve J, Beg AM. Design and development of ligand-appended polysaccharidic nanoparticles for the delivery of oxaliplatin in colorectal cancer. Nanomedicine 2010; 6(1): 179-90.
[74] Soppimath KS, Kulkarni AR, Aminabhavi TM. Chemically modified polyacrylamide- *g*-guar gum-based crosslinked anionic microgels as pH-sensitive drug delivery systems: preparation and characterization. J Control Release 2001; 75(3): 331-45.

Subject Index

A

B

C

Viness Pillay, Yahya E. Choonara and Pradeep Kumar (Eds.)

D

E

F

G

H

I

O

P

R

S

T

U

V

X

Made in the USA
Las Vegas, NV
04 November 2026

57437164R00278